Lecture Notes in Computer Science   9877

Commenced Publication in 1973
Founding and Former Series Editors:
Gerhard Goos, Juris Hartmanis, and Jan van Leeuwen

More information about this series at http://www.springer.com/series/7409

Muhammad Aamir Cheema · Wenjie Zhang
Lijun Chang (Eds.)

Databases Theory and Applications

27th Australasian Database Conference, ADC 2016
Sydney, NSW, September 28–29, 2016
Proceedings

 Springer

Editors
Muhammad Aamir Cheema
Monash University
Clayton
Australia

Lijun Chang
University of New South Wales
Sydney, NSW
Australia

Wenjie Zhang
School of Computer Science and
  Engineering
University of New South Wales
Sydney
Australia

ISSN 0302-9743   ISSN 1611-3349 (electronic)
Lecture Notes in Computer Science
ISBN 978-3-319-46921-8   ISBN 978-3-319-46922-5 (eBook)
DOI 10.1007/978-3-319-46922-5

Library of Congress Control Number: 2016952520

LNCS Sublibrary: SL3 – Information Systems and Applications, incl. Internet/Web, and HCI

This Springer imprint is published by Springer Nature
The registered company is Springer International Publishing AG
The registered company address is: Gewerbestrasse 11, 6330 Cham, Switzerland

Preface

It is our pleasure to present to you the proceedings of the 27th Australasian Database Conference (ADC2016), which took place in Sydney, Australia. The Australasian Database Conference is an annual international forum for sharing the latest research advancements and novel applications of database systems, data-driven applications, and data analytics between researchers and practitioners from around the globe, particularly Australia and New Zealand. The mission of ADC is to share novel research solutions to problems of today's information society that fulfill the needs of heterogeneous applications and environments and to identify new issues and directions for future research. ADC seeks papers from academia and industry presenting research on all practical and theoretical aspects of advanced database theory and applications, as well as case studies and implementation experiences. All topics related to database are of interest and within the scope of the conference. ADC gives researchers and practitioners a unique opportunity to share their perspectives with others interested in the various aspects of database systems.

ADC 2016 was held during September 26–29, 2016 in Sydney, Australia. As in previous years, ADC 2016 accepted all papers that the Program Committee considered as being of ADC quality without setting any predefined quota. The conference accepted 33 research papers including some invited papers, and 11 demo papers. Each paper including every invited submission was peer reviewed in full by at least three independent reviewers, and in some cases four referees produced independent reviews. A conscious decision was made to select the papers for which all reviews were positive and favorable. The Program Committee that selected the papers comprised 53 members from around the globe including Australia, Bangladesh, Canada, China, Finland, Germany, Japan, New Zealand, Qatar, Singapore, Taiwan, and the UK, who were thorough and dedicated to the reviewing process.

We would like to thank all our colleagues who served on the Program Committee or acted as external reviewers. We would also like to thank all the authors who submitted their papers and the attendees. We hope that with these proceedings, you can have an overview of this vibrant research community and its activities. We encourage you to make submissions to the next ADC conference and contribute to this community.

August 2016
Muhammad A. Cheema
Wenjie Zhang
Lijun Chang

General Chair's Welcome Message

Welcome to the proceedings of the 27th Australasian Database Conference (ADC2016)! ADC is a leading Australia- and New Zealand-based international conference on research and applications of database systems, data-driven applications, and data analytics. In the past 10 years, ADC has been held in Melbourne (2015), Brisbane (2014), Adelaide (2013), Melbourne (2012), Perth (2011), Brisbane (2010), Wellington (2009), Wollongong (2008), Ballarat (2007), and Hobart (2006). This year, the ADC conference came to Sydney.

In the past, the ADC conference series was held as part of the Australasian Computer Science Week (ACSW). Starting from 2014, the ADC conferences have departed from ACSW as the database research community in Australasia has grown significantly larger. Now the new ADC conference has an expanded research program and focuses on community building through a PhD School. ADC 2016 was the third of this new ADC conference series.

The conference this year had four eminent speakers to give keynote speeches: Michael Stonebraker from Massachusetts Institute of Technology, USA, Mark Sanderson from RMIT University, Australia, Stephan Winter from the University of Melbourne, Australia, and Mohamed F. Mokbel from the University of Minnesota, USA. In addition to 33 full research papers and 11 demo papers carefully selected by the Program Committee, we were also very fortunate to have two invited talks presented by world-leading researchers: Zi (Helen) Huang from the University of Queensland, Australia, and Ying Zhang from the University of Technology Sydney, Australia. We had a two-day PhD School program as part of this year's ADC with great support from invited speakers: Alan David Fekete from the University of Sydney, Australia, Jeffrey Xu Yu from the Chinese University of Hong Kong, China, Zi (Helen) Huang from the University of Queensland, Australia, Ying Zhang from the University of Technology Sydney, Australia, and Kai (Alex) Qin from RMIT University, Australia.

We wish to take this opportunity to thank all speakers, authors, and organizers. I would especially like thank our Organizing Committee members: Program Committee co-chairs Muhammad A. Cheema and Wenjie Zhang, for their dedication in ensuring a high-quality program, proceedings chair Lijun Chang, for his effort in delivering the conference proceedings timely, local co-chairs Jianbin Qin, Shiyu Yang, and Xiaoyang Wang, for their efforts in covering every detail of the conference logistics, and webmaster Xiang Wang for his effort in maintaining the conference website. I would also like thank the University of New South Wales (UNSW) for the generous financial support. Without them, this year's ADC would not have been a success.

Sydney is a multi-cultural city and ADC 2016 was held on the main campus of the University of New South Wales located on a 38-hectare site in the suburb of Kensington. We trust all ADC 2016 participants had wonderful experience with the conference, the campus, and the city.

Xuemin Lin

Organization

General Chair

Xuemin Lin   University of New South Wales, Australia

Program Co-chairs

Muhammad A. Cheema   Monash University, Australia
Wenjie Zhang   University of New South Wales, Australia

Local Co-chairs

Jianbin Qin   University of New South Wales, Australia
Shiyu Yang   University of New South Wales, Australia
Xiaoyang Wang   University of Technology Sydney, Australia

Proceedings Chair

Lijun Chang   University of New South Wales, Australia

Steering Committee

Rao Kotagiri   University of Melbourne, Australia
Timos Sellis   Swinburne University of Technology, Australia
Gill Dobbie   University of Auckland, New Zealand
Alan Fekete   University of Sydney, Australia
Xuemin Lin   University of New South Wales, Australia
Yanchun Zhang   Victoria University, Australia
Xiaofang Zhou   University of Queensland, Australia

Program Committee

Muhammad Ahmad   McGill University, Canada
Shafiq Alam   The University of Auckland, New Zealand
Mohammed Ali   Bangladesh University of Engineering and Technology,
                                       Bangladesh
Zhifeng Bao   RMIT University, Australia
Ljiljana Brankovic   University of Newcastle, Australia
Junbin Gao   Charles Sturt University, Australia
Dimitrios   RMIT University, Australia
  Georgakopoulos

Janusz Getta	University of Wollongong, Australia
Yusuke Gotoh	Okayama University, Japan
Yu Gu	Northeastern University, China
Tanzima Hashem	Bangladesh University of Engineering and Technology, Bangladesh
Michael Houle	National Institute of Informatics, Japan
Guangyan Huang	Deakin University, Australia
Arijit Khan	Nanyang Technological University, Singapore
Khalid Latif	Aalto University, Finland
Jianxin Li	University of Western Australia, Australia
Jianzhong Li	Harbin Institute of Technology, China
Jiuyong Li	University of South Australia, Australia
Xue Li	University of Queensland, Australia
Zhixu Li	Soochow University, China
Ghulam Murtuza	Intersect, Australia
Muhammad Naeem	University of Auckland, New Zealand
Mourad Ouzzani	Qatar Computing Research Institute, Qatar
Weining Qian	East China Normal University, China
Jianbin Qin	University of New South Wales, Australia
Lu Qin	University of Technology, Sydney, Australia
Goce Ristanoski	National ICT Australia, Australia
Shazia Sadiq	University of Queensland, Australia
Sherif Sakr	University of New South Wales, Australia
Flora Salim	RMIT University, Australia
Haichuan Shang	National Institute of Information and Communications, Japan
Quan Sheng	University of Adelaide, Australia
Bela Stantic	Griffith University, Australia
Farhan Tauheed	Oracle
Anwaar Ulhaq	Victoria University, Australia
Junhu Wang	Griffith University, Australia
Lei Wang	University of Wollongong, Australia
Yan Wang	Macquarie University, Australia
Yang Wang	University of New South Wales, Australia
Lin Wu	University of Adelaide, Australia
Chuan Xiao	Nagoya University, Japan
Guandong Xu	University of Technology, Australia
Mi-Yen Yeh	Academia Sinica, Taiwan
Daisaku Yokoyama	University of Tokyo, Japan
Weiren Yu	Imperial College London, UK
Nayyar Zaidi	Monash University, Australia
Xianchao Zhang	Dalian University of Technology, China
Xiuzhen Zhang	RMIT University, Australia
Ying Zhang	University of Technology Sydney, Australia
Kai Zheng	University of Queensland, Australia
Aoying Zhou	East China Normal University, China

Xiangmin Zhou   Victoria University, Australia
Guido Zuccon   Queensland University of Technology, Australia
Andreas Zufle   Ludwig-Maximilians-Universität, Germany

External Reviewers

Maitraye Das   Bangladesh University of Engineering and Technology,
                           Bangladesh
Wei Kang   University of South Australia, Australia
Ali Shemshadi   University of Adelaide, Australia
Mehnaz Tabassum   Bangladesh University of Engineering and Technology,
                           Bangladesh
Xin Wang   Tianjin University, China
Wei Zhang   University of Adelaide, Australia
Yihong Zhang   University of Adelaide, Australia

ADC Keynotes

Big Data, Technological Disruption and the 800 Pound Gorilla in the Corner

Michael Stonebraker

Massachusetts Institute of Technology, Cambridge, USA

Abstract. This talk will focus on the current market for "Big Data" products, specifically those that deal with one or more of "the 3 V's". I will suggest that the **volume** problem for business intelligence applications is pretty well solved by the data warehouse vendors; however upcoming "data science" tasks are poorly supported at present. On the other hand, there is rapid technological progress, so "stay tuned". In the **velocity** arena recent "new SQL" and stream processing products are doing a good job, but there are a few storm clouds on the horizon. The **variety** space has a collection of mature products, along with considerable innovation from startups. I will discuss opportunities in this space, especially those enabled by possible disruption from new technology. Also discussed will be the pain levels I observe in current enterprises, culminating in my presentation of "the 800 pound gorilla in the corner".

Short Biography. Professor Stonebraker has been a pioneer of data base research and technology for more than forty years. He was the main architect of the INGRES relational DBMS, and the object-relational DBMS, POSTGRES. These prototypes were developed at the University of California at Berkeley where Stonebraker was a Professor of Computer Science for twenty five years. More recently at M.I.T. he was a co-architect of the Aurora/Borealis stream processing engine, the C-Store column-oriented DBMS, the H-Store transaction processing engine, the SciDB array DBMS, and the Data Tamer data curation system. Presently he serves as Chief Technology Officer of Paradigm4 and Tamr, Inc. Professor Stonebraker was awarded the ACM System Software Award in 1992 for his work on INGRES. Additionally, he was awarded the first annual SIGMOD Innovation award in 1994, and was elected to the National Academy of Engineering in 1997. He was awarded the IEEE John Von Neumann award in 2005 and the 2014 Turing Award, and is presently an Adjunct Professor of Computer Science at M.I.T, where he is co-director of the Intel Science and Technology Center focused on big data.

Getting Rid of the Ten Blue Links

Mark Sanderson

RMIT University, Melbourne, Australia

Abstract. In this talk, I will first give a brief overview of the IR group at RMIT. Then I will describe the work we are doing at RMIT to change one of the commonest web pages we all look at: the Search Result Page (SERP). In our work we are looking to replace the SERP with a set of answer passages that address the user's query. In the context of general web search, the problem of finding answer passages has not been explored extensively. Previous studies have found that many informational queries can be answered by a passage of text extracted from a retrieved document, relieving the user from having to read the actual document. While current passage retrieval methods that focus on topical relevance have been shown to be not effective at finding answers, the result shows that more knowledge is required to identify answers in the document. We have been formulating the answer passage extraction problem as a summarization task. We initially used term distributions extracted from a Community Question Answering (CQA) service to generate more effective summaries of retrieved web pages. An experiment was conducted to see the benefit of using the CQA data in finding answer passages. We analyze the fraction of answers covering a set of queries, the quality of the corresponding result from the answering service, and their impact on the generated summaries. I will also talk about recent work where we re-rank retrieved passages according to the summary quality and incorporating document summarizability into the ranking function.

Short Biography. Prof. Mark Sanderson is the deputy head of the School of Computer Science and IT at RMIT University in Melbourne, Australia. According to a range of international ranking systems RMIT is in the top 8 of CS schools in Australia. Prof Sanderson is head of the RMIT Information Retrieval (IR) group, which is regarded as the leading IR group in Australia. He is co-editor of Foundations and Trends in Information Retrieval, which is currently the highest impact rated IR journal. He is also an associate editor of IEEE TKDE and of ACM TWeb. Prof. Sanderson was co-PC chair of ACM SIGIR in 2009 and 2012, and general chair of the conference in 2004. Prof Sanderson is also a visiting professor at NII in Tokyo.

ADC Invited Talks

Location Understanding in Social Media

Zi (Helen) Huang

The University of Queensland, St Lucia, Australia

Abstract. Location data has been playing an important role in many social media applications, particularly the location-based services. Unfortunately, location information is often missing in social media data, such as online images. In this talk, we introduce novel methods to estimate missing locations for social images by effectively fusing multi-modalities of social media data. Interestingly, by integrating visual data and location data, such as geo-tagged images and check-ins, important location proximities can be better understood and discovered to users.

Short Biography. Dr Zi Huang received her BSc degree from Tsinghua University, China, in 2001, and her PhD in Computer Science from the University of Queensland, Australia, in 2007. She is currently an ARC Future Fellow with the School of Information Technology and Electrical Engineering, University of Queensland. Her research interests include multimedia indexing and search, social data analysis and knowledge discovery.

Continuous Spatial-Keyword Queries over Streaming Data

Ying Zhang

University of Technology Sydney, Ultimo, Australia

Abstract. As the prevalence of social media and GPS-enabled devices, a massive amount of geo-textual data has been generated in a stream fashion, leading to a variety of applications such as location-based recommendation and information dissemination. For example, a location-based e-coupon system may allow potentially millions of users to register their continuous spatial-keyword queries (e.g., interests in nearby sales) by specifying a set of keywords and a spatial region; the system then delivers each incoming spatial-textual object (e.g., a geo-tagged e-coupon) to all the matched queries (i.e., users) whose spatial and textual requirements are satisfied. In this talk, I will introduce our recent work on continuous spatial-keyword queries over streaming data. Novel indexing structures, which seamlessly and effectively integrate keyword and spatial information, will be presented to support various continuous spatial-keyword queries.

Short Biography. Ying Zhang is a senior lecturer and ARC DECRA research fellow (2014–2016) at QCIS, the University of Technology Sydney (UTS). He received his BSc and MSc degrees in Computer Science from Peking University, China, and PhD in Computer Science from the University of New South Wales, Australia. His research interests include query processing on spatial data, spatial-textual data, streaming data, uncertain data and graphs. He has published 40+ papers on prestigious conferences and journals such as SIGMOD, SIGIR, VLDB, ICDE, TODS, VLDBJ, and TKDE. He was an Australian Research Council Australian Postdoctoral Fellowship (ARC APD) holder during 2010 and 2013.

PhD School Keynotes

Where am I? Where Do I Want to Go?

Stephan Winter

The University of Melbourne, Parkville, Australia

Abstract. Sensors and the related information and communication technology get ever smarter in localizing people, vehicles, events or goods. And yet, at the end always a person is consuming this information, or even producing it (sometimes called "people as sensors"). In this talk I will focus on the gap between the concepts of people about their environment, and the concepts of sensors, spatial databases and geographic information systems. This gap is a major impediment for communication between people and systems; some examples are emergency calls, tracking bushfires, guiding an autonomously driving car, planning a trip through a city by public transport, helping people to evacuate, or simply the general search in a search engine (Ed Parsons, Geospatial Chief Technologist of Google, indicated that "about 1 in 3 of queries that people just type into a standard Google search bar are about places"). The talk will identify the issues with this gap, and show some steps to overcome this gap, including novel, complementary ways of representing spatial information in databases.

Short Biography. Stephan Winter is Professor in Spatial Information Science at the Department of Infrastructure Engineering, The University of Melbourne. He holds a PhD (Dr.-Ing.) from the University of Bonn (1997), and a habilitation from the Technical University Vienna (2001). Within spatial information science Stephan Winter is specializing on human wayfinding and navigation, with a vision of developing intelligent spatial machines. He has contributed to topics such as spatial human-computer interaction, network analysis, routing heuristics, and collaborative transportation and evacuation.

The Era of Big Spatial Data

Mohamed F. Mokbel

University of Minnesota, Minneapolis, USA

Abstract. In recent years, there has been an explosion in the amounts of spatial and spatio-temporal data produced from several devices including smart phones, space telescopes, medical devices. Unfortunately, managing and analyzing such big spatial data is hampered by the lack of specialized systems, techniques, and algorithms. While big data is well supported with a variety of distributed systems and cloud infrastructure, none of these systems or infrastructure provide any special support for spatial or spatio-temporal data. This talk presents our efforts in indexing, querying, and visualizing big spatial and spatio-temporal data. We will describe our efforts within SpatialHadoop; our full-fledged MapReduce framework with native support for spatial data, including support for basic spatial operations, computational geometry, and spatial visualization.

Short Biography. Mohamed F. Mokbel (Ph.D., Purdue University, MS, B.Sc., Alexandria University) is Associate Professor in the Department of Computer Science and Engineering, University of Minnesota. His research interests include the interaction of GIS and location-based services with database systems and cloud computing. His research work has been recognized by five Best Paper Awards and by the NSF CAREER award. Mohamed was the program co-chair for the ACM SIGSPATIAL GIS conference from 2008 to 2010, IEEE MDM Conference 2011 and 2014, and the General Chair for SSTD 2011. He is an Associate Editor for ACM TODS, ACM TSAS, VLDB journal, and GeoInformatica. Mohamed is an elected Chair of ACM SIGSPATIAL 2014-2017. For more information, please visit: www.cs.umn.edu/~mokbel.

PhD School Tutorials

Consistency Properties for Distributed Storage Platforms

Alan David Fekete

University of Sydney, Sydney, Australia

Abstract. A scalable and fault-tolerant data storage layer is extremely useful when constructing scalable fault-tolerant application software. The application developer is a consumer of a service provided by the storage layer, and the interface between these parties needs to be precise. This tutorial reflects on several bodies of research that relate to understanding the implications for the consumer, of the consistency aspects of that interface. We cover in turn how consistency properties can be defined, how the consumer can measure consistency, and how to reason about applications when they must run over storage with consistency that is weaker-than-ideal.

Short Biography. Alan Fekete is Professor of Enterprise Software Systems within the School of Information Technologies at the University of Sydney. His undergraduate education was at the University of Sydney, and his doctorate was earned in the mathematics department of Harvard University. He has been recognized as a Distinguished Scientist by ACM, and he serves as Trustee for the VLDB Endowment. He is particularly known for a body of research on transaction management. He is also active in CS Education.

Large Graph Processing: Algorithms and Systems

Jeffrey Xu Yu

The Chinese University of Hong Kong, Hong Kong, China

Abstract. The real applications that need graph processing techniques to handle a large graph can be found from many real applications including online social networks, biological networks, ontology, transportation networks, etc. In this talk, we will discuss some selected research topics on graph processing over large graphs from the algorithm perspectives and the systems perspectives.

Short Biography. Dr Jeffrey Xu Yu is a Professor in the Department of Systems Engineering and Engineering Management, The Chinese University of Hong Kong. His current main research interests include graph mining, graph query processing, graph pattern matching, keywords search in databases, and online social networks. Dr. Yu served as an Information Director and a member in ACM SIGMOD executive committee (2007–2011), an associate editor of IEEE Transactions on Knowledge and Data Engineering (2004–2008), and an associate editor in VLDB Journal (2007-2013). Currently he servers as an associate editor of WWW Journal, the International Journal of Cooperative Information Systems, the Journal on Health Information Science and Systems (HISS), and Journal of Information Processing. Dr. Yu served/serves in many organization committees and program committees in international conferences/workshops including PC Co-chair of APWeb'04, WAIM'06, APWeb/WAIM'07, WISE'09, PAKDD'10, DASFAA'11, ICDM'12, NDBC'13, ADMA'14, and CIKM'15.

Social Event and Behavior Modelling

Zi (Helen) Huang

The University of Queensland, St Lucia, Australia

Abstract. Social media data has provided great opportunities for many challenging data mining tasks. Its value has been widely exhibited in real-world applications. In this tutorial, we are focused on the impact of social media on public social event and individual online user behavior using heterogeneous social media data. We will review recent research advances in social event detection and prediction, and online user behavior modelling and prediction. As a step further, we will also discuss the effect of public events on individual behavior and explore their potential sequential correlations for new research opportunities.

Short Biography. Dr Zi Huang received her BSc degree from Tsinghua University, China, in 2001, and her PhD in Computer Science from the University of Queensland, Australia, in 2007. She is currently an ARC Future Fellow with the School of Information Technology and Electrical Engineering, University of Queensland. Her research interests include multimedia indexing and search, social data analysis and knowledge discovery.

Querying and Mining of Geo-Textual Data

Ying Zhang

University of Technology Sydney, Sydney, Australia

Abstract. Proliferation of geo-position technologies (e.g., smart phones, general mobile devices and sensor networks) and online social media (e.g., Twitter, Foursquare and Facebook) has resulted in a huge flood of location data being integrated with various textual data (e.g., tweets and news), leading to the "geo-textual" data. The ever increasing amounts of geo-textual data have tremendous potential for the discovery of new and useful knowledge in many key applications such as location-based services (LBS), e-marketing and social networks. In this tutorial, we first highlight the importance of geo-textual data management and the unique challenges that need to be addressed. Subsequently, we provide an overview of the existing research on geo-textual data, covering modelling, ad-hoc spatial-keyword queries, continuous spatial-keyword queries, mining of geo-textual data and other relevant topics. Finally, we discuss the future research directions in this important and growing research area.

Short Biography. Ying Zhang is a senior lecturer and ARC DECRA research fellow (2014–2016) at QCIS, the University of Technology Sydney (UTS). He received his BSc and MSc degrees in Computer Science from Peking University, China, and PhD in Computer Science from the University of New South Wales, Australia. His research interests include query processing on spatial data, spatial-textual data, streaming data, uncertain data and graphs. He has published 40+ papers on prestigious conferences and journals such as SIGMOD, SIGIR, VLDB, ICDE, TODS, VLDBJ, and TKDE. He was an Australian Research Council Australian Postdoctoral Fellowship (ARC APD) holder during 2010 and 2013.

Data-Driven Evolutionary Optimisation

Kai (Alex) Qin

RMIT University, Melbourne, Australia

Abstract. Optimisation aims at finding the best solution from numerous feasible ones, which is demanded in nearly every field when resolving various problems arising therein. Evolutionary optimisation represents a family of optimisation techniques based on Darwinian principles, characterized by a population of candidate solutions which will be evolved via nature-inspired operators to search for the optimum. Intrinsically, it belongs to a generate-and-test problem solver which incrementally produces a large volume of "data" (i.e. candidate solutions) as search progresses with search experience encoded by such "data". In the past few decades, a lot of efforts had been made to enhance evolutionary optimisation techniques via exploiting (e.g. using data analytics techniques) the "data" generated in the course of search. However, modern optimisation problems, featured with the fast-growing scale, complexity and uncertainty, can seldom be tackled by simply hybridizing evolutionary optimisation with some off-the-shelf data analytics techniques, and therefore call for an in-depth investigation on how to leverage the "data" generated during search to facilitate optimisation. This tutorial aims to introduce a unified perspective on evolutionary optimisation techniques that adopts data analytics as an indispensable component, describe how to identify and address various data analytics tasks during the search process, and discuss an emerging research trend which makes use of search experience gained by solving some problems to facilitate solving other problems via knowledge transfer. The audience is expected to get to know the fundamentals and recent developments in data-driven evolutionary optimisation, and be inspired to employ such techniques to deal with their encountered optimisation problems.

Short Biography. Dr Kai Qin received his BEng degree from Southeast University, China, in 2001, and his PhD from Nanyang Technological University, Singapore, in 2007. He is now a lecturer in Computer Science and Information Technology at RMIT University. His research interests include evolutionary computation, machine learning, image processing, GPU computing and service computing. He has published 60+ papers and received two best paper awards. Two of his co-authored papers are the 1st and 3rd most cited papers (Thomson Reuters) in IEEE Transactions on Evolutionary Computation (ERA A*) over the last 10 years. He is currently chairing the IEEE Computational Intelligence Society task force on collaborative learning and optimisation, promoting research on synergizing machine learning and intelligent optimisation techniques to resolve challenging real-world problems which involve learning and optimisation as indispensable and interwoven tasks.

Contents

Uncertain Data and Trajectory

Data Mining and Analytics

Miscellaneous

Demo Papers

Social Network and Graphs

Finding Influencers in Temporal Social Networks Using Intervention Analysis

Maximilian Franzke[1], Janina Bleicher[1(✉)], and Andreas Züfle[2]

[1] Ludwig-Maximilians-Universität München, München, Germany
{franzke,bleicher}@dbs.ifi.lmu.de
[2] George Mason University, Fairfax, VA, USA
azufle@gmu.edu

Abstract. People influence, inspire and learn from each other. The result of such a latent cooperation can be observed in social networks, where interacting users are connected with each other. In previous work, the influence casted by a single individual is measured purely quantitatively, by analyzing the topology of a social network. In this work, we take a step towards analyzing the quality of an influencer. For this purpose, we analyze how attributes describing an individual change over time, and put this change in the context of the social network topology changing over time. Each social node is associated with time-series of their attribute values, such as their number of citations. For each individual, we apply the concept of intervention analysis, to identify points in time, so-called interventions, when these attributes have been affected significantly. Such interventions can be of either positive or negative type, and either short-term or long-term. For each intervention, we use the temporal social network's topology to identify candidate individuals, potentially responsible for this intervention. We use these interventions to score their influence on others. This allows us to find users, who have a significant bias towards affecting the performance of other users, in a positive or negative way. We evaluate our solution using a data set obtained from the ACM digital library – containing both a temporal collaboration network, as well as time-dependent attributes such as the number of citations. Our resulting most-positively influential researches is quite different from traditional purely quantitative metrics such as citation count and h-index.

1 Introduction

Human beings are social. They tie and break social links over time. Such social links include family, friends, colleagues, idols and teachers. Through these social links and interactions, individuals influence and change each other: Parents educate their children, friends form common interests and opinions, colleagues share the same projects and ideas, fans want to be like their idol, and students learn from their teachers. All of these examples of social interactions coin the notion of social learning, a well-established concept in social sciences [4,5,14]. For a process to qualify as social learning, it has to *"occur through social interactions*

© Springer International Publishing AG 2016
M.A. Cheema et al. (Eds.): ADC 2016, LNCS 9877, pp. 3–16, 2016.
DOI: 10.1007/978-3-319-46922-5_1

and processes between actors within a social network, either through direct inter-action, e.g., conversation, or through other media, e.g., mass media, telephone, or Web 2.0 applications" [16]. In this work, we want to quantify this learning effect. By using historical social network data, such as citation networks, we measure how a social link established between two users A and B at some time t affects their attributes, such as their citation count, in the future. We call this measure *influence*. The task of this work is to find vertices of an attributed temporal social network, which have the highest influence on other nodes.

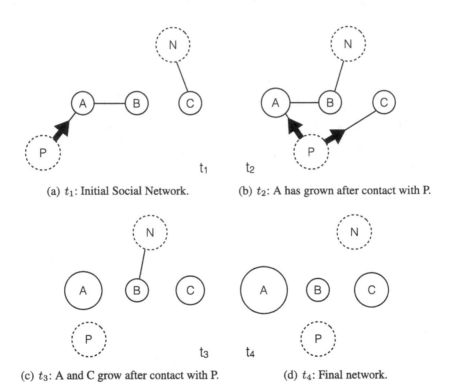

(a) t_1: Initial Social Network.

(b) t_2: A has grown after contact with P.

(c) t_3: A and C grow after contact with P.

(d) t_4: Final network.

Fig. 1. Toy Social Network: Users P and N are making connections to other nodes over time. While every node that has been in touch with P grows in size afterwards, a connection with N seems to have no effect.

Example 1. As an abstract toy example of social learning, consider Fig. 1, which depicts a social network of five individuals $\{A, B, C, P, N\}$ at four consecutive points in time t_1, t_2, t_3, and t_4. For illustration in this example, let individuals A, B and C be high-school students, and let P and N be two teachers. The size of each node corresponds to one of its attributes, such as their current score in mathematics tests. Initially, at time t_1 shown in Fig. 1(a), all three students A, B and C are struggling, having a low score. Thus at this time, A and C are seeking private lessons by teachers P and N respectively. At time t_2 shown in Fig. 1(b), we can observe that the score of A has increased, illustrate by the increases size of the node of A, whereas the score of B and C remained constant. At time t_2, teacher P helps both A and C, leading to an improvement for both of these at time t_3 in Fig. 1(c). At the same time, we see that teacher N seems to have no imminent influence on the scores of the students that N connects with. Finally, in Fig. 1(d), we see that A has obtained a strong improvement in his math score, C has improved slightly, and B remained at their low initial score.

The challenge tackled in this work is to find users of a social network which have an influence on the attributes of other users. A main challenge here is to find influencers that are stochastically significant. In the above example, we have no way to guarantee that the increase in math-score of user A can indeed be contributed to the social link to teacher P: there might be other, unobservable and latent factors which might affect the math score of user A, that are not captured in the data. However, in a real social network, such as the citation network that we employ in our experimental evaluation, a user is involved in a plethora of social connections over time. If a user positively affects a significant fraction of his social links, then this might be a result of the users influence, rather than a result of chance.

To quantify *influence* in a temporal social network, we propose to measure the degree at which individuals of a social network influence the attributes of each others. To find influencers, i.e., nodes of a social network which significantly affect the attributes of others over time, our challenge is to find significant attribute changes over time. However, the influence of one individual may not result in an instant change of attributes of the ones that he affects. Rather, the effect may be delayed and continuous. For instance, an inspiring researcher may improve the research skills of PhD students that he supervises. Thus, the number of papers and the number of citations, may, on average, continuously increase significantly after having interacted in the corresponding collaboration network. To allow detection of delayed and continuous effects on the time series of attributes of an individual, we choose to employ the concept of intervention analysis [6,13] on time-series of attributes of nodes of a temporal social network. We summarize the contributions of this work as follows

- We give the first definition of *influence* in time-dependent attributed graphs. Previous work on influence estimation in social networks only considered the graph topology. But arguably, for a more practical definition of influence, we need to consider attributes (such as citation count) that are influenced, and we need to consider how these attributes change over time.
- We define the problem of finding the top-k most influential users of a social network.
- We utilize the concept of intervention analysis, to find points in time in the past, where a user's attribute values have been influenced significantly.
- We propose an algorithm to efficiently find the top-k most influential users.

The remainder of this work is organized as follows. In the following section, we formally define our notion of influence and define the problem of finding the top-k most influential users. After reviewing the state-of-the-art in Sect. 2, we define the problem of finding influential users of a temporal social network in Sect. 3. To solve the defined problem, we propose two solutions to estimate the influence of a user in Sect. 4. In Sect. 6 we provide an experimental evaluation of our solutions. The experiments were applied to synthetic data and to a large real citation network. Our results on synthetic data show that we can find artificially added influencers with high precision. The experiments on real data show the trade-off between our two proposed solutions in terms of run-time.

2   Related Work

Identifying important nodes in a graph which influence others has long been the focus of numerous studies. An extensive introduction into the field of diffusion models and influence maximization can be found in [7]. Our approach combines traditional maximization problems with intervention analysis with the novel consideration of the temporal behaviour of the influenced nodes. The following sections summarize representative work of both fields.

Influence Maximization. Selecting a subset of vertices that maximizes influence in the overall non-temporal network while restraining the subset to a constraint (e.g. subset size, marketing cost to approach the subset) has first been researched by [9,17]. This is an NP-hard problem, for which [12] first gave an efficient approximation; while [8] provide a scalable solution that allows to set a compromise between runtime and influence spread. Solving the influencer maximization problem generally assumes that the magnitude of the influence from one node to another is already known or trivial [15].

Information and Influence Propagation. Since it is crucial for several application purposes, like the previously mentioned influence maximization, quantifying the influence of one node to another is a necessary step. When working with social graphs, it is handy to derive this information from the graph itself. [11] proposes a list of models for the probability that one node influences another based on an action propagation approach: Static models, which are independent of the temporal domain (Bernoulli distribution, Jaccard Index, Partial Credits), Continuous Time models that assume a different probability within various time intervals, and Discrete Time models, that assume a propagated action will have a lasting effect for a set time interval and end after it.

Intervention Analysis. In the field of Statistics, intervention analysis is a common problem. The effect of an external action, called an intervention, on a variable has been studied by [6], while both providing a model for noise and seasonal frequency. This is done by modelling the intervention as a timeseries on its own, and then estimating the likelihood that this impact can explain the time series. [1] furthermore extended the concept by analysing multiple interventions and multiple variables simultaniously. In the context of these works, it is usually assumed that the timepoint of the intervention is known; although the type of intervention can be modelled through an external variable; e.g. a singular impact (like a terror attack), or an ongoing impact (sustained economic subsidies). The ARIMA-model is best suited to model and to analyze time series [10], as it provides variables for the autoregressive part (AR), which represents a time-series dependency on its previous values (thus for example allowing for an exponential decay), the integration of the data I and the moving average (MA), which models the random noise part in the data. Extending the ARIMA model, ARIMAX furthermore supports a deeper analysis of external influencing variables on the time series. Besides the three integer values another fourth variable – a vector – encodes when the intervention is happening. Therefore, the vector holds an entry for every timepoint, thus representing a timeseries itself. It encodes, how strong the intervention is affecting the model. A simple X vector would consist of just zeros, with a single one (which could for example represent a terror attack). A change in environmental policies on the other would for example be represented by a series of zeros, followed by a series of ones.

3 Problem Definition

Traditional research on social networks (e.g. [2,3]) uses static social networks, that is a snapshot without any notion of time. Our following definitions generalize social networks to consider a notion of time, allowing links between individuals to bond and break over time, and allowing the attribute values of individuals to change over time.

Definition 1 (Temporal Social Graph). *Let T be a time domain. Let $G = (V, E(t), A)$ be a temporal social graph, where $V = \{v_1, v_2, \ldots, v_{|V|}\}$ is the set of vertices, $E(t) \subseteq V \times V \times T$ is the set of edges changing for each time $t \in T$, and $A = \{a_1(v, t), \ldots, a_k(v, t)\}$ a set of time-dependent attribute functions, where each attribute function $A \ni a_i(v, t) : V \times T \mapsto \mathcal{R}$ maps each vertex v and point of time t to an attribute value. More intuitive, using a curried representation of function $a_i(v, t)$, the function $a_i(v)(t) : V \mapsto (T \mapsto \mathcal{R})$ maps each vertex to a time series $T \mapsto \mathcal{R}$, describing the change of the i'th attribute of node v over time.*

The above definition allows edges to appear and disappear depending on the time t. Furthermore, each attribute value $a_i(v)$ of a node $v \in V$ is dependent on t, and thus, is a time-series.

Detecting trends and significant turn-around-points in a temporal performance graph is not trivial. Consider the temporal social toy network depicted in Fig. 2 having three users $V = \{A, B, C\}$. The edge-labels of Fig. 2(a) show the time intervals at which the corresponding pair of users had been interacting. For example, the edges existing at time $t = 8$ are $E(8) = \{(A, B), (B, C)\}$. In addition, Fig. 2(b) shows the time series of the attribute "Number of New Citations", which, formally, is described by attribute function $a(v, t)$, and here for user A by the time-series $a(A, t)$. Consider the yearly publications of a researcher depicted in Fig. 2(b): The overall trend is clear, but dedicating the gain to an actual single (or multiple) time point is not easy. Therefore, a formal definition

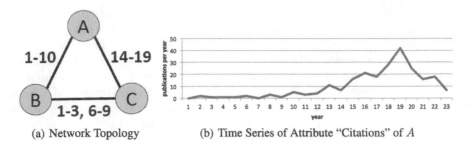

(a) Network Topology   (b) Time Series of Attribute "Citations" of A

Fig. 2. A toy example temporal social network.

of such trend definition is necessary. To express the trend at which the attribute of a node changes over time, a performance function is required. A performance function gives a numeric indication of the temporal change of an attribute a of vertex v at time t.

Definition 2 (Performance Function). *Let $v \in V$ be a vertex, let $a \in A$ be an attribute, and let $a(v)(t)$ be the time-series describing the change of attribute a of v over time. A performance function $P : V \times A \times T \mapsto \mathcal{R}$ maps the times series $a(v)(t)$ and a given point of time $t \in T$, to a score value describing the quality of the intervention of $a(v)(t)$ at time t.*

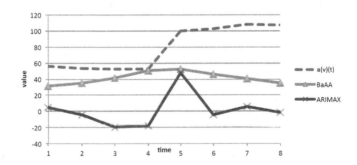

Fig. 3. A synthetic timeseries of a single performance attribute, which experiences a severe boost at $t = 5$. The plot of several performance functions are displayed accordingly.

Intuitively, a performance function assesses the quality of a possible intervention of a point of time in a time series. As an example, consider the time-series in Fig. 3, which shows a fairly constant trend, except for a significant increase at time $t = 5$. Intuitively, a good performance function should be able to map time $t = 5$ to a high performance value, while mapping each other point of time to a low performance value. We will propose two performance functions in Sect. 4. Once we are able to identify the performance of an individual vertex $v \in V$ at a time t, our challenge is to contribute this performance to other vertexes v' that have been connected to v at or around time t. As a first simplistic approach to credit the change in value of an attribute to social neighbour vertices, we propose the following approach: For each point in time, the performance is credited to all current neighbour vertices connected to v at time t:

Definition 3 (Social Influencer Score). *Let $v \in V$ be a node in a temporal social graph G. Its influencer score S^v is defined as*

$$S^v := \sum_{t \in T} \sum_{a \in A} \sum_{(v,v',t) \in E(t)} P(v', a, t),$$

where $P(v', a, t)$ is a performance function as introduced in Definition 2.

As an example, reconsider Fig. 3 showing the attribute times series $a(v)$ of a node v. The high performance increase at time 5 will be contributed to all users connected to v at time t. Furthermore, the slight negative performance at time $t = 3$ and $t = 4$ will be added to all friends of v at these times, this reduces their overall social influence score. Summarizing, the influence score of an individual vertex v is defined as the sum of all interventions of all users that v has been connected to. This allows us to define the problem of finding the top-k most influencial users.

Definition 4 (Top-k Influencer Query (TkIQ)). *Given a temporal social network G, a performance function $P := P(v, a, t)$ and a positive integer k, the Top-k Influencer Query retrieves the subset $TkIQ(G, P, k)$ from V that contains k objects for which the following condition holds:*

$$\forall v \in TkIQ(G, P, k), \forall v' \in V \setminus TkIQ(G, P, k) : S^v > S^{v'}$$

In informal words, the Top-k Influencer Query returns the k users having the highest social influencer score.

4   Performance Functions

In the previous section, we have abstractly defined a performance function, which maps each point of time t of a time-series $a_i(v)$ to a value which captures the quality of the intervention at time t. Yet, no specific implementation has been specified. In the following, we propose two implementation of a performance function, i.e., two approaches to detect interventions that increase an individual's performance. That is, we try to find points in time in the time-series of a node's attributes where a significant improvement in attributes can be suspected. The first approach is a straight-forward approach, which is elegant in terms of computation cost. The second approach applies a state-of-the-art solution in intervention analysis [6,13], which promises to yield better results, but at a much higher computational cost.

Before and After Average (BaAA): We define a temporal window of size r. For a given intervention at time t and a time series $a_i(v)$ of attribute a_i and vertex v, we consider the average values $\overline{a(v,t)}$ of the attribute a of v in the r time points before the intervention t, and analogously we consider the average values $\underline{a(v,t)}$ in the r time points after t. We define the Before-and-After-Average score $\overline{BaAA}(v,a,t)$ as follows:

$$BaAA(v,a,t) := \overline{a(v,t)} - \underline{a(v,t)} = \frac{1}{r}\sum_{j=1}^{r}(a(v,t-j) - a(v,t+j)).$$

Auto-Regression Integrated Moving Average (ARIMA): We propose to model an intervention by an ARIMAX(0,1,1) model [10]. This assumes that the impact of the intervention is assumed to be continuous, having zero decay, using the first derivation (i.e. the trend), with a moving average window size of 1. The value returned by this model is the intervention impact estimated using maximum likelihood estimation following these assumptions. Details of this maximum likelihood estimation can be found in [18]. Figure 3 uses an artificial time series of $a(v)$ with a severe performance increasement at $t = 5$. We also see the results of our proposed performance functions BaAA and ARIMAX. BaAA has a maximum value around the boost, but fails to detect the precise time point. This is due to the window size r, which acts as a blur filter on hard changes of the plot. In this simple example, ARIMAX manages to detect the negative growths at $t = 3$ and $t = 4$. It clearly detects the large step which represents the boost in performance. Intuitively, ARIMAX appears to be more powerful, as it exploits existing research on intervention analysis. However, as we will see in our experimental evaluation, the ARIMAX model does not scale to large temporal social networks, where an extremely large number of models have to be computed. As we will see, the simple BaAA model sacrifices model quality for scalability.

5   Algorithm

This section shows how to handle $TkIQ(G,P,k)$ queries efficiently. For this purpose, we present Algorithm 1, which is able to answer TkIQ queries. Since for an accurate response it is necessary to determine the score-value of every node, a first iteration over the influencee nodes ensures that fewer calls of the performance function $P(v,a,t)$ are necessary, as the computed performance value can be applied to all relevant influencers at once. In terms of complexity, the above algorithm scales linear in the number of temporal edges in G.

6   Experiments

In this section, we evaluate the performance of Algorithm 1, using our two performance functions BaAA and ARIMA, in terms of run-time, and in terms of accuracy of detecting ground-truth influencers. In the next section, Sect. 6.1, we describe the ACM citation dataset that we employ, in Sect. 6.2 we perform a

---

Algorithm 1. TkIQ

Require: G, A, P, k

1: $Score = key \rightarrow value$
2: **for** $v_j \in V$ **do**
3:    **for** $t \in time$ **do**
4:       $perf = 0$
5:       **for** $i = 1$; $i \le |A|$; $i = i + 1$ **do**
6:          $perf = perf + P(v, a_i, t)$
7:       **end for**
8:       **for** $(v_i, v_j, t) \in E$ **do**
9:          $Score_{v_i} = Score_{v_i} + perf$
10:      **end for**
11:    **end for**
12: **end for**
13: *Sort Score by keys ascending*
14: **return** *First k entries of sorted Score*

---

purely qualitative analysis of our query result followed by a quantitative run-time evaluation in Sect. 6.3. Finally, in Sect. 6.4, we employ artificial data where we can define ground-truth influencers, to evaluate how accurately our algorithm can identify these.

6.1   Dataset

To apply our concept of detecting influencers through intervention analysis, we apply it to the ACM citation graph. It consists of 373'830 researchers, that correspond to the set of vertices V of our temporal attributed graph G. For each paper, an edge is added between all co-authors in the corresponding year of the publication (the timeline is discretized to a yearly interval). Thefore, the dataset's 541'230 papers result in 2'900'444 individual cooperations. We choose a one-dimensional vector A for a node's performance, and the performance we measure is *'publications per year'*; so that $a(v, t)$ gives the number of papers v has published in year t. The goal of our Top-k Influencer Query is now to find those people in the graph, that can increase the productivity of others through a cooperation. Imagine for example a PhD student that visits another university for a short period and works with an skillful instructor. Ideally, the student will learn practical methods from their influencer and apply the concepts again back home, thus experiencing an enormous boost in productivity, which can be credited to the instructor. As we present a novel approach to quantify an individual's influence, we will discuss the dataset's performance as we evaluate several performance functions. The underlying social graph is the ACM library. Every time a paper has been listed there, we assume a collaboration between all the authors in that year. Since the timeline has be discretized, we choose yearly intervals as time points. For the sake of simplicity, we only consider a one-dimensional attribute vector A with just a single value per node. In the following, the only attribute $a_1 = a$ denote a users number of publications in year t.

Table 1. TkIQ on the ACM dataset. Microsoft field ranking score and h-index from Google Scholar are denoted as well.

TkIQ Position	Name	Field ranking score (Microsoft)	h-index (Google)
1	Hector Garcia-Molina	69	125
2	Jim Gray	43	79
3	David Maier	38	58
4	Jeffrey D. Ullman	60	105
5	Michael J. Carey	57	n/a
6	Abraham Silberschatz	33	73
7	Serge Abiteboul	52	77
8	Elisa Bertino	39	83
9	David J. DeWitt	62	81
10	Michael Stonebraker	60	n/a
11	John Mylopoulos	35	80
12	Jennifer Widom	66	92
13	Jiawei Han	51	136
14	Sushil Jajodia	36	93
15	Moshe Y. Vardi	24	90
16	Christos H. Papadimitriou	24	111
17	W. Bruce Croft	19	93
18	Joseph M. Hellerstein	41	78
19	Bruce G. Lindsay	31	41
20	Gerhard Weikum	42	74

6.2   TkIQ Qualitative Evaluation

We ran the Top-k Influencer Query on the ACM graph and selected $k = 20$ using only a single attribute a which corresponds to a person's yearly new publications (i.e. their 'productivity'). Finding the top influencers in the database community (only showing people in the table who had published a paper at SIGMOD) applying our approach using the ARIMAX performance function yields the results shown in Table 1. For comparison reasons, we listed the persons' h-indices (from Google Scholar) and field ranking scores (Microsoft Academic) as well. Finding well-known people in the top-k-ranking was expected; however interestingly there seems not to be just a strong correlation with h-index or field ranking, but the influencer ranking, which measures the positive influence on an author's co-authors, seems to be deviating from these existing measures. As an example, author Jim Gray ranks second using our influence score, whereas his ranking is significantly lower using traditional rankings. This implies that Jim Gray's coauthors have been performing exceptionally well, which can possibly be explained by inspiration of their co-author Jim Gray.

6.3   TkIQ Quantitative Evaluation

To answer any Top-k Influencer Query, a large amount of performance function values $P(v, a, t)$ need to be considered – one for each edge in the temporal social graph G. As described in Sect. 5, most of the data can be precomputed in advance to speed up the query. Therefore, we evaluate the overall query time for the two presented performance functions BaAA and ARIMAX. More specificially, we evaluate the total time required to answer a TkIQ query. Using the described ACM dataset, a total of 1'221'809 performance functions needs to be determined. The results of this run-time analysis can be found in Table 2.

Table 2. Detailed information about runtime costs of ARIMA and BaAA performance functions, along a calculation of the total cost of a TkIQ without any precalculations.

	BaAA	ARIMA
Average runtime per performance function calculation	0.0038 ms	238 ms
Total runtime for performance function calculation	4642 ms	290790542 ms
Query overhead	22427 ms	22427 ms
Total TkIQ runtime	27069 ms	290812969 ms

6.4 Hero-Experiment

Using the ranking of authors using real data as shown in Table 1, it is hard to judge the quality of this ranking. Clearly, there is no ground-truth to assess if the co-authors of Hector Garcia-Molina and Jim Gray have truly inspired their co-authors, or if our results is merely a result of random noise captured by our models. To obtain such a ground-truth, we have to fall back to a synthetic data set. In this data set, the (only) attribute value of each user is chosen uniformly in the interval $[0, 1]$. In other words, each node now is associated with a completely random attribute time series, and there is no correlation between collaboration. The network topology of our temporal social graph is the same as the real-data set employed in the previous experiments. To obtain a ground-truth of influences, we choose a random node v_{Hero} uniformly from the network. For every other node v that v_{Hero} interacts with at a time t, the influenced time series $a(v, t+i)$ $(0 \leq i \leq 2)$ will be increased by an 'impact bonus' b to represent the impact at time t and two subsequent points in time. After this adjustment, the Top-k Influencer Query is restarted and v_{Hero}'s position within the internal ranking is analyzed. Since this dataset is generated in a way such that v_{Hero} is the only positive influencer, while all other authors are independent of each other, we hope that v_{Hero} is ranked highly in our query result.

For this experiment, we choose $|V| = 1000$ and a hundred time points $|Time| = 100$. Edges and temporal intensity function are taken from a local cluster of the ACM citation network graph. A hundred different v_{Hero} were chosen randomly to be the influencer in each iteration of this experiment, and results were averaged over those hundred runs.

Figure 4 shows the outcome for a varying impact bonus b. On the y-axis, the node's global ranking position is depicted. For a bonus $b = 0$, node v_{Hero} has

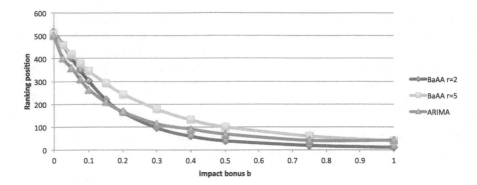

Fig. 4. Varying the size of the bonus a user has on its neighbours. On the y-axis, this node's global influencer ranking is denoted.

no impact on its influencees, just like all other nodes. In this case, the ranking position of v_{Hero} is completely random, such that all approaches yield an average ranking position of about 500/1000. As the impact bonus increases, we see that the average ranking position of v_{Hero} decreases quickly for all approaches. Converging towards the perfect results of a ranking position of one. We note that the performance function BaAA having a window size parameter value of $r = 2$ appears to perform the best. This is contributed to the fact that our artificial interventions affect exactly three points of time, for which the window size of $r = 2$ is perfectly adjusted to. However, in practice, it is not possible to estimate how long and how continuous the impact of a collaboration affects a coauthor. In particular, some authors might be inspired for a longer time than others. Consequently, the BaAA approach for $r = 2$ has an unfair advantage in this setting. For the case where the parameter r is non-optimally set to $r = 5$, we see that BaAA performs much worse. Looking at the performance of ARIMA, we see that the performance is in-between both approaches. This indicates that ARIMA is able to yield high quality influencer results without making any assumptions on the form and shape of an influence. Furthermore, ARIMA further allows to estimate different kinds of interventions, having a higher or lower decay factor, i.e., being more or less lasting. Yet, we can see that this increase in result quality comes at a price. Table 2 shows that ARIMA has an extremely high run-time, making this model inapplicable for large datasets.

7   Conclusions

In this work we apply the concept of intervention analysis on temporal social networks, to find the most influential users. For this purpose, we defined an attributed temporal social network, where connections between users, as well as attributes of users may change over time. To find influential users, we use a performance function to estimate the quality of an intervention of the attribute time-series of a user v at each point of time. Specifically, we present two performance functions, one straight-forward function which compares the attribute-average before and after the intervention, and a function which uses state-of-the-art intervention analysis using the ARIMA model. Regardless of the employed performance function, the resulting performance is projected to all users having been connected to v at the time of the intervention. Thus, if a user incurs positive interventions to his co-authors, this user will have a high total social influencer score. Note that a performance function may also yield negative values, such that a collaboration which yield a significant drop in one users productivity, will have a negative impact on the social influencer score of the coauthors of the user. Our experiments show that our simple model can be used to find the most influential users efficiently. In contrast, applying state-of-the-art intervention analysis yields a higher quality in terms of accuracy of finding ground-truth influencers, but these models do not scale to large networks.

The models that we proposed in this work are simple, and should be seen as a seed for future research. In particular our very straight-forward allows a multitude of optimizations, which are out of scope of this paper. Also, the ARIMA model has a multitude of parameters that can be tuned for specific datasets. In particular, the ARIMA model allows to specify the time-window of interest, as well as a decay function to model a loss of influence after a collaboration.

References

1. Abraham, B.: Intervention analysis and multiple time series. Biometrika **67**(1), 73–78 (1980)
2. Armenatzoglou, N., Ahuja, R., Papadias, D.: Geo-social ranking: functions and query processing. VLDB J. - Int. J. Very Large Data Bases **24**(6), 783–799 (2015)
3. Armenatzoglou, N., Papadopoulos, S., Papadias, D.: A general framework for geo-social query processing. Proc. VLDB Endowment **6**(10), 913–924 (2013)
4. Bandura, A., McClelland, D.C.: Social learning theory (1977)
5. Bandura, A., Walters, R.H.: Social learning and personality development, vol. 14, JSTOR (1963)
6. Box, G.E., Tiao, G.C.: Intervention analysis with applications to economic and environmental problems. J. Am. Stat. Assoc. **70**(349), 70–79 (1975)
7. Chen, W., Lakshmanan, L.V., Castillo, C.: Information and influence propagation in social networks. Synth. Lect. Data Manage. **5**(4), 1–177 (2013)
8. Chen, W., Wang, C., Wang, Y.: Scalable influence maximization for prevalent viral marketing in large-scale social networks. In: Proceedings of the 16th ACM SIGKDD International Conference on Knowledge Discovery and Data Mining, pp. 1029–1038. ACM (2010)
9. Domingos, P., Richardson, M.: Mining the network value of customers. In: Proceedings of the Seventh ACM SIGKDD International Conference on Knowledge Discovery and Data Mining, pp. 57–66. ACM (2001)
10. Fleming, N.S., Gibson, E., Fleming, D.G.: The use of proc arima to test an intervention effect. In: SAS Conference Proceedings: South-Central SAS Users Group, pp. 91–98 (1997)

11. Goyal, A., Bonchi, F., Lakshmanan, L.V.: Learning influence probabilities in social networks. In: Proceedings of the Third ACM International Conference on Web Search and Data Mining, pp. 241–250. ACM (2010)
12. Kempe, D., Kleinberg, J., Tardos, É.: Maximizing the spread of influence through a social network. In: Proceedings of the Ninth ACM SIGKDD International Conference on Knowledge Discovery and Data Mining, pp. 137–146. ACM (2003)
13. Khandker, S.R., Koolwal, G.B., Samad, H.A.: Handbook on Impact Evaluation: Quantitative Methods and Practices. World Bank Publications, Washington, D.C (2010)
14. Miller, N.E., Dollard, J.: Social learning and imitation (1941)
15. Pham, H., Shahabi, C.: Spatial influence - measuring followship in the real world. In: IEEE 32nd International Conference on Data Engineering (ICDE), pp. 529–540. IEEE (2016)
16. Reed, M., Evely, A.C., Cundill, G., Fazey, I.R.A., Glass, J., Laing, A., Newig, J., Parrish, B., Prell, C., Raymond, C., Stringer, L.: What is social learning? Ecology and Society (2010)
17. Richardson, M., Domingos, P.: Mining knowledge-sharing sites for viral marketing. In: Proceedings of the Eighth ACM SIGKDD International Conference on Knowledge Discovery and Data Mining, pp. 61–70. ACM (2002)
18. Sowell, F.: Maximum likelihood estimation of stationary univariate fractionally integrated time series models. J. Econometrics **53**(1), 165–188 (1992)

Retrieving Top-k Famous Places
in Location-Based Social Networks

Ammar Sohail[1]([⊠]), Ghulam Murtaza[2], and David Taniar[1]

[1] Faculty of Information Technology, Monash University, Melbourne, Australia
{ammar.sohail,david.taniar}@monash.edu
[2] Intersect, Melbourne, Australia
ghulam@intersect.org.au

Abstract. The widespread proliferation of location-acquisition techniques and *GPS*-embedded mobile devices have resulted in the generation of geo-tagged data at unprecedented scale and have essentially enhanced the user experience in location-based services associated with social networks. Such location-based social networks allow people to record and share their location and are a rich source of information which can be exploited to study people's various attributes and characteristics to provide various Geo-Social (GS) services. In this paper, we propose a new type of query called *Top-k famous places T_kFP* query, which enriches the semantics of the conventional spatial query by introducing a social relevance component. In addition, three approaches namely, (1) Social-First (2) Spatial-First and (3) Hybrid are proposed to efficiently process T_kFP queries. Finally, we conduct an exhaustive evaluation of the proposed schemes using real and synthetic datasets and demonstrate the effectiveness of the proposed approaches.

1 Introduction

The fusion of social and geographical information has given rise to the notion of online social media known as location-based social networks (LBSNs) such as Facebook and Foursquare. An LBSN is usually represented as a complex graph where nodes represent various entities in the social network (such as users, places or pages) and the edges represent the relationships between different nodes. These relationships are not only limited to friendship relations but also contain other types of relationships such as `works-at`, `born-in`, and `studies-at` etc. In addition, the nodes may also contain spatial information such as a user's check-ins at different locations. Consider the example of a Facebook user Alice who was born in Germany, works at Monash University and checks-in at a particular restaurant. Facebook records this information by creating Facebook pages for *Monash University* and *Germany* and adding their relationships with the node corresponding to Alice [1], e.g., Alice and Monash University are connected by an edge labelled `works-at` and Alice and Germany are connected with an edge labelled `born-in`. The check-in information records the places the user has visited.

© Springer International Publishing AG 2016
M.A. Cheema et al. (Eds.): ADC 2016, LNCS 9877, pp. 17–30, 2016.
DOI: 10.1007/978-3-319-46922-5_2

Spatial data and social relationships in LBSNs provide a rich source of information which can be exploited to offer many interesting services. Consider the example of a German tourist visiting Melbourne. She may want to find a nearby pub which is popular (e.g., frequently visited) among people from Germany. This involves utilizing spatial information (i.e., near by pub, check-ins) as well as social information (i.e., people who were born-in Germany). Similarly, a user may want to find near by places that are most popular among her friends, e.g., the places most frequently visited by her friends.

Although various types of queries have been studied on LBSNs [2–7], to the best of our knowledge, none of the existing techniques can be applied to answer the queries like the above that aim at finding near by places that are popular among a particular group of users satisfying a social constraint. Motivated by this, in this paper, we formalize this problem as a *Top-k famous places T_kFP* query and propose efficient query processing techniques. Specifically, a T_kFP query retrieves top-k places (points of interest) ranked according to their spatial and social relevance to the query user where the spatial relevance is based on how close the place is to a given location and the social relevance is based on how frequently it is visited by the one-hop neighbors of the query user in the social graph. A formal definition is provided in Sect. 2.1.

We present three approaches to answer T_kFP query processing called, (1) Social-First, (2) Spatial-First and (3) Hybrid. The first two approaches separately process the social and spatial components of the query and do not require a specialized index. The third approach (*Hybrid*) is capable of processing social and spatial components simultaneously by utilizing a hybrid index specifically designed to handle T_kFP queries.

We make the following contributions in this paper.

1. To the best of our knowledge, we are the first to study the T_kFP queries that retrieves near by places popular among a particular group of users in the social network.
2. We propose three approaches to process the query which enable flexible data management and algorithmic design.
3. We conduct an exhaustive evaluation of the proposed schemes using real and synthetic datasets and demonstrate the effectiveness of the proposed approaches.

2   Preliminaries

2.1   Problem Definition

Location Based Social Network (LBSN): A *location-based social network* consists of a set of entities U (e.g., users, Facebook Pages etc.) and a set of places P. The relationship between two entities u and v are indicated by a labeled edge where the label indicates the type of relationship (e.g., friend, lives-in). LBSN also records check-ins where a check-in of a user $u \in U$ at a particular place $p \in P$ indicates an instance that u had visited the place P.

Score of a place p: Given a query user q, and a range r, the score of a place $p \in P$ is 0 if $||q, p|| \geq r$ where $||q, p||$ is the Euclidean distance between query location and p. If $||q, p|| \leq r$, the score of p is a weighted sum of its spatial score (denoted as $p_{spatial}$) and its social score (denoted as p_{social}).

$$p.score = \alpha \times p_{spatial} + (1 - \alpha) \times p_{social} \tag{1}$$

where α is a parameter used to control the relative importance of spatial and social scores. The social score $p.social$ is computed as follows. Let F_q denote the one-hop neighbors of the query user considering a particular relationship type, e.g., if the relationship is **born-in** and the query entity is the Facebook Page named Germany, then F_q is a set of users born in Germany. Although our techniques can be used on any type of relationship, for the ease of presentation, in the rest of the paper we only consider the friendship relationships. In this context, F_q contains the friends of the query user q. Let $p.visitors$ denote the set of all users that visited (i.e., checked-in at) the place p. The social score p_{social} is computed as follows.

$$p_{social} = \frac{|F_q \cap p.visitors|}{|F_q|} \tag{2}$$

where $|X|$ denote the cardinality of a set X. Intuitively, p_{social} is the proportion of the friends of q who have visited the place p.

The spatial score $p_{spatial}$ is based on how close the place is to the query location. Formally, given a range r, $p_{spatial} = 0, (r - ||q, p||)$ where $||q, p||$ indicates Euclidean distance between the query location and p. Note that p_{social} is always between 0 to 1 and we normalize $p_{spatial}$ such that it is also within the range 0 to 1, e.g., the data space is normalized such that $||q, p|| \leq 1$ and $r \leq 1$.

Top-k Famous Places ($T_k FP$) Query: Given a LBSN, a $T_k FP$ query q returns k places with the highest scores where the score $p.score$ of each place p is computed as described above.

Example 2.1: Fig. 1(a) illustrates the locations of a set of places $P = \{p_1, p_2, p_3, p_4\}$. The query q shown in Fig. 1(a) with $k = 2$ and range $r = 0.15$, has a set of friends $F_q = \{u_1, u_2, u_3, u_4, u_5, u_6, u_7, u_8, u_9, u_{10}\}$. The number in bracket next to each place is the check-in count made by q's friends. Figure 1(b) shows the Euclidean distances and the visitors of each place amongst q's friends. Let us assume $\alpha = 0.5$, the score of the p_1 w.r.t. q is computed as $0.025 + 0 = 0.025$. Similarly, we have $Score(p_2) = 0.185$, $Score(p_3) = 0.205$ and $Score(p_4) = 0.115$. The result of the query q is (p_2, p_3) according to scoring function in Eq. 1.

2.2   Related Work

Since Geo-Social query processing is an emerging field, there is only limited literature available about it [3,8–10]. In [11], an Algebric model is proposed to process geo-social query. This model consists of set of operators to query the geo-social data. They replicate the graphs to represent spatial and social components

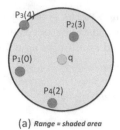

P	$\|q, p_i\|$	P_i.visitors	Final Score
P_1	0.10	-	0.025
P_2	0.08	u_1, u_3, u_9	0.185
P_3	0.14	u_5, u_6, u_2, u_7	0.205
P_4	0.12	u_7, u_{10}	0.115

(a) *Range = shaded area*

(b) $\alpha = 0.5, k = 2, r = 0.15$
$F_q = \{u_1, u_2, u_3, u_4, u_5, u_6, u_7, u_8, u_9, u_{10}\}$

Fig. 1. Top-k Query example

which can be very large in terms of social networks thus, making query processing more cumbersome. Huang et al. [12] studied a Geo-Social query that retrieves the set of nearby friends of a user that share common interests, without providing concrete query processing algorithms. In [6], they defined a new query namely, *Geo-Social Circle of Friends* to retrieve the group of friends in geo-social settings whose members are close to each other based on their geographical and social circumstances such as for group sports, social gathering and community services. Yang et al. [7] introduced another type of group query by extending the work presented in [6] namely, *Social-Spatial Group Query* (SSGQ) which is useful for impromptu activity planning.

In addition, nearest neighbour queries have been widely applied in location-based social networks recently [13–17]. Wu et al. [18] proposed a new query named as social-aware $top-k$ spatial keyword query (SkSK) which retrieves a list of k objects ranked according to their spatial proximity, textual (e.g. restaurant has menu and different facilities) and social relevance. Another work is presented by Jiang et al. [19] in which they proposed a method to find *top-k* local users in geo-social media data. First, they compute spatial score of a user who has posted tweets in given region followed by computing social score of her tweets based on number of replies or forwards to her tweets. Subsequently, by fusing both scores, the user is ranked accordingly.

To the best of our knowledge, none of the existing algorithms can be applied or trivially extended to answer $T_k FP$ queries studied in this paper.

2.3 Framework Overview

The proposed framework consists of three approaches to answer $T_k FP$ query: (I) Social-First (II) Spatial-First and (III) Hybrid. The *Social-First* approach first processes the social component (e.g., friendship relations and their check-ins) and then processes the spatial component (e.g., places in given range) where as *Spatial-Fisrt* initially processes the spatial component followed by processing the social component. In contrast, the *Hybrid approach* is capable of processing both social and spatial components simultaneously to answer such queries. More specifically, it leverages two types of pre-processed information associated with each user $u \in U$, her check-in information (check-ins) and summary of her friends' check-ins information.

To the best of our knowledge, there is no unanimously accepted social or spatial storage implementation. Specifically, Facebook uses adjacency lists stored in Memcached [20] which is a distributed memory caching system, on the other hand, Twitter leverages the R*-Tree [21] spatial index. Further, Foursquare uses MongoDB [22], a document oriented database. Similarly, academics research has been adopting various kind of approaches such as [5] uses adjacency list stored in Neo4j which is a graph based database, where as [11] utilizes relational tables for storing the friendship relations.

Similar to the existing work on KNN queries, we tailored the storage implementation for our technique in a way that suites our requirements. More specifically, we index places and users' check-in information by adopting the R-Tree [23] spatial Index structure. Before presenting our technique, we present the definition of *Facility R-Tree, Check-in R-Tree and Friendship Index*.

Facility R-Tree: Stores all places ($p \in P$) in a given data set. A node of *Facility R-Tree* is represented by a minimum bounded rectangle (MBR) that constitutes all places in its sub-tree.

Check-in R-Tree: For the sake of efficiency, we store the check-in information of each user by indexing all visited places by her in a separate index based on R-Tree.

Friendship Index: To index each user $u \in U$ and their social relationships, we build an index structure by employing B^+-*Tree* based on *Unicorn* [1] built to search the *facebook* social graph.

3 Proposed Technique

3.1 Social-First Based Approach

Intuitively, *Social-First* approach, first processes the social component of given query q and computes the score of all the places $p \in P$ which have been checked-in by her friends $u \in F_q$. Next, using given range r, it processes the spatial component of the q and computes the score of the remaining places $p \in P$ which are not checked-in and returns the set of *top-k* places based on their score and q's defined preference criteria α. We next describe the technique in detail with pseudocode given in Algorithm 1.

Initially, in the first loop of the algorithm, we compute the score of each place p in given range r, that has been visited by the friends $u \in F_q$ of query q while maintaining the score of current k_{th} place p based on their social and spatial scores by exploiting the *Check-in R-Tree* of each friend u. In-addition, in the second loop, the *Facility R-Tree* is exploited to compute the score of those places $p \in P$ in range r which are not visited by q's friends hence, their respective score only comprises of spatial component and their $(P_{social}) = 0$ which subsequently yields the *top-k* result set. Let us assume, the score of current k_{th} place p is

$Score_k$, next lemma shows that if the $||q,p|| \geq (r - \frac{Score_k}{\alpha})$, we can prune that place p. Next, we introduce our first pruning rule in Lemma 1.

Lemma 1. *Every place p that has a distance $||q,p||$ from query q greater than current $(r - (Score_k/\alpha))$, cannot be in the Top-k places.*

Proof. Given a query user q, a range r, preference factor α, a place p which is not checked-in by any user $u \in F_q$ has social component $(p_{social}) = 0$, by using Eqs. 1 and 2 we get,

$$Score(p) = \alpha(r - ||q,p||) + 0 \tag{3}$$

To be the candidate for the *Top-k places*, a place p's score must be greater than current $Score_k$, hence

$$Socre_k \leq Socre(p) \tag{4}$$

By substituting the value of $Socre(p)$ from Eq. 3,

$$Socre_k \leq \alpha(r - ||q,p||)$$
$$||q,p|| \leq (r - (Score_k/\alpha)) \tag{5}$$

□

Algorithm 1. Social-First

   Input   : Query q, range r, weight-age constant α, integer k,
   Output: Result set R
1 **foreach** *friend u in F_q* **do**
2   Traverse check-in R-Tree of u ; // `Accessing Check-in R-Tree by branch and bound method`
3   **foreach** *place p in r in Check-in R-Tree* **do**
4   update social score and score of p;
5   update $Score_k$;
6   **end**
7 **end**
8 Traverse Facility R-Tree ;   // `Accessing Facility R-Tree by branch and bound method`
9 **foreach** *place p in range r in Facility R-Tree* **do**
10   **if** $p.dist \leq (r - (Score_k/\alpha))$;   // `from Lemma 1`
11   **then**
12   compute $Score(p)$;
13   update $Score_k$;
14   **end**
15 **end**
16 **return** *Return R*

3.2   Spatial-First Based Approach

Initially, this approach starts with the processing of spatial component of the query q by computing the score of each place p in given range r regardless of the fact whether it is checked-in by any friend $u \in F_q$ of q or not. Moreover, it then computes the social score of each place p by first computing the number of friends checked-in to it by performing an intersection of the set of q's friends F_q and the set of visitors of the place V_p followed by computing the social score p_{social} of p and then yields the result set. We next elaborate the technique in detail with pseudocode given in Algorithm 2.

Specifically, for each place $p \in P$ in range r, we compute the $Score(p)$ in ascending order of the distance of the place p from q. To achieve this, a $Heap$ is initialized with the root entry of $Facility$ $R\text{-}Tree$ with $||q, e||$ as a key to process spatial component first. Each entry is iteratively retrieved from $Heap$ and processed as follows. For each place p in range r, we first, compute social component of the score by counting the number of friends $u \in F_q$ who have visited the place p followed by the computation of the final score using Eq. 1. Let us assume, the score of current k_{th} place p is $Score_k$, next lemma shows that if the $||q, p|| \geq (r - \frac{(Score_k - (1-\alpha))}{\alpha})$, the process stops since every subsequent place p entry in $Heap$ is further than the current place p entry from q. Next, we introduce our second pruning rule in Lemma 2.

Lemma 2. *Every place p that has distance $||q, p||$ from query q greater than current $Score_k$, cannot be in the Top-k places.*

Proof. Given a query user q, a range r, preference factor α, to be the candidate for the $Top\text{-}k$ $places$, a place p's score must be greater than current $Score_k$, by using Eqs. 1 and 2, we get,

$$Socre_k \leq Socre(p) \tag{6}$$

By substituting the value of $Socre(p)$, we get,

$$Socre_k \leq \alpha(r - ||q, p||) + (1 - \alpha)\left(\frac{F_q \cap V_p}{|F_q|}\right) \tag{7}$$

Since the maximum possible social score of given place can be 1, we get

$$Socre_k \leq \alpha(r - ||q, p||) + (1 - \alpha) * 1$$
$$||q, p|| \leq (r - \frac{(Socre_k - (1 - \alpha))}{\alpha}) \tag{8}$$

$\square$

3.3   Hybrid Approach

Friends Check-Ins R-Tree: To Optimize, we propose a spatial indexing structure, the *Friends Check-ins R-tree*, that supports the simultaneous pruning of

Algorithm 2. Spatial-First

Input   : Query q, range r, weight-age constant α, integer k,
Output: Result set R

1 Traverse Facility R-Tree ;   `// Accessing Facility R-Tree by branch and`
  `bound method`
2 **foreach** *place p in r in Facility R-Tree* **do**
3 | **if** $||q,p|| \geq (r - ((Score_k - (1 - \alpha))/\alpha))$;   `// from Lemma 2`
4 | **then**
5 | | **return** *Result set R*
6 | **end**
7 | count friends $\leftarrow F_u \cap V_p$;
8 | compute $Score(p)$;
9 | update $Score_k$;
10 **end**
11 **return** *Result set R*

---

friends and places. It is an R-Tree based structure which is constructed for each user $u \in U$ and is able to prune the search space. *FCR-Tree* stores check-in information of each friend $u \in F_q$ of q, thus representing the check-in summary of all friends of q. The objects of *FCR-Tree* are the root MBRs of each friend's *Check-in R-Tree*. The update of the index in case of new check-in entry of any friend u, is not costly since these are being bulk updated after certain period of time.

Let us assume a query $q \in U$ where the friends of q are $F_q = \{u_2, u_3, u_4, u_5, u_6\}$. Figure 2 shows the conceptual view of the *FCR-Tree* of u_1. In following Sect. 3 we describe our proposed technique in detail.

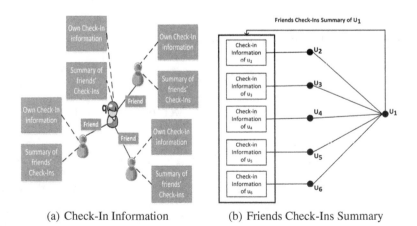

(a) Check-In Information   (b) Friends Check-Ins Summary

Fig. 2. Check-Ins Example

In this approach, the score of each place p in range r is computed by processing the social and spatial components of query q together. To answer the $T_k FP$ queries efficiently, this approach leverages the *Friends Check-ins R-Tree* of query q to prune the friends which have not visited the top-k places and *Grid* Spatial Index to prune the places in given range r which cannot be the candidate for the top-k result set. More specifically, to compute the social and spatial scores of a place p, this approach supports simultaneous pruning of q's friends set F_q and places $p \in P$ in given range r. We next elaborate the technique in detail with pseudocode given in Algorithm 3.

To achieve this, initially, grid portioning approach is employed to divide the area formed by given range r into small cells. Similarly, for each grid cell g_c, a set of places $P_{gc} \in P$ which lie inside the cell is maintained by using the *Facility R-Tree* and distance of the closest place p to q in a cell is recorded as the cell distance from q. In addition, a set of friends who might have visited a cell denoted as V_{cell} is computed for each cell by exploiting *Friends Check-ins R-Tree* ($FCR-Tree$) of q and counting the number of overlapping objects of $FCR-Tree$ with the cell. Figure 3 illustrates an example of a cell and overlapping objects in which four objects are overlapping with the cell and therefore, the overlap count of the cell is 4.

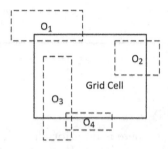

Fig. 3. Cell Overlap

Once the overlap count and distance to the q for each cell g_c is computed, a ranking score of each cell is computed using Eq. 9 which serves as an upper bound on the score of any place in the cell. Moreover, to compute the score of a place p in range r, the places $p \in P_{gc}$ of the cell g_c with highest score are processed first. If the current k_{th} score is greater than the next cell's score, the process stops since all subsequent cells can not contain a place with higher ranking score.

$$Score_{cell} = \alpha(r - cell.distance) + (1 - \alpha)\left(\frac{OverlapCount_{cell}}{|F_q|}\right) \quad (9)$$

Algorithm 3. Hybrid Algorithm

Input   : Query q, range r, weight-age constant α, integer k,
Output: Result set R
1 Traverse Facility R-Tree ;   // Using branch and bound method
2 **foreach** *place p in r in Facility R-Tree* **do**
3 | compute $Score(p)$;
4 | insert p in corresponding cell's places set;
5 | update the distance of corresponding cell;
6 | update $Score_k$;
7 **end**
8 Traverse FCR-Tree of q ;   // Using branch and bound method
9 **foreach** *cell g_c* **do**
10 | compute $overlapCcount(g_c)$ and $Score(g_c)$;
11 **end**
12 **foreach** *cell g_c* **do**
13 | **if** $Score_{gc} \leq Score_k$ **then**
14 | | return Result set R;
15 | **end**
16 | **foreach** *place p in cell g_c* **do**
17 | | compute check-in count of $p \leftarrow V_{cell} \cap V_p$;
18 | | update $Score(p)$ and $Score_k$;
19 | **end**
20 **end**
21 **return** *Result set R*

---

4   Experiments

4.1   Experimental Setup

To the best of our knowledge, there is no prior algorithm to solve $T_k FP$ queries therefore, we compare the three proposed algorithms with each other to evaluate their performance.

All algorithms were implemented in C++ and experiments were run on Intel Core $I3$ 2.4 GHz PC with 8GB memory running on 64-bit Ubuntu Linux. Specifically, we use same real data set of *Gowalla* as of [24]. *Gowalla* is a location based social network which later was acquired by *Facebook*. It contains 196,591 users, 950,327 friendships, 6,442,890 check-ins and 1,280,956 checked-in places across the world. The page size of each *Facility R-Tree* index is set to 4096 Bytes and 1024 Bytes for *Check-in R-Tree* and *FCR-Tree* indexes. We randomly select 100 users and treat them as query points. The cost in the experiments correspond to the average cost of 100 $T_k FP$ queries. The default value of range r is 100 Km and the default value of k is set to 10 unless mentioned otherwise.

4.2   Performance Evaluation

Effect of Range: We analyse the performance of our algorithms for various range values ranging from $10 - 400$ kms. The size of the area formed by given range determines the number of places it contains (ranging from 1500 − 94000). Figure 4 shows that *Spatial-First* algorithm is most affected at bigger range values hence its performance deteriorates due to large number of places. Note that, the *Hybrid* algorithm performs better for bigger range since it is more likely to find *top-k* places after processing fewer cells. Figure 4(b) shows that I/O cost increases for bigger range values due to large number of places in range which result in higher index access rate.

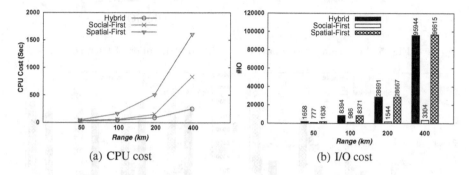

(a) CPU cost   (b) I/O cost

Fig. 4. Effect of varying range (number of places)

Effect of Average number of Friends: In Fig. 5, we study the effect of the average number of friends of each query. Note that the size of *FCR-Tree* depends on the size of friends set of each user in data set which essentially affects the Hybrid algorithm. Further, *Spatial-First* algorithm is greatly affected by it because the intersection of two large sets i.e. visitors set and friends set is more expensive. Specifically, Fig. 5(a), shows the CPU cost and Fig. 5(b) shows the I/O cost of each method for varying average number of friends. The average number of places in given range r is 38319. Note that when average number of friends increases, the CPU and I/O cost of all three algorithms increases since each friend's check-in information is required to verify candidate places.

Effect of concurrent number of Queries: Geo-Social services seek to answer large number of incoming queries simultaneously due to the enormous size of registered users. Therefore, the number of concurrent queries ranging from 50 to 200 are analyzed for all three algorithm. In addition, each experiment involves average number of friends ranging from 750 − 1350 and approximately 10,000 average number of places in given range r. All three algorithms need to traverse the *Facility-RTree* every time a T_kFP query is issued to verify candidate place. The *Social-First* algorithm also needs to traverse the *Check-in R-Tree*. On the other hand, *Hybrid* algorithm leverages the *FCR-Tree* and both *Spatial-First and Hybrid* greatly rely on the visitors set of the places. In Fig. 6, we report the CPU and I/O cost of each algorithm on *Gowalla* data set for different number of queries. As expected, the I/O cost of *Social-First* algorithm is less than the other two due to low dependency on indexes. Note that *Hybrid* is up to four times better than *Social-First* and *Spatial-First* algorithms.

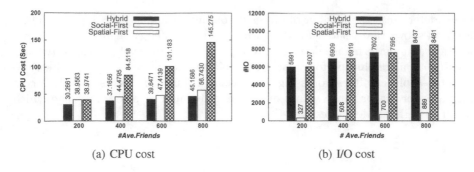

(a) CPU cost   (b) I/O cost

Fig. 5. Performance comparison on different number of friends

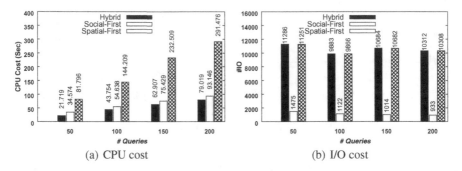

(a) CPU cost   (b) I/O cost

Fig. 6. Effect of number of Queries

Effect of Grid Size: In Fig. 7(a), we study the effect of the size of grid partitioning ranges from $2 - 64$ on Hybrid algorithm. The size of grid affects the CPU cost since the size of a cell defines how many places will be processed/pruned at once. Similarly, it also affects the termination condition on the algorithm. Note that the best CPU performance can be achieved by dividing the area into grid of size 4×4.

Effect of k: In previous experiments, the value of k is set to 10. Next, we analyze the performance of three algorithms for various values of k. Note that in Fig. 7(b), all three algorithms are nearly independent of k. The reason is that, we have to update the result set every time we update the score of a place. Therefore, the size of the result set does not impose great computation load. In terms of I/O cost, Fig. 7(c) shows that all three algorithms do not get affected by the value of k.

Effect of Data Set Size: In Fig. 8(a) and (b), we study the effect of data set size on the performance of the three algorithms. Specifically, we conduct experiments on synthetic data sets of different sizes containing places ranging from $100\,k$ to $500\,k$. In Fig. 8(a), note that the *Spatial-First* algorithm is most effected by number of places. Similarly, in Fig. 8(b), *Hybrid and Spatial-First* have higher I/O cost due to the intersection performed on visitors set of places and friends set of query q.

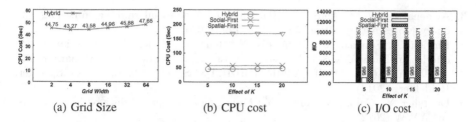

(a) Grid Size   (b) CPU cost   (c) I/O cost

Fig. 7. Effect of Grid Size and varying number of requested places (k)

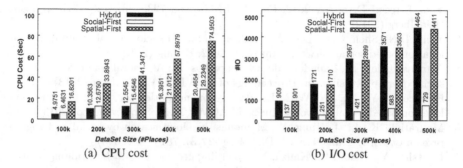

(a) CPU cost   (b) I/O cost

Fig. 8. Effect of varying data set sizes (number of places)

5 Conclusions

This paper studies *Top-k famous places* $T_k FP$ query, which enriches the semantics of the conventional spatial query by introducing a social relevance component. We propose three approaches namely, (1) Social-First (2) Spatial-First and (3) Hybrid to efficiently process such queries. In addition, an *FCR-Tree* index structure is proposed that integrates social information with spatial information. Results of empirical studies with an implementation demonstrate the effectiveness of the proposed approaches using real and synthetic datasets.

In the future, it will be interesting to explore the implications of $T_k FP$ query, i.e., aiming to retrieve a result that is as beneficial as possible for an important use case based on dependable ground truth by collecting real data including query workloads.

References

1. Curtiss, M., Becker, I., Bosman, T., Doroshenko, S., Grijincu, L., Jackson, T., Kunnatur, S., Lassen, S., Pronin, P., Sankar, S., Shen, G., Woss, G., Yang, C., Zhang, N.: Unicorn: a system for searching the social graph. PVLDB **6**(11), 1150–1161 (2013)
2. Ahuja, R., Armenatzoglou, N., Papadias, D., Fakas, G.J.: Geo-social keyword search. In: Claramunt, C., Schneider, M., Wong, R.C.-W., Xiong, L., Loh, W.-K., Shahabi, C., Li, K.-J. (eds.) SSTD 2015. LNCS, vol. 9239, pp. 431–450. Springer, Heidelberg (2015)

3. Armenatzoglou, N., Ahuja, R., Papadias, D.: Geo-social ranking: functions and query processing. VLDB J. **24**(6), 783–799 (2015)
4. Emrich, T., Franzke, M., Mamoulis, N., Renz, M., Züfle, A.: Geo-social skyline queries. In: Bhowmick, S.S., Dyreson, C.E., Jensen, C.S., Lee, M.L., Muliantara, A., Thalheim, B. (eds.) DASFAA 2014, Part II. LNCS, vol. 8422, pp. 77–91. Springer, Heidelberg (2014)
5. Doytsher, Y., Galon, B., Kanza, Y.: Managing socio-spatial data as large graphs. In: WWW (2012)
6. Liu, W., Sun, W., Chen, C., Huang, Y., Jing, Y., Chen, K.: Circle of friend query in geo-social networks. In: Lee, S., Peng, Z., Zhou, X., Moon, Y.-S., Unland, R., Yoo, J. (eds.) DASFAA 2012, Part II. LNCS, vol. 7239, pp. 126–137. Springer, Heidelberg (2012)
7. Yang, D.N., Shen, C.Y., Lee, W.C., Chen, M.S.: On socio-spatial group query for location-based social networks. In: KDD 2012, Beijing (2012)
8. Armenatzoglou, N., Papadopoulos, S., Papadias, D.: A general framework for geo-social query processing. In: Proceedings of the VLDB Endowment (2013)
9. Ference, G., Ye, M., Lee, W.C.: Location recommendation for out-of-town users in location-based social networks. In: 22nd ACM, CIKM, San Francisco (2013)
10. Mouratidis, K., Li, J., Tang, Y., Mamoulis, N.: Joint search by social and spatial proximity. IEEE Trans. Knowl. Data Eng. **27**(3), 781–793 (2015)
11. Doytsher, Y., Galon, B., Kanza, Y.: Querying geo-social data by bridging spatial networks and social networks. In: LBSN, San Jose (2010)
12. Huang, Q., Liu, Y.: On geo-social network services. In: 17th International Conference on Geoinformatics, 2009, pp. 1–6. IEEE (2009)
13. Ye, M., Yin, P., Lee, W.C.: Location recommendation for location-based social networks. In: Proceedings of the 18th SIGSPATIAL International Conference on Advances in Geographic Information Systems, pp. 458–461. ACM (2010)
14. Sarwat, M., Levandoski, J.J., Eldawy, A., Mokbel, M.F.: Lars*: an efficient and scalable location-aware recommender system. IEEE Trans. Knowl. Data Eng. **26**, 1384–1399 (2014)
15. Gao, H., Liu, H.: Data analysis on location-based social networks. In: Chin, A., Zhang, D. (eds.) Mobile Social Networking, pp. 165–194. Springer, New York (2014)
16. Li, J., Cardie, C., Timeline generation: tracking individuals on twitter. In: 23rd International World Wide Web Conference, WWW 2014, Seoul (2014)
17. Li, G., Chen, S., Feng, J., Tan, K. L., Li, W.S.: Efficient location-aware influence maximization. In: SIGMOD, Snowbird (2014)
18. Wu, D., Li, Y., Choi, B., Xu, J.: Social-aware top-k spatial keyword search. In: IEEE MDM, 2014, Brisbane (2014)
19. Jiang, J., Lu, H., Yang, B., Cui, B.: Finding top-k local users in geo-tagged social media data. In: 31st IEEE ICDE 2015, Seoul (2015)
20. Memcached. http://memcached.org/
21. Twitter: Real-time Geo. http://slideshare.net/raffikrikorian/rtgeo-where-20-2011
22. GeoSpatial indexes in MongoDB. http://docs.mongodb.org/manual/core/geospatial-indexes/
23. Guttman, A.: R-trees: a dynamic index structure for spatial searching. In: SIGMOD 1984, pp. 47–57 (1984)
24. Cho, E., Myers, S.A., Leskovec, J., Friendship, mobility: user movement in location-based social networks. In: KDD. ACM (2011)

Comprehensive Graph and Content Feature Based User Profiling

Peihao Tong[1], Junjie Yao[1(✉)], Liping Wang[1], and Shiyu Yang[2]

[1] School of Computer Science and Software Engineering,
East China Normal University, Shanghai, China
10122510148@student.ecnu.edu.cn, {junjie.yao,lipingwang}@sei.ecnu.edu.cn
[2] School of Computer Science and Engineering,
The University of New South Wales, Sydney, Australia
yangs@cse.unsw.edu.au

Abstract. Nowadays, users post a lot of their ordinary life records to online social sites. Rich social content covers discussion, interaction and communication activities etc. The social data provides insights into users' interest, preference and communication aspects. An interesting problem is how to profile users' occupation, i.e., professional categories. It has great values for users' recommendation and personalized delivery services. However, it is very challenging, compared to gender or age prediction, due to the multiple categories and complex scenarios.

This paper takes a new perspective to tackle the occupation prediction. We propose novel methods to transfer the commonly used social network feature and textual content feature into vector space representation. Specifically, we use the embedding method to transfer the social network feature into a low dimensional space. We then propose an integrated framework that combines the graph and content feature for the occupation classification problem. Empirical study on a large real social dataset verifies the effectiveness and usefulness of the proposed approach.

Keywords: User profiling · Graph embedding · Prediction model

1 Introduction

With the stunning growth of the social media fever, the amount of user contributed data is growing exponentially. Nowadays, ordinary users usually spend much more time surfing on the social media applications. Take Facebook, Twitter, and Weibo as examples, Facebook has more than 1.65 billion monthly active users every day; Twitter has around 500 millions registered users and generates several hundreds of million messages every day; Weibo from Sina has around 214 million registered users and 93 million active users every day[1]. Lots of users' life

---

[1] http://facebook.com
http://twitter.com
http://weibo.com
http://www.sootoo.com/content/654707.shtml.

© Springer International Publishing AG 2016
M.A. Cheema et al. (Eds.): ADC 2016, LNCS 9877, pp. 31–42, 2016.
DOI: 10.1007/978-3-319-46922-5_3

record is collected and stored. It is becoming valuable to profile users' interests and preferences from social data [6,16]. The user profile mining has been a hot topic in recent years.

An important problem in user recommendation and content delivery is the occupation prediction [7]. Occupation describes users' professional area and work interest. However, occupation prediction is very challenging due to the lacking of enough data collection. The social media provides a great opportunity to dive into users' life record to extract occupation indicants. The occupation prediction from social data can be modeled as a classification problem of social features. Possible features include users' posted messages, his/her social networks and other kinds of activity features. The large social network data and the heterogeneous features inside social data domain hinder the effective processing [3,13,16]. Though there has already exist several attempts towards occupation inference from social data [4,7,13], the performance is still not comparable with other user attribute prediction.

This paper focuses on effective methods to process the social features for a better prediction performance. We propose a new framework that combines the representative content and graph features in order to analyze social data. For the content information, we propose a text classification approach. For the graph features, we resort to effective embedding way to process these features properly. Recently, many efficient techniques for analyzing the real-world network were proposed. One of the techniques is graph embedding [15]. Graph embedding is a representation learning process, aiming to transform the large-scale network to low dimensionality representation. It makes the network data much simpler and easier to handle. The most fantastic part is that, representation learning can combine different kinds of features in a unified space, enabling machine learning techniques more general and easier to tune. This paper exploits the graph embedding power and the corresponding unified occupation prediction approach.

This paper chooses a unique dataset from Sina Weibo, one of the Chinese largest social media platform. It has a special feature that some verified users' occupation information has already been properly annotated with editors. It alleviates our annotation cost and provides accurate ground truth for the prediction algorithm training. Weibo dataset has posts, social networks and users' other kind of information. In the preprocessing stage, we extract a network from the Weibo dataset via bi-follower relationship.

The proposed approach's framework is shown in Fig. 1. We use the graph embedding method to process the bi-follower network and deliver a low dimensional latent representation for every user. We concentrate on the messages' content of the users and merge the all the messages posted by each user as his/her content representation. On top of the new representation layer, we utilize the classification methods to process the graph feature and content feature respectively. We then merge the individual prediction results in a novel fusion way to finally deliver user's occupation estimation. We would illustrate the processing details of this unified occupation prediction in the later parts of this paper.

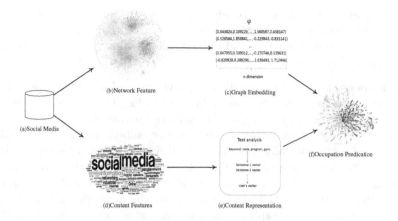

Fig. 1. The unified occupation prediction approach

Our main contributions of this paper are listed as follow:

1. First, we systematically analyze the feature representation of users and dive into the network structure to capture users' latent affiliations representation. This simplifies the feature extraction work and provides better foundation for the prediction task.
2. Second, We propose a multi-source supervised classification framework combined with both content-based and graph-based features. It is very versatile and able to cope with rich indicants of social domain.
3. At last, we conduct extensive experiments and the results clearly show the good performance of the proposed approach. The features are useful, especially the graph ones.

The rest of this paper is organized as follows. In Sect. 2 we discuss related work. Section 3 introduces the problem definition and approach framework. Section 4 illustrates the feature processing steps and the inference model. Section 5 presents the experiments on a real large dataset and finally we conclude this work in Sect. 6.

2   Related Work

The work in this paper is related to several areas. Here we briefly review the corresponding literature.

User Profiling: It is of great importance to a variety of user oriented applications. Lots of features and applications are investigated in recent years [6]. For example, [16] proposes a framework which utilizes the friendship to predict user's interests and friends. [4] takes the Twitter as its dataset and measures user's influence in different aspects. [7] combine the graph and text features to infer users' occupation and improve the performance. Other works also include the mobility and social network prediction. However, they are either limited in the feature processing or the prediction accuracy. In this work, we focus on the novel low dimensional reduction and unified approach to guarantee the occupation prediction performance.

Graph Representation: Graph data is pervasive and ubiquitous across a lot of domains. Effective graph representation is vital for the processing. There are already many different processing methods, including not limited heterogeneous modeling [13], the joint propagation [16] and the embedding [2,11]. Traditional embedding algorithms usually suffer from the difficulties in processing large-scale network due to the complexity of these algorithms quadratic to the network's vertices [12]. SocioDim [14] is a graph embedding algorithm based on community detection and it generates the representation by the network's eigenvectors. In recent years, many novel graph embedding algorithms are proposed, performing well in handling large-scale network. [11] naturally combines graph modeling with language models. This paper focuses on effective graph representation methods for the social media prediction.

3 Preliminaries

In this section, we define the problem of occupation prediction and introduce the used datasets and the features.

3.1 Problem Definition

Definition 1. *We define graph $G = (V, E, O)$, where V is the set of the vertices in the network and the E is the set of vertices' correlation in the network and $E \in V \times V$. O is the set of the text information in vertex. In this work these text information usually implies vertex's occupation.*

In our work, we use the users' friendship network for analysis. For example, there is an edge between user a and user b if there is a bi-follower between a and b.

Definition 2. *Graph embedding aims to learn every vertex's low-dimensional representation which $V \in R^d$ where d is the dimension of the representation and $d << |V|$. In other word, it tries to generate a function ϕ which $\phi(V)$ representing the low-dimensional vector representation where $\phi(V) \in R^d$.*

This paper discusses effective methods which process the network features and produce the latent low dimensional representation. Based on it, this paper then adopts the prediction technique to infer the user's occupation.

Definition 3. *Given a group set that contains I group labels, it can be represented $O = \{o_1, o_2, \ldots, o_I\}$. Every vertex in the network G has a label and the occupation inference is to infer the missing label of o_i. We model the occupation inference as a classification problem.*

3.2   Datasets

This paper uses one of China's largest social media platforms, i.e., Sina Weibo[2] as our dataset. In Sina Weibo, users can post, comment and re-share messages and interact with other users. This paper uses the users' verified occupation information, since it is more credible, acting as the ground truth for the prediction model training and testing.

After the pre-processing, we get the most common top groups. Their portions are shown in the following Table 1. We find that media accounted for the largest proportion, followed by entertainment, real estate and finance.

Table 1. Occupation distribution of verified users on Sina Weibo

Fashion	Government	Finance	Arts	Car	Real Estate	Construction
5.22%	5.59%	9.65%	4.39%	1.06%	10.67%	1.30%
Education	Media	Service	Entertainment	Press	Sport	
5.18%	25.82%	4.05%	11.32%	3.54%	7.51%	

3.3   Features

Content Feature: Users in Sina Weibo can post the messages that called status. These messages are allowed within 140 words. A user can post any message that he/she likes. However, although users can post anything, users usually post the message that he/she is interested in.

We observe that users are usually interested in the things which are related to them, such as the things related to their occupations. For example, a programmer is more interested in IT filed than others and he is more likely to post a message that contains the words just like JAVA or other words in IT filed. So the status may imply the information that is related to the users' occupation. This paper takes the content of status as a source of occupation prediction features.

Network Feature: There are several networks in Sina Weibo. As we all know, users in Sina Weibo could comment and reweet others' messages. Relying on these relationships we could construct two types of networks. However, sometimes users will reweet or comment a message posted by a stranger, which leads the relationships being weak in practical. So we take the bi-follower relationship as the criterion to construct the network, since is more stable and general. In bi-follower network, the relationship is strong because users usually follow the people they are interested in or know well. The community feature in bi-follower network is more clear and it is easier for us to extract valuable information.

---

[2] http://weibo.com.

4   Prediction Approach

In this section, we discuss the used features and the unified approach for the occupation inference.

4.1   Content Features

This paper utilizes the text information of the user to infer the user's occupation. It concentrates on the content text and tries to transform the text information to latent representation. However, it's infeasible to embed the text directly because of the complexity of the content. So we resort to handle the content and pick a data structure for the content in order to process the content accurately and efficiently.

A user posts several messages. What we get is the representation of the content of the posted message, so we still need to do some process to transform it into the users' latent representation. In this paper, we choose to merge the representation by aggregating the messages' content representation.

There are many meaningless words in the text just like the preposition, conjunction and so on. Meanwhile, there are also some words that we do not concern about. We extract the key words via the term frequency and inverse document frequency criterion. In this way, similar content can be merged and then we use the feature selection metrics to reduce the representation length for each post.

4.2   Network Features

The objective of graph embedding is to learn a latent representation of a graph, which means transforming the large scale network to a low dimension vector representation. After embedding, the new format will be much easier for handling and we can adopt some machine learning techniques to process the data.

However, graph embedding has a risk that may lead to the loss of the structure's data. So it is important to select a suitable graph embedding algorithm to embed this data. DeepWalk [11] is a graph embedding algorithm which adopts to random walk to capture the structure of the network. Facts prove that random walk has good performance in capturing local structure and can adapt to the changed network easily. Besides, using random walk in the network, the frequency of the vertices appearing in the walks follows the power law distribution, while the word frequency in the natural language follows the same distribution. This discovery leads to adopting the natural language model to analyze graph data.

After generating the random walks, this approach continues to utilize the language model word2vec [9,10] to process the intermediate result. Language model aims to estimate the probability of the specific words' occurrence in the corpus. For Word2Vec, it is usually used to maximize the probability of the specific words sequence's occurrence in the corpus. Word2vec takes the words' corpus as the input and produces the distributed representation which called word embedding as output. There are two common models in word2vec. One is

CBOW and the other is called skip-gram. CBOW utilizes the context to predict a missing word. However, with the words sequence's length becoming longer, facts show that calculating the probability by CBOW model becomes difficult and even unfeasible. In contrast, skip-gram model removes the order constraint and uses a word to predict its context within window size rather than use the context to predict a missing word. This model is proved to be friendly to calculating.

Motivated by the content process, we regards a walk as a sentence and selects skip-gram because of the long length of the walk in the network. Its goal is to maximize the probability, targeting vertex's contexts make a co-occurence.

DeepWalk can be used to generate the latent representation of the vertices. We replace the vertex with the vector representation and use $\phi(V_i)$ represents the vector representation of V_i. In practice, it is a common way to adopt the log-likelihood function to handle the Eq. 1. We can get the new representation as following:

$$\max log(p(\{v_1, v_2, \cdots, v_{i-1}\}/v_i|\phi(v_i))) \tag{1}$$

Calculating the probability directly is still difficult and it is unfeasible in practical experiment. In order to derive the probability with guaranteed performance, we adopt a way called Hierarchical Softmax. It treats the occurrence frequency of the vertex in the walk as the weight of the leaf node in order to construct a Huffman tree for factorizing the conditional probability. After Huffman tree constructed, if we set the v_i as our target vertex and intend to compute the v_i's context v_j's occurrence probability $P(v_j|v_i)$, we should find the path from root to the leaf v_j in the Huffman tree.

In the path, non-leaf nodes should be given a vector representation θ as the undetermined parameter for the sake of the later calculation. And We set the left node as positive class and the right node as negative class. So According to the sigmoid function, we could get the probability of every level and thus we can get the final probability. Then by means of adjusting the vectors, we could maximize the probability and at the same time we get the vector representation.

4.3   Prediction Model

The framework we proposed combines the graph features and content features together. Therefore, there is a key problem that we have to solve is how to fuse the network feature and the text content feature.

There are two main fusion methods which are early fusion and late fusion. Early fusion fuses the features before processing this data while the late fusion concentrates on fusing the processed results of the different features' data. In our work, to explicitly compare each feature's contribution and guarantee the generality, we choose the late fusion method to fuse the graph feature and content feature.

In late fusion, there are still several strategies. The most common strategy is averaging. Averaging takes the results of the different learning machines and accumulates these results to calculate the average of the sum result. However, when a big gap exists among the learning machines' performances, the prediction performance may be impacted and even be worse than some of the original learning machines. So in order to solve the problem, an improved version of averaging called weighted averaging is proposed which can be represented as $H(x) = \sum_{i=1}^{T} w_i h_i(x)$, where h is the individual learning machine and w is its weight. However, facts show that finding suitable weights for different learning machines is still difficult because of the noise of the data. There is another fusion strategy called stacking. Stacking is a method that needs an extra learning process to get the fusion result. For example, if the original learning machines take the training data x_i as input and produce the probability lists as the predict result $h_i(x_i)$, stacking combines the result as $H(x)$ which can be represented as Eq. 2

$$H(x) = h^{'}(h_1(x_1), h_2(x_2), \ldots, h_n(x_n)) \tag{2}$$

The fusion result could be regarded as the new input of the secondary learning machine and the prediction of this machine is the result of the fusion. Many studies proved that stacking is an efficient method in feature fusion. This paper compares averaging and stacking and chooses the stacking as our final fusion method. The result will be given in the experiment section.

5   Experiments

In this section, we present the empirical studies conducted on the real large dataset, i.e., Sina Weibo. We first introduce the baselines and the dataset description. We then discuss the prediction performance and reveal illustrate cases.

5.1   Dataset Setup

The network of the users in Sina Weibo is based on the relationship of bi-follower. There are 93,264 vertices and 1,688,681 edges used in this experiments. We choose the first 5 occupation groups as the evaluation set.

5.2   Baseline and Evaluation Metrics

To systematically compare the proposed methods' performance, we choose several baseline according to their features and methods:

Graph Partitioning: Graph Partitioning is an unsupervised partition method. It takes network G as input and decomposes the graph into different communities C. For every $c \in C$, we label the community with the most common group appears in c. In this experiment, we use the METIS library [1].

SocioDim: SocioDim algorithm is a representation learning algorithm based on community detection process [14]. SocioDim takes the network's communities in use and utilizes the network's eigenvectors to generate a latent representation in R^d where d is a small dimension. SocioDim usually takes advantage of the modular graph partitions of the network.

DeepWalk: We use Deepwalk approach [11] to get the low dimensional vector representation of users' social network information. It acts as the baseline to test the potential of the social network features.

Content Classification: Different from the methods above, this method takes the content of the messages as the input. After several processing just like extracting keywords, calculating weights and so on, we get the vector representation of the content for users and use the boost classification based on XGBoost package[3] to estimate users' occupation [8].

5.3 Effectiveness

We conduct the five-fold cross validation experiment, on top of scikit-learn package[4].

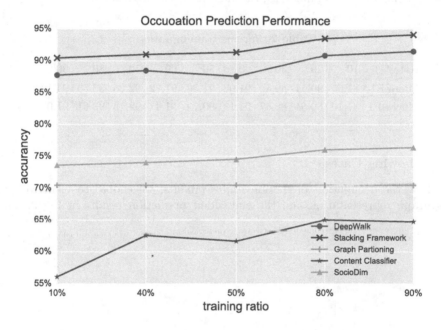

Fig. 2. Occupation prediction performance

Figure 2 shows the accuracy score of different methods. We can find that content classification method is not comparable with graph ones. Under five class prediction tasks, content feature is useful, though not competitive enough to provide satisfying performance. It is largely due to the short text collection. In the categories of graph based methods, the unsupervised partition method

---

[3] https://xgboost.readthedocs.io/en/latest/.
[4] http://scikit-learn.org/.

lags behind. The SocDim performance also can not compare with the new low dimensional space learning approaches. With the help of feature fusion, the proposed approach can gain a further increase on top of Deepwalk's already very high performance.

The result demonstrates that the unified framework with graph feature embedding and content feature fusion can improve the occupation prediction in social media domains. The classification based inference model is effective and able to cope with the social media challenges.

5.4   Training Ratio

To test the sensitivity of the proposed approach, we extend the ratio scope and report the Micro/Macro F1 measures in Table 2. It is revealing that the accuracy is growing gradually with more training data. The network effect enables that a small portion of the training data can deliver satisfying results.

Table 2. Varying training ratio

Rate(%)	10	20	30	40	50	60	70	80	90
Micro-F1	87.65	90.44	89.80	91.00	91.33	91.72	92.25	93.50	94.10
Macro-F1	87.10	89.54	88.87	91.17	90.74	91.47	92.13	93.61	91.09

5.5   Showing Cases

To show the occupation prediction improvements, we choose to visualize the raw graph representation and the embedded prediction results in Fig. 3. The

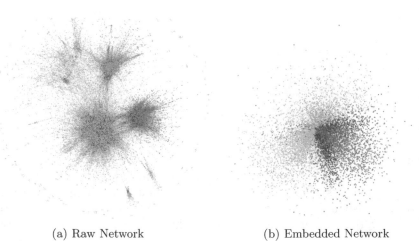

(a) Raw Network   (b) Embedded Network

Fig. 3. Top five occupation categories of users.

raw graph presentation is generated with the help of Gephi[5], and the embedded result is transformed using MDS algorithm [5]. We can clearly find the network coherence in the occupation categories are vivid, and the embedding process guarantees the occupation distinct and easy to interrupt.

6 Conclusions

In this paper, we discuss the problem of user profiling from social media data. We choose the occupation prediction to illustrate the features and models used in this new scenario. Based on a unique micro-blog dataset, we discuss the content and graph features for occupation prediction.

We propose to utilize the graph embedding and text classification for the prediction model. The fused approach also improve the individual based baselines. Extensive experiments on the large real dataset demonstrate the improvement and effectiveness of our new approach.

It is also remarkable that the comprehensive occupation prediction method can illustrate the users' professional interaction and posting activities. The future work of this paper can be focused on the occupation interpretation and friendship recommendation, with the help of efficient occupation pattern mining.

Acknowledgements. The research is supported by the National Natural Science Foundation of China under Grant No. 61502169, 61401155 and NSFC-Zhejiang Joint Fund for the Integration of Industrialization and Informatization Grant No. U1509219.

References

1. Abou-Rjeili, A., Karypis, G.: Multilevel algorithms for partitioning power-law graphs. In: 20th International Parallel and Distributed Processing Symposium, IPDPS 2006, p. 10-pp. IEEE (2006)
2. Bengio, Y., Courville, A., Vincent, P.: Representation learning: a review and new perspectives. IEEE Trans. Pattern Anal. Mach. Intell. **35**(8), 1798–1828 (2013)
3. Cao, S., Lu, W., Xu, Q.: GraRep: Learning graph representations with global structural information. In: Proceeding of CIKM, pp. 891–900 (2015)
4. Cha, M., Haddadi, H., Benevenuto, F., Gummadi, P.K.: Measuring user influence in twitter: The million follower fallacy. In: Proceeding of ICWSM, pp. 10–17 (2010)
5. Cox, T.F., Cox, M.A.: Multidimensional Scaling. CRC Press, Boca Raton (2000)
6. Farseev, A., Nie, L., Akbari, M., Chua, T.S.: Harvesting multiple sources for user profile learning: a big data study. In: Proceeding of ACM Multimedia, pp. 235–242 (2015)
7. Huang, Y., Yu, L., Wang, X., Cui, B.: A multi-source integration framework for user occupation inference in social media systems. World Wide Web **18**(5), 1247–1267 (2015)
8. Manning, C.D., Raghavan, P., Schütze, H., et al.: Introduction to Information Retrieval, vol. 1. Cambridge University Press, Cambridge (2008)

---

[5] https://gephi.org.

9. Mikolov, T., Chen, K., Corrado, G., Dean, J.: Efficient estimation of word representations in vector space. arXiv preprint arXiv:1301.3781 (2013)
10. Mikolov, T., Sutskever, I., Chen, K., Corrado, G.S., Dean, J.: Distributed representations of words and phrases and their compositionality. In: NIPS, pp. 3111–3119 (2013)
11. Perozzi, B., Al-Rfou, R., Skiena, S.: DeepWalk: Online learning of social representations. In: Proceeding of SIGKDD, pp. 701–710 (2014)
12. Roweis, S.T., Saul, L.K.: Nonlinear dimensionality reduction by locally linear embedding. Science **290**(5500), 2323–2326 (2000)
13. Sun, Y., Norick, B., Han, J., Yan, X., Yu, P.S., Yu, X.: Integrating meta-path selection with user-guided object clustering in heterogeneous information networks. In: Proceeding of SIGKDD, pp. 1348–1356 (2012)
14. Tang, L., Liu, H.: Relational learning via latent social dimensions. In: Proceeding of SIGKDD, pp. 817–826 (2009)
15. Yan, S., Xu, D., Zhang, B., Zhang, H.J., Yang, Q., Lin, S.: Graph embedding and extensions: a general framework for dimensionality reduction. IEEE Trans. Pattern Anal. Mach. Intell. **29**(1), 40–51 (2007)
16. Yang, S.H., Long, B., Smola, A., Sadagopan, N., Zheng, Z., Zha, H.: Like like alike joint friendship and interest propagation in social networks. In: Proceeding of WWW, pp. 537–546 (2011)

Efficient Maximum Closeness Centrality Group Identification

Chen Chen[1]([✉]), Wei Wang[1], and Xiaoyang Wang[2]

[1] The University of New South Wales, Sydney, Australia
{cchen,weiw}@cse.unsw.edu.au
[2] University of Technology, Sydney, Australia
xiaoyang.wang@uts.edu.au

Abstract. As a key concept in the social networks, closeness centrality is widely adopted to measure the importance of a node. Many efficient algorithms are developed in the literature to find the top-k closeness centrality nodes. In most of the previous work, nodes are treated as irrelevant individuals for a top-k ranking. However, in many applications, it is required to find a set of nodes that is the most important as a *group*. In this paper, we extend the concept of closeness centrality to a set of nodes. We aim to find a set of k nodes that has the largest closeness centrality as a whole. We show that the problem is NP-hard, and prove that the objective function is monotonic and submodular. Therefore, the greedy algorithm can return a result with $1 - 1/e$ approximation ratio. In order to handle large graphs, we propose a baseline sampling algorithm (BSA). We further improve the sampling approach by considering the order of samples and reducing the marginal gain update cost, which leads to our order based sampling algorithm (OSA). Finally, extensive experiments on four real world social networks demonstrate the efficiency and effectiveness of the proposed methods.

Keywords: Closeness centrality · Random sampling · Top-k · Shortest path distance

1 Introduction

As a subject of broad and current interest, social networks have been widely studied for decades. A social network is usually represented as a graph $G = (V, E)$ where V denotes the set of nodes (users) and E denotes the set of edges (relationships among users). Centrality, which measures the importance of a node in a social network, has been a fundamental concept investigated in the social networks. There are different measurements of centrality developed for various purposes, such as closeness centrality [4], betweenness centrality [1], eigenvector centrality [5], etc. In this paper, we focus on the classic closeness centrality, which is defined as the inverse of the average distance from a node to all the other nodes in the social network. The distance between two nodes is calculated by the shortest path distance. The smaller the average distance of a node is, the

© Springer International Publishing AG 2016
M.A. Cheema et al. (Eds.): ADC 2016, LNCS 9877, pp. 43–55, 2016.
DOI: 10.1007/978-3-319-46922-5_4

more important or more influential the node will be. To find the influential nodes (users) in a social network, many research efforts have been made to find the k nodes with the largest closeness centrality [10,13,14]. However, in many real applications, such as team formation, we may need to find a set of k users which has large closeness centrality as a *group*, instead of returning the k independent users in the top-k ranking. In this paper, we extend the definition of closeness centrality for a single node to a set of nodes. Specifically, the closeness centrality of a set S of nodes is defined as the inverse of the average distance from S to the nodes in G. And the distance from S to a node $u \in V$ is defined as the minimum distance from u to the nodes in S. The maximum closeness centrality group identification problem is to find a set of k nodes in the social network with the largest closeness centrality.

The challenges of the problem lie in two aspects. First, we show that the problem is NP-Hard, thus there is no polynomial time solution unless $P = NP$. Fortunately, we prove that the objective function is monotonic and submodular. It means we can obtain a result with $1 - 1/e$ approximation ratio by adopting a greedy framework. Second is that we still need the information of the all pairs shortest path distances even for the simple greedy algorithm, which is prohibitive to compute ($O(|V|^3)$ time) and store ($O(|V|^2)$ space) when the graph is large. There are some efficient index such as the network structure index techniques [15]. It partitions the graph into zones and approximates the shortest path distance between two nodes by using their distances to the same zone. However, this approximation offers no guarantee for the quality of the returned distance. In order to scale to large graphs, we propose a sampling based approach by extending the traditional sampling method for estimating the closeness centrality of a single node. In addition, we bring order into the samples, such that the nodes can be identified incrementally. Then we utilize the selected nodes to reduce the cost of computing the distances from the nodes to the samples. To further accelerate the process, we develop optimization techniques to reduce the updating cost for the less important nodes. Through experiments on real world social networks, we verify the efficiency and effectiveness of the proposed techniques.

The rest of the paper is organized as follows. Section 2 formally introduces the problem studied in this paper as well as the greedy algorithm based framework. Section 3 surveys related work. Section 4 presents the proposed sampling based algorithms. We demonstrate the efficiency and effectiveness of the proposed techniques on four real social networks in Sect. 5 and conclude the paper in Sect. 6.

2   Preliminary

We formally define the problem in this section. Then we analyze the properties of the objective function and introduce our greedy algorithm based framework.

2.1   Problem Definition

We consider a social network $G = (V, E)$ as an undirected connected graph, where V denotes the set of nodes, and E denotes the set of edges. $|V| = n$ and $|E| = m$. For nodes $u, v \in V$, the distance $d(u, v)$ between the two nodes is calculated as the shortest path distance and $d(u, u) = 0$. Then the classic closeness centrality of a node u is defined by the inverse average distance from u to the nodes in G.[1] By extending this definition, we can define the closeness centrality for a set of nodes as follows.

Definition 1 (Closeness Centrality for a Set of Nodes). *Given a social network G and a set S ($S \subseteq V$) of nodes, the closeness centrality of S is denoted by $c(S)$, which is measured by the inverse average distance from S to all the nodes in G, i.e.,*

$$c(S) = \frac{n}{\sum_{v \in V} d(S, v)}$$

where $d(S, v) = \min\{d(u, v)\}$ for $u \in S$.

Problem Statement. Based on Definition 1, we define the *maximum closeness centrality group identification* (**MCGI**) problem as follows. Given a social network G and a positive integer k, the MCGI problem aims to find a set S^* of k nodes that has the largest closeness centrality, *i.e.,*

$$S^* = \arg \max_{S \subseteq V} \{c(S) \mid |S| = k\}. \tag{1}$$

The selected node set is call *seed set* and each node in the set is called a *seed*. According to Lemma 1, the MCGI problem is NP-Hard.

Lemma 1. *The maximum closeness centrality group identification problem is NP-Hard.*

Proof Sketch. The correctness of Lemma 1 can be proved by reducing from a known NP-hard problem "k-means clustering problem in the Euclidean space" [2]. The polynomial time reduction can be summarized as follows. Each object in the k-means clustering corresponds to a node in the social network. The k centers in the k-means clustering problem correspond to the selected k nodes. The Euclidean distance between objects corresponds to the shortest path distance between nodes in MCGI. Even if we can compute the shortest path distance in constant time, the hardness of the problem still remains the same.

---

[1] Note that there are different variants of the definition of closeness centrality, in this paper we focus on the classic closeness centrality.

2.2　Objective Function Analysis

According to Lemma 1, we know that there is no polynomial time solution for the MCGI problem. Fortunately, based on Lemma 2, the closeness centrality function for a set of nodes shown in Definition 1 has the following two properties.

- Monotonicity. Given any two sets S, $T \subseteq V$ with $S \subseteq T$, a function $f(x)$ is monotonic, if $f(S) \leq f(T)$.
- Submodularity. Given any two sets S, $T \subseteq V$ with $S \subseteq T$ and $v \in V \backslash T$, a function $f(x)$ is submodular, if $f(S \cup \{v\}) - f(S) \geq f(T \cup \{v\}) - f(T)$.

Lemma 2. *The closeness centrality function for a set of nodes is monotonic and submodular.*

Proof Sketch. Assume there are two sets satisfying S, $T \subseteq V$ with $S \subseteq T$.

- Based on the definition of shortest path distance from a set of nodes to a single node, we have $d(S, u) \geq d(T, u)$ for $u \in V$, which means $c(S) \leq c(T)$. Thus the monotonic property is correct.
- Given two nodes u, v, where $v \in V \backslash T$ and $u \in V$, if $d(T, u) > d(T \cup \{v\}, u)$, we must have $d(T \cup \{v\}, u) = d(S \cup \{v\}, u) = d(v, u)$. Then we have $d(S \cup \{v\}, u) - d(S, u) \leq d(T \cup \{v\}, u) - d(T, u)$. If $d(T, u) = d(T \cup \{v\}, u)$, we have $d(S \cup \{v\}, u) - d(S, u) \leq d(T \cup \{v\}, u) - d(T, u) = 0$. Thus the submodular property holds based on the definition of closeness centrality. □

2.3　Greedy Algorithm Framework

Since the objective function is monotonic and submodular, we can utilize the greedy algorithm to iteratively select the node with the largest marginal gain (*i.e.*, the node will increase the closeness centrality of the set most by adding it). Through k iterations it can return a set S of k nodes with $1 - 1/e$ approximation ratio, *i.e.*, $c(S) \geq (1 - 1/e)c(S^*)$, where S^* is the optimal result. The details of the greedy algorithm framework are shown in Algorithm 1.

In Algorithm 1, S maintains the set of nodes already selected in the previous iterations. M is a matrix that stores the distance $d(S \cup \{u\}, v)$ $(v \in V)$ for each node u. *Score* is a vector that maintains the closeness centrality of adding u to S, *i.e.*, $c(S \cup \{u\})$ for each node $u \in V$. And $Score[u]$ equals $c(u)$ at the beginning. After initialization, we iteratively select the node with the largest marginal gain from $V \backslash S$ in Line 5. After selecting the node v with the largest marginal gain, we add it to S and update the distance matrix M as well as the *Score* value for each node $u \in V \backslash S$ from Line 7 to 10. The procedure is repeated until we find k nodes.

Analysis. The space complexity of Algorithm 1 is $O(n^2)$, since we need to store the all pairs distances in M. To compute the all pairs shortest paths, we can use the Floyd-Warshall algorithm which needs $O(n^3)$ time. Note that we can also use other algorithms to compute the all pairs shortest paths but it is not the major concern in this paper. In the node selection phase, we need $O(kn^2)$ time to select the k nodes as we need to update the distance matrix and the marginal gain for all the unselected nodes after each iteration.

Algorithm 1. Greedy Framework

    Input   : G : a social network, k : a positive integer.
    Output : S : a set of k nodes
1 $S \leftarrow \emptyset$;
2 $M \leftarrow$ all pairs shortest path distances ;
3 $Score \leftarrow \{c(u) \mid u \in V\}$;
4 **while** $|S| < k$ **do**
5      $v = \arg\max_{w \in V \setminus S} Score[w]$;
6      $S \leftarrow S \cup \{v\}$;
7   **foreach** $u \in V \setminus S$ **do**
8   **foreach** $w \in V$ **do**
9   **if** $d(u, w) > d(v, w)$ **then**
10               $M[u, w] = d(v, w)$;
11         $Score[u] \leftarrow c(S \cup \{u\})$;

12 **return** S;

---

3 Related Work

As a key concept in the social networks, centrality has been widely used to measure the importance of the nodes in a social network. Many centrality metrics [6,11,17] are proposed for different considerations to measure the importance of a node. For example, Google's PageRank is a variant of the eigenvector centrality. In this paper, we focus on the closeness centrality. The concept of closeness centrality is first considered in [3]. The definition of closeness centrality is formalized in [4] as the inverse of the average distance of a given node to all the other nodes in the graph. A major problem in calculating the closeness centrality is the scalability issue for large graphs. In [12,16], the authors use sampling approaches to find the top-1 node by utilizing the average distance to the sampled nodes. In [10], the authors extend the sampling approach for estimating the closeness centrality for all the nodes. Okamoto et al. [13] further extend the sampling framework to efficiently find the top-k nodes. To further improve the estimation performance, authors take advantage of a hybrid framework and weighted sampling techniques in [7–9]. [14] aims to find the top-k nodes exactly by leveraging the hierarchy structure of the graphs. In [18], the authors also study the group closeness centrality maximization problem. However, the techniques proposed are for disk-based graphs and only consider the case when the weight of edge is 1. While our method is memory-based and suitable for any weight setting.

4 Sampling Based Algorithms

Although the greedy approach in Algorithm 1 can return a result with a bounded approximation ratio, it suffers from serious limitations when scaling to large graphs. As analyzed previously, it needs $O(n^2)$ space to store the all pairs distances, which is prohibitive for large graphs. If we do not store the all pairs distances in memory and only compute them on the fly when needed, the computation time is still not affordable. Moreover, we need considerable amount of time to update the marginal gain for each node in each iteration. Therefore, we propose two sampling based approaches to make it possible to scale to large graphs in this section.

4.1   Baseline Sampling Method

To make it possible for processing large graphs, a natural consideration is to utilize the sampling techniques. It is able to obtain a high quality result by accessing only a small part of the graph's information. In the previous works, sampling based approaches are developed to estimate the closeness centrality of a single node, or to identify the top-k closeness centrality nodes. The idea of estimating the closeness centrality of a single node u can be summarized as follows. We first randomly sample several nodes without replacement from V. Then we calculate the shortest path distances from u to all the samples. Next we use the average distance from u to all the samples as an estimation of the average distance of u to all the nodes. It is easy to see that the estimator is unbiased.

Motivated by the idea, we can use the average distance from a set S of nodes to all the samples as the estimation of the average distance of S to all the nodes. It can be formally expressed in Eq. (2).

$$\hat{c}(S) = \frac{l}{\sum_{v \in \mathcal{L}} d(S, v)}, \tag{2}$$

where $\mathcal{L}$ denotes the sample set of size l. It is easy to verify $1/\hat{c}(S)$ is an unbiased estimation of $1/c(S)$, i.e., $1/c(S) = \mathbf{E}\left[1/\hat{c}(S)\right]$.

Based on the estimator, we can modify Algorithm 1 to apply the sampling technique and solve the MCGI problem. The idea is to use the estimated closeness centrality and marginal gain to replace the exact calculated value in the greedy framework. To be more specific, we first sample l nodes from V. For the distance matrix M, it stores the distances from each node $u \in V$ to all the samples, i.e., $d(S \cup \{u\}, v)$ $(v \in \mathcal{L})$. Initially, it only stores the distances from all the nodes to the l samples. This can be done by running a single source shortest path algorithm from every sample. Similarly, $Score[u]$ stores $\hat{c}(S \cup \{u\})$. Then we iteratively select k nodes with the largest marginal gain based on the sampling method. The pseudo-code is omitted due to the space limitation.

Analysis. The space cost is $O(nl)$ for the baseline sampling algorithm, since only the distances from all the nodes to the l samples are maintained. In order to compute the initial matrix M, we can run l times single source shortest path algorithm (e.g., Dijkstra algorithm) which requires $O(l(m + n \log n))$ time. We also need $O(kln)$ time to select the k nodes. Due to the sampling techniques, the result will have $1 - 1/e - \epsilon$ approximation ratio, where ϵ is the error introduced by sampling. From previous studies [10,12,16] we know the estimation procedure converges quickly with the sample size l.

4.2   Order Based Sampling Method

Although the baseline sampling algorithm can greatly reduce the space cost and time complexity, there is still room for improvement.

- The first limitation is that it strictly separates the shortest path distance calculation phase and the node selection phase. As introduced later, we can further reduce the cost if incrementally doing the sampling and the seed selection.
- The second limitation is that after selecting a node, it needs to update the marginal gain for all the nodes, which is costly when n is large.

Based on the observations, we present two optimized techniques to further reduce the cost of baseline sampling algorithm.

Incremental Sampling and Node Selection. We first explain the motivation. Suppose we already select certain nodes by using $l' < l$ samples, then we can use these selected nodes (seed set) to reduce the calculation cost of shortest path distances for the rest $l - l'$ samples. Following is a motivating example.

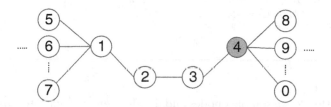

Fig. 1. Motivating example of incremental sampling and node selection

Example 1. As shown in Fig. 1, assume $k = 2$ and the weight of each edge is 1. We use half of the samples to select the first seed v_4 hence currently $S = \{v_4\}$. Then we examine the rest samples to find the second seed. Suppose v_3 is the next sample, then we start running a Dijkstra's algorithm from v_3. The property of Dijkstra's algorithm is that it will incrementally reach the nodes close to the source node. So in the figure, v_3 first reaches nodes v_2 and v_4 with shortest path distance of 1. Remember that v_4 is the node already selected. Then we do not need to further compute the distance from all the other nodes to v_3, as their distances are greater than that of v_4. The reason is that adding any of them to S will not change the distance from S to the sample v_3. Consequently, their distances to v_3 are all recorded as $1 = d(v_3, v_4)$, that is for $v' \in V \backslash \{v_2, v_3, v_4\}$, it holds $d(S \cup \{v'\}, v_3) = 1$.

According to the motivating example, if we can incrementally do the sampling and node selection, we will significantly reduce the cost of computing shortest path distances from nodes to the samples. To fulfill it, we firstly divide the sample into k partitions $\{P_1, P_2, ..., P_k\}$. For the i-th node selection, we compute the distances from all the nodes to the samples in P_i by considering the selected $i-1$ nodes. Then we can obtain the node with the largest marginal gain by considering the estimation due to samples in $\cup_{j=1}^{i} P_j$. The details of the algorithm are shown in Algorithm 2.

Algorithm 2. Order Based Sampling

Input   : G : a social network, k : a positive integer, l : sample size.
Output : S : a set of k nodes
1 $S \leftarrow \emptyset$; $Score \leftarrow \emptyset$; $M \leftarrow \emptyset$;
2 Get samples and partitions $\{P_1, P_2, ..., P_k\}$; /* partition of l samples */;
3 **for** i from 1 to k **do**
4 $\quad$ Compute the distance from all nodes to P_i with S as constraint ;
5 $\quad$ Update $Score$ for $u \in V\backslash S$ based on the samples in $\cup_{j=1}^{i} P_j$;
6 $\quad$ $v = \arg\max_{w \in V\backslash S} Score[w]$;
7 $\quad$ $S \leftarrow S \cup \{v\}$;
8 $\quad$ **foreach** $u \in V\backslash S$ **do**
9 $\quad\quad$ **foreach** $w \in \cup_{j=1}^{i} P_j$ **do**
10 $\quad\quad\quad$ **if** $d(u,w) > d(v,w)$ **then**
11 $\quad\quad\quad\quad$ $M[u,w] = d(v,w)$;

12 **return** S;

In Algorithm 2, we sample l nodes and divide them into k partitions in Line 2. To get l samples without replacement, we can do a random permutation on all the nodes and select the first l nodes. In such case, each sample has a rank induced by the permutation. Then each partition can store the samples based on the order, i.e., P_i stores the $(i-1)l/k$-th to il/k-th sampled nodes.

In the i-th iteration, we compute the distances from all the nodes to samples in P_i with S as the constraint. It means when running Dijkstra's algorithm from a sample v', we can early stop if we get the shortest path distance from v' to a seed node u' in S. In Line 5, we update the $Score$ value for nodes based on the samples in $\cup_{j=1}^{i} P_j$. Then we select the node v with the largest marginal gain and add it to S. Finally, we update the distance matrix, i.e., $d(S \cup \{u\}, w)$ for $u \in V\backslash S$ and $w \in \cup_{j=1}^{i} P_j$. The algorithm terminates after k iterations.

Correctness of Algorithm 2. The aim of the algorithm is to select the node with the largest marginal gain from $V\backslash S$ in each iteration. The input of the iteration is the seed set S from previous $i-1$ iterations. Since the proposed optimization only prunes the unnecessary calculation, the distance matrix M and $Score$ vector are exactly calculated based on the samples in $\cup_{j=1}^{i} P_j$ given S. Moreover, $\cup_{j=1}^{i} P_j$

consists of the first $| \cup_{j=1}^{i} P_j |$ samples of the l samples, as we partition the samples based on the sample order (*i.e.*, permutation order). Consequently, the case in iteration i amounts to selecting the node with the largest marginal gain, based on the already selected $i - 1$ nodes with $| \cup_{j=1}^{i} P_j |$ samples. Therefore the algorithm is guaranteed to be correct. Although the incremental selection strategy may affect the accuracy of estimation, the quality drop is negligible as shown in the experimental evalutions.

Update Optimization. In the baseline sampling algorithm, after selecting one node in an iteration, we need to update the *Score* vector for all the other nodes. However, the closeness centrality tends to be very small for most of the nodes in a social network. Usually we have $k << n$. As a result, the nodes with small centralities will never be selected into the results. If we can avoid updating their marginal gains we will save a large amount of computational cost. Following is a motivating example.

Example 2. Suppose node v_1 is a less important node with $c(v_1) = 0.0001$. In the i-th iteration, we find a node v_2 with marginal gain of 0.05. Due to the submodular property of the objective function, the marginal gain of v_1 must be smaller than $c(v_1) < 0.05$. Thus we can safely prune v_1 from the i-th node selection without updating its marginal gain.

Based on the motivating example, we come up with the *update when needed strategy*. The idea is that in the i-th iteration, we do not calculate the marginal gain for node $v \in V$ unless necessary. It is easy to adopt this strategy by modifying Algorithm 2. For each node $u \in V \backslash S$, $Score[u]$ stores its centrality or values calculated in the previous iteration, so does for the distance matrix M. In the i-th iteration, we first calculate the distance of nodes to the samples in P_i, then update the matrix and the score vector based on P_i. Next we sort the nodes decreasingly based on their values in *Score*. We continuously pop nodes from the queue and calculate their true marginal gains, until the largest marginal gain found is greater than the top value of the queue. Then we can safely prune all the untouched nodes from this iteration and select the node with the largest marginal gain as the next seed. Through this way, we can reduce the update cost for many unpromising nodes, which will greatly improve the efficiency when n is large.

5   Experiment

In this section, we present the results of a comprehensive performance study to evaluate the efficiency and effectiveness of the proposed techniques in this paper.

5.1   Experiment Setup

In the experiments, we report the response time of finding k nodes to evaluate the efficiency of the algorithms. We also report the closeness centrality of the returned node set to measure the effectiveness.

Algorithms and Workload. We evaluate the performance of two proposed algorithms, baseline sampling algorithm (**BSA**) and order based sampling algorithm (**OSA**). We omit the greedy algorithm in Algorithm 1, since it needs to compute and store all pairs shortest path distances. Even for a small graph with 50,000 nodes, it will need more than one day to return the results. We set the sample size for both BSA and OSA as 1000 to make a tradeoff between the accuracy and efficiency. In the experiments, we vary k from 1 to 50 with 50 by default. For each algorithm, we run 20 times and report the average performance.

Datasets. To demonstrate the effectiveness and efficiency of our methods, we conduct experiments on 4 real world social networks. The parameters of the datasets [2] are reported in Table 1. The diameter is defined as the longest shortest path distance in the graph.

Table 1. The summary of datasets

Dataset	n	m	Diameter
Gowalla	196,591	950,327	14
Amazon	334,863	925,872	44
Youtube	1,134,890	2,987,624	20
LiveJournal	3,997,962	34,681,189	17

Implementation Environment. All experiments are carried out on a PC with Intel Xeon 2.30 GHz and 96 G RAM. The operating system is Redhat. All algorithms are implemented in C++ and compiled with GCC 4.8.2 with -O3 flag.

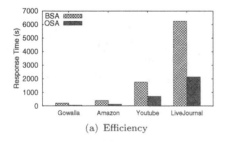

(a) Efficiency

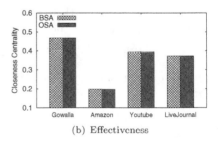

(b) Effectiveness

Fig. 2. Performance evaluation on all the datasets

---

[2] https://snap.stanford.edu/data/.

5.2 Efficiency Evaluation

In this section we evaluate the efficiency of the algorithms through response time. In the first set of experiments, Fig. 2a reports the response time of BSA and OSA on all the datasets under the default settings. As the increase of dataset size, the response time grows for both algorithms. Because the cost of computing the distance from nodes to the samples and calculating the marginal gain will increase with the growing graph size. OSA constantly outperforms BSA by up to 4 times acceleration, due to the pruning in computing shortest path distances and updating the marginal gains.

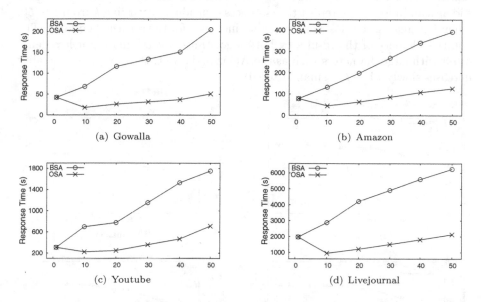

Fig. 3. Efficiency evaluation by varying k

In Fig. 3, we report the response time by varying k from 1 to 50 on four datasets. When k equals 1, the response time of both algorithms is the same. It is because when k equals 1, there is no incremental node selection optimization and the updating optimization for OSA. Under the same sample size and $k = 1$, the procedure of OSA is the same as that of BSA. When k increases, the response time of BSA increases because of the increase of cost in the node selection. For OSA, the response time firstly drops then increases, because when k is larger than 1, OSA can take advantage of incremental node selection to reduce the cost of calculating shortest path distances to the samples. However, when k becomes larger, the node selection cost and marginal gain updating cost increase, which leads to the increase of the running time.

5.3   Effectiveness Evaluation

In this section, we evaluate the effectiveness of the proposed algorithms. In Fig. 2b, we report the closeness centrality on all the datasets under default settings. As can be seen, the quality of the results returned by OSA is almost the same as that of BSA. Only in few cases there is a very slight drop in the closeness centrality returned by OSA. Since for each node selection in OSA, it only utilizes part of the samples for estimating, $i.e.$, $\cup_{j=1}^{i} P_j$ for selecting the i-th node. While BSA always uses the full samples to do the estimation. Also note that the closeness centrality of returned nodes is decided by the diameter of the graph. For graphs with small diameters, it tends to return a set of nodes with large closeness centrality when k is identical. In Fig. 4, we report the closeness centrality by varying k from 1 to 50. When k increases, the closeness centrality increases for the returned node set. The quality difference of the results by both algorithms is very small. For less dense graph with large diameters such as the Amazon dataset, the closeness centrality increases slowly when k is small (1 to 10).

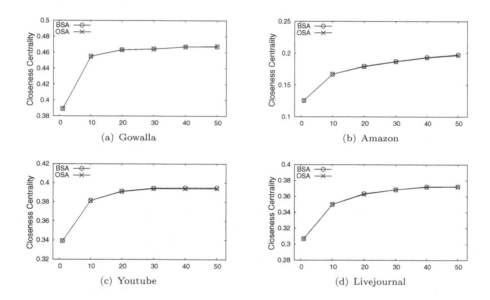

Fig. 4. Effectiveness evaluation by varying k

6   Conclusion

In this paper, we consider closeness centrality for a set of nodes and aim to find the set with the largest closeness centrality. We show the problem is NP-Hard. By proving the monotonic and submodular properties of the objective function, we present a greedy framework which can achieve $1 - 1/e$ approximation ratio. Unfortunately, naïve implementation of the greedy framework will result in large space and time cost. To be able to scale to large graphs, we present two sampling based algorithms, BSA and OSA, respectively. OSA significantly accelerates BSA due to the optimizations in the shortest path distance computation and the updating procedure. By conducting extensive experiments on four real social networks, we demonstrate the efficiency and effectiveness of the proposed techniques.

References

1. Abboud, A., Grandoni, F., Williams, V.V.: Subcubic equivalences between graph centrality problems, APSP and diameter. In: SODA, pp. 1681–1697 (2015)
2. Aloise, D., Deshpande, A., Hansen, P., Popat, P.: NP-hardness of euclidean sum-of-squares clustering. Mach. Learn. **75**(2), 245–248 (2009)
3. Bavelas, A.: A mathematical model for small group structures. Hum. Organ. (1948)
4. Bavelas, A.: Communication patterns in task oriented groups. J. Acoust. Soc. Am. **22**(6), 726–730 (1950)
5. Bonacich, P., Lloyd, P.: Eigenvector centrality and structural zeroes and ones: when is a neighbor not a neighbor? Soc. Netw. **43**, 86–90 (2015)
6. Brandes, U.: A faster algorithm for betweenness centrality. J. Math. Sociol. **25**, 163–177 (2001)
7. Chechik, S., Cohen, E., Kaplan, H.: Average distance queries through weighted samples in graphs and metric spaces: high scalability with tight statistical guarantees. In: APPROX/RANDOM (2015)
8. Cohen, E.: All-distances sketches, revisited: HIP estimators for massive graphs analysis. In: PODS (2014)
9. Cohen, E., Delling, D., Pajor, T., Werneck, R.F.: Computing classic closeness centrality, at scale. In: COSN (2014)
10. Eppstein, D., Wang, J.: Fast approximation of centrality. In: SODA (2001)
11. Freeman, L.C.: Centrality in social networks conceptual clarification. Soc. Netw. **1**, 215–239 (1978)
12. Indyk, P.: Sublinear time algorithms for metric space problems. In: STOC (1999)
13. Okamoto, K., Chen, W., Li, X.-Y.: Ranking of closeness centrality for large-scale social networks. In: Preparata, F.P., Wu, X., Yin, J. (eds.) FAW 2008. LNCS, vol. 5059, pp. 186–195. Springer, Heidelberg (2008)
14. Olsen, P.W., Labouseur, A.G., Hwang, J.: Efficient top-k closeness centrality search. In: ICDE (2014)
15. Rattigan, M.J., Maier, M., Jensen, D.: Graph clustering with network structure indices. In: ICML, pp. 783–790 (2007)
16. Thorup, M.: Quick k-Median, k-Center, and facility location for sparse graphs. In: Orejas, F., Spirakis, P.G., van Leeuwen, J. (eds.) ICALP 2001. LNCS, vol. 2076, p. 249. Springer, Heidelberg (2001)
17. Wasserman, S., Faust, K.: Social network analysis: methods and applications. Cambridge University Press, Cambridge (1994)
18. Zhao, J., Lui, J.C., Towsley, D., Guan, X.: Measuring and maximizing group closeness centrality over disk-resident graphs. In: WWW 2014 Companion, pp. 689–694 (2014)

Analyzing EEG Signal Data for Detection of Epileptic Seizure: Introducing Weight on Visibility Graph with Complex Network Feature

Supriya[1(✉)], Siuly[1], Hua Wang[1], Guangping Zhuo[2], and Yanchun Zhang[1]

[1] Centre for Applied Informatics, College of Engineering and Science, Victoria University, Melbourne, VIC 8001, Australia
Supriya.Supriya@live.vu.edu.au,
{Siuly.Siuly, Hua.Wang, Yanchun.Zhang}@vu.edu.au
[2] Department of Computer Science, Taiyuan Normal University, Taiyuan, China
zhuoguangping@163.com

Abstract. In the medical community, automatic epileptic seizure detection through electroencephalogram (EEG) signals is still a very challenging issue for medical professionals and also for the researchers. When measuring an EEG, huge amount of data are obtained with different categories. Therefore, EEG recording can be characterized as big data due to its high volume. Traditional methods are facing challenges to handle such Big Data as it exhibits non-stationarity, chaotic, voluminous, and volatile in nature. Motivated by this, we introduce a new idea for epilepsy detection using complex network statistical property by measuring different strengths of the edges in the natural visibility graph theory. We conducted 10-fold cross validation for evaluating the performance of our proposed methodology with support vector machine (SVM) and Discriminant Analysis (DA) families of classifiers. This study aims to investigate the effect of segmentation and non-segmentation of EEG signals in the detection of epilepsy disorder.

Keywords: EEG · Epilepsy · Complex network · Visibility graph · Average weighted degree · SVM and LDA

1 Introduction

Around 50 million people in world-wide is effected by one of the most widely existing chronic neurological syndrome named as epilepsy [1]. In epilepsy, recurrent seizure attacks due to malfunctioning of the electrophysiological part of the brain come to pass at any stage of life. Electroencephalogram (EEG) is one of electro-physiological technique to study and measure the voltage fluctuation of the brain and helps in diagnose the epilepsy disorder as epilepsy leaves their signature in the EEG signals [2]. There are wide-ranging of existing methods from linear to non-linear for the detect epilepsy disorder [3]. However, these techniques do not preserve all characteristics of

© Springer International Publishing AG 2016
M.A. Cheema et al. (Eds.): ADC 2016, LNCS 9877, pp. 56–66, 2016.
DOI: 10.1007/978-3-319-46922-5_5

EEG time series data such as, non-stationarity, chaotic nature [4]. Hence there is an ever-increasing need to develop new techniques that can detect epilepsy disorder by preserving the relevant information and provide additional information about epileptic EEG signals.

In the current era, complex network and graph theory approach is becoming the emergent field to detect various brain disorders [5]. In 2010, Ahmadlou et al. [6] firstly applied visibility graph algorithm for the detection of Alzheimer disorder and obtained promising results. After that many researchers and clinicians applied visibility graph algorithm for the detection of epilepsy disorder [7, 8] but their proposed methods have some limitations as they have not considered an important fact that in network, the links exhibit different strengths and all the nodes of network are connected with each other on the basis of this strength. Therefore, by focusing the limitation of the existing (especially visibility graph) methods to detect the epilepsy disorder, our proposed technique has explore the idea of, the importance of edge weight in epilepsy detection through visibility graph by constructing weighted complex network. It is our believe that, this proposed methodology is really new and will be very useful in the field of epilepsy and other brain disorder detection.

1.1 Contribution and Organization of the Paper

In this paper, we perform several experiments to discriminate between different kinds of EEG signals and make the following contributions.

- In this paper we developed new edge weight calculation method which helps to record the sudden changes happen in EEG signal for the duration of seizure activity. As during seizure activity the amplitude of EEG signals are too much fluctuating with time and our proposed method helps to recognize this fluctuation easily.
- We investigate the effect of segmentation and non-segmentation process on EEG signals in detection of epilepsy disorder with our proposed methodology.
- Several experiments performed for different classification problems and the outcomes results for all the test cases also suggests that our proposed technique is best appropriate to differentiate between different kinds of EEG signals. Moreover, it is quite promising for the classification of EEG signals of epileptic seizure activity set (E) and healthy person with eye open (A) with 100 % accuracy.

This remaining paper has been structured as: Sect. 2 comprises complete description of the proposed methodology. Section 3 presents the detailed discussion about the experimentation procedure and results. In Sect. 4, conclusions along with the future work have been mention.

2 Proposed Methodology

In this work, a novel algorithm based on new edge weight method for visibility graph is proposed to detect epilepsy disorder from EEG brain signals. The schematic diagram of proposed methodology is presented in Fig. 1. The approach is effective to distinguish between different EEG signals and epileptic EEG signals. The entire procedure of this methodology is composed of various sections: *conversion of time series EEG signals into weighted complex network, statistical feature extraction of weighted complex network, classification of epileptic EEG signals* from different kinds of EEG signals.

Fig. 1. The schematic diagram of proposed methodology for epileptic seizure detection.

2.1 Conversion of Time Series Data into Complex Network

For mapping the time series data into weighted complex network we have used lucasa [9] visibility graph algorithm. Following steps are used for the construction of weighted complex network:

  I. Consider each sample point of a time series $x(t_i)$, $i = 1,2,\ldots\ldots N$ of N sampling points as a node n_i of graph G(N,E), where N represents the node set i.e. N = $\{n_i\}$, $i = 1,2,\ldots\ldots N$, and E = e_i, $i = 1,2,3,\ldots\ldots\ldots N$, are the edges of graph.
 II. The edges between the nodes of the graph are determined on the basis of [9] the following Eq. (1).

$$n_j < n_i + (n_k - n_i)\frac{t_j - t_i}{t_k - t_i}, k > j > i \qquad (1)$$

where, n_i, n_j and n_k are the nodes corresponds to the data sample points $x(t_i)$, $x(t_j)$ and $x(t_k)$ with the time events t_i, t_j and t_k.

 III. In this study, we develop the below Eq. (2) to calculate the edge weight between two nodes:

$$w_{ij} = \frac{n_j - n_i}{t_j - t_i}, j > i \qquad (2)$$

where, w_{ij} is the edge weight between node i and j and also directional in nature from i to j. Also absolute value of edge weight has been considered in all cases.

 IV. Finally a weighted complex network has been constructed from the EEG time series data by utilizing the edges weight values.

Figure 2 presents an example of the weighted complex network build upon the above four steps using EEG time series data sample points = {100, 124, 153, 185, 210, 220, 216, 222, 240, 265, 298, 330, 362, 381, 391}. The edge weight showing in this figure is constructed on the basis of our newly developed Eq. (2). The thickness of edges in the below figure is according to the edge weight values.

2.2   Feature Extraction

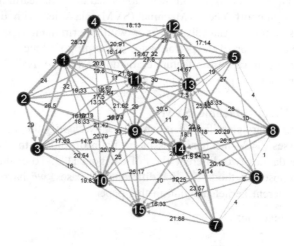

Fig. 2. Weighted complex network of EEG signal

Feature extraction plays an important role for classification of EEG signal data. It helps to make the analysis process easier as extracted features compress the huge amount of EEG signal data into feature vector set by minimizing the loss of information in the original EEG signals. In this paper, we have extracted one statistical property of network named as average weighted degree of network as feature from weighted complex network. This statistical property helps to determine the underlying pattern of hidden information from brain EEG signals.

If a $A_{NxN} = \{a_{ij}\}$ is an adjacency matrix with N number of nodes is used to represent the weighted complex network then $a_{ij} = 1$ if there is an edge from node i to j otherwise it's 0. According to [10] the weighted degree of the node i is the total weights of all the edges attached to node i which is represented by:

$$wd_i = \sum_{j \in B(i)} w_{ij} \qquad (3)$$

where, B(i) represents the neighborhood of node i and w_{ij} represents the weight of the edges between nodes i and j. And the average weighted degree of the network is the average mean of the total weights of the incident links on all the vertices in the network. It is important to note that due to sudden fluctuations in the epileptic EEG signals, the edge weight will show a discrepancy and different kinds of EEG signals exhibit different edge weight among their nodes and thus their resultant networks has different average weighted degree values.

2.3   Classification

In this paper, we have used two well-known supervised machine learning classification method named as Support Vector Machine (SVM) classifier with different kernel function and Discriminant Analysis classifier with discriminant type as linear and quadratic, for the evaluation of the performance of the proposed technique by utilizing the resulting feature extracted from feature extraction technique. As LDA and SVM is easy to implement and has fast prediction speed.

2.3.1   Support Vector Machine (SVM)

SVM is basically a binary classifier i.e. it can efficiently classify the data that belong to two different classes. SVM mechanism is based upon of finding the best hyperplane that separates the data of two different class of category [11]. To evaluate the performance of our proposed methodology for different test cases, we have employed the following three different kernel function in this paper.

I. Linear kernel function:

$$K(x,y) = (x^T y) \tag{4}$$

II. Polynomial kernel function with degree d:

$$K(x,y) = (x^T y + 1)^d \tag{5}$$

III. Radial basis kernel function with width σ:

$$K(x,y) = e^{\left(\frac{(-\|x-y\|)^2}{2\sigma^2} \right)} \tag{6}$$

2.3.2   Discriminant Analysis

In this paper, we have used two Discriminant Analysis (DA) classification methods named as Linear Discriminant Analysis (LDA) and Quadratic Discriminant Analysis (QDA). In case of LDA, each class has same covariance but the means vary whereas in case of QDA, the covariance and means parameters varies for each class. The detailed description about DA classifier are available in [12].

3 Experimental Results and Discussion

In this section, the proposed method is tested on the online available epileptic benchmark (http://epileptologiebonn.de/cms/front_content.php?idcat=193&lang=3) database: Bonn university epileptic EEG data. The whole database comprises of five EEG datasets (denoted as Set A–Set E). EEG signal in Set A and Set B were recorded from surface EEG recordings of five healthy volunteers with eyes open and eyes closed, respectively. Set C and Set D were collected in seizure-free intervals from five epileptic patients from the hippocampal formation of the opposite hemisphere of the brain and from within the epileptogenic zone, respectively. Set E contains the EEG records of five epileptic patients during seizure activity. Each channel of every set contains 4097 data sample points of 23.6 s. For detail description of this database please refer Andrzejak et al. [13]. The proposed technique is tested on the below four different classification problems named as test cases build upon this data set:

Test Case I: Set A versus Set E
Test Case II: Set B versus Set E
Test Case III: Set C versus Set E
Test Case IV: Set D versus Set E

The proposed technique has been implemented with the help of MATLAB R2015b (version 8.6, 64 bit). In this research study, we wanted to examine the effect of segmentation of EEG signals in the detection of epilepsy disorder. In order to include more data, the segmentation of a signal can provide more meaningful information and can be considered as a part of the entire data set [7]. Moreover, this will also help to make computation task faster. By considering this information into account, the experimentation of the proposed technique has been conducted for the following two objectives:

1 First objective is to check the performance of our proposed method by considering whole data samples per channel i.e. by considering 4097 data sample points per channel. Here each channel is considered as independent samples. The implementation process of this method as discussed in Sect. 2.
2 Second objective is to check the performance of our proposed methodology with segmentation of each channel of EEG signals. During second approach, we divided each channel into four segment i.e. Seg1 = 1024, Seg2 = 1024, Seg3 = 1024, Seg4 = 1025 data sample points. Then these four segments are further used as a four independent samples. As in each data subset, there are 100 channels data with 4097 data points therefore after segmentation; we have 400 segments with 1024 data sample points. Afterward the proposed method (Sect. 2) is implemented on these segments.

To evaluate the performance of proposed epilepsy detection method and also to achieve more reliable results, k-fold cross validation method is applied on all the four test cases. In this paper, we have considered k = 10 i.e. in 10 fold cross validation, the feature sets are randomly partitioned into 10 groups. The classification model utilize 9 groups for training purposes and the remaining 10th group is used for testing. This whole procedure has been repeated 10 times. The performance of extracted feature vector sets is analyzed with the help of SVM and LDA families of classifiers. The results of different test cases after applying the proposed technique with LDA and QDA classifier is presented in Tables 1 and 2.

Table 1. Overall performance of the LDA Classifier considering with segmentation and without segmentation process of EEG signals

Different Test Cases	LDA with segmentation			LDA without segmentation		
	Sensitivity (%)	Specificity (%)	Accuracy (%)	Sensitivity (%)	Specificity (%)	Accuracy (%)
Test Case 1	100	73.75	86.88	100	76	88
Test Case 2	100	72	86	100	74	87
Test Case 3	98	72	85.13	98	75	86.5
Test Case 4	100	73.25	86.62	100	76	88

Table 2. Overall performance of the Quadratic LDA Classifier considering with segmentation and without segmentation process of EEG signals

Different Test Cases	Quadratic DA with segmentation			Quadratic DA without segmentation		
	Sensitivity (%)	Specificity (%)	Accuracy (%)	Sensitivity (%)	Specificity (%)	Accuracy (%)
Test Case 1	100	100	100	100	100	100
Test Case 2	96.25	90.75	93.5	97	93	95
Test Case 3	96	83	89.62	96	82	89
Test Case 4	98	98.5	98.25	98	99	98.5

It is clear from both the tables, that the QDA is more efficient for our proposed methodology as compared to LDA. Moreover in case of QDA during segmented and non-segmented of EEG signals, the accuracy results are very close to each other except for test case 2, which is also showing only slight increase in case of non-segmentation. It can be seen that quadratic LDA classifier is demonstrating 100 % efficiency to distinguish between healthy (set A) and epileptic seizure activity (set E) EEG signals in both segmented and non-segmented process.

Our methodology is further analysed with other different kernel functions of SVM named as Linear, RBF and Polynomial kernel function. Table 3 illustrates the results after applying SVM classifier using linear kernel function on all the four test cases. Tables 4 and 5 lists the experimental results of all the four test cases with segmentation and without segmentation approach using SVM Rbf and SVM Polynomial kernel functions.

The experimental outcomes of Table 5 signifies that SVM Polynomial kernel function outperforms as compared to other SVM classifiers and achieve promising results with 100 % accuracy for distinguish between healthy person and epileptic seizure activity EEG signals. Moreover during segmented and non-segmented EEG

Table 3. Overall performance of the SVM Linear Classifier considering with segmentation and without segmentation process of EEG signal

Different Test Cases	SVM Linear with segmentation			SVM Linear without segmentation		
	Sensitivity (%)	Specificity (%)	Accuracy (%)	Sensitivity (%)	Specificity (%)	Accuracy (%)
Test Case 1	100	98.5	99.25	100	91	95.5
Test Case 2	98.5	87.75	93.13	100	83	91.5
Test Case 3	94.75	90.5	92.63	96	82	89
Test Case 4	99	93.75	96.38	100	85	92.5

Table 4. Overall Performance of the SVM classifier using RBF kernel function considering with segmentation and without segmentation process of EEG signal

Different Test Cases	SVM RBF with segmentation			SVM RBF without segmentation		
	Sensitivity (%)	Specificity (%)	Accuracy (%)	Sensitivity (%)	Specificity (%)	Accuracy (%)
Test Case 1	100	98.25	99.12	100	91	96.5
Test Case 2	97.25	90	93.63	100	85	92.5
Test Case 3	95	91.5	93.25	96	86	91
Test Case 4	99	95.5	97.25	99	91	95

Table 5. Overall performance of the SVM classifier using Polynomial kernel function considering with segmentation and without segmentation process of EEG signal

Different Test Cases	SVM Poly with segmentation			SVM Poly without segmentation		
	Sensitivity (%)	Specificity (%)	Accuracy (%)	Sensitivity (%)	Specificity (%)	Accuracy (%)
Test Case 1	100	100	100	100	100	100
Test Case 2	94	92.75	93.37	96	94	95
Test Case 3	93.75	95.25	94.5	94	95	94.5
Test Case 4	97.75	99	98.38	99	98	98.5

signals, the accuracy results are very close to each other except for test case 2, which is also showing only slight increase in case of non-segmentation.

In order to provide a clear scenario, Fig. 3(a)–(c) presents a comparison of obtained performances for the different classifiers in both segmentation and non-segmentation process.

As can be observed from Fig. 3(a) SVM polynomial has higher accuracy and specificity classification performance as compared to other classifiers for all the four test cases problems whereas sensitivity of the LDA achieved higher performance results as compared to other classifiers. Table 6 presents the comparative analysis of the classification accuracy of the proposed method with different methods in the literature that perform experimentation on the same EEG data set and illustrate that the proposed methodology is more accurate for detection of epileptic seizure (set A vs set E) as

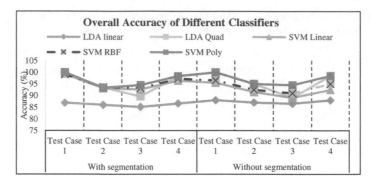

Fig. 3(a)

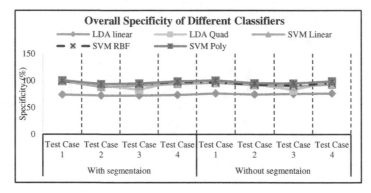

Fig. 3(b)

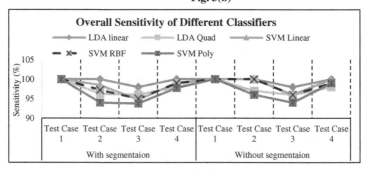

Fig. 3(c)

Fig. 3. Illustration of classification performance for all the four test cases with different classfiers. (a) Accuracy, (b) Specificity, and (c) Sensitivity

compared to them. Moreover provide promising results for different test cases except for test case 3.

As for the duration of seizure activity, there is sudden change in neural discharge in the brain. Thus these consequences increase in variation of EEG signals and henceforth results with the sudden fluctuation in EEG signals. Due to this sudden fluctuation, the

Table 6. Comparative analysis of the accuracy of the proposed work with existing work that used the same data set for their experimentation

Researchers	Data set	Dimension of features	Accuracy (%)
Siuly et al. 2011 [14]	A vs E	9	99.9
	B vs E	9	93.6
	C vs E	9	96.20
	D vs E	9	93.60
Nicolaou et al. 2012 [15]	A vs E	1	93.42
	D vs E	1	83.13
Guohun Zhu et al. 2014 [16]	A vs E	2	100
	B vs E	2	93.0
	C vs E	2	97
	D vs E	2	93
The proposed technique	A vs E	1	100
	B vs E	1	95
	C vs E	1	94.5
	D vs E	1	98.5

edge weight of the complex network build from this seizure EEG signals starts varying and helps to reveal the hidden information of brain functionality. This study also explores that different nodes of EEG weighted complex network interact with each other with different strengths. Therefore when complex network theory is used to detect epilepsy from brain EEG signals, the edge weight play an important role to detect the sudden fluctuation during seizure activity. This study also investigate that our proposed methodology produce very close results during segmentation and without segmentation approaches with SVM classifier. So there is no huge effect of segmentation and without segmentation of EEG signals on our proposed technique. Moreover segmentation approach takes fast computation time due to less amount of analysis data.

4 Conclusion

In this research study, we presented an efficient technique to detect epilepsy disorder from EEG brain signals. This methodology introduces new method to calculate the edge weight of complex network. We then constructed weighted complex network with the help of our newly developed edge weight method. Statistical property of complex network named as average weighted degree is used as extracted feature to compare the classification performance of SVM and DA families of classifiers. The outcomes of the experiments yield that in DA family, QDA provides higher accuracy performance results as compared to LDA but overall SVM polynomial is most promising classifier for our proposed methodology with higher performance results as compared to other classifiers. Moreover in this study, we investigate that, in case of segmentation and without segmentation approaches of EEG signals, the performance results for different test cases are very close to each and does not varies a lot. The pilot study in this paper has examined that the proposed methodology is best suitable to discriminate between different EEG signals. The classification accuracy performance result for ictal (set E) and normal healthy person EEG (set A) is achieved by100 % with SVM polynomial and quadratic discriminant analysis when considering both segmentation and without segmentation approach. Moreover the sensitivity performance for set A versus set E is 100 % with the family of SVM and LDA classifier. It is our believed that this research study will support the technicians to build a software system that will provides support for automatic detection of epileptic seizure. We are currently planning to extend this proposed methodology to detect other brain disorders through EEG signals and also to multi-class EEG classification in the epilepsy detection from brain signals.

References

1. Siuly, S., Li, Y.: Designing a robust feature extraction method based on optimum allocation and principal component analysis for epileptic EEG signal classification. Comput. Methods Programs Biomed. **119**, 29–42 (2015)
2. Supriya, S., Siuly, S., Zhang, Y.: Automatic epilepsy detection from EEG introducing a new edge weight method in the complex network. Electron. Lett. (2016)
3. Donner, R., Small, M., Donges, J., Marwan, N., Zou, Y., Xiang, R., Kurths, J.: Recurrence-based time series analysis by means of complex network methods. Int. J. Bifurcat. Chaos **21**, 1019–1046 (2011)
4. Campanharo, A., Sirer, M., Malmgren, R., Ramos, F., Amaral, L.: Duality between Time Series and Networks. PLoS ONE **6**, e23378 (2011)
5. van Stam, C., Straaten, E.: The organization of physiological brain networks. Clin. Neurophysiol. **123**, 1067–1087 (2012)
6. Ahmadlou, M., Adeli, H., Adeli, A.: New diagnostic EEG markers of the Alzheimer's disease using visibility graph. J. Neural Transm. **117**, 1099–1109 (2010)
7. Tang, X., Xia, L., Liao, Y., Liu, W., Peng, Y., Gao, T., Zeng, Y.: New approach to epileptic diagnosis using visibility graph of high-frequency signal. Clin. EEG Neurosci. **44**, 150–156 (2013)
8. Ni, Y., Wang, Y., Yu, T., Li, X.: Analysis of epileptic seizures with complex network. Comput. Math. Methods Med. **2014**, 1–6 (2014)
9. Lacasa, L., Luque, B., Ballesteros, F., Luque, J., Nuno, J.: From time series to complex networks: The visibility graph. Proc. Nat. Acad. Sci. **105**, 4972–4975 (2008)
10. Antoniou, I., Tsompa, E.: Statistical analysis of weighted networks. Discrete Dyn. Nat. Soci. **2008**, 1–16 (2008)
11. Andrew, A.: An Introduction to Support Vector Machines and Other Kernel-based Learning Methods (2001)
12. Mikat, S., Fitscht, G., Weston, J., Scholkopft, B., Muller, K.-R.: Fisher discriminant analysis with kernels. Neural Net. Signal Proc. **IX**, 41–48 (1999)
13. Andrzejak, R., Lehnertz, K., Mormann, F., Rieke, C., David, P., Elger, C.: Indications of nonlinear deterministic and finite-dimensional structures in time series of brain electrical activity: Dependence on recording region and brain state. Phys. Rev. E **64**, 61907 (2001)
14. Siuly, L.: Y., Wen, P.: Clustering technique-based least square support vector machine for EEG signal classification. Comput. Methods Programs Biomed. **104**, 358–372 (2011)
15. Nicolaou, N., Georgiou, J.: Detection of epileptic electroencephalogram based on Permutation Entropy and support vector machines. Expert Syst. Appl. **39**, 202–209 (2012)
16. Zhu, G., Li, Y., Wen, P.: Epileptic seizure detection in EEGs signals using a fast weighted horizontal visibility algorithm. Comput. Methods Programs Biomed. **115**, 64–75 (2014)

Spatial Database

Spatial Textual Top-k Search in Mobile Peer-to-Peer Networks

Thao P. Nghiem[1(✉)], Cong Ma[1], J. Shane Culpepper[1], and Timos Sellis[2]

[1] RMIT University, Melbourne, Australia
{jessie.nghiem,shane.culpepper}@rmit.edu.au, s3496858@student.rmit.edu.au
[2] Swinburne University, Melbourne, Australia
tsellis@swin.edu.au

Abstract. Mobile hardware and software is quickly becoming the dominant computing model for technologically savvy people around the world. Nowadays, mobile devices are commonly equipped with GPS and wireless connections. Users have also developed the habit of regularly checking into a location, and adding comments or ratings for restaurants or any place of interest visited. This work explores new approaches to make data available from a local network, and to build a collaborative search application that can suggest locations of interest based on distance, user reviews and ratings. The proposed system includes light-weight indexing to support distributed search over spatio-textual data on mobile devices, and a ranking function to score objects of interest with relevant user review content. From our experimental study using a Yelp dataset, we found that our proposed system provides substantial efficiency gains when compared with a centralised system, with little loss in overall effectiveness. We also present a methodology to quantify efficiency and effectiveness trade-offs in decentralized search systems using the Rank-based overlap (RBO) measure.

1 Introduction

Location-aware services are becoming increasing popular in advanced database applications. One of the most important fields in location-aware services is local business search using associated user reviews and ratings. For example, a person moves to a new suburb, and wishes to find an affordable Chinese restaurant. They can go to Zomato to search for local restaurants, and read user reviews and ratings[1]. This type of search involves both spatial and textual search. The expected results of this type of search are a list of the k highest ranking objects according to some spatial and textual similarity metric. To rank the result, many different scoring functions could be applied. One example is a linear combination of the spatial relevance of the location of the objects and the query point, along with the user-rating and the textual relevance between query keywords and user review contents.

---

[1] https://www.zomato.com/.

© Springer International Publishing AG 2016
M.A. Cheema et al. (Eds.): ADC 2016, LNCS 9877, pp. 69–81, 2016.
DOI: 10.1007/978-3-319-46922-5_6

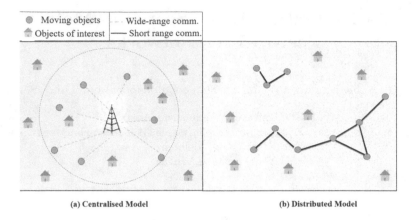

(a) Centralised Model   (b) Distributed Model

Fig. 1. Centralised Systems versus P2P Systems. Circles represent moving objects; house symbol: objects of interest; dashed lines: wide-range communication; continuous lines: short-range communication; dots: the base station network range.

In the vast majority of previous research, search systems combining spatial and keyword queries are centralised, which requires a single server to store data and process queries [7,13,14,19]. This model is shown in Fig. 1(a). However, a single point of failure makes the system susceptible to too much traffic, natural disasters, and/or denial of service attacks, which can lead to widespread disruptions [11]. Moreover, with the continual growth of data, update costs and storage is a persistent issue.

This work proposes a collaborative P2P search framework with spatial and textual indexing. It does not rely on a centralised server to store data or process queries; instead the queries are processed by each mobile device. Specifically, the overall contributions of this research are as the following:

1. We introduce a new direction in mobile distributed location-aware search for spatial and textual data.
2. We propose an indexing structure for textual data (user reviews) that can be searched for cached locations of interest associated with reviews from other mobile users.
3. A ranking function is also developed to score objects of interest with relevant user review content.
4. From our experimental study using a real-dataset, we found that our proposed system is substantially more efficient than a centralised system. We also show that the effectiveness of the proposed distributed workload model is comparable to the centralised approach using the RBO measure [18].

2   Background

2.1   Top-k kNN Queries

Top-k search in an important application in geographical information retrieval. This type of query returns a ranked list of the top-k documents, ranked by spatial and textual similarity. These queries are supported through a variety of spatial and textual indexing data structures.

There have several recent studies on new approaches to combining spatial and textual indexes [3,6,7,14,19]. In general, they can be classified into two categories. The first approach maintains two independent indexing structures; one for text (for example, inverted files) and another for spatial data (such as: R-tree and variants [9]). For example, the work in [20] loosely combined R*-tree indexing. Other spatial indexes like grid-based and space filling based structures are also possible [6,17]. The queries are answered by using a spatial index as an initial filter, and reranking the remaining items with an inverted index, or vice versa. The second approach focuses on a more tightly combined scoring reqime by combining both data representations into a single, hybrid index. The hybrid indexes simultaneously handle spatial and textual pruning to produce the final top-k result set. The IR-tree [7,12] is the most widely used indexing structure in this group. Conceptually, an IR-tree is an R-tree, where each node is augmented with an inverted file.

To estimate the relevance between documents and user queries, a *scoring function* must be defined. In particular for geographical textual search, the scoring function can be composed of two main components: textual relevance and spatial relevance [2,12,14]. Textual relevance can be measured using variants of the TF-IDF model such as BM25, or the language model. Spatial relevance is often measured using a distance metric such as Euclidean or network distance.

2.2   Mobile Collaborative Caching and Local Distributed Query Processing

With the development of the state-of-the-art wireless communication technologies, such as IEEE 802.11 and Bluetooth, mobile collaborative caching has increasingly drawn attention as an alternative for information sharing among mobile hosts over standard centralized models. In general, the technique can improve data retrieval performance by allowing moving objects to access local caches on peers [4]. The first on-demand distributed data sharing algorithm for kNN queries was introduced by Ku and Zimmermann [11]. The scenario is shown as follows. The query node collects and verifies information from peers. If results cannot be verified, they are sent to the server or base station (BS). The BS will complete the task, and send the result back to the query node. This approach is efficient in reducing server workload, and alleviating traffic congestion in the BS.

A distributed multi-dimensional index structure, called P2PRdNN was introduced by Chen et al. [1] to efficiently support reverse nearest neighbour queries. Other related work [5,16] proposed a framework to find an approximate answer for spatial-only range and kNN queries. Another solution for nearest neighbour queries in static sensor networks called a *peer-tree* was proposed by Demirbas and Ferhatosmanoglu [8] The approach is not amenable to mobile P2P environments due to the fixed communication infrastructure.

The novelty of our approach is that query processing and indexing is accomplished using a purely distributed spatial and textual search model. Top-k range queries are answered only by harnessing the power of peer collaboration without any central supervision.

3   Proposed Model

3.1   System Model and Assumptions

We assume a mobile network with no central supervision, where query objects and peers are dynamic as is commonly found in a mesh network. The environment is a symmetric system where each moving object, such as a smart mobile phone or tablet, can be both a query node and a peer of other nodes. Moving objects are also self-aware of their current location through an equipped GPS. The location of moving objects and objects of interest mentioned in this paper are physical locations.

Moving objects are equipped to support ad-hoc communication with other moving neighbours via Bluetooth, Wireless Local Networks (WLANs), Wireless Local Personal Networks (WPANs), or WiFi Direct – an emerging form of P2P communication. In addition, points of interest are randomly distributed in the network. To enhance the P2P query processing, a memory cache is assigned to store spatial data for points of interest from previous queries. A priority queue manages requests based on the distance from the cached point of interest to the moving object. When the cache (priority queue) is full, new points of interest will be cached only if their ranking score is high enough to displace the k-th ranked object in the cache. This deletion strategy assures that cached data is the most useful answer for future queries, and increases the accuracy of the query results.

3.2   Query Models and Message Types

The proposed system is designed to answer top-k kNN queries based on the available information from peers. The query point is always at the same location as the moving object issuing the query. The information from peers is the result from previous queries stored in memory cache of peers.

There are four distinct message types between the query node q and a peer p_j:

1. A beacon message (*beacon_msg*) is broadcast from the query node to detect peers within communication range.
2. An acknowledgement message (*ack_msg*) from peers to the query node to assign a location to the responding peers.

3. A query message (*query_msg*) from the query node to the selected peers asking for points of interest cached locally by those peers.
4. A query reply (*reply_msg*) from peers to the query node with a possible answer from the cache of the peer. The answer consists of the location, and the type of the point of interest.

Here the system is working with cached data from moving objects; therefore, it is expected that the number of points of interest stored in each cache is relatively small. Hence, a spatial index is not necessary. Instead, indexing text data (user reviews) associated to the objects of interest is most important.

3.3   Indexing User Reviews and Ranking Function

Indexing Structure. To support text search, an indexing structure for cached user reviews at each moving object is required. As the storage and computation for the moving objects is limited, a simple inverted index is used. First, the reviews are read from the dataset file. Then all the terms minus stop words are extracted, formatted into lower case and registered in the indexing file [15]. The indexing structure consists of a lexicon and a posting list, as shown in Fig. 2.

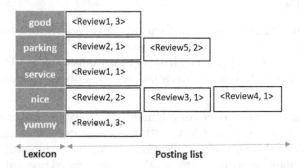

Fig. 2. Indexing structure for user reviews

Ranking Function. For a query q with location l_q, a set of keywords $t \in q$, and a candidate user review r with location l_r, the combined score of q and r, $s(q, r)$ is computed as:

$$s(q, r) = w_1 \times ls(l_q, l_r) + w_2 \times ts(q, r) + w_3 \times rs \qquad (1)$$

where $w_1, w2, w3 \in (0, 1)$ are the parameters used to weight the importance of the spatial, textual or rating components, $w_1 + w_2 + w_3 = 1$, ls is the spatial relevance component, and ts is textual relevance component. The normalised user rating rs is also associated with the object for each user (node).

$$rs = \frac{user\_rating}{rating_{max}} \qquad (2)$$

where $rating_{max}$ is the maximum rating allowed.

Spatial Relevance Component. In this model, Euclidean distance is used to measure the distance between the objects of interest and the query point. The spatial relevance score ls is computed as below:

$$ls(l_q, l_r) = 1 - \frac{distance_E(l_q, l_r)}{distance_{max}} \tag{3}$$

where $distance_{max}$ is the maximum distance from two unique points in the geographical space.

Textual Relevance Component. Variants of TF-IDF are the most commonly used textual similarity metric, and are used in this work. Specifically, the calculation of this score is as the following:

$$ts(q, r) = \frac{\sum_{t \in q} tf_{r,t} * log \frac{N_r}{df_{r,t}}}{ts_{max}} \tag{4}$$

where $tf_{r,t}$ is the number of times the term occurs in each review, N_r is the total number of reviews, and $df_{r,t}$ is the total count of the term in the collection.

3.4   System Details

Overview. Our ultimate goal is to harness the collaborative power of mobile devices to process spatial and keyword queries locally. Overall, the proposed system is divided into two primary phases: (1) Initialisation and Peer Discovery Phase; and (2) Query Processing. Each phase is described in detail below.

Initialisation and Peer Discovery Phase. Each moving object maintains a default map of the associated objects with user reviews and ratings in a cache. This can be loaded during the initialisation phase, or downloaded from a local provider. Since mobile users move frequently, the associated peers also change. As a result, before starting to send queries, a query node q needs to discover which moving objects are in communication range by sending a one-hop broadcast message. Moving objects receiving the broadcast message send an acknowledgement message which contains their ID and location information. More specifically, this phase is described in Algorithm 1. The query node q collects all acknowledgement messages from the surrounding nodes to construct a peer list. Note that q is assigned an acknowledgement time-out period. Therefore, q waits to receive acknowledgement messages from peers for a fixed period of time.

Query Processing Phase. After the first stage, q is aware of all peers close enough to query. When a peer receives the query, data is retrieved from the local cache, followed by pruning and ranking which is computed as the follows.

Pruning objects and user reviews at peers. For the user reviews associated with the candidate set, if the reviews contain the keywords, they will be ranked using the ranking function described in Eq. (1). The ranked lists of reviews and objects of interest from the peers are sent to the query object. Each query object

Data: Query node q, transmission range R
Result: At node q, a priority queue of peers P
begin
 | $P \leftarrow \varnothing$
 | Node q broadcasts a one-hop *beacon_msg* to every peer.
 | **foreach** p_i in range R **do**
  | **if** a peer p_i receives *beacon_msg* **then**
   | p_i sends *ack_msg* with a location and ID to q.
  | **end**
 | **end**
 | **if** q receives an *ack_msg* from p_i **then**
  | $P \leftarrow P \cup \{p_i\}$.
 | **end**
end

Algorithm 1. Initialisation and Peer Discovery

then collects objects of interest sorted by similarity, and return the top k objects of interest along with the relevant user reviews. Overall, the query processing phase is summarised in Algorithm 3.

4   Performance Evaluation

4.1   Simulation Setup and Configuration

All results are computed using the MiXiM simulation environment, which is derived from an OMNeT++-based framework to model and analyse Mobile P2P Query Processing Systems [10]. Each moving object contains 8 modules as shown in Fig. 3. Here we use Nic80211 for the Wi-Fi connection. According to the configuration for the network interface cards in MiXiM, a transmission current $txCurrent = 153$ mA and a receiving current $rxCurrent = 200$ mA are used. The communication to the server is conducted via 3G (WCDMA), band I-2100 which is used by Vodaphone and Optus in Australia. Data rate for high speed moving objects in this network is 128 kbps[2] and current consumption in connected state is 365.6 mW[3]. All queries are generated using a Poisson arrival model. A universal λ is assigned to all moving objects to represent the average number of queries arriving per unit time; or the expected number of queries generated by each moving object is $E(N) = \lambda T$ where T is the simulation time. Initially, each moving object is assigned random objects of interest and the corresponding user review and added to the local cache. The query expiry time is set to 30 s. After this time, even if there are still peers to query, the query object aborts communications, and processes the current results.

---

[2] http://www.silicon-press.com/briefs/brief.3g/index.html.
[3] http://www.option.com/en/newsroom/media-center/white-papers/.

Data: A query node q initialised with the number of ranked results required k,
a range *range*, a set of keywords QK. On each peer p, a set of cached
objects of interest IO_p, and an indexing structure I of user reviews R

Result: To node p, a set of sorted objects of interests $Result_p$ with relevant user
reviews and ratings

begin

    foreach IO_i in C **do**

        if $distance(l_{IO_i}, l_q) > range$ **then**

            | $C \leftarrow C - \{IO\}_i$

        end

        else if IO_i has no review containing any $k_i \in QK$ **then**

            | $C \leftarrow C - \{IO\}_i$

        end

    end

    foreach IO_i in C **do**

        $score \leftarrow 0$

        $review_{no} \leftarrow 0$

        foreach r_i in R **do**

            Calculate score $s(r_i)$ using the ranking function in Section 3.3.

            Increment $review_{no}$.

            $score+ = s(r_i)$

        end

        $score = score/review_{no}$

    end

    Return k objects in C with the highest score.

end

Algorithm 2. Query Processing Algorithm

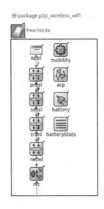

Fig. 3. Moving objects' modules

The simulation was ran using both real and synthetic datasets for a large-scale network. Empirically, we set w_1 (distance weight) to 0.8, w_2 (text relevance weight) to 0.15 and w_3 (rating weight) to 0.05 in the ranking function as locality

Data: Query node q, on q, a set of peers P, k value, range *range*, set of
     keywords QK, and *expiry_time*

Result: To node q, a set of sorted objects of interest IO_q with relevant user
     reviews and ratings

begin
    Node q sends a query set of keywords QK to every peer in P.
    Node q starts a timer *waiting_time*.
    foreach p_i in P **do**
        Node p calls **PeerRankingFunction** (Algorithm 2) for local cached
        data.
        Node p return the top k results to q.
    end
    if *waiting_time* > *expiry_time* **then**
        Node q keeps only the top k IO_i from peer with highest score.
        Return IO_q with relevant user reviews and ratings.
        Node q updates cached data.
    end
end

Algorithm 3. Query Processing Algorithm

Table 1. Simulation parameters

Parameter	Value
Playground	$5\,km \times 5\,km$
Number of reviews	10000
Number of moving objects	600
k	10
Expected number of queries generated	1000
Cache Size	1000 reviews
Simulation time	600 s

is the most important feature. Data is sampled from Yelp dataset[4]. This dataset includes data for businesses in America, including location, attributes, user ratings and reviews. In this simulation, we consider a subset of restaurants in Las Vegas with total 10,000 user reviews. To prevent reviews from popular restaurants dominating the dataset, a maximum of 20 reviews for each restaurant is used. In the initial stage, each moving object caches a number of restaurants in the neighbourhood area with the associated user reviews, which is defined by the cache size parameter. Other parameters are described in Table 1.

---

[4] https://www.yelp.com/academic_dataset/.

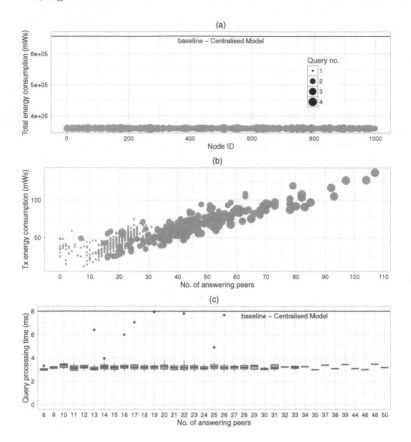

Fig. 4. Efficiency

4.2 Simulation Results and Discussions

In this section, we evaluate the performance of the proposed model in term of efficiency and effectiveness when varying the parameter values. The evaluation is based three different measures: processing time (from the time a query object discovers peers to the time the query is answered), energy consumption (energy spent at each moving object) and RBO (a similarity measure between incomplete rankings that handles non-conjointness, and gives higher weight to higher ranking objects). It is noted that to calculate RBO, the query results from the proposed method are in comparison with that of the central method.

Figure 4 compares the efficiency of the proposed P2P Model to the centralised model. In particular, energy consumption and query processing time are measured at the node level. Figure 4(a) clearly shows that the total energy consumption of the centralised model is higher than in the P2P model. In this simulation, mobile nodes do not go into sleep mode. Therefore, the receiving energy consumption at each node is stable, and much greater than the transmission energy consumption. That is why the transmission energy consumption

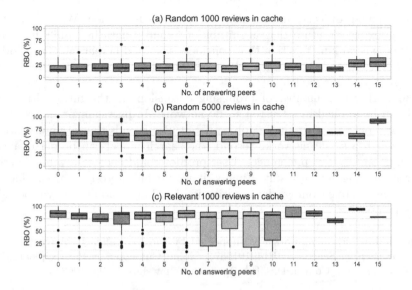

Fig. 5. Effectiveness

has little fluctuation in Fig. 4(b). As expected, increasing the number of peers queried or queries result in higher energy consumption. In addition to the energy efficiency, Fig. 4(c) indicates that the processing time of the P2P model can save up to 25 % over the centralised model. This is due to the difference costs in wide-range communications between the server and mobile objects, versus short-range P2P communications.

Figure 5 shows the accuracy of top-k results in the P2P model. The cen-tralised model is the ground truth as this system can exhaustively process the entire dataset, while moving objects in the P2P model only cache a subset of review lists. This trade-off in query processing time and power consumption is the core idea exploited in our model. Increasing the size of the dataset sub-set cached in moving objects, the RBO increases as expected. There is a three fold increase in RBO when the initial number of reviews in the memory cache changes from 1,000 to 5,000. Another possible way to improve the accuracy is to select the most relevant objects, and load them into the initial cache. In this simulation at the initialisation stage, reviews related to the nearest restaurants to the moving objects are randomly chosen during cache initialisation. Figure 5 shows the effects on RBO when relevance caching is used, achieving an RBO of 75 % when the cache size is 1,000 reviews (10 % of the collection).

5   Conclusions and Future Work

We present a peer-to-peer solution to solve the problem of retrieving the top-k spatial-textual objects of interest that are associated with a list of relevant user reviews and ratings. The proposed model harnesses the power of collaboration between moving objects, and requires no central supervision. We also apply a simple indexing structure that is suitable for shared data in mobile networks, and develop a ranking function that considers several different factors including distance, rating, and user review relevance. The simulation results demonstrate the feasibility of our model.

The work has a number of promising extensions in the future. First, the model can be applied in a trusted social network. In particular, a query object can ask not only the surrounding objects, but also friends that are not spatially close through a social network. Second, the search model can be personalised, depending on the user preferences. User profiles could become a valuable criteria for improving search result. Finally, it would be interesting to consider an incentive model to encourage data sharing in distributed mobile query processing environments.

Acknowledgment. This work was supported by the Australian Research Council's *Discovery Projects* Scheme (DP140101587). Shane Culpepper is the recipient of an Australian Research Council DECRA Research Fellowship (DE140100275).

References

1. Chen, D., Zhou, J., Le, J.: Reverse nearest neighbor search in peer-to-peer systems. In: Larsen, H.L., Pasi, G., Ortiz-Arroyo, D., Andreasen, T., Christiansen, H. (eds.) FQAS 2006. LNCS, vol. 4027, pp. 87–96. Springer, Heidelberg (2006)
2. Chen, L., Cong, G., Jensen, C.S., Wu, D.: Spatial keyword query processing: an experimental evaluation. PVLDB **6**(3), 217–228 (2013)
3. Choudhury, F.M., Culpepper, J.S., Sellis, T., Cao, X.: Maximizing bichromatic reverse spatial and textual k nearest neighbor queries. PVLDB **9**(6), 456–467 (2016)
4. Chow, C., Leong, H.V., Chan, A.T.S.: GroCoca: group-based peer-to-peer cooperative caching in mobile environment. J. Sel. Areas Commun. **25**(1), 179–191 (2007)
5. Chow, C., Mokbel, M., Leong, H.: On efficient and scalable support of continuous queries in mobile peer-to-peer environments. IEEE Trans. Mob. Comput. **10**, 1473–1487 (2011)
6. Christoforaki, M., He, J., Dimopoulos, C., Markowetz, A., Suel, T.: Text vs. space: efficient geo-search query processing. In: Proceedings of CIKM, pp. 423–432 (2011)
7. Cong, G., Jensen, C.S., Wu, D.: Efficient retrieval of the top-k most relevant spatial web objects. PVLDB **2**(1), 337–348 (2009)
8. Demirbas, M., Ferhatosmanoglu, H.: Peer-to-peer spatial queries in sensor networks. In: Proceedings of P2P, pp. 32–39 (2003)
9. Guttman, A.: R-trees: a dynamic index structure for spatial searching. SIGMOD Rec. **14**, 47–57 (1984)
10. Köpke, A., Swigulski, M., Wessel, K., Willkomm, D., Haneveld, P.T.K., Parker, T.E.V., Visser, O.W., Lichte, H.S., Valentin, S.: Simulating wireless, mobile networks in OMNeT++ the MiXiM vision. In: Proceedings of STTCNS (2008)
11. Ku, W., Zimmermann, R.: Nearest neighbor queries with peer-to-peer data sharing in mobile environments. Pervasive Mob. Comput. **4**(5), 775–788 (2008)
12. Li, Y., Chen, H., Xie, R., Wang, J.Z.: Bgn: a novel scatternet formation algorithm for bluetooth-based sensor networks. Mob. Inf. Syst. **7**, 93–106 (2011)

13. Li, Z., Lee, K.C.K., Zheng, B., Lee, W.-C., Lee, D., Wang, X.: IR-Tree: an efficient index for geographic document search. TKDE **23**(4), 585–599 (2011)
14. Mackenzie, J., Choudhury, F.M., Culpepper, J.S.: Efficient location-aware web search. In: Proceedings of ADCS, pp. 4: 1–4: 8 (2015)
15. Manning, C.D., Raghavan, P., Schütze, H.: Introduction to Information Retrieval. Cambridge University Press, New York (2008)
16. Nghiem, T.P., Maulana, K., Green, D., Waluyo, A.B., Taniar, D.: Peer-to-peer bichromatic reverse nearest neighbors in mobile ad-hoc networks. JPDC **74**(11), 3128–3140 (2013)
17. Vaid, S., Jones, C.B., Joho, H., Sanderson, M.: Spatio-textual indexing for geographical search on the web. In: Medeiros, C.B., Egenhofer, M., Bertino, E. (eds.) SSTD 2005. LNCS, vol. 3633, pp. 218–235. Springer, Heidelberg (2005)
18. Webber, W., Moffat, A., Zobel, J.: A similarity measure for indefinite rankings. ACM Trans. Inf. Syst. **28**(4), 20: 1–20: 38 (2010)
19. Zhang, D., Chan, C.-Y., Tan, K.-L.: Processing spatial keyword query as a top-k aggregation query. In: Proceedings of SIGIR (2014)
20. Zhou, Y., Xie, X., Wang, C., Gong, Y., Ma, W.-Y.: Hybrid index structures for location-based web search. In: Proceedings of CIKM (2005)

Continuous Maximum Visibility Query for a Moving Target

Ch. Md. Rakin Haider[1(✉)], Arif Arman[1], Mohammed Eunus Ali[1],
and Farhana Murtaza Choudhury[2]

[1] Bangladesh University of Engineering and Technology, Dhaka, Bangladesh
{rakinhaider,eunus}@cse.buet.ac.bd, arman@cse.uiu.ac.bd
[2] RMIT University, Melbourne, Australia
farhana.choudhury@rmit.edu.au

Abstract. Opportunities to answer many real life queries such as *"which surveillance camera has the best view of a moving car in the presence of obstacles?"* have become a reality due to the development of location based services and recent advances in 3D modeling of urban environments. In this paper, we investigate the problem of continuously finding the best viewpoint from a set of candidate viewpoints that provides the best view of a moving target in presence of visual obstacles in 2D or 3D space. We propose a query type called k *Continuous Maximum Visibility* *(kCMV)* query that ranks k query viewpoints (or locations) from a set of candidate viewpoints in the increasing order of the visibility measure of the target from these viewpoints. We propose two approaches that reduce the set of query locations and obstacles to consider during visibility computation and efficiently update the results as target moves. We conduct extensive experiments to demonstrate the effectiveness and efficiency of our solutions for a moving target in presence of obstacles.

Keywords: Visibility query · Moving object · Spatial database

1 Introduction

Recent development in 3D modeling of urban environments and popularity of location based services have increased the opportunity to answer different real life queries involving the visibility of objects in the presence of 3D obstacles [1,5]. For example, in a smart city, surveillance cameras may need to track a moving car continuously, and then a security officer may want to find the cameras that produce the best view of the car at different timestamps in the presence of obstacles. Consider another scenario, where a motor racing is telecasted, and the director may want to continuously track the leading car and display the best view of the leading racer continuously. In each of these cases, the objective is to continuously find the viewpoint or camera location from a set of candidate locations that has the maximum visibility of the moving target at particular time instance. We call this new type of query *Continuous Maximum Visibility Query*

© Springer International Publishing AG 2016
M.A. Cheema et al. (Eds.): ADC 2016, LNCS 9877, pp. 82–94, 2016.
DOI: 10.1007/978-3-319-46922-5_7

(*CMVQ*). In this paper, we investigate efficient techniques to answer the CMVQ query. At each timestamp and for the current position of the target, the *CMVQ* returns the query point that provides the maximum visibility. A generalization of *CMVQ* is *k-Continuous Maximum Visibility Query* or *kCMVQ* that finds k best query points (or cameras) in increasing order of their visibility of T.

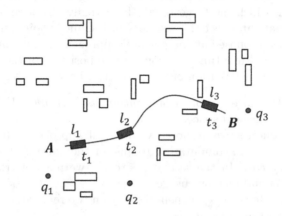

Fig. 1. A target object T moving along the trajectory AB

Figure 1 shows an example of a CMVQ query, where $Q = \{q_1, q_2, q_3\}$ are the locations of three security cameras, and O is a set of obstacles in a city. A security officer wants to track a moving car, T, by using the installed security cameras. At time t_1, target T is at location l_1 and q_1 provides the maximum visibility (or the best view) of T. But as the target changes its location and moves from l_1 to l_2 and then from l_2 to l_3, the camera that provides maximum visibility changes from q_1 to q_2 and then from q_2 to q_3, respectively.

The visibility of an object from a viewpoint depends on the distance, angle and existence of obstacles between the target and the viewpoint. Thus the main challenge of solving a CMVQ query is the computation of visibility of a target object for a given set of viewpoints by considering a large set of obstacles. Visibility computation has been extensively studied in computational geometry and computer graphics, and different algorithms have been proposed to solve the visibility problem. However, they rely on accessing all the obstacles [3,14,15], and propose in-memory algorithms. Moreover, they only consider the visibility in terms of binary measure, i.e., visible or non-visibile. An efficient approach was suggested in [10] that overcame these drawbacks for visibility queries for a static object. They compute visibility of a static target object from a set of query points and find relative ranking of these query points. This approach needs to update visibility of all the query points based on the obstacles that are inside the view of those query points. Applying this technique to solve the *kCMVQ* requires considering all the query points and obstacles within a certain range for each position of the target repetitively, which will incur high query processing and I/O overhead.

We have proposed techniques to solve $kCMVQ$ queries with reduced processing time and I/O. The novelty of our approach is that we do not need to access all the obstacles and also do not require to update the visibility of the target with respect to all the query points. Moreover, our solution relies on some pre-computed data structure based on obstacles and query datasets to enhance the query processing time. At each position of the target object we only use those query points from which the target might be actually visible and consider only those obstacles that can affect the visibility of the query points under consideration. Also we have proposed an incremental approach that will avoid retrieving the same obstacle multiple times for consecutive positions of the target.

In summary, the contribution of the paper can be stated as follows.

- We introduce the novel problem of *Continuous Maximum Visibility (CMV)* query in a d-dimensional space.
- To solve CMV queries, we propose two different approaches. In the first approach, we resort to pre-computation to answer the query fast at the expense of high memory cost. In the second approach, we make a tradeoff between pre-computation and memory usage.
- We conduct an extensive experimental study on two real datasets that demonstrate efficiency of our algorithms.

2   Related Works

Visibility in Computational Geometry and Computer Graphics. In computational geometry, several approaches have been proposed to construct visibility graph and visibility polygon [3,4,14,15]. The problem of computing visbility of input polygon P from a query point q, $V(q)$ was first addressed for simple polygons in [6]. A visibility graph is defined by a set P of n points inside a polygon Q where two points $p, q \in P$ are joined by an edge if the segment $pq \subset Q$ [4]. They have given a nearly optimal algorithm for simple polygons and introduced a notion of *robust* visibility for polygons with holes (non-simple polygons). Different works [2,15] showed that query time for simple polygons can be reduced to logarithmic bound using polynomial time preprocessing. In [3], Asano et al. presented an algorithm with $O(n)$ query time with $O(n^2)$ pre-processing time and space. These algorithms efficiently constructs the visibility polygon with the expense of heavy preprocessing and/or accessing all obstacles present in dataset, which makes this technique inappropriate for many spatial database applications that handles a large number of obstacles.

Although computation of visibility is also an active study topic [13] in computer graphics, their main focus is rendering a scene while we focus on calculating the visibility of a specific target object from various query points.

Visibility in Spatial Queries. kNN queries have been extensively studied in spatial databases. Variants of kNN query consider the effect of obstacles while measuring the distance between two objects. Visible NN (VNN) query [11] finds the NN that is visible to a query point. Continuous Obstructed NN (CONN) query [7] retrieves the nearest neighbor of each query point according to the obstructed distance. Continuous Visible NN (CVNN) query [8] retrieves visible nearest neighbor along a query line segment in the presence of obstacles.

A new type of query called the k Maximum Visibility (kMV) query was introduced in [10], where for a given fixed target T, a set of obstacles O, and a set of query points Q, the kMV query returns k query points in the increasing order of their visibilities to T. Another study [5] quantified the visibility of a target object from the surrounding area with a visibility color map (VCM). But their approach has no notion of partial visibility and has not considered the case of a moving target. A method is presented in [12] that considers partial visibility of the target and support incremental updates of the VCM if the target moves to near-by positions.

In this paper, we focus on answering kMV query for a moving target, that is, we need to continuously find k query points based on the visibility of the target T for each time instance.

3   Problem Formulation

Let O be a set of obstacles stored in R^*-Tree, Q be a set of query points or candidate locations and T be a moving target object in a dataspace. In this scenario Continuous Maximum Visibility Query (CMVQ) can be defined as follows:

Definition 1 (CMVQ). Given a set Q of n query points $\{q_1, q_2, ..., q_n\}$, a set O of m obstacles $\{o_1, o_2, ..., o_m\}$ and a moving target object T that is at location l_i at time instance t_i in a d-dimensional space R^d, the Continuous Maximum Visibility Query (CMVQ) for T continuously returns the query point $q \in Q$ at each time instance where $visibility(q) \geq visibility(q_j)$ where $q_j \in Q \setminus q$.

The generalization of $CMVQ$ is $kCMVQ$, where k query points $Q'\{q'_1, q'_2, ..., q'_k\}$ are returned at each time instance, where $visibility(q'_i) \geq visibility(q'_j)$, $1 \leq i < j \leq k$, and $visibility(q') \geq visibility(q'')$, for each $q' \in Q'$, $q'' \in Q \setminus Q'$.

3.1   Preliminaries

The visibility of a target from a viewpoint varies with the distance and the angle between the viewpoint (or camera) and the target. Similar to [5], in this paper, by considering both of the above factors, we quantify the visibility as visual angle that varies with both the angle and distance between the target and the viewpoint. We assume each camera q has a *field of view* (FOV), a region beyond which nothing is visible to that camera. If we use the visual angle measure, a FOV produces a circular sector in 2D and a spherical sector in 3D which we call the *visible region* (VR). Any target object lying in this region is *point to point visible* from q in the absence of obstacles. For simplicity, we assume that an FOV is triangular in 2D and conical in 3D.

Aggregated Visible Region and Potentially Visible Query Point Set.
An AVR is a region formed by overlapping VRs of a set of query points and is
disjoint from all other $AVRs$. Any point within an AVR is visible only from a
specific set query points. We call this set of query points *potentially visible query
point set* (PVQS). While computing the visibility of a target object T inside
an AVR, we only need to consider query points in the corresponding $PVQS$.
The maximum number of $AVRs$ generated by VRs of n query points in 2D is
$\frac{3n(3n+1)}{2} + 1$ since each VR is bounded by 3 lines [9].

Blocking Set and Aggregated Blocking Set. The visibility of a target
object T from a query point q_i is affected by the obstacles that lie in the *visible
region* V_i of q_i. All other obstacles can be safely pruned while computing the
visibility of T from q_i. We call this set of obstacles a *blocking set* (BS) of that
query point. Each AVR A_i has an associated $PVQS$ P_i. Since the target will be
visible only from the query point $q \in P_i$, the only obstacles to consider are the
blocking sets of each q. Thus, by combining BSs of each query point $q \in P_i$, we
get an *aggregated blocking set (ABS)* for the AVR.

4   Our Approach

In this section we present an algorithm to efficiently solve $kCMVQ$. The key
challenge of solving $kCMVQ$ is that since the target object is moving, we need
to re-compute kMV query and update the ranking of query points for every
position change of the target. A straightforward approach is to run the $kMVQ$
[10] every time when T moves. This approach result in high processing and
I/O overhead as for each position change of the target we need to consider all
obstacles and query points that fall within the visible range. Since there can
be many common obstacles that fall within the range of two consecutive target
locations, the retrieval of same object occur multiple times.

To overcome the above limitation, we introduce an approach with reduced
I/O and processing overhead. To answer a $kCMVQ$ we rely on pre-computed
data structure based on our obstacle and query datasets.

Preprocessing. A moving target may change its position fast. For each position
update, finding the query points from which T becomes visible and retrieving the
required obstacles from the database by traversing the R^*-*Tree* is costly and time
consuming. Since in our problem setting, we assume query points and obstacles
are static, we can precompute certain steps of our approach. Specifically, we
precompute (i) the set of AVR and their corresponding $PVQS$, as when T is in
an AVR, we only need to consider the query points of its corresponding $PVQS$,
and (ii) the BS of each query point $q \in Q$, so that we do not need to consider
the obstacles that has no affect on the visibility of T while processing a query.

As the first step, we construct the AVR set for the dataspace from the VRs
of the queries in Q. The query points whose overlapping VRs form an AVR are

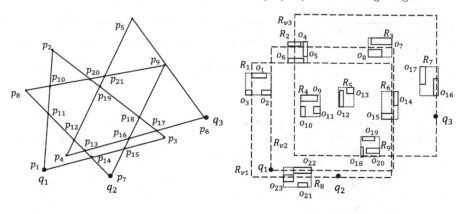

(a) Aggregated Visible Regions   (b) Computing Blocking Sets

Fig. 2. Building aggregated visible regions

recorded as its *PVQS*. We do not consider obstacles while computing *AVRs*.
Second, we generate the set of *BS* for every query. To reduce the complexity, we
maintain the *minimum bounding rectangle (MBR)* of *VR*, and use it to compute
the set of obstacles that are inside the *VR*. Note that, for each obstacle in a *BS*,
we store block number and entry number of the *R*-Tree*, and thus we can avoid
the *R*-Tree* traversal.

Example 1. We will next show the construction of *AVR*, *PVQS*, and *BS* using
an example. Figure 2(a) shows locations of a set Q of three query points
$\{q_1, q_2, q_3\}$. First we take q_1 and its *VR* ,V_1. Since *AVR* set A is empty at
this point, V_1 (p_1, p_2, p_3) itself is added as an *AVR*. Next q_2 is considered and
its *VR* V_2 (p_7, p_8, p_9) intersects with existing *AVRs* and creates new *AVRs*.
For example, V_2 intersects with (p_1, p_2, p_3) and divides it into four regions
(p_1, p_{11}, p_{14}), (p_2, p_{10}, p_{20}), (p_3, p_{15}, p_{18}) and ($p_{10}, p_{11}, p_{14}, p_{15}, p_{18}, p_{20}$). For each
AVR its *PVQS* is updated, e.g., the region bounded by ($p_{10}, p_{11}, p_{14}, p_{15}, p_{18}, p_{20}$)
has *PVQS* $\{q_1, q_2\}$ since overlapping *VRs* of these query points form the
AVR. This process is repeated until all query points are considered. To
compute *BS*, we start with q_1 and compute *MBR* of *VR* V_1. Let this
MBR be R_{V1}. We perform a range query in the *R*-tree* to find out the
obstacles inside R_{V1}. We can see from Fig. 2(b) that *BS* of q_1 is $B_1=$
$\{o_1, o_2, o_9, o_{10}, o_{11}, o_{12}, o_{13}, o_{15}, o_{18}, o_{19}, o_{20}, o_{22}, o_{23}\}$. We compute *BS* of q_2 and
q_3 in a similar manner.

4.1   AVR-BS Incremental Approach

We divide the dataspace into *AVRs* such that when T is inside an *AVR*, the set of
query points from which T is visible and set of obstacles that affect the visibility
of T remain same. Thus we do not need any extra I/O as long as the target does
not change its current *AVR*. Since neighboring AVRs are likely to have many
queries in common in their PVQS, we re-use the already retrieved obstacles for
those common queries when T moves to a neighboring AVR.

Algorithm 1. KCMVQ-INCREMENTAL(T,k)

    Input: T a target object, k
    Output: L a set of k query points ordered by visibility
1 $A_c \leftarrow getCurrentAVR(T)$; $Q_c \leftarrow getCurrentQs(A_c)$;
    $B_a \leftarrow getBlockingSets(Q_c)$
2 **while** *true* **do**
3   **if** $getTargetMovement(T) < th$ **then** *continue*
4   **if** $isAVRChanged(T, A_c)$ **then**
5         $A_c \leftarrow updateAVR(T)$; $Q_n \leftarrow getCurrentQs(A_c, T)$
6         $Q_{com} \leftarrow Q_n \cap Q_c$
7         $Q_{new} \leftarrow Q_n \setminus Q_{com}$
8   **for** $q \in Q_{new}$ **do** $B_a \leftarrow B_a \cup q.B$
9         $Q_{obs} \leftarrow Q_c \setminus Q_{com}$
10   **for** $q \in Q_{obs}$ **do** $UpdateABS(B_a, q, Q_n)$
11       $Q_c \leftarrow Q_n$
12     $L \leftarrow ComputeVisibility(T, Q_c, B_a, k)$

Algorithm 1 shows our incremental approach for answering a *kCMV* query. It takes target object T and positive integer k as input and reports k best query points in set L ordered by the visibility of T. We update L only when T has moved at least a threshold amount th. Algorithm starts with retrieving current *AVR* A_c where T resides, PVQS Q_c of A_c and *ABS* B_a for all query points in Q_c (Line 1). If T resides on the border of multiple *AVRs*, A_c stores the set of *AVRs*. Q_c is constructed by union of *PVQS* of each *AVR* in A_c. The *while* loop iterates as long as the target continues querying in the dataspace. If target changes its position within the current *AVR*, we compute visibility in Line 12. If T moves to another *AVR*, we update current *AVR* A_c and compute a new query set Q_n by combining *PVQS* of *AVRs* that overlap with T. Line 6 calculates Q_{com} whose corresponding *blocking set*s are already retrieved in previous iteration. At Line 7 Q_{new} holds the new query points from which T is now visible. Line 9 updates *ABS* B_a. With the region change, T may also become invisible to some query points from which it was visible earlier. Set of such query points Q_{obs} is computed in Line 9. For each such query point *ABS* is updated, i.e., we remove obstacles from B_a that do not affect the visibility anymore. This is not a set minus operation since an obstacle can be in multiple *blocking sets*. So removing *BS* of a query point may remove some obstacles that are part of *BS* of some other query points from which T is still visible. Line 12 then uses Q_c and A_b to compute visibility.

Example 2. We explain the incremental approach using *AVRs* A_1 $(p_1, p_2, p_3, p_4, p_5)$ and A_2 $(p_4, p_5, p_6, p_7, p_8)$. In Fig. 3(a) at time t_1 target T is completely inside A_1. The corresponding *PVQS* is $P_1 = \{q_1, q_2\}$ and *ABS* is $B_{a(1)} = \{B_1 \cup B_2\}$. We retrieve all the objects of $B_{a(1)}$ and compute visibility of the target from the query points in P_1. Then at t_2, T moves to the region boundary between the *AVRs* A_1 and A_2. At this point we need to consider query

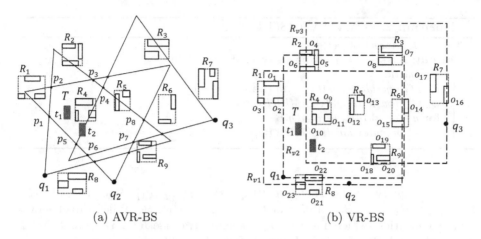

(a) AVR-BS   (b) VR-BS

Fig. 3. An example scenario of approaches

point sets $P_{1\cup2} = P_1 \cup P_2$ and set of obstacles $B_{a(1\cup2)} = B_{a(1)} \cup B_{a(2)}$. T is now visible from a new query point q_3 that was not in P_1 and thus the obstacles in its BS B_3 are not in the ABS that was computed in the previous step. We can now update ABS as $B_{a(1\cup2)} = B_{a(1)} \cup B_3$. We do not need to retrieve obstacles that are already in B_1 and B_2.

4.2   VR-BS Approach

The incremental approach computes AVRs based on the given set of query locations. If a large number of query points are densely placed in the dataspace, the number of AVRs can be huge. In some scenario, it may happen that the number of AVRs may outnumber the number of obstacles in the dataset. For such cases, the previous approach may not perform well. Thus, in this section, we introduce another approach that preprocesses obstacle set to compute BS of each query point and avoids computation of AVRs.

In this approach, we maintain an R^*-Tree that holds $MBRs$ of VRs (VRMBR) of query points. We search the R^*-Tree to find $VRMBRs$ with which target T overlap, and then we add the corresponding query point to *potentially visible query point set* (PVQS).

Algorithm 2 shows the steps of VR-BS approach. Lines 1–5 iterate as long as the target continues querying. If T does not move a threshold distance, we do not update existing ranking of query points in L. Line 3 computes Q_c by traversing the R^*-Tree of VRMBRs and finding query points whose $VRMBRs$ have non-empty intersecting region with T. The ABS B_a is reconstructed in Line 4. With the reduced set of query points Q_c and the reduced set of obstacle B_a, we compute visibility of target in Line 5.

Example 3. Figure 3(b) shows $VRMBRs$ of query point set $Q = \{q_1, q_2, q_3\}$. For target position at time t_1 we traverse the tree to find that T lies inside $VRMBR$

Algorithm 2. κCMVQ-VRBS(T,k)

Input: T a target object, k
Output: L a set of k query points ordered by visibility
1 **while** *true* **do**
2   **if** *getTargetMovement(T)* $< th$ **then** *continue*
3      $Q_c \leftarrow findActiveQ(T, VRMBRTree); B_a \leftarrow \emptyset$
4   **for** $q \in Q_c$ **do** $B_a \leftarrow B_a \cup q.B$
5      $L \leftarrow ComputeVisibility(T, Q_c, B_a, k)$

---

of all three query points. Therefore $PVQS\ P = \{q_1, q_2, q_3\}$. Since $P \subset Q$, ABS $B_a = B_1 \cup B_2 \cup ... \cup B_{|P|}$. Hence at t_1 $B_a = B_1 \cup B_2 \cup B_3$. When target moves to a new location depicted at time t_2, we again traverse R^*-*Tree* to find that T lies inside $VRMBR$ of q_3 only. Hence $PVQS\ P = \{q_3\}$ and ABS $B_a = B_1$.

5 Experimental Evaluation

We evaluate performance of our proposed algorithms for answering the *kCMV* query with two real datasets, for both *Uniform*(U) and *Zipf*(Z) distribution of query point locations. Two real datasets are, the British[1] dataset, representing 5985 data objects obtained from British ordnance survey[2] and Boston dataset represents 130,043 data objects in Boston downtown[3]. Random paths in search space are generated to simulate movement of a target. We assume a total of 300 queries requested by target while it is moving in the dataspace. All obstacles are stored in a R^*-*Tree* with the disk page size fixed at 1 KB and block size fixed at 256B. The target size is kept fixed at 2 units. The algorithms are implemented in C++, and the experiments are conducted on a core i5 2.67 GHz PC with 4 GB RAM, running 64 bit Microsoft Windows 8.1 (Table 1).

Table 1. Parameters

Parameter	Range	Default
dataset size	British (6 K), Boston(130 K)	
number of query points n_q	100, 200, 300, 400, 500	400
field of view fov	30, 45, 60, 75, 90	30
maximum distance of visibility D	300, 400, 500, 600, 700	500

---

[1] http://www.citygml.org/index.php?id=1539.
[2] http://www.ordnancesurvey.co.uk/oswebsite/indexA.html.
[3] bostonredevelopmentauthority.org/BRA_3D_Models/3D-download.

5.1 Performance Evaluation

We conduct four sets of experiments to evaluate the performance of AVR-BS and VR-BS, which are referred as A and V respectively, in the following figures. In each set of experiments, one parameter is varied while all other parameters are set to their default values. In all cases AVR-BS outperforms VR-BS in terms of execution time and I/O cost with the expense of high pre-computation cost. For each experiment we have evaluated the results of 10 iterations and reported average performance.

Straightforward Approach. We have compared our proposed approaches with the straightforward approach, which finds *kMVQ* at each time instance. The straightforward approach is not scalable for a moving target and is outperformed by both *AVR-BS* and *VR-BS* algorithms. *AVR-BS* is about 250 times (British) and 1500 times (Boston) faster than the straightforward approach, whereas *VR-BS* is about 125 times (British) and 128 times (Boston) faster than the straightforward approach. *AVR-BS* has 48 times (British) and 28 times (Boston) less I/O cost than the straightforward approach. *VR-BS* has similar I/O cost to straightforward approach. *VR-BS* first traverses a R^*-*Tree* to find out the query points whose *VRs* intersect T and then retrieves obstacles that belong to *BS* of these query points. This results in high I/O cost. But *VR-BS* uses reduced set of obstacles and set of query points during visibility computation, which explains its lower processing time than the straightforward approach (Table 2).

Table 2. Comparison with straightforward approach

	British		Boston	
	Time (ms)	I/O (avg)	Time (ms)	I/O (avg)
Straightforward	778.78	24.92	21946.42	365.82
AVR-BS	3.31	0.54	13.93	13.91
VR-BS	6.36	47.05	170.31	354.72

Preprocessing. Table 3 shows that precomputation time in milliseconds (ms) for both *AVR* and *BS* for default values of parameters. Both costs increase with the increase in n_q, *fov* or D. For more query points, more polygon intersections take place and more *AVRs* are generated. For *AVR* both datasets show that the cost is more for Z as query points are clustered nearby and their *VRs* are more likely to overlap and is similar for both datasets.

Varying No. of Query Points. Figure 4 shows that with the increase of number of query points, execution time and I/O cost both increase for both *AVR-BS* and *VR-BS*. *VR-BS* approach has higher execution time than *AVR-BS* as *PVQS* and *ABS* needs to be recomputed at each position change of the target. On average *AVR-BS* is 10 times (Boston) and 2 times (Bristish) faster than *VR-BS*. *VR-BS*

Table 3. Preprocessing time for AVR and BS

	British		Boston	
	Uniform	Zipf	Uniform	Zipf
AVR	56135.66	85149.0	58718.67	92671.67
BS	1438.0	1734.33	12320.33	14827.67

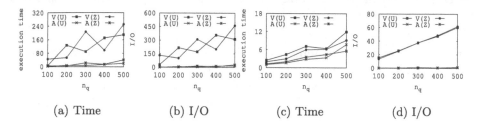

(a) Time   (b) I/O   (c) Time   (d) I/O

Fig. 4. Effect of varying n_q for Boston (a–b) and British (c–d) datasets

incurs higher I/O cost than *AVR-BS* as building *PVQS* each time requires the *R*-Tree* traversal. Execution time of *VR-BS* is similar to that of *AVR-BS*, for British dataset. This is because British dataset has a smaller search space and with the increase in number of query points in this smaller space, *VRs* overlap more frequently and hence number of *AVRs* increases. Therefore, a moving target changes *AVR* more often, which requires more computation in AVR-BS.

Varying Field of View. Figure 5 shows that execution time increases for both approaches with increase in *fov*. More obstacles come into consideration for each query point as *fov* increases. Effect of varying *fov* is similar to the effect of varying n_q. On an average *AVR-BS* performs approximately 10 times and 1.5 times faster than *f* for Boston and British datasets, respectively.

Varying Maximum Distance of Visibility. Figure 6 shows that execution time increases for both approaches in both Boston and British datasets. *AVR-BS* outperforms *VR-BS* in all cases. Execution times for both approaches are similar in British dataset; the underlying reason can be high density of query points.

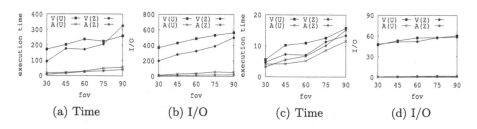

(a) Time   (b) I/O   (c) Time   (d) I/O

Fig. 5. Effect of varying *fov* for Boston (a–b) and British (c–d) datasets

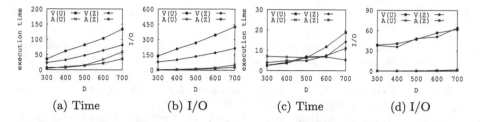

(a) Time   (b) I/O   (c) Time   (d) I/O

Fig. 6. Effect of varying D for Boston (a–b) and British (c–d) datasets

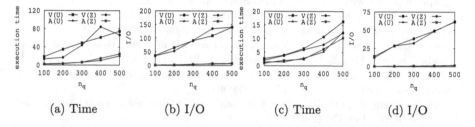

(a) Time   (b) I/O   (c) Time   (d) I/O

Fig. 7. Effect of varying paths for Boston (a–b) and British (c–d) datasets

Varying Paths. Figure 7 shows the result of average execution time and I/O cost of 10 different paths. In all cases I/O cost for VR-BS approach is higher than that of AVR-BS as expected. On average *AVR-BS* performs 7 times and 2 times faster than *VR-BS* for Boston and British datasets, respectively.

6   Conclusion

In this paper, we have introduced a new type of query, namely k *Continuous Maximum Visibility Query* that continuously finds the best viewpoint for a moving target in the presence of obstacles. To efficiently answer a *kCMV* query, we have proposed two approaches. The first approach, *AVR-BS*, relies on pre-computation to provide fast answer to queries. On the other hand, the second approach, *VR-BS*, makes a tradeoff between pre-computation and memory usage while processing queries. Our experimental results show that *AVR-BS* is two to three orders of magnitude faster than the straightforward approach, and *VR-BS* is at least one order of magnitude faster than the straightforward approach.

Acknowledgements. This research is supported by the ICT Division - Government of the People's Republic of Bangladesh.

References

1. Ali, M.E., Tanin, E., Zhang, R., Kulik, L.: A motion-aware approach for efficient evaluation of continuous queries on 3d object databases. VLDB J. **19**(5), 603–632 (2010)
2. Aronov, B., Guibas, L.J., Teichmann, M., Zhang, L.: Visibility queries and maintenance in simple polygons. DCG **27**(4), 461–483 (2002)
3. Asano, T., Guibas, L.J., Hershberger, J., Imai, H.: Visibility of disjoint polygons. Algorithmica **1**(1), 49–63 (1986)
4. Ben-Moshe B., Hall-Holt, O., Katz, M.J., Mitchell, J.S.B.: Computing the visibility graph of points within a polygon. In: SCG, pp. 27–35 (2004)
5. Choudhury, F.M., Ali, M.E., Masud, S., Nath, S., Rabban, I.E.: Scalable visibility color map construction in spatial databases. Inf. Syst. **42**, 89–106 (2014)
6. Davis, L.S., Benedikt, M.L.: Computational models of space: Isovists and isovist fields. Comput. Graph. Image Process. **11**(1), 49–72 (1979)
7. Gao, Y., Zheng, B.: Continuous obstructed nearest neighbor queries in spatial databases. In: SIGMOD, pp. 577–590 (2009)
8. Gao, Y., Zheng, B., Lee, W., Chen, G.: Continuous visible nearest neighbor queries. In: EDBT, pp. 144–155 (2009)
9. Graham, R., Knuth, D., Patashnik, O.: Concrete Mathematics. Addison-Wesley, Boston (1994)
10. Masud, S., Choudhury, F.M., Ali, M.E., Nutanong, S.: Maximum visibility queries in spatial databases keys. In: ICDE, pp. 637–648 (2013)
11. Nutanong, S., Tanin, E., Zhang, R.: Visible nearest neighbor queries. In: Kotagiri, R., Radha Krishna, P., Mohania, M., Nantajeewarawat, E. (eds.) DASFAA 2007. LNCS, vol. 4443, pp. 876–883. Springer, Heidelberg (2007)
12. Rabban, I.E., Abdullah, K., Ali, M.E., Cheema, M.A.: Visibility color map for a fixed or moving target in spatial databases. In: Claramunt, C., Schneider, M., Wong, R.C.-W., Xiong, L., Loh, W.-K., Shahabi, C., Li, K.-J. (eds.) SSTD 2015. LNCS, vol. 9239, pp. 197–215. Springer, Heidelberg (2015)
13. Shou, L., Huang, Z., Tan, K.-L.: Hdov-tree: the structure, the storage, the speed. In: ICDE (2003)
14. Suri, S., ORourke, J.: Worst-case optimal algorithms for constructing visibility polygons with holes. In: SCG, pp. 14–23 (1986)
15. Zarei, A.R., Ghodsi, M.: Efficient computation of query point visibility in polygons with holes. In: SCG, pp. 6–8 (2005)

An Index-Based Method
for Efficient Maximizing Range Sum Queries
in Road Network

Xiaoling Zhou$^{(\boxtimes)}$ and Wei Wang

University of New South Wales, Sydney, Australia
{xiaolingz,weiw}@cse.unsw.edu.au

Abstract. Given a set of positive weighted points, the Maximizing Range Sum (maxRS) problem finds the placement of a query region r of given size such that the weight sum of points covered by r is maximized. This problem has long been studied since its wide application in spatial data mining, facility locating, and clustering problems. However, most of the existing work focus on Euclidean space, which is not applicable in many real-life cases. For example, in location-based services, the spatial data points can only be accessed by following certain underlying (road) network, rather than straight-line access. Thus in this paper, we study the maxRS problem with road network constraint, and propose an index-based method that solves the online queries highly efficiently.

Keywords: Maximizing range sum · Road network · Query processing

1 Introduction

In recent years, location-based services that answer queries on spatial databases have drawn much attention, due to the proliferation of mobile computing devices. One of the common queries is the *Maximizing Range Sum* (maxRS) problem [1–6]. Given a set of positive weighted spatial points and a query shape(eg. rectangular or circular) r of user specified size, the maxRS problem finds an optimal placement of r such that the total weight of all points covered by r is maximized. The maxRS problem is widely applied in facility locating problems [8] for finding the best facility location with maximum number of potential clients, spatial data mining for extracting interesting locations from log data [9], and point enclosing problems.

However, most of the existing work adopt Euclidean distance metric in their method. This is not applicable in many real-life location-based services, where the spatial data points can only be accessed by following certain underlying (road) network. For example a tourist service that answers user queries of finding the most attractive places in a city in the sense that it is close to as many sightseeing spots as possible within a given range (eg. 5 km walking distance). In such scenarios, the ways to access scenic spots are constrained by the road

© Springer International Publishing AG 2016
M.A. Cheema et al. (Eds.): ADC 2016, LNCS 9877, pp. 95–109, 2016.
DOI: 10.1007/978-3-319-46922-5_8

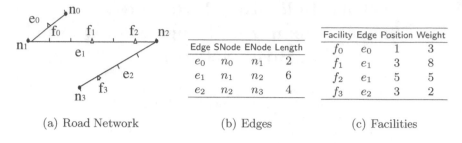

Edge	SNode	ENode	Length
e_0	n_0	n_1	2
e_1	n_1	n_2	6
e_2	n_2	n_3	4

Facility	Edge	Position	Weight
f_0	e_0	1	3
f_1	e_1	3	8
f_2	e_1	5	5
f_3	e_2	3	2

(a) Road Network   (b) Edges   (c) Facilities

Fig. 1. Road network example

network, hence the actual distance between two locations is probably significantly different from their Euclidean distance, for example the distance between
facilities f_0 and f_3 in Fig. 1(a).

Therefore, in this paper, we study the maxRS problem with a road network
constraint, where the distance between two points in the network is determined
by the length of the shortest path connecting them (i.e. network distance [17]).
That is, given a road network, a set of positive weighted facilities on it, and a
network radius r, we find a location p on the network that maximizes the total
weight of facilities whose network distance to p is no larger than r. Figure 1
shows an example of a road network containing 4 nodes, 3 edges, and 4 weighted
facilities. Given a query radius $r = 2$, any position between f_1 and f_2 can be the
answer of the maxRS query, since the total weight of facilities that can reach
these positions within distance 2 is the maximum (i.e. 13).

Currently, the only existing work, as far as we know, that deals with the
maxRS problem in a road network is [14]. The authors proposed an external-
memory algorithm based on segments generation and linesweeping on the road
network. However, their method is not efficient for large radius or road network
databases, as shown in the experimental results in Sect. 6. They take tens or hundreds of seconds to answer one query with specified radius, which is undesirable
for a mobile or web service that copes with the needs of answering millions of
concurrent user queries, each with a different radius parameter. Thus, we propose
a new solution to this problem that answer online queries much more efficient
(around 6–8 orders of magnitude faster) than previous method, by making use
of a tiny precomputed index which is of size linear to the facility number, and
therefore is superior for dealing with large number of concurrent or batch query
workloads.

In the rest of the paper, we formally define the problem in Sect. 2, and introduce the details of our proposed method in Sects. 3 and 4. Two optimizations for
index construction are presented in Sect. 5. Experimental results on various parameters demonstrate the superiority of our method compared with existing work
in Sect. 6, and finally followed the related work review in Sect. 7 and conclusion
in Sect. 8.

2   Problem Definition

We follow the definition in [14] to formally describe the problem. A road network is represented as an undirected graph $G = (V, E)$, where V is a set of nodes, and E is a set of edges. We use F to denote the set of facilities, each of which, denoted as f, is located on an edge and is associated with a positive weight $w(f)$.

Definition 1 (Network Radius and Network Range). *The network range $p(r)$ of a point p in a road network contains all points in the network whose network distance to p is no greater than r, where r is called the network radius.*

Definition 2 (MaxRS query in road network). *Given G, a set of positive-weighted facilities F, and a network radius value r, let $p(r)$ be the network range of a point p in the network, and $F_{p(r)}$ be the set of facilities covered by $p(r)$. A Maximizing Range Sum (maxRS) query in a road network finds an optimal point (i.e. position) p in G that maximizes: $\sum_{f \in F_{p(r)}} w(f)$.*[1]

3   The Proposed Method

The previous method [14] generates segments from each facility f along the network until distance reaches query radius r, then sorts the segments and uses line-sweeping method to find final answer. This method takes time $O(|E||F|\log|F|)$ in worst case. The performance is undesirable, especially when r is large, which can be observed from the experimental results in Sect. 6.

Therefore, we hope to accelerate online query via some precomputed index. A direct idea is to compute the optimal location p and its weight sum w for each possible query radius r and store $<r, w, p>$ as index entries. This is obviously impossible since there are infinite number of distinct radius values. However, we can substantially reduce the index to a feasible size by the following way: (1) When radius r increases, the maximum weight sum w is non-decreasing. Thus, we can categorize the naive index entries into equivalent classes, where each class contains entries whose weights are the same. (2) For each class, store only the entry with minimum r into index. For example, if for both radiuses r and $r'(r' > r)$, the maximum weight sum can be achieved is w, we store only $<r, w, p>$, and leave out $<r', w, p'>$. The reduced index of previous example (Fig. 1) is shown in Table 1.[2]

The reduced index size is upper bounded by $\sum_{f \in F} w(f)$ since this is the maximum number of distinct weight sum values. Then we have $\sum_{f \in F} w(f) \le \max_{f \in F} w(f) \cdot |F| = O(|F|)$ when $\max_{f \in F} w(f)$ is constant, which is usually the case.

The above index structure immediately leads to a simple binary search query processing method with complexity $O(\log|F|)$.

---

[1] W.l.o.g., we assume r and edge length are real numbers, and $w(f)$ is constant integer.

[2] Note that there may exist multiple optimal locations, but we store only one in the index.

Table 1. Example index

r	MaxWeight	OptPosition
0	8	$[e_1, 3]$
1	13	$[e_1, 4]$
3	16	$[e_1, 2]$
5	18	$[e_1, 4]$

Theorem 1 *Given query radius r, assume $r*$ is the largest radius in the index that is smaller than or equal to r, then $maxRS(r) = maxRS(r*)$.*

Therefore, according to Theorem 1, given query r, we use binary search to find $r*$ in the index, and retrieve the index entry as maxRS result for r. For example, if given $r = 4$ and the index in Table 1, we will return the entry $<3, 16, [e_1, 2]>$.

4 Index Construction

In this section, we introduce the details of index construction.

Lemma 1 *Given radius r, if a point p is in the network range of facility f (i.e. $f(r)$), then f is in $F_{p(r)}$, and vice versa.*

Using Lemma 1, we can transfer the problem of finding an optimal location in the network that can *reach* facilities with maximum total weight within radius r to the problem of finding a position that can *be reached* by facilities with maximum total weight within radius r.

Unlike the previous method [14] where r is known in advance, we do not assume r during index construction, thus the aggressively generating segments until reach length r is not applicable to our case. Therefore, we design an event-driven algorithm which is essentially a simulation of the following process.

From each facility f, we generate directed cursors carrying weight $w(f)$, and go to every possible way along the network. We move all the cursors simultaneously. Whenever there are cursors meet together, we check at the meet position (p) the total weight (w) of cursors have passed this position (we call it **Location Weight** of p). If w is larger than the last indexed entry, we add new index entry $<R, w, p>$, where R is the current total moving distance for each cursor.

The above moving process is driven by the following two kinds of events:

- **Meet Event:** There are cursors meet, i.e. two or more cursors having different directions reach the same position on an edge.
- **End Event:** There are cursors reach the edge ends, i.e. nodes in the network.

Algorithm 1 shows the main construction process. Here we assume all facilities are on edges but not on nodes, for ease of presentation. Thus we generate two cursors for each f, one goes left (towards startnode) and another goes right (towards endnode), we call them **reverse cursor** of each other. A cursor c is denoted as $[facilityId, direction, position]$. After generating all cursors, we organize them by the edge they belong to, and sort them according to their positions (Lines 4–5). We call all the edges who have cursors on it as *active edges* ($\mathcal{A}$). Then we compute for each active edge the minimum distance required to reach one of the above events (denoted as **dmin**), and choose among all the *dmins* the global minimum one as the moving distance d (Lines 6–8) to move all the cursors.

The while loop simulates the continuous moving process. During each iteration, an optimization applied here is that (1) if an edge entry ee's $dmin(ee) > d$, we say its a general case edge. We just accumulate d to its *need move distance m*, rather than actually move all the cursors on it. The new $dmin(ee)$ is computed as $dmin(ee) - d$ (Lines 13–14); (2) if the edge's $dmin(ee) = d$, we say it's an exact case edge. We know there must be at least one event occur on this edge after moving all the cursors. Therefore, we check after moving:

- All the meet positions to see if a new index entry has to be added (Lines 17–19).
- All the cursors that reach edge ends, to see if they have fully passed through this edge. If so, we add the facility this cursor belongs to to the edge's fully passed list FP (Line 22). We also need to erase the end cursors, and add new ones onto the adjacent edges (Lines 23).
- Remove this edge from the active edges list if there's no cursors on it.

Lemma 2 *We say a facility f has fully passed edge e only if the total track of all cursors generated from f has fully covered edge e. For example in Fig. 2(a), c and c' are the two cursors generated from f, and the dotted line represents the total track of the two cursors. Although c' has reached the edge end, the track has not fully covered edge e yet, so we do not add f to e's fully passed list until c reaches edge end.*

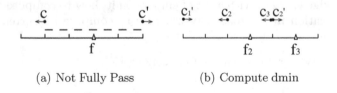

(a) Not Fully Pass   (b) Compute dmin

Fig. 2. Examples

Lemma 3 *To avoid redundant edge visiting, we only add a facility f's cursor c to an edge e if the following conditions meet:*

1. *f is not in e's fully passed list $FP[e]$.*
2. *There is no other cursors of f that has same direction with c on e.*

Algorithm 1. IndexConstruction(E, V, F)

Input : The list of edges E, vertices V, and facilities F.
Output: The list of index entries I.

1 $R \leftarrow 0$; $I, \mathcal{A}, FP \leftarrow \emptyset$;
2 $d \leftarrow \infty$;   /* the global dmin */;
3 $I \leftarrow I \cup <0, fmax.w, fmax.pos>$;   /* fmax:the facility with max weight */;
4 **for** *each f in F* **do**
5     $\mathcal{A} \leftarrow \mathcal{A} \cup <f.eid, \infty, 0, [f.id, \text{left}, f.pos], [f.id, \text{right}, f.pos]>$;

6 **for** *each edge entry ee in $\mathcal{A}$* **do**
7     $ee.dmin \leftarrow$ GetDmin(ee) ;
8     $d \leftarrow ee.dmin < d$? $ee.dmin : d$;   /* update global dmin */;

9 **while** $I.back.weight < \sum_{f \in F} w(f)$ **do**
10     $R \leftarrow R + d$; $d \leftarrow \infty$;
11   **for** *each ee in $\mathcal{A}$* **do**
12   **if** $ee.dmin > d$ **then**
13         $ee.dmin \leftarrow ee.dmin - d$;
14         $ee.m \leftarrow ee.m + d$;   /* accumulate need move dist */;
15   **else**   /* exact case edge */
16   move all cursors on ee with distance $(ee.m + d)$; $ee.m \leftarrow 0$;
17   **for** *each meet position p on ee* **do**
18   **if** *GetLocationWeight*$(p, ee) > I.back.weight$ **then**
19             $I \leftarrow I \cup <R, p.weight, p.pos>$;
20   destroy meet cursors at p belonging to common facility ;
21   **for** *each end cursor c on ee* **do**
22   add $c.fid$ to ee's fully passed list FP[ee] with precheck ;
23   add c to adjacent edges with precheck; destroy c ;
24   destroy ee if no cursor on it; $ee.dmin \leftarrow$ GetDmin(ee) ;
25         $d \leftarrow ee.dmin < d$? $ee.dmin : d$;   /* update global dmin */;

26 **return** I;

In the above processes, there are two technical problems:

– Given an edge entry ee with a list of cursors on it, how to compute $dmin(ee)$.
– Given a position in the road network, how to compute its current location weight.

We solve these two problems in the following sections.

4.1   Compute $dmin$ for Each Edge

Algorithm 2 shows the process to compute $dmin$ for an active edge. Recall that $dmin$ is the minimum distance required to reach one of the above events. We first consider the distances that will trigger End Event: (1) the distance between the leftmost go-left cursor and startnode (Line 4); (2) the distance between the rightmost go-right cursor and endnode (Line 5).

Next, we consider the distances that will trigger Meet Event: Choose the shorter list between go-left cursors list and go-right cursors list as S. For each cursor c in S, find its closest but not yet meet cursor c' in another list. The meet distance is computed as $dist(c, c')/2$ (Lines 7–9).

From all the distances considered above, we choose the minimum one as $dmin(ee)$ and return it. Consider an illustrative example in Fig. 2(b). To compute its $dmin$, we look at:

(1) the distance to reach left end: $dist(startnode, c_2) = 2$.
(2) the distance to reach right end: $dist(c'_2, endnode) = 2$.
(3) the distance for meet event(assume S is go-right cursors): $dist(c'_1, c_2)/2 = 1$.

Thus, the final $dmin$ is 1.

Algorithm 2. GetDmin(ee)

   Input   : The edge entry ee.
   Output: The minimum moving distance $dmin$ to trigger an event.
1 $d \leftarrow \infty$;
2 $GL \leftarrow$ sorted list of go-left cursors on ee ;
3 $GR \leftarrow$ sorted list of go-right cursors on ee ;
4 $d \leftarrow GL[first].dist < d$? $GL[first].dist : d$;   /* trigger End Event */;
5 $d' \leftarrow ee.length - GR[last].dist$; $d \leftarrow d' < d$? $d' : d$; /* trigger End Event */;
6 $S \leftarrow$ the shorter list between GL and GR ;
7 **for** *each cursor c in S* **do**
8 | $c' \leftarrow$ the closest but not meet yet cursor in the other list ;
9 | $d' \leftarrow dist(c, c')/2$; $d \leftarrow d' < d$? $d' : d$;   /* trigger Meet Event */;

10 **return** d;

4.2   Compute Location Weight

Algorithm 3 addresses the problem of retrieving the current location weight of a given position in the network. Given a position p on edge entry ee, its location weight is composed by two parts:

1. The weights of facilities that fully passed edge ee. This can be obtained by checking ee's fully passed list $FP[ee]$ (Line 1).
2. The weights of facilities that passed p, but not in ee's fully passed list, which means they must have active cursors on ee. Thus, we retrieve them by looking at all the go-left cursors on the left of p, and all the go-right cursors on the right of p. For each such cursor c, assume its facility is f, we check if they actually passed p in the following way:
    – If f is not on edge ee, then we know c must get on ee from one end and moving towards another end, so c must have passed p.
    – If f is on edge ee: (1) if p is in between the positions of f and c, then c must have passed p; (2) If p is not in between, then the possible positions of p must be in case p_1 or p_2 shown in Fig. 3(a). We can judge whether c or

its *reverse cursor* c' passed p by checking whether $dist(c, f) \geq dist(f, p)$, if so, we add f as a passed facility. (This comparison is to address the special case in Lemma 2 where f is added to fully passed list only if its cursors' moving track fully covers the edge.)
We use set F' to store passed facilities to avoid redundant weight accumulation.

For example, the location weight of p in Fig. 3(b) is computed as $w(f_2) + w(f_3)$.

Algorithm 3. GetLocationWeight(p, ee)

Input : The edge entry ee, and a position p on it.
Output: The current location weight of p.
1 $F' \leftarrow \emptyset; w \leftarrow \sum_{f \in FP[ee]} w(f)$;
2 **for** *each cursor c on ee* **do**
3   **if** $c.pos \leq p.pos$ **and** $c.dir = left$ **then**
4   **if** f *not on ee* **or** $f.pos \geq p.pos$ **or** $dist(c, f) \geq dist(f, p)$ **then**
5           $\lfloor\ F' \leftarrow F' \cup f$;   /* f: the facility c belongs to */;
6   **else if** $c.pos \geq p.pos$ **and** $c.dir = right$ **then**
7   **if** f *not on ee* **or** $f.pos \leq p.pos$ **or** $dist(c, f) \geq dist(f, p)$ **then**
8           $\lfloor\ F' \leftarrow F' \cup f$;
9 $w \leftarrow w + \sum_{f \in F'} w(f)$;
10 **return** w;

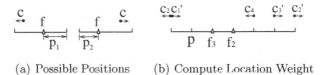

(a) Possible Positions   (b) Compute Location Weight

Fig. 3. Examples

4.3 Running Example

The running of Algorithm 1 on the road network in Fig. 1 is shown in Table 2. At the beginning (Round 0), the total moving distance R is 0, and the initial index entry $<0, 8, [e_1, 3]>$ is added into index I. Two opposite cursors from each facility are generated and organized by edges (shown in column 2). For each edge entry, its need move distance m is initialized as 0, and $dmin$ is computed by Algorithm 2 (equal to 1 for all edges in this example). Thus, the global dmin d is 1.

In Round 1, we move all the cursors with distance 1 and show the current active edges. The underlined cursors represent a meet event, and a new index

Table 2. Index construction: running Algorithm 1 on example in Fig. 1

Round#	Active Edges ($\mathcal{A}$)	FP	Index(I)	R	d
0	$< e_0, 1, 0, [f_0, l, 1], [f_0, r, 1] >$ $< e_1, 1, 0, [f_1, l, 3], [f_1, r, 3], [f_2, l, 5], [f_2, r, 5] >$ $< e_2, 1, 0, [f_3, l, 3], [f_3, r, 3] >$	$\emptyset$	0, 8, $[e_1, 3]$	0	1
1	$< e_1, 1, 0, [f_0, r, 0], [f_1, l, 2], [f_1, r, 4], [f_2, l, 4] >$ $< e_2, 1, 0, [f_2, r, 0], [f_3, l, 2] >$	$e_0 : f_0$	0, 8, $[e_1, 3]$ 1, 13, $[e_1, 4]$	1	1
2	$< e_1, 1, 0, [f_0, r, 1], [f_1, l, 1], [f_2, l, 3], [f_1, r, 5] >$ $< e_2, 1, 0, [f_2, r, 1], [f_3, l, 1] >$	$e_0 : f_0$	0, 8, $[e_1, 3]$ 1, 13, $[e_1, 4]$	2	1
3	$< e_0, 2, 0, [f_1, l, 2] >$ $< e_1, 2, 0, [f_0, r, 2], [f_2, l, 2], [f_3, l, 6] >$ $< e_2, 2, 0, [f_1, r, 0], [f_2, r, 2] >$	$e_0 : f_0$ $e_1 : f_1$ $e_2 : f_3$	0, 8, $[e_1, 3]$ 1, 13, $[e_1, 4]$ 3, 16, $[e_1, 2]$	3	2
4	$< e_0, 2, 0, [f_2, l, 2] >$ $< e_1, 2, 0, [f_0, r, 4], [f_3, l, 4] >$ $< e_2, 2, 0, [f_1, r, 2] >$	$e_0 : f_0$ $e_1 : f_1$ $e_2 : f_3$	0, 8, $[e_1, 3]$ 1, 13, $[e_1, 4]$ 3, 16, $[e_1, 2]$ 5, 18, $[e_1, 4]$	5	-

entry $<1, 13, [e_1, 4]>$ is added. f_0 as a fully passed facility of e_0, is added to FP. The new global $dmin$ is computed as 1.

We repeat the above moving process, until the weight of last indexed entry equal to $\sum_{f \in F} w(f)$. After 4 rounds, the final index is shown in column 4.

5 Optimizations

Next, we present optimization techniques to improve index construction.

The Upper Bound Filter (Opt1). Given the network G and the set of facilities F, the index entries (i.e. the radius r required to reach each distinct weight sum) are fixed. To obtain the r values, in each while round in Algorithm 1, we choose the global $dmin$ as the moving distance. Therefore, the total indexing time is closely related to $dmin$ in each round. The larger moving distance in each round, the less number of rounds to go through, and the earlier we finish the index construction. So the goal is to move as long as possible in each round. Another observation is that around 90 % of edge entries (statistics from experimental results) achieve $dmin$ at meet events.

Based on these observations, we propose a simple filtering method used during computing $dmin$ for each edge e: before checking the meet cases from Line 6 in Algorithm 2, we compute the total weight of all cursors on edge e and all facilities that have fully passed e. If this weight is no larger than the last indexed entry, we know even there are meet cases occur in this edge, there will be no new index entries to be added. Therefore, we can safely skip the Lines 6–10. To be more specific, we add the following line before Line 6 in Algorithm 2: return d if $(\sum_{c \in e} w(c) + \sum_{f \in FP[e]} w(f) \leq I.back.weight)$.

For example, at the Round 1 in Table 2, after checking end event distances on edge e_2, we compute $w(f_2) + w(f_3) = 7 < 13$, we know any meet event occur on e_2 will not generate new index entry, so we directly return 2 rather than 1 as $dmin(e_2)$.

The Exact Filter (Opt2). The above filtering method uses an upper bound to estimate the location weight when meet event occurs. Although it already achieves good performance improvement (saves 21 % construction time), the bound is not tight and can be further improved with a bit more expenses.

Using the same example in Table 2, after checking end event distances on edge e_1 in Round 1, if the upper bound filter is applied, we get $w(f_0) + w(f_1) + w(f_2) = 16 > 13$, therefore the filter has no effect. However, when f_0's go-right cursor and f_1's go-left cursor meet, the actual location weight at meet position is $w(f_0) + w(f_1) = 11 < 13$, so we can still skip this meet distance 1.

Thus, instead of using total weight of cursors and fully passed facilities as an estimation, we compute the exact location weight at future meet positions, and compare it with last indexed entry to decide whether to use this $dmin$ or not. Algorithm 4 presents the details. It is similar to Algorithm 3 with the difference that Algorithm 3 computes the current location weight given a position p, while Algorithm 4 computes the future location weight at the meet position given two will meet cursors.

Continue with the above example, if exact filter is applied, we get $dmin(e_1) = 2$, therefore, the global $dmin$ after Round 1 is increased to 2, which effectively reduces one round. This optimization technique reduces nearly half of the total round number and index time, and if used together with upper bound filter, achieves further speed-up as shown in Sect. 6.

Algorithm 4. ExactFutureLocationWeight(ee, c_1, c_2)

Input : The edge entry ee, and two will meet cursors c_1 and c_2 on it.
Output: The location weight of middle position p between c_1 and c_2 after meet.

1 $F' \leftarrow \emptyset$; $w \leftarrow \sum_{f \in FP[ee]} w(f)$;
2 **for** *each cursor c on ee* **do**
3   **if** $c.pos < c_1.pos$ **and** $c.dir = left$ **then**
4   **if** f *not on ee* **or** $f.pos \geq c_1.pos$ **or** $dist(c,f) \geq dist(f,c_1)$ **then**
5           $F' \leftarrow F' \cup f$;   /* f: the facility c belongs to */;
6   **else if** $c_1.pos \leq c.pos$ **and** $c.pos \leq c_2.pos$ **then** /* cursors in between */
7        $F' \leftarrow F' \cup f$;
8   **else if** $c.pos > c_2.pos$ **and** $c.dir = right$ **then**
9   **if** f *not on ee* **or** $f.pos \leq c_2.pos$ **or** $dist(c,f) \geq dist(f,c_2)$ **then**
10          $F' \leftarrow F' \cup f$;
11 $w \leftarrow w + \sum_{f \in F'} w(f)$;
12 **return** w;

---

6   Experiment

In this section, we perform empirical experiments to confirm the substantial query performance improvement by our proposed method in practice.

Experiment Setting. In the experiment, we use two real network datasets, the North American (NA) and San Francisco (SF) road network, same as the previous work [14]. The NA dataset is obtained from [15] and SF from [16]. The facilities are generated randomly in the network with uniform distribution in terms of road network distance, and their weights are within range $(0, 50]$.[3] The cardinalities of datasets are shown in Table 3. The default facility number is 12500 if not explicitly mentioned. The maximum moving radius is set to 1000 while index construction.

We compare our method with the segment generation based algorithm [14] (denoted as SEG) mentioned before; our index-based algorithm is denoted as IND. Both methods are implemented in C++, and experiments are conducted on a server with Quad-Core 2.4 GHz Processor and 96 GB RAM. Although the method described in [14] generates segments in a DFS fashion, we implemented both the DFS and BFS versions. The results show that the DFS version is much slower than BFS due to the significant amount of redundant segments generation, therefore, in the following demonstrations, we compare our method with the BFS version result.

Table 3. Dataset statistics

Dataset	Nodes	Edges	Avg.EdgeLength	Facilities
NA	175813	179179	4.028	12500, 25000, 50000, 100000
SF	174956	223001	8.782	12500, 25000, 50000, 100000

Varying Query Radius Size. The first set of experiments compares the performance of both methods on various query radiuses. The four groups of radius ranges are $(0, 50]$, $(50, 100]$, $(100, 200]$ and $(200, 400]$. We produce 100 queries with lengths generated uniformly from each group, and show the *total query time* for each group in Fig. 4.

It can be seen that our method is around 6+ orders of magnitude faster than SEG. This demonstrates the substantial advantage of our index-based method to efficiently support large or batch query workloads. When query range increases, the running time of our method stays steady due to the high performance of simple binary search, while SEG takes much more time when query radius rises since they need to generate and process more segments with longer radius. This is more obvious on the SF dataset as the density of edges in SF is higher than NA.

---

[3] We also test on another popular real dataset(California road network) from [15], where the 87635 facilities are carefully generated using the real-life facility distribution. The result trend is similar with other datasets hence omitted due to space limitation.

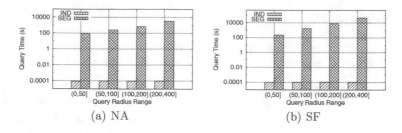

(a) NA

(b) SF

Fig. 4. Vary query radius

Vary Facility Number. The second group tests method scalability. We start with a set of 12500 facilities, and increase its size till 4x, i.e. 100000, and measure the total query time of both methods on 100 queries randomly generated within length (0,100]. The result is plotted in Fig. 5. Clearly, our method has much better scalability than SEG as SEG's query time climbs up quickly when the data size increases. This is expected as our query time increases only logarithmically with facility number, while SEG's increases superlinearly.

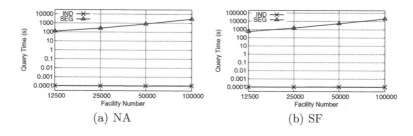

(a) NA

(b) SF

Fig. 5. Vary facility number

Index Construction. We show statistics about index construction in Table 4 to demonstrate the performance of different optimizations introduced in Sect. 5. The results are obtained on NA dataset with 12500 facilities. Results for other settings are similar and therefore omitted. In the table, Original means no optimization, and Opt12 means both optimizations are adopted. It is shown that Opt1 itself reduces 25 % of the Original round number, and achieves a 21 % acceleration of the index construction. While Opt2, if used alone, leads to 50 % reduction of the round number and 46 % of the total construction time. This is expected since Opt2 uses a tighter bound and thus achieves a larger global dmin in each round compared with Opt1. Finally, if both optimizations are applied, although the round number is still decreased by 50 % (as it depends on the tighter bound), the total time is further reduced to 51 % of the Original time. This is because in the cases where both Opt1 and Opt2 take effects, Opt12 uses Opt1 which requires less processing time than Opt2.

Table 4. Index construction optimizations

	Original	Opt1	Opt1/orig	Opt2	Opt2/orig	Opt12	Opt12/orig
Round#	199928	150421	75 %	101246	50 %	101246	50 %
Time(s)	37661	29893	79 %	20404	54 %	19274	51 %
Size	76 KB						

The index size is 76 KB despite what optimizations are used to construct it. The small index size further confirms the superiority and practicability of our method.

7 Related Work

Facility location optimization problem finds the optimal location by maximizing/minimizing some objective functions such as location influence (i.e. total weight of its RNNs) [10], average min-dist (distance from each object to its nearest facility) [11], total weighted distance to RNNs [12], and so on (see [13] for survey).

The maxRS problem can been seen as another instance of the facility location problem. Considering the axis-parallel rectangular query range, Nandy et al. [2] proposed an $O(n \log n)$ time algorithm to solve it using the plane-sweeping technique [1] with interval trees. Choi et al. [3] proposed an external memory solution following the distribution sweep paradigm [7], and the work was further extended in [5] by providing solutions to the AllMaxRS problem, which retrieves all optimal locations achieving the maximum total covered weight. Tao et al. [4] studied the approximate maxRS problem, and obtained a $(1 - \epsilon)$-approximate answer with high confidence in time $O(n \log \frac{1}{\epsilon} + n \log \log n)$ via grid sampling.

Another variation of maxRS problem is maximizing circular range sum(maxCRS) problem, meaning that query range is a circle. Chazelle et al. [18] solved the maxCRS problem in time $O(n^2)$. Aronov et al. [19] proposed a $(1-\epsilon)$-approximate algorithm with complexity $O(n\epsilon^{-2} \log n)$ for unweighted points, and $O(n\epsilon^{-2} \log^2 n)$ for weighted case. Choi et al. [3] solved the maxCRS problem by first converting it to the maxRS problem.

All the above maxRS related work assume Euclidean space. The only existing work, as far as we know, that studied the maxRS problem in road network is [14]. Their proposed method finds answer for a particular query radius r in time $O(|E||F| \log |F|)$, which is very time-consuming when the network or r is large. We devise an index-based method that answers queries in time $O(\log |F|)$ with $O(|F|)$ index size, hence provides significant speed-up to existing method and is beneficial to frequent queries.

8 Conclusion

In this paper, we study the maximizing range sum problem in road network. The only existing work [14] proposed an external-memory algorithm that solves one specific query with $O(|E||F| \log |F|)$ time. It is not satisfactory for mobile or web services dealing with millions of concurrent user queries or batch queries. We propose an index-based method that results in $O(\log |F|)$ online query time (6+ orders of magnitude faster than existing method in practice), with a tiny index of size linear in facility number $|F|$. Besides, we propose optimization techniques that achieve around 50 % reduction of the offline index construction time. Experiments on various settings verify the efficiency and scalability of our method.

References

1. Imai, H., Asano, T.: Finding the connected components and a maximum clique of an intersection graph of rectangles in the plane. J. Algorithms **4**(4), 310–323 (1983)
2. Nandy, S.C., Bhattacharya, B.B.: A unified algorithm for finding maximum and minimum object enclosing rectangles and cuboids. Math. Appl. **29**(8), 45–61 (1995)
3. Choi, D.W., Chung, C.W., Tao, Y.: A scalable algorithm for maximizing range sum in spatial databases. Proc. VLDB Endow. **5**(11), 1088–1099 (2012)
4. Tao, Y., Hu, X., Choi, D.W., Chung, C.W.: Approximate MaxRS in spatial databases. PVLDB **6**(13), 1546–1557 (2013)
5. Choi, D.W., Chung, C.W., Tao, Y.: Maximizing range sum in external memory. ACM Trans. Database Syst. **39**(3), 21: 1–21: 44 (2014)
6. Mukherjee, M., Chakraborty, K.: A polynomial time optimization algorithm for a rectilinear partitioning problem with applications in VLSI design automation. Inf. Process. Lett. **83**, 41–48 (2002)
7. Goodrich, M.T., Tsay, J.-J., Vengroff, D.E., Vitter, J.S.: External-memory computational geometry (preliminary version). In: FOCS, pp. 714–723 (1993)
8. Abellanas, M., Hurtado, F., Icking, C., Klein, R., Langetepe, E., Ma, L., Palop, B., Sacristán, V.: Smallest color-spanning objects. In: Meyer auf der Heide, F. (ed.) ESA 2001. LNCS, vol. 2161, p. 278. Springer, Heidelberg (2001)
9. Tiwari, S., Kaushik, H.: Extracting region of interest (roi) details using lbs infrastructure and web databases. In: MDM 2012, pp. 376–379 (2012)
10. Du, Y., Zhang, D., Xia, T.: The optimal-location query. In: Medeiros, C.B., Egenhofer, M., Bertino, E. (eds.) SSTD 2005. LNCS, vol. 3633, pp. 163–180. Springer, Heidelberg (2005)
11. Zhang, D., Du, Y., Xia, T., Tao, Y.: Progressive computation of the min-dist optimal-location query. In: VLDB (2006)
12. Xiao, X., Yao, B., Li, F.: Optimal location queries in road network databases. In: ICDE (2011)
13. Farahani, R.Z., Hekmatfar, M.: Facility Location: Concepts, Models, Algorithms and Case Studies, 1st edn. Physica-Verlag, Heidelberg (2009)
14. Phan, T.K., Jung, H.R., Kim, U.M.: An efficient algorithm for maximizing range sum queries in a road network. Sci. World J. **2014** (2014). Article ID 541602
15. http://www.cs.fsu.edu/lifeifei/SpatialDataset.htm
16. Brinkhoff, T.: A framework for generating network-based moving objects. GeoInformatica **6**(2), 153–180 (2002)
17. Yiu, M.L., Mamoulis, N.: Clustering objects on a spatial network. In: SIGMOD (2004)

18. Chazelle, B.M., Lee, D.T.: On a circle placement problem. Computing **36**, 1–16 (1986)
19. Aronov, B., Har-Peled, S.: On approximating the depth and related problems. In: Proceedings of the Sixteenth Annual ACM-SIAM Symposium on Discrete Algorithms, SODA 2005, pp. 886–894 (2005)

Trip Planning Queries for Subgroups in Spatial Databases

Tanzima Hashem[1]([⊠]), Tahrima Hashem[2], Mohammed Eunus Ali[1], Lars Kulik[3], and Egemen Tanin[3]

[1] Department of Computer Science and Engineering,
Bangladesh University of Engineering and Technology,
Dhaka, Bangladesh
{tanzimahashem,eunus}@cse.buet.ac.bd
[2] Department of Computer Science and Engineering,
Dhaka University, Dhaka, Bangladesh
tahrimacsedu14@gmail.com
[3] Department of Computing and Information System,
University of Melbourne, Melbourne, Australia
{lkulik,etanin}@unimelb.edu.au

Abstract. In this paper, we introduce a novel type of trip planning queries, a subgroup trip planning (SGTP) query that allows a group to identify the subgroup and the points of interests (POIs) from each required type (e.g., restaurant, shopping center, movie theater) that have the minimum aggregate trip distance for any subgroup size. The trip distance of a user starts at the user's source location and ends at the user's destination via the POIs. The computation of POI set for all possible subgroups with the straightforward application of group trip planning (GTP) algorithms would be prohibitively expensive. We propose an algorithm to compute answers for different subgroup size concurrently with less query processing overhead. We focus on both minimizing the total and maximum trip distance of the subgroup. We show the efficiency of our algorithms in experiments using both real and synthetic datasets.

1 Introduction

Location-based services that involve more than one user have become common in recent years [11,13,16]. A group of users located at different places may want to visit a number of points of interests (POIs) such as a restaurant, a shopping center, or a movie theater together, and then go towards their own individual destinations. Group trip planning (GTP) queries [6,7,15] return points of interests (POIs) of required types (e.g., restaurant, shopping center, movie theater) that minimize the aggregate trip distance with respect to the source and destination locations of the group members. The trip distance of a user starts at the user's source location and ends at the user's destination via the returned POIs and the aggregate trip distance is computed as the total or the maximum trip distance of group members.

© Springer International Publishing AG 2016
M.A. Cheema et al. (Eds.): ADC 2016, LNCS 9877, pp. 110–122, 2016.
DOI: 10.1007/978-3-319-46922-5_9

Sometimes it may happen that the majority of the users in the group are located nearby, whereas a few users may be located far away from them. These distant users, considered as *outliers*, may change the query answer for the group and increase the trip distances of the other group members. In such scenarios, if the aggregate trip distance of a subgroup improves significantly compared to a complete group, the group may prefer to select the POIs for the trip that minimize the aggregate trip distance for a subgroup. Other group members who are not included in the subgroup can join the trip in return of traveling a little bit more for the overall interest of the group. In Fig. 1, for the entire group (of size 4), the pair of POIs (p_1'', p_2'') minimizes the maximum travel distance, while (p_1'', p_2') minimizes the maximum travel distance if we consider a subgroup of size 3. We observe that the aggregate trip distance significantly improves if the group does not take (s_4, d_4) into account as (s_4, d_4) is located far away from other source-destination pairs. To address the above mentioned scenario, in this paper, we propose a *subgroup trip planning (SGTP) query* that returns for every subgroup size, the subgroup and the POIs from each required type that have the minimum aggregate trip distance.

Evaluating an SGTP query requires identifying the subgroup and the POIs from each required type that have the minimum aggregate trip distance for every subgroup size. A straightforward approach to evaluate an SGTP query is to apply our GTP query algorithms [6,7,15] for all possible subgroups independently, and select the best subgroup that results in a minimum aggregate distance. This approach incurs high processing overhead and is not scalable as it requires to combinatorially enumerate all possible subgroups and apply our GTP query algorithm for each of these subgroups. We propose an efficient solution to process an SGTP query that can identify the query answer in a single search on the database. Our approach gradually refines the POI search space based on the smallest aggregate trip distance computed using already retrieved POIs, and does not retrieve those POIs from the database that cannot be part of the answer for any subgroup. Furthermore, we develop an efficient technique to identify subgroups of different sizes with the minimum aggregate trip distances for a set of POIs, without enumerating all subgroups.

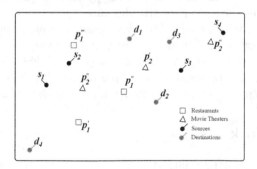

Fig. 1. An example SGTP query, where s_i and d_i represent the source and destination locations of a user u_i, and p_1 and p_2 represent POI types restaurant and movie theater, respectively.

In summary, the contributions of this paper are summarized as follows:

- We introduce subgroup trip planning (SGTP) queries in spatial databases.
- We propose a hierarchical algorithm to evaluate SGTP queries. We consider minimizing both the total trip distance and the maximum trip distance of the members in the subgroup.
- We develop a technique to compute aggregate trip distances for different subgroup sizes with respect to a set of POIs with reduced computational overhead.
- We evaluate the performance of our approach in experiments using both synthetic and real datasets.

2   Problem Overview

A subgroup trip planning (SGTP) query is formally defined as follows:

Definition 1 *(SGTP Queries). Given a group G of n users $\{u_1, u_2, \ldots, u_n\}$, the minimum subgroup size n', a set of source locations S, a set of destination locations D, sets of m types of data points $\{D_1, D_2, \ldots, D_m\}$, and an aggregate function f the SGTP query returns for every subgroup size $n'' \in [n', n]$, a subgroup $G' \in G$ of n'' users and a set of data points $P = \{p_1, p_2, \ldots, p_m\}$, $p_t \in D_t$, that minimizes f.*

In this paper, an aggregate function f can be either SUM or MAX, where SUM and MAX return the total or maximum trip distance of group members, respectively. The total trip distance for a subgroup is the summation of each user's trip distance in the subgroup, where a user's trip distance starts from a user u_i's source location s_i and ends at u_i's destination location d_i via the returned data points. On the other hand, the maximum trip distance for a subgroup is computed as the maximum of the trip distances of users in the subgroup.

For an ordered SGTP query, the group fixes the order of the types of data points in which a subgroup visits the data points (e.g., the group may want to visit a shopping center before a restaurant) and the aggregate trip distance is computed accordingly. For a flexible SGTP query, the order of visiting data point types is not fixed, i.e., the group is happy to visit the data points in any order that minimizes the aggregate trip distance.

An SGTP query can be extended to a k *subgroup trip planning (kSGTP) query* that returns for every subgroup size, k subgroups and sets of data points that have k smallest aggregate distances for the group trip. A kSGTP query enables a group to select data points based on other parameters like cost or preference for restaurants. In this paper, we develop solution for processing kSGTP queries.

3   Related Work

GTP queries have been first introduced in [7], where the authors developed algorithms to determine group trips with the minimum total trip distance. In [1, 15], algorithms have been proposed to process ordered GTP queries. Recently, in [6], the authors have proposed an algorithm that can minimize both total and maximum distances of the group members for both ordered and flexible GTP queries. In this paper, we focus on an SGTP query, a new variant of a GTP query. A GTP query identifies only a set of data points whereas an SGTP query returns the subgroup and the set of data points that together minimize the aggregate group trip distance.

Group nearest neighbor (GNN) queries [10,12–14] and variants [2,4,9,11] have been extensively studied in the literature. A GNN query returns a single type of data point that minimize the aggregate travel distance with respect to the current locations of group members. On the other hand, a GTP query returns a set of data points of different types that minimize the aggregate travel distance with respect to the source and destination locations of group members. It has been already shown in the literature [7] that extending GNN algorithms for processing SGTP queries would be an exhaustive search and prohibitively computationally expensive.

In [11], Li et al. have proposed a flexible aggregate similarity search that finds the nearest data point and the corresponding subgroup for a fixed subgroup size; for example, a group may query for the nearest data point to 50 % of group members. In [2], the authors have developed an approach for spatial consensus queries that find subgroups of different size and corresponding data points that together minimize the aggregate travel distance. In this paper, we introduce a subgroup trip planning query, which is different from the above mentioned variants of GNN queries.

4 Our Approach

In this section, we present our solution to process kSGTP queries. Though it is possible to evaluate an SGTP query by applying the GTP query algorithm [6,7,15] independently for all subgroups, this straightforward solution would be an exhaustive search and require to access same data multiple times. More specifically, for every subgroup of size j from a group of n users, this approach requires to apply kGTP query algorithm $\binom{n}{j}$ times to evaluate the answers for $\binom{n}{j}$ groups independently and then select k sets of data points from the evaluated answers and corresponding k sets of j users that together provide k smallest aggregate trip distances. The limitation of this straightforward approach is that the combined query processing overhead is very high and does not scale. We assume that data points of the same type (e.g., restaurants) are indexed using a single R^*-tree [3], and develop an approach to evaluate the query answers for different subgroup sizes *concurrently* with a single traversal on the R^*-trees $R_1, R_2, \ldots, R_m$. Our approach prunes those POIs that cannot be part of the SGTP query answer using the smallest aggregate trip distance computed based on the already retrieved POIs from the database. We develop a technique to identify the subgroups of different sizes with minimum aggregate trip distances without computing aggregate trip distances for all possible subgroups, which further reduces the query processing overhead.

Algorithm 1 shows the steps to evaluate an kSGTP query. The input to the algorithm are a group of n users $G = \{u_1, u_2, \ldots, u_n\}$, source locations $S = \{s_1, s_2, \ldots, s_n\}$, destination locations $D = \{d_1, d_2, \ldots, d_n\}$, m types of data points for $m > 0$, the number k of required sets of data points, the minimum subgroup size n', and an aggregate function f. The output of the algorithm is the answer set $A = \{A_j\}$, where $n' \leq j \leq n$, represents the query answer for subgroup size j. A_j contains k subgroups consisting of j members from G and corresponding data points, $\{G^1, \{p_1^1, p_2^1, \ldots, p_m^1\}\}, \{G^2, \{p_1^2, p_2^2, \ldots, p_m^2\}\}, \ldots, \{G^k, \{p_1^k, p_2^k, \ldots, p_m^k\}\}$ having the k smallest aggregate distances for the group trip.

We use the following notations in our algorithm:

- r_i: a data point or a minimum bounding rectangle (MBR) of a node of R_i.
- $Dist_{min}(.,.)(Dist_{max}(.,.))$: the minimum (maximum) distance between two parameters.
- $d_{min}^j(w_1, w_2, \ldots, w_m)$: the distance computed as the smallest minimum aggregate trip distances computed for all possible subgroups of size j, where the minimum aggregate distance for a subgroup G' of size j is determined as $f_{u_i \in G'}(Dist_{min}(s_i, w_1) + \sum_{t=2}^{m-1} Dist_{min}(w_t, w_{t+1}) + Dist_{min}(d_i, w_m))$, where $w_1, w_2, \ldots, w_m$ are data points or MBRs of R^*-tree nodes.
- $d_{max}^j(w_1, w_2, \ldots, w_m)$: the distance computed as the smallest maximum aggregate trip distances computed for all possible subgroups of size j, where the maximum aggregate distance for a subgroup G' of size j is determined as $f_{u_i \in G'}(Dist_{max}(s_i, w_1) + sum_{t=2}^{m-1} Dist_{max}(w_t, w_{t+1}) + Dist_{max}(d_i, w_m))$, where $w_1, w_2, \ldots, w_m$ are data points or MBRs of R^*-tree nodes.
- $MinDist[j][k]$: The kth smallest distance for subgroup size j from already computed $d_{max}^j(w_1, w_2, \ldots, w_m)$s.

The algorithm works in a hierarchical manner. The algorithm maintains a priority queue Q_p, which is initially ordered based on the minimum $d_{min}^{n'}$ and reordered based on the minimum $d_{min}^{n'+1}$ when the query answer for subgroup size $n' + 1$ has been identified. The reordering continues through $d_{min}^{n'+2}, d_{min}^{n'+3}, \ldots, d_{min}^n$ until the query answer for subgroup size n, i.e., the entire group has been identified.

The algorithm first initializes all entries of A with $\emptyset$, $Mindist$ with ∞, cur with n', and end with 0 (Line 1.1). The search for the query answer starts from the root nodes of R^*-trees, $R_1, R_2, \ldots, R_m$; the algorithm inserts the root nodes of $R_1, R_2, \ldots, R_m$ together with $\bigcup_{j=n'}^n d_{min}^j$ into a priority queue Q_p. The elements of Q_p are stored in order of $d_{min}^{n'}$. In each iteration of the search, the algorithm dequeues $r_1, r_2, \ldots, r_m$ from Q_p. If all $r_1, r_2, \ldots, r_m$ are data points then the algorithm updates the answers for a subgroup size j, $cur \leq j \leq n$, if the condition $d_{min}^j(r_1, r_2, \ldots, r_m) < MinDist[j][k]$ is satisfied (Lines 1.8–1.9). The algorithm checks this condition because at any point in time Q_p is ordered based on a d_{min}^j and it may happen that $d_{min}^{j'}(r_1, r_2, \ldots, r_m)$ for $j' > j$ is greater than current $MinDist[j'][k]$ and cannot be part of the answer for subgroup size j'.

Algorithm 1. kSGTP(G, S, D, n', k, f)

Input : $G = \{u_1, u_2, \ldots, u_n\}$, $S = \{s_1, s_2, \ldots, s_n\}$, $D = \{d_1, d_2, \ldots, d_n\}$, n', k, and f

Output: $A = \{A_{n'}, A_{n'+1}, \ldots, A_n\}$

1.1 $Initialize(A, MinDist, cur, end)$;

1.2 $Enqueue(Q_p, root_1, root_2, \ldots, root_m, \bigcup_{j=n'}^n d_{min}^j (root_1, root_2, \ldots, root_m))$;

1.3 **while** Q_p is not empty and $end = 0$ **do**

1.4 $\quad \{r_1, r_2, \ldots, r_m, \bigcup_{i=n'}^n d_{min}^i (r_1, r_2, \ldots, r_m)\} \leftarrow Dequeue(Q_p)$;

1.5 $\quad$ **if** $r_1, r_2, \ldots, r_m$ are data points **then**

1.6 $\quad\quad j \leftarrow cur$;

1.7 $\quad\quad$ **while** $j \leq n$ **do**

1.8 $\quad\quad\quad$ **if** $d_{min}^j (r_1, r_2, \ldots, r_m) < MinDist[j][k]$ **then**

1.9 $\quad\quad\quad\quad$ $Update(G, S, D, k, f, A_j, d_{min}^j, j)$;

1.10 $\quad\quad\quad$ $j \leftarrow j + 1$;

1.11 $\quad\quad$ **if** $d_{min}^{cur} (r_1, r_2, \ldots, r_m) > MinDist[cur][k]$ **then**

1.12 $\quad\quad\quad$ $cur \leftarrow cur + 1$;

1.13 $\quad\quad\quad$ **if** $cur \leq n$ **then**

1.14 $\quad\quad\quad\quad$ $Reorder(Q_p, cur)$;

1.15 $\quad\quad\quad$ **else**

1.16 $\quad\quad\quad\quad$ $end \leftarrow 1$;

1.17 $\quad$ **else**

1.18 $\quad\quad$ $W \leftarrow FindSets(r_1, r_2, \ldots, r_m)$;

1.19 $\quad\quad$ **for** each $(w_1, w_2, \ldots, w_m) \in W$ **do**

1.20 $\quad\quad\quad$ $\bigcup_{j=n'}^n \{d_{min}^j (w_1, w_2, \ldots, w_m), d_{max}^j (w_1, w_2, \ldots, w_m)\} \leftarrow$ $CompTripDist(S, D, n', f, w_1, w_2, \ldots, w_m)$;

1.21 $\quad\quad\quad$ $j \leftarrow n'$;

1.22 $\quad\quad\quad$ $entry \leftarrow 0$;

1.23 $\quad\quad\quad$ **while** $j \leq n$ **do**

1.24 $\quad\quad\quad\quad$ **if** $d_{min}^j (w_1, w_2, \ldots, w_m) \leq MinDist[j][k]$ **then**

1.25 $\quad\quad\quad\quad\quad$ **if** $entry = 0$ **then**

1.26 $\quad\quad\quad\quad\quad\quad$ $Enqueue(Q_p, w_1, w_2, \ldots, w_m, \bigcup_{i=n'}^n d_{min}^i (w_1, w_2, \ldots, w_m))$;

1.27 $\quad\quad\quad\quad\quad\quad$ $entry \leftarrow 1$;

1.28 $\quad\quad\quad\quad$ **if** $d_{max}^j (w_1, w_2, \ldots, w_m) \leq MinDist[j][k])$ **then**

1.29 $\quad\quad\quad\quad\quad$ $Update(MinDist[j], d_{max}^j (w_1, w_2, \ldots, w_m))$;

1.30 $\quad\quad\quad\quad$ $j \leftarrow j + 1$;

1.31 **return** A;

After updating the answer set, the algorithm checks whether kSGTP answer for subgroup size cur have been already found, i.e., $d_{min}^{cur} (r_1, r_2, \ldots, r_m) > MinDist[cur][k]$. If the next subgroup size, $cur + 1$ is greater than n then the algorithm terminates by assigning 1 to the variable end (Line 1.16). Otherwise, the remaining elements of priority queue is reordered based on $d_{min}^{cur+1} (r_1, r_2, \ldots, r_m)$.

On the other hand, if $r_1, r_2, \ldots, r_m$ are not data points, the algorithm computes W using the function $FindSets$. $FindSets$ determines all possible sets $\{w_1, w_2, \ldots, w_m\}$s, where w_j represents either one of the child node of r_j or the data point r_j. In case of an ordered SGTP query, $FindSets$ only computes ordered set of data points/R^*-tree nodes, whereas in case of a flexible SGTP query, $FindSets$ considers all combination of data points/R^*-tree nodes.

For each set $\{w_1, w_2, \ldots, w_m\}$ in W, the algorithm computes $\bigcup_{j=n'}^{n} \{d_{min}^j$ $(w_1, w_2, \ldots, w_m), d_{max}^j(w_1, w_2, \ldots, w_m)\}$ using function $CompTripDist$ (please see Sect. 4.1). Before inserting $\{w_1, w_2, \ldots, w_m\}$ into Q_p, the algorithm checks whether it is possible to prune $\{w_1, w_2, \ldots, w_m\}$. The algorithm prunes $\{w_1, w_2, \ldots, w_m\}$ if for every $n' \leq j \leq n$, $d_{min}^j(w_1, w_2, \ldots, w_m) > MinDist[j][k]$.

4.1   Function $CompTripDist$

The purpose of the function $CompTripDist$ is to compute d_{min}^js and d_{max}^js for a set of R^*-tree nodes or data points $\{w_1, w_2, \ldots, w_m\}$ with respect to a set of source and destination locations for subgroup size j varying from n' to n. In a straightforward approach to determine these distances, we can first compute each user's individual minimum and maximum trip distances and then for every subgroup size j, we have to determine the minimum and maximum aggregate trip distances for all possible subgroups of size j and select the minimum and maximum from the computed aggregate trip distances as d_{min}^j and d_{max}^j, respectively. We develop an algorithm that avoids computing aggregate distances for every subgroup and thus, reduces computational overhead.

Algorithm 2. ComputeTripDist$(S, D, n', f, w_1, w_2, \ldots, w_m)$

Input : $S = \{s_1, s_2, \ldots, s_n\}$, $D = \{d_1, d_2, \ldots, d_n\}$, n', f, and $w_1, w_2, \ldots, w_m$
Output: $\bigcup_{j=n'}^{n} \{d_{min}^j(w_1, w_2, \ldots, w_m), d_{max}^j(w_1, w_2, \ldots, w_m)\}$

2.1 $InterMinDist \leftarrow Dist_{min}(w_1, w_2) + Dist_{min}(w_2, w_3) + \cdots + Dist_{min}(w_{m-1}, w_m)$;
2.2 $InterMaxDist \leftarrow Dist_{max}(w_1, w_2) + Dist_{max}(w_2, w_3) + \cdots + Dist_{max}(w_{m-1}, w_m)$;
2.3 $i \leftarrow 1$;
2.4   **while** $i \leq n$ **do**
2.5 | $Enqueue(DMinQ_p, Dist_{min}(s_i, w_1) + Dist_{min}(d_i, w_m))$;
2.6 | $Enqueue(DMaxQ_p, Dist_{max}(s_i, w_1) + Dist_{max}(d_i, w_m))$;
2.7 |_ $i \leftarrow i + 1$;

2.8 $j \leftarrow 1$;
2.9 $EndMinDist, EndMaxDist \leftarrow 0$;
2.10 **while** $j \leq n$ **do**
2.11 | $EndMinDist \leftarrow Compf(EndMinDist, Dequeue(DMinQ_p), f)$;
2.12 | $EndMaxDist \leftarrow Compf(EndMaxDist, Dequeue(DMaxQ_p), f)$;
2.13 | **if** $j \geq n'$ **then**
2.14 | | **if** $f =$ SUM **then**
2.15 | | | $d_{min}^j(w_1, w_2, \ldots, w_m) \leftarrow EndMinDist + j \times InterMinDist$;
2.16 | | |_ $d_{max}^j(w_1, w_2, \ldots, w_m) \leftarrow EndMaxDist + j \times InterMaxDist$;

2.17 | | **else**
2.18 | | | $d_{min}^j(w_1, w_2, \ldots, w_m) \leftarrow EndMinDist + InterMinDist$;
2.19 | |_ |_ $d_{max}^j(w_1, w_2, \ldots, w_m) \leftarrow EndMaxDist + InterMaxDist$;

2.20 |_ $j \leftarrow j + 1$;
2.21 **return** $\bigcup_{j=n'}^{n} \{d_{min}^j(w_1, w_2, \ldots, w_m), d_{max}^j(w_1, w_2, \ldots, w_m)\}$;

Algorithm 2 shows pseudocode for the function $CompTripDist$. The input to the algorithm are source locations $S = \{s_1, s_2, \ldots, s_n\}$, destination locations $D = \{d_1, d_2, \ldots, d_n\}$, the minimum subgroup size n', an aggregate function f, and a set of R^*-tree nodes or data points $\{w_1, w_2, \ldots, w_m\}$. The output of the algorithm is the set of distances $\bigcup_{j=n'}^{n} \{d_{min}^j(w_1, w_2, \ldots, w_m), d_{max}^j(w_1, w_2, \ldots, w_m)\}$.

For $d_{min}^j(w_1, w_2, \ldots, w_m)$, we need to consider j users who have j smallest minimum (maximum) trip distances via $(w_1, w_2, \ldots, w_m)$. The minimum (maximum) distances to travel from w_1 to w_m via $w_2, w_3, \ldots, w_{m-1}$ remain constant for all users' minimum (maximum) trip distances. The trip distances of individual users differ due to their different source and destination locations, i.e., the summation of the minimum (maximum) distance to travel from a user's source location to w_1 and the minimum (maximum) distance to travel from w_m to the user's destination. Based on this observation, the steps of $CompTripDist$ are as follows.

To compute $\bigcup_{j=n'}^n \{d_{min}^j(w_1, w_2, \ldots, w_m)\}$, the algorithm uses two variables, $InterMinDist$ and $EndMinDist$, where $InterMinDist$ stores the minimum distance from w_1 to w_m via $w_2, w_3, \ldots, w_{m-1}$ and $EndMinDist$ stores the summation of the users' minimum aggregate distance from their sources to w_1 and the users' minimum aggregate distance from w_m to their destinations. The algorithm computes $InterMinDist$ in Line 2.1 and $InterMinDist$ remains constant for all subgroups of size j.

On the other hand, $EndMinDist$ changes based on the source and destination locations of the users in a subgroup. To compute $EndMinDist$ for a subgroup size j, we need to determine j smallest values from n distances, where n is the number of users in the group and each distance represents the summation of the minimum distance from a user u_i's source to w_1 and the minimum distance from w_m to u_i's destination, i.e., $Dist_{min}(s_i, w_1) + Dist_{min}(d_i, w_m)$. The algorithm uses a priority queue $DMinQ_p$ to store these n distances in a sorted manner; $DMinQ_p$ is ordered based on $Dist_{min}(s_i, w_1) + Dist_{min}(d_i, w_m)$. We can have j smallest distances for computing $EndMinDist$ by dequeuing first j elements from $DMinQ_p$.

In every iteration, the algorithm starts to dequeue a distance from $DMinQ_p$ and updates a variable $EndMinDist$ using function $Compf$ based on the aggregate function f. If f is SUM, $Compf$ adds the dequeued distance to $EndMinDist$, and if it is MAX, $Compf$ assigns the dequeued distance to $EndMinDist$, since the current dequeued distance is always greater than previous one dequeued from $DMinQ_p$ (Line 2.11).

In addition, if n' or more distances have been dequeued from $DMinQ_p$, i.e., $j \geq n'$, the algorithm computes corresponding $d_{min}^j(w_1, w_2, \ldots, w_m)$ as $EndMinDist + i \times InterMinDist$, if f is SUM. Otherwise, for $f = $ MAX, $d_{min}^j(w_1, w_2, \ldots, w_m)$ is computed as $EndMinDist + InterMinDist$.

Similarly, the algorithm computes $\bigcup_{j=n'}^n \{d_{max}^j(w_1, w_2, \ldots, w_m)\}$. In summary, our algorithm avoids redundant computations, because we do not need to consider all subgroups while finding the subgroups that provide smallest minimum aggregate trip distances for different subgroup sizes.

5 Experiments

In this section, we present experiments to show the performance of our proposed approach for SGTP queries using both real and synthetic datasets. The real dataset C consists of 62,556 points of interests (i.e., data points) of California.

The synthetic datasets U and Z are generated using uniform and a Zipfian distribution, respectively. The total space is normalized into a span of $10,000 \times 10,000$ square units. We use a desktop with a Intel Core 2 Duo 2.40 GHz CPU and 4 GBytes RAM to run our experiments.

We measure the performance of our algorithm in terms of IO costs and the query processing time. We run the experiments for 10 sample SGTP queries and show the average performance. For every SGTP query sample, we randomly select an area in the total space as a query area, and then generate the source and destination locations of group members in the selected query area using a uniform random distribution. We vary the group size, the subgroup size, the number of required sets of data points (k), the query area, i.e., the minimum bounding rectangle covering the source and destination locations (M), and the dataset size in different sets of experiments. Table 1 shows the range and default value of each parameter. The subgroup size is measured in terms of percentage of the group size; e.g., 60 % subgroup size means 38 group members with respect to a default group size 64. We fix the number of required POI types to 2 as a group normally plans a trip for a limited number of POI types. We run the experiments for ordered kSGTP queries as it is a common scenario that the group predefines the order of visiting POI types. Note that our approach is applicable for both flexible and ordered kSGTP queries for any number of POI types.

Table 1. Experiment setup

Parameter	Range	Default
Group size	4, 16, 64, 256	64
Subgroup size	60 %, 70 %, 80 %, 90 %	70 %
Query area M	2 %, 4 %, 8 %, 16 %	4 %
k	2, 4, 8, 16	4
Data set size (Synthetic)	5 K, 10 K, 15 K, 20 K	-

Effect of Query Area (M): The query area M is varied as 2 %, 4 %, 8 %, and 16 % of the data space. The IO cost and processing time for processing SGTP queries are measured for both aggregate SUM and MAX functions (Figs. 2(a) and (b)). Both the IO cost and the processing time increase with the increase of the area M as for a larger M we need to access more data points from R*-trees than that of a smaller M for the aggregate SUM and MAX functions.

Effect of Group Size: We vary the group size as 4, 16, 64, and 256, and measure IOs and processing time for SGTP queries for aggregate functions SUM and MAX. Figure 3(b) shows that the processing time increases with the increase of the group size for both MAX and SUM functions. The reason behind this is for a larger group size, SGTP requires to process a large number of subgroups. Thus, it computes the aggregate trip distances for these increasing number of subgroups. However, the IOs remain almost constant (MAX) or slightly increase (SUM) with the increase of the group size (see Fig. 3(a)).

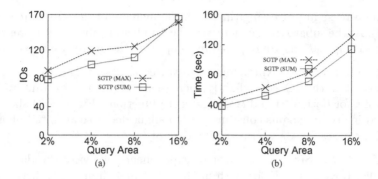

Fig. 2. Effect of query area for kSGTP queries (dataset C)

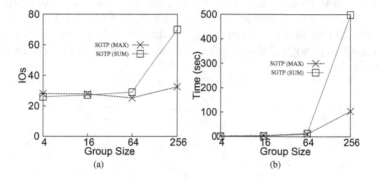

Fig. 3. Effect of group size for kSGTP queries (dataset C)

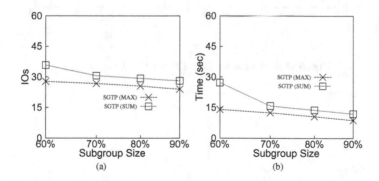

Fig. 4. Effect of subgroup size for kSGTP queries (dataset C)

Effect of Subgroup Size: In this set of experiments, we vary the subgroup size as
60 %, 70 %, 80 %, and 90 % of the default group size of 64, and measure IOs and
processing time for both aggregate functions SUM and MAX. In Fig. 4(a), we can
see that the IOs remain constant with the increase of the subgroup size. However,

there is a decrease in processing time with the increase of the subgroup size (see Fig. 4(b)). As the subgroup size increases, the number of subgroups decreases and hence, less computations are required in computing the aggregate trip distances.

Effect of k: In this set of experiments, k is varied as 2, 4, 8, and 16. In Fig. 5(a), we observe that the IO cost slightly increases or remains constant with the increase of k for both SUM and MAX aggregate functions. Figure 5(b) shows that the processing time remains constant (with a slight decrease at initial phase) for SUM and increases for MAX with the increase of k.

Effect of Data Set Size: In this set of experiments, we vary the dataset size as 5K, 10K, 15K, and 20 K for both uniform (U) and Zipfian (Z) distributions. Figures 6(a) and (b) show the IO cost and processing time, respectively, for different data set sizes with U distribution for aggregate functions SUM and MAX. The experimental results show that SGTP requires more IOs and processing time for SUM aggregate function than MAX aggregate function. Figures 7(a) and (b) show the IO cost and processing time required by SGTP, respectively, for different data set sizes with Z distribution. The results show similar characteristics of U distribution.

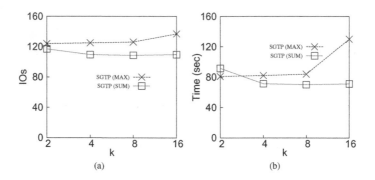

Fig. 5. Effect of k for kSGTP queries (dataset C)

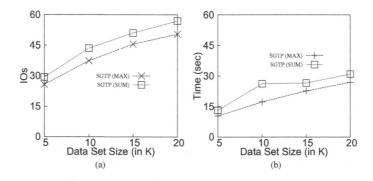

Fig. 6. Effect of dataset size for kSGTP queries (dataset U)

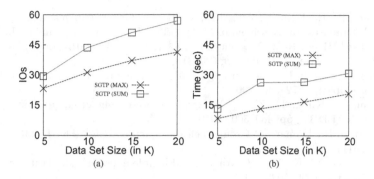

Fig. 7. Effect of dataset size for kSGTP queries (dataset Z)

6 Conclusion

In this paper, we introduced subgroup trip planning (SGTP) queries that enable a group to identify the subgroups and POI sets that together minimize the total or maximum trip distance. We have developed a hierarchical approach to process SGTP queries. Our approach avoids the computation of aggregate trip distances independently for different subgroups and thus, reduces computational overhead. Our experiments also show that our algorithm can evaluate a SGTP query with reduced processing time. In the future, we plan to develop algorithms for processing SGTP queries for road networks and the obstructed space. We also aim to protect location privacy [5,8] of group members for SGTP queries.

Acknowledgments. This research is partially supported by the ICT Division - Government of the People's Republic of Bangladesh.

References

1. Ahmadi, E., Nascimento, M.A.: A mixed breadth-depth first search strategy for sequenced group trip planning queries. In: MDM, pp. 24–33 (2015)
2. Ali, M.E., Tanin, E., Scheuermann, P., Nutanong, S.: Spatial consensus queries in a collaborative environment. ACM Trans. Spat. Algorithms Syst. **2**(1), 3 (2016)
3. Beckmann, N., Kriegel, H.-P., Schneider, R., Seeger, B.: The R*-tree: an efficient and robust access method for points and rectangles. SIGMOD Rec. **19**(2), 322–331 (1990)
4. Deng, K., Sadiq, S.W., Xiaofang Zhou, H., Xu, H., Fung, G.P.C., Yansheng, L.: On group nearest group query processing. IEEE TKDE **24**(2), 295–308 (2012)
5. Hashem,T., Ali, M.E., Kulik, L., Tanin, E., Quattrone, A.: Protecting privacy for group nearest neighbor queries with crowdsourced data and computing. In: UbiComp, pp. 559–562 (2013)
6. Hashem, T., Barua, S., Ali, M.E., Kulik, L., Tanin, E.: Efficient computation of trips with friends and families. In: CIKM, pp. 931–940 (2015)

7. Hashem, T., Hashem, T., Ali, M.E., Kulik, L.: Group trip planning queries in spatial databases. In: Nascimento, M.A., Sellis, T., Cheng, R., Sander, J., Zheng, Y., Kriegel, H.-P., Renz, M., Sengstock, C. (eds.) SSTD 2013. LNCS, vol. 8098, pp. 259–276. Springer, Heidelberg (2013)
8. Hashem, T., Kulik, L.: Safeguarding location privacy in wireless ad-hoc networks. In: Ubicomp, pp. 372–390 (2007)
9. Hashem, T., Kulik, L., Zhang, R.: Privacy preserving group nearest neighbor queries. In: EDBT, pp. 489–500 (2010)
10. Li, F., Yao, B., Kumar, P.: Group enclosing queries. IEEE TKDE **23**(10), 1526–1540 (2011)
11. Li, Y., Li, F., Yi, K., Yao, B., Wang, M.: Flexible aggregate similarity search. In: SIGMOD, pp. 1009–1020 (2011)
12. Namnandorj, S., Chen, H., Furuse, K., Ohbo, N.: Efficient bounds in finding aggregate nearest neighbors. In: Bhowmick, S.S., Küng, J., Wagner, R. (eds.) DEXA 2008. LNCS, vol. 5181, pp. 693–700. Springer, Heidelberg (2008)
13. Papadias, D., Shen, Q., Tao, Y., Mouratidis, K.: Group nearest neighbor queries. In: ICDE, p. 301 (2004)
14. Papadias, D., Tao, Y., Mouratidis, K., Hui, C.K.: Aggregate nearest neighbor queries in spatial databases. TODS **30**(2), 529–576 (2005)
15. Samrose, S., Hashem, T., Barua, S., Ali, M.E., Uddin, M.H., Mahmud, M.I.: Efficient computation of group optimal sequenced routes in road networks. In: MDM, pp. 122–127 (2015)
16. Yan, D., Zhao, Z., Ng, W.: Efficient processing of optimal meeting point queries in euclidean space and road networks. Knowl. Inf. Syst. **42**(2), 319–351 (2015)

A Sketch-First Approach for Finding TSP

Weihuang Huang[✉], Jeffrey Xu Yu, and Zechao Shang

The Chinese University of Hong Kong, Hong Kong, China
{whhuang,yu,zcshang}@se.cuhk.edu.hk

Abstract. Travel planning is one of the most important issues in location-based services (LBS), and TSP (traveling salesman problem) is to find the shortest tour that traverses all the given points exactly once. Given the hardness of TSP as an NP-hard problem, a large number of heuristic methods are proposed to find a tour efficiently. Here, the heuristics proposed are based on a similar idea that is to expand a partial tour by adding points one by one in different ways until all points are visited. In this paper, we study TSP query with a given set of points Q. We propose a new heuristic called Sketch-First, which is different from the existing approaches. By Sketch-First, we select a set of points out of Q, forming a sketch of Q, and add the points that are not in the sketch back to the sketch to obtain the answer for Q. The sketch gives a global picture on the points, and can be used to guide to add the other points back effectively. We discuss the heuristics to find a sketch for Q. Our approach is based on the observation that a better sketch with the same number of points is the sketch over which its optimal tour is larger in length. In addition, as the number of such points is to be small, we can find the optimal tour for the sketch. We discuss our methods, and conduct extensive experiments to show the effectiveness and efficiency of our methods.

1 Introduction

Travel planning has been recently studied as an important issue of location-based services (LBS), which are to find tours among points of interest (POI), where POIs are with latitude and longitude in a 2-dimensional space or in a road network, and might be associated with keywords. In the literature, the work includes finding the shortest tour connecting two POIs [1,2] and searching the optimal meeting point for a set of POIs [3,4]. Traveling salesman problem (TSP), which is to find a tour that traverses all given points exactly once and also has the minimum overall distance, is also an important issue in travel planning. However, given the hardness of TSP, as a well known NP-hard problem, there are less investigations on TSP in LBS, even though there are TSP-like problems being studied. Li et al. [5] investigates a problem that aims to find the shortest tour passing through a set of points, and those points will cover all the keywords given by the user. Other works [6,7] investigate similar problems with extra constraints given a specific order and/or time window. The problems studied in [5–7] are solved to expand a tour by adding nearest neighbours one by one, as an answer by heuristics.

© Springer International Publishing AG 2016
M.A. Cheema et al. (Eds.): ADC 2016, LNCS 9877, pp. 123–136, 2016.
DOI: 10.1007/978-3-319-46922-5_10

There are numerous heuristics proposed to find the shortest tour for a given set of points Q. They are usually categorized into two categories [8]. One category is constructive heuristics. The approaches in this category try to find a tour by expanding the partial answer tour one by one until all the points in Q are visited. The other category is improvement heuristics. The approaches in this category try to refine a given answer by for example inserting k pair of direct visits followed by deleting k pair of direct visits. However, the improvement heuristics are less effective since it is unknown whether they can improve the answer by the cost they spend. We will not discuss them in this paper.

In this paper, we study a new heuristic which we call Sketch-First. By Sketch-First, first, we pick a subset from Q of a TSP query, and find the optimal tour for these points, as a sketch of the answer for all points in Q. Second, we insert all the points that are not in the sketch back into the sketch. It is important to note that the sketch guides the insertion. This heuristic is different from the existing approaches, since it is to find a sketch first. The main contributions of this paper are summarized as follows.

- We revisit TSP with a set of points, Q, which has not been well discussed in the recent studies related to LBS, and explore new heuristics to find a better answer.
- We propose Sketch-First as one to find a sketch first followed by refinement of the sketch in a global-to-local manner. We discuss heuristics to select a subset of points from Q to form the sketch. In order to select a set of points which may end up such an optimal TSP, we propose three deletion based heuristics to remove points from Q by which the points left are in a better position to form the sketch. We find that such an order by deletion plays an important role to find a better answer efficiently.
- We discuss two Sketch-First algorithms when dealing with the refinement. We propose an algorithm that is to use a splice operation during the refinement and remove any possible intersection (or crossover) in the TSP tour following the fact that the optimal tour does not intersect itself if the triangle inequality is held [9].
- We have conducted extensive experimental studies. We compare our Sketch-First approaches with the existing constructive heuristics. We show that our approaches can give an answer with a better accuracy and with a shorter execution time.

The rest of the paper is organized as follows. Section 2 provides the background of TSP. In Sect. 3, we discuss the representative heuristics for solving TSP query. In Sect. 4, we highlight the main idea of Sketch-First and heuristics to select a subset of points to form the sketch. In Sect. 5, we show another Sketch-First algorithm with a new splice operation to refine insertion while inserting the points back to the sketch. In Sect. 6, we report our experiment results. We discuss the related works in Sect. 7, and conclude our paper in Sect. 8.

2 Preliminaries

Consider a set of points V in a 2-dimensional space. The distance between two points v_i and v_j in V is the Euclidean distance, denoted as $d(v_i, v_j)$, such that $d(v_i, v_j) = \sqrt{(v_i.x - v_j.x)^2 + (v_i.y - v_j.y)^2}$, where the two coordinates of a point v are $v.x$ and $v.y$. An edge-weighted complete undirected graph $G = (V, E)$ can be constructed for the set of points, where V is the set of nodes representing the points, and E is a set of edges for every pair of points (v_i, v_j) in V with an edge weight $d(v_i, v_j)$ (Fig. 2).

Let Q be a set of nodes of size $n = |Q|$ in V. A Hamilton path over Q is a path that visits every node exactly once, and a Hamilton circuit over Q is a Hamilton path that forms a circle with an additional visit from the last node to the first node in the Hamilton path. Both Hamilton path and Hamilton circuit can be regarded as a permutation of nodes (or points) in Q. Here a permutation π over Q is a one-to-one mapping. Below, we use $\pi_i(Q)$ to denote a node v at the i-th position in the permutation π, and for simplicity we use π_i to indicate a specific node v in Q at the i-th position. Note that a node can only appear at a specific position in a permutation. We indicate a permutation over Q as $T = (\pi_1, \pi_2, \cdots, \pi_n)$. Given a permutation T over Q, the distance of a Hamilton path by T is defined as $d(T) = \sum_{1 \leq i < n} d(\pi_i, \pi_{i+1})$, and the distance of a Hamilton circuit by T is defined as $d(T) = \sum_{1 \leq i < n} d(\pi_i, \pi_{i+1}) + d(\pi_n, \pi_1)$. Let $\mathcal{T}$ be the set of all possible paths (circuits) for Q. The size of $\mathcal{T}$ is $|\mathcal{T}| = n!$ for Hamilton paths, and the size is $|\mathcal{T}| = \frac{(n-1)!}{2}$ for Hamilton circuits. In this paper, we focus on Hamilton circuit, and use "circuit", "tour" and "route" interchangeably. Among all possible permutations in $\mathcal{T}$, the optimal Hamilton circuit over Q is the shortest Hamilton circuit, denoted as T^*, such that $d(T^*) = \min_{T \in \mathcal{T}} d(T)$. The problem of finding the optimal Hamilton circuit is known as Traveling Salesman Problem (TSP), which is known to be NP-hard. The error-ratio for an approximate T is defined as $\frac{d(T) - d(T^*)}{d(T^*)}$. It is worth mentioning that the TSP problem we study in this paper is the symmetric and metric TSP. Here, by symmetric it implies $d(v_i, v_j) = d(v_j, v_i)$, and by metric it implies $\forall v_i, v_j, v_k \in V, d(v_i, v_j) + d(v_j, v_k) \geq d(v_i, v_k)$.

The Problem: In this paper, we compute TSP for a TSP query over the user-given set of points, Q.

In the following, in computing TSP over Q, we have to deal with a subset of points $P (\subset Q)$. We call a Hamilton path/circuit for P a partial path/circuit for Q.

Figure 1 shows an example of a TSP query Q with 12 points. The optimal tour for Q is $T^* = (v_1, v_2, v_3, v_4, v_5, v_6, v_7, v_8, v_9, v_{10}, v_{11}, v_{12})$, which is shown with black solid lines, where $d(T^*) = 3,018.11$.

3 Heuristics

One reason for the hardness of computing TSP is that the local optimum property does not hold, which is of importance to many combinatorial optimization problems. This is known as *knowing all optimal solutions does not help for TSP*

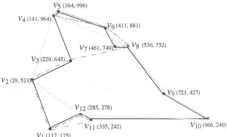

Heuristics	Expansion order	Run time	Approximate ratio
NN	from-near-to-far	fast	$O(\log n)$
RI	random	medium	$O(\log n)$
NI	from-near-to-far	slow	2
CI	from-near-to-far	slow	2
FI	from-far-to-near	slow	1.5
NA	from-near-to-far	medium	2
FA	from-far-to-near	medium	-

Fig. 1. A TSP example   **Fig. 2.** Properties of heuristics

reoptimization [10]. There are many reported heuristics in the literature to compute an approximate answer. We discuss 7 representative heuristics using the same example given in Fig. 1. The results of them are shown in Table 1.

The Nearest Neighbor Heuristic (NN) [11]: As the simplest and fastest heuristic to solve TSP query, it randomly picks a point as the start point, and then expands the circuit to visit one unvisited point, which is the nearest neighbor to current end, in every iteration.

The Insertion Methods [8]: Like the nearest neighbor heuristic, the insertion methods start from a randomly selected point and add unvisited points to current solution one by one. Unlike the nearest neighbor heuristic which generates a partial path in every iteration, the insertion methods find a partial circuit in every iteration by taking the length increment of inserting a new point into consideration. Here, the increment of the total length of the tour when adding a point v into an edge (v_i, v_j), as the cost of this insertion is given as Eq. (1).

$$c(v_i, v_j, v) = d(v, v_i) + d(v, v_j) - d(v_i, v_j) \tag{1}$$

The insertion methods are to insert points between two points (an edge) with minimum cost. Let the current partial circuit be T_p, and assume v is the next point for insertion. The insertion methods insert v between v_i^* and v_j^* in T_p following Eq. (2).

$$(v_i^*, v_j^*) = \operatorname*{argmin}_{(v_i, v_j) \in T_p} c(v_i, v_j, v) \tag{2}$$

There are 4 strategies to choose the next point v to be inserted.

– **Random Insertion (RI)**: It picks the next point randomly.
– **Nearest Insertion (NI)**: It chooses $v^* = \operatorname{argmin}_{v \notin T_p} \{d(v, v_i), \forall v_i \in T_p\}$.
– **Cheapest Insertion (CI)**: It picks $v^* = \operatorname{argmin}_{v \notin T_p} \{c(v_i, v_j, v), \forall (v_i, v_j) \in T_p\}$.

Table 1. Heuristics, Expansion order, and the Result

Heuristic	Start	1	2	3	4	5	6	7	8	9	10	11	Circuit	Length
NN	v_1	v_{12}	v_{11}	v_2	v_3	v_7	v_8	v_6	v_5	v_4	v_9	v_{10}	$(v_1, v_{12}, v_{11}, v_2, v_3, v_7, v_8, v_6, v_5, v_4, v_9, v_{10})$	3674.52
RI	v_1	v_2	v_3	v_4	v_5	v_6	v_7	v_8	v_9	v_{10}	v_{11}	v_{12}	$(v_1, v_2, v_4, v_5, v_6, v_8, v_7, v_{10}, v_9, v_3, v_{12}, v_{11})$	3593.27
NI	v_1	v_{12}	v_{11}	v_2	v_3	v_7	v_8	v_6	v_5	v_4	v_9	v_{10}	$(v_1, v_2, v_3, v_4, v_5, v_6, v_7, v_8, v_9, v_{10}, v_{12}, v_{11})$	3099.69
CI	v_1	v_{12}	v_{11}	v_2	v_3	v_7	v_8	v_6	v_9	v_4	v_5	v_{10}	$(v_1, v_2, v_3, v_4, v_5, v_6, v_7, v_8, v_9, v_{10}, v_{12}, v_{11})$	3099.69
FI	v_1	v_{10}	v_5	v_4	v_2	v_6	v_3	v_9	v_{11}	v_{12}	v_8	v_7	$(v_1, v_2, v_3, v_4, v_5, v_6, v_7, v_8, v_9, v_{10}, v_{11}, v_{12})$	3018.11
NA	v_1	v_{12}	v_{11}	v_2	v_3	v_7	v_8	v_6	v_5	v_4	v_9	v_{10}	$(v_1, v_2, v_3, v_4, v_5, v_6, v_7, v_8, v_9, v_{10}, v_{12}, v_{11})$	3099.69
FA	v_1	v_{10}	v_5	v_4	v_2	v_6	v_3	v_9	v_{11}	v_{12}	v_8	v_7	$(v_1, v_8, v_9, v_{11}, v_{12}, v_5, v_4, v_2, v_6, v_3, v_7, v_{10})$	5440.71

– **Furthest Insertion (FI) [12]:** It picks up points which are far away from each other in a higher priority. It chooses the point $v^* = \mathrm{argmax}_{v \notin T_p}\{d(v, v_i), \forall v_i \in T_p\}$.

The Addition Methods [13]: For the point v^* to be added, the addition methods pick its nearest neighbor v_i^* already on the current circuit T_p, and insert v^* into the position between v_i^* and one of its neighbors, v_i and v_j on T_p. The addition methods calculate $c(v_i^*, v_i, v^*)$ and $c(v_i^*, v_j, v^*)$ following Eq. (1) and choose the smaller one. There are 2 strategies to select the next point v^*: **Nearest Addition (NA)** and **Furthest Addition (FA)**. They take the identical insertion orders of points as **NI** and **FI**, respectively.

4   A Sketch-First Approach

Following human cognition [14,15], TSP can be solved in a global sketch first manner, followed by local refinement. In this way, a sketch of Q can be an optimal TSP for a small number of points in Q that outlines the big picture on the tour to be, and the remaining points are inserted into the global sketch to obtain a TSP answer for Q. In other words, the Sketch-First approach we study in this paper is to find a small subset of Q as the sketch to get the global picture followed by local refinement. In the following, we use Q_g and Q_l for the sketch and for the remaining points, respectively, such that $Q = Q_g \cup Q_l$ where $Q_g \cap Q_l = \emptyset$. We call Q_g and Q_l a sketch and a local-set.

Given a TSP query Q, assume Q_g can be computed as the sketch of Q. The first question is whether we need to compute the optimal TSP for Q_g. Our answer is positive since the points in Q_g decide the sketch of TSP based on which the remaining points in local-set are inserted to find the answer for Q. In terms of the cost to compute such an optimal TSP for Q_g, it is affordable if the size of Q_g is small.

Algorithm 1. Sketch-First (Q)

Input: Q: a TSP query of a set of points
Output: T: the answer for Q
1 **begin**
2   Select Q_g out of Q ;
3     $Q_l = Q \setminus Q_g$;
4   Compute optimal tour T_g^* for Q_g ;
5   **for** $v \in Q_l$ **do**
6 | Insert v into T_g^* by local refinement;
7   **end**
8     $T = T_g^*$;
9 **end**

Our algorithm is shown in Algorithm 1. First, it selects Q_g out of Q to construct the sketch (line 2), Let Q_l be $Q \setminus Q_g$ (line 3). In line 4, it finds the optimal tour, i.e. Sketch for Q_g. Finally, it inserts all the points in Q_l (line 6) by local refinement.

For Sketch-First framework, we will insert points in Q_l with local refinement. In order to search for a better sketch efficiently, we consider the heuristics mentioned in Sect. 3 and give two properties based on the insertion methods.

Observation 1. The further a point is to the current circuit, the larger error the insertion may introduce.

We explain it for a current partial circuit T with a new v to be inserted. The closer v is to the edge to be inserted, denoted as (v_i, v_j), the smaller error it might introduce. Assume there are two points, v_1 and v_2, that should be inserted into the same edge (v_i, v_j) in the partial circuit T, according to an insertion strategy. Let the two new circuits be T_1 and T_2 for v_1 and v_2, respectively. With a cost function, assume $c(v_i, v_j, v_1) < c(v_i, v_j, v_2)$ and $d(T_1) \leq d(T_2)$, the upper bounds of error rate for the two circuits are: $\frac{d(T_1)}{d(T)} = \frac{d(T)+c(v_i,v_j,v_1)}{d(T)}$ and $\frac{d(T_2)}{d(T)} = \frac{d(T)+c(v_i,v_j,v_2)}{d(T)}$. The former one is smaller. Moreover, suppose that $T = (v_1, v_2, ..., v_s)$ is the optimal TSP for current point set, and v is the new point to be inserted. Let a new circuit be T' by inserting v into an edge $(v_k, v_{k+1}) \in T$ with $c(v_k, v_{k+1}, v) = \min_{1 \leq i < s} c(v_i, v_{i+1}, v)$. We have a property (Property 1), which suggests that the smaller $c(v_k, v_{k+1}, v)$ is, the higher probability that T' is optimal. Due to space limit, we omit all the proofs in this paper.

Property 1. If $\forall v_i, v_j \in T$, $c(v_k, v_{k+1}, v) \leq c(v_i, v_j, v)$, $T' = (v_1, ..., v_k, v, v_{k+1}, ..., v_s)$ is optimal.

Observation 2. For all the optimal partial circuits with size s, the longer the circuit is the more information it carries.

This is obvious. For the subset $Q_g \subset Q$ where $|Q_g| = s$, if the points in Q_g are concentrated, the optimal tour connected will be short, and it only carries some limited local information. On the contrary, it will be long and reflects the global distribution. Next, consider the length of a circuit. Let the optimal answer be T^* for Q. Suppose there are subsets Q_1 and Q_2 of Q, where $|Q_1| = |Q_2|$, and suppose the optimal circuits as T_1^* and T_2^*, respectively, such that $d(T_1^*) < d(T_2^*)$. As the length of T^* is a certain number, the points in $Q - Q_2$ has higher probability to be near to T_2^*.

In general, if a point is far away from the other points, its position is more important in the Sketch. On the other hand, if a point is near enough to other points, it is easy to find the position to insert this point. We explain why we cannot use the furthest insertion (**FI**) to find the sketch for Q, using an example.

Example 1. As shown in Table 1, consider the 3rd iteration of the furthest insertion (**FI**), it selects v_4 since $d(v_4, v_{10})$ is the largest among all pairwise distances. However, v_4 is near to v_5 which is on the current partial circuit.

This example suggests that by the furthest insertion (**FI**), it does not necessarily mean it can be a good sketch since it may be close to some points on the partial circuit. Our Sketch-First approach is to avoid such problem. It is important to mention that we use a local approach to find the sketch, and find a sketch by deletion. Intuitively, if a point is near to many other points, it means we can insert it back easily based on Observation 1, and the cluster of local points can be used to get a better answer. Therefore, we do not consider the point in the sketch.

Algorithm 2. select(Q, s)

Input: Q: a TSP query of a set of points
$\quad\quad\quad$ s: the size constraint of Q_g
Output: Q_g: the sketch of Q
1 **begin**
2 $\quad$ Initialize $Q_g = Q$ and $Q_l = \emptyset$;
3 $\quad$ **while** $|Q_g| > s$ **do**
4 $\quad\quad$ $v^* \leftarrow \text{argmin}_{v \in Q_g}\, d(v, Q_g - \{v\})$;
5 $\quad\quad$ Remove v^* from Q_g ;
6 $\quad\quad$ Append v^* into Q_l ;
7 $\quad$ **end**
8 **end**

Algorithm 2 shows how it works to select Q_g with size constraint s. There are two inputs: the TSP query Q and the size constraint s. In line 2, we initialize Q_g to be Q and Q_l to be an empty list to keep all the removed points. We keep removing points from Q_g until it reaches the size constraint s. In every iteration, we find a point v which is the nearest to others in Q_g, delete v from Q_g, and append it to Q_l. The s points left in Q_g form the sketch of Q. We discuss $d(v, Q_g - \{v\})$ (line 4), which is the distance from one point to a set of points. To measure whether a point is near to a set of points, we propose 3 different heuristics.

Nearest Neighbor Deletion (NND): Following the idea of the nearest insertion, we define $d(v, V)$ in Eq. (3), where v is a point and V is a set of points.

$$d(v, V) = \min_{v' \in V} d(v, v') \qquad (3)$$

Minimum Cost Deletion (MCD): We define the distance based on cost like the cheapest insertion heuristics (**CI**), as shown in Eq. (4).

$$d(v, V) = \min_{v_i, v_j \in V, v_i \neq v_j} c(v_i, v_j, v) \qquad (4)$$

Two Nearest Neighbor Deletion (2NND): Comparing with **NND**, **MCD** takes the cost of inserting v into a possible circuit for V. In order to get the smallest cost, **MCD** needs to check all possible pairs. Suppose the value of $d(v, V)$ is the cost for inserting v into V, and v_i, v_j are corresponding points. When v is to be inserted back into the sketch as a point in the local-set Q_l, it may not lead to a better answer since at that stage v_i and v_j might not be connected (next to each other) in the sketch to be. Therefore, we relax the condition of **MCD** for getting a better answer. As given in Eq. (5), the distance is defined by considering the shortest path for combinations of 3 points, where v is the middle one. In other words, it is the sum of distances from v to its 2 nearest neighbors, v_i and v_j.

$$d(v, V) = \min_{v_i, v_j \in V, v_i \neq v_j} (d(v_i, v) + d(v_j, v)) \qquad (5)$$

For the expanding order or the choice of next point to be inserted, as an alternative, we can insert points that are not in the sketch back to the sketch according to the existing heuristics in Sect. 3. In our approach, the order of removing points from the sketch Q_g is the order for us to insert them back.

Example 2. Consider Algorithm 2. With **NND**, the nearest neighbor for v_1 is v_{12} with $d(v_1, Q - \{v_1\}) = d(v_1, v_{12}) = 197.06$. Following Algorithm 2, the first point to be removed from the sketch with the minimum distance is either v_4 or v_5, with distance 39.41. With **MCD**, we need to compute the cost for $12 * 11 * 10 = 1,320$ triples. v_3 has the minimum value with $d(v_3, Q - \{v_3\}) = c(v_4, v_{11}, v_3) = 0.01$. Here, v_3 will be removed from Q_g and added to Q_l. With **2NND**, the minimum distance from v to $Q - \{v\}$ is v_7 with $d(v_7, Q - \{v_7\}) = d(v_6, v_7) + d(v_7, v_8) = 216.21$.

5   Splice-Based Insertion

In this work, we explore to insert the points in Q_l into the sketch following the reverse order of the points being appended into Q_l. There is a new issue. When we remove a point v from Q_g (**2NND**), the point v is removed regarding its 2 nearest neighbors v_i and v_j. The best is that the error introduced is small by inserting v back between v_i and v_j on the partial circuit. However, in this setting, the two points v_i and v_j may not be next to each other on the partial circuit. Connecting (v, v_i) and (v, v_j) might lead to an invalid or bad tour. To address this issue, we give a new algorithm SBSF for Splice-Based Sketch-First in Algorithm 4 following the same main idea given in Sketch-First (Algorithm 1). In brief, as shown in Algorithm 4, we compute the sketch of size s for Q, and then insert the points following the reverse order of deletion. In every iteration, when v_i and v_j are not connected in current partial circuit, we apply a new operation called "splice" (Algorithm 3). The splice inserts v by reconnecting the points in the partial circuit, and the resulting partial circuit is a Hamilton circuit.

Algorithm 3. splice(T_p, v, v_i, v_j)

    Input: T_p: a Hamilton circuit for a set of point V
               v: a new point for insertion
               v_i, v_j: points in T_p
    Output: T': a Hamilton circuit for $V \cup \{v\}$, with v between v_i and v_j

1 **begin**
2   **if** $(v_i, v_j) \in T_p$ **then**
3   Remove (v_i, v_j) from T_p ;
4         $T' =$ connect (v_i, v) and (v_j, v) ;
5   **end**
6   **else**
7   Fix the order for T_p ;
8         $v_i^*, v_j^* =$ the predecessors of v_i, v_j in T_p ;
9   Remove edge (v_i^*, v_i) and (v_j^*, v_j) from T_p ;
10       $T' =$ connect (v_i^*, v_j^*), (v_i, v) and (v_j, v) ;
11   **end**
12 **end**

There are 4 inputs in Algorithm 3. T_p is the current circuit with point set V, and v is the new point to be inserted. Here, v_i and v_j are the two nearest neighbors of v in V. If (v_i, v_j) is next each other, as an edge appearing in the current circuit in T_p, we can simply insert v between them (line 4), since the resulting circuit is still a Hamilton circuit for $V \cup \{v\}$. Otherwise, we fix an order for T_p, such as clockwise or counterclockwise (line 7). Based on the order, we pick up one neighbor for v_i and v_j each, as v_i^* and v_j^* (line 8). By deleting (v_i, v_i^*) and (v_j, v_j^*) from T_p and adding (v_i^*, v_j^*) into T_p, v_i and v_j will be two endpoints of a Hamilton path for V. Connecting (v_i, v) and (v_j, v) will form the Hamilton circuit for $V \cup \{v\}$. After each splice operation, we will detect and remove crosses, in order to decrease the length of circuit.

Example 3. Assume $T_p = (v_1, v_2, v_4, v_5, v_6, v_{10})$ and the next point to be inserted is v_3, with 2 nearest neighbors as v_2 and v_6, which are are not next to each other on T_p. Inserting v_3 between v_2 and v_6 will introduce an invalid tour. Assume we adopt the clockwise order for T_p. The predecessors for v_2 and v_6 will be v_1 and v_5. After the deletion and addition step, we can get a new circuit as $(v_1, v_5, v_4, v_2, v_3, v_6, v_{10})$.

The splice operation helps to obtain a new circuit with one more point based on existing partial circuit. But, it may lose some accuracy, since it greedily finds the optimal choice for v and sacrifices the order of the points already in the circuit. Theorem 1 shows the incremental length of splice is not larger than $2(d(v_i, v) + d(v_j, v))$.

Theorem 1. *For Algorithm 3, $d(T') - d(T_p) \leq 2(d(v_i, v) + d(v_j, v))$.*

Theorem 1 demonstrates if v is near to current circuit T_p, the length of the new circuit generated by splice is not large. If T_p is optimal for V, we can get an approximate circuit for $V \cup \{v\}$ by simply applying Algorithm 3.

Corollary 1. *If T_p is the optimal circuit for V, T' is an approximate answer for $V \cup \{v\}$ with approximation ratio as $1 + \frac{2(d(v_i,v)+d(v_j,v))}{d(T_p)}$.*

Theorem 1 and Corollary 1 suggest that, if a point is near to other points, we can put it aside and consider others first. After getting the circuit for other points, it can be safely inserted to the right position. It is the main idea of our Sketch-First approach.

Algorithm 4. SBSF (Q, s)

Input: Q: a TSP query of a set of points
       s: the size constraint of Q_g
Output: T: an approximate answer for Q
1 **begin**
2   **if** $|Q| \leq s$ **then** $T =$ the optimal answer for Q ;
3   ;
4   let $\mathcal{L}$ be a priority queue ;
5   **for** each v in Q **do**
6   v_i and $v_j =$ the 2 nearest neighbors for v in Q ;
7   Insert $(v, d(v_i, v) + d(v_j, v))$ into $\mathcal{L}$;
8   **end**
9    $v = \mathcal{L}.\text{top}()$;
10   Let v_i and v_j be the corresponding neighbors of v ;
11   $T' =$ SBSF $(Q \setminus \{v\}, s)$;
12   $T = \text{splice}(T', v, v_i, v_j)$;
13   Remove cross for T ;
14 **end**

Algorithm 4 finds an approximate answer for a TSP query Q recursively. If the size of current query is less than or equal to the size constraint s, these points will form the sketch, and we should get the accurate answer for them (line 2). Otherwise, it uses the distance of **2NND** to find the next point to be removed. We remove v from Q and call this function recursively (line 10). This step decomposes Algorithm 2 to facilitate the insertion. T' will be a partial circuit for $Q - \{v\}$, and we will insert v back. As proved in Theorem 1, we can use splice operation to get a new Hamilton circuit with quality guarantee (line 11). At last, we will remove the crosses in current circuit (line 12).

Example 4. Reconsider Q in Fig. 1. We find the 2 nearest neighbors for each point and calculate corresponding distances. Among all the triples, (v_6, v_7, v_8) has the shortest length. Then, we remove v_7 from Q, and execute the algorithm recursively. Suppose $s = 6$, we will keep deleting points in the following order: v_{12}, v_5, v_6, v_3, and v_1. The left point set is $\{v_2, v_4, v_8, v_9, v_{10}, v_{11}\}$. Here, the optimal TSP for the left points in the Sketch is $(v_2, v_4, v_8, v_9, v_{10}, v_{11})$. Then we will insert those removed points back in the reverse order. The first one to be inserted back is the last one being removed, v_1, which we can directly get the new circuit as $(v_1, v_2, v_4, v_8, v_9, v_{10}, v_{11})$. The next one to be considered is v_3 with its two nearest neighbors in current circuit as v_2 and v_8. v_2 and v_8 are not connected, by splice operation, the new circuit should be $(v_1, v_2, v_3, v_8, v_4, v_9, v_{10}, v_{11})$. We can further remove the cross of (v_3, v_8) and (v_4, v_9).

6 Experiment

We have conducted extensive experiments on a PC with two Intel Xeon X5550@2.67 GHz CPU and 48 GB main memory. The algorithms are implemented in C++. Like previous works [8,13], we generate our queries by sampling from real datasets: we crawled millions of real POIs of two cities, Beijing (BJ) and New York (NY). As the complexity of a TSP query is only related to the query size n, we generate TSP queries by randomly sampling points from the two datasets. All experiments are repeated 100 times and the average value is reported. For heuristics listed in Sect. 3, we randomly pick the start point. We do not report the performance for the addition methods, which cannot outperform the others as confirmed in [13] and our experimental studies.

Effect of approximate algorithms: We compare two Sketch-First algorithms. One is Algorithm 1. We denote it as **SF**. It is important to note that by **SF** we insert the points in Q_l back to the Sketch T_g^* following the reverse order obtained, and insert every point at the best position following the cost function given in Eqs. (1) and (2). The other is **SBSF** (Algorithm 4) using the splice. We set $s = |Q_g| = 3$. Note that when the size of sketch is 3, there is no need to compute the optimal TSP since any permutation will lead to the same triangle. For both **SBSF** and **SF**, we take **2NND** as discussed in Sect. 4 to delete points from the set to form Q_g. We compare **SBSF** and **SF** with **NN**, **RI**, **NI**, **CI**, and **FI** by randomly selecting points from 20 to 100. **SF** finds the best results for BJ (Fig. 3(a)). For NY, as shown in Fig. 3(b), **FI** is the best while **SF** is different from **FI** marginally. **SBSF** is more accurate than **NI**, **CI**, **NN**, and worse than **FI**.

Efficiency of approximate algorithms: We compare **SBSF** and **SF** with the other heuristics as well as the exact algorithm (**Concorde**), which is the state-of-the-art technique. The results are shown in Fig. 3(c) and (d). **Concorde** takes much longer processing time. Here, **NN** is the fastest followed by **RI**. Our **SBSF** is next fastest, as it only follows the deletion order and does not spend time on finding the insertion position. **SF** is marginally slower than **SBSF**. **SF** becomes faster than **SBSF** when n is larger than 60. In conclusion, **NN** is the fastest

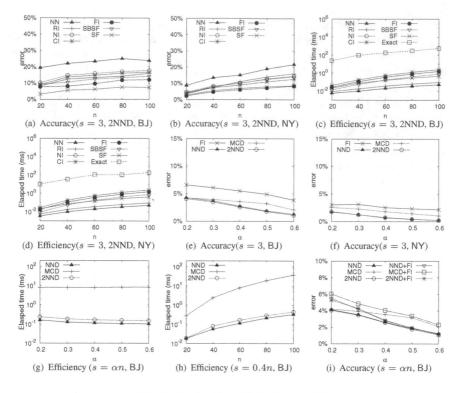

Fig. 3. Evaluation (Default: $n = 60$)

approximate algorithm with high error rate. **SBSF** finds a result with less error rate and with less execution time than **CI** and **NI**. **SF** finds tours with highest quality in relatively short time.

Selection of Q_g: We compare the 3 methods, **NND**, **MCD**, and **2NND** used in **SF** (Algorithm 2), with **FI** which can be considered as a way to catch the global sketch by selecting the furthest s points. Figure 3(e) and (f) illustrate the accuracy of answers by them. **2NND** produces the shortest tour. The results confirm that **SF** (Algorithm 2) can find the set of points to form a sketch that helps to generate approximate answers with high quality. Our approaches outperform **FI**. In Fig. 3(g), we show the efficiency of **SF** (Algorithm 2) by varying the value of α, where $n = 60$. The **MCD** takes the longest time to delete $(1-\alpha)n$ points. **2NND** considers one more nearest neighbor than **NND**, and therefore takes a little bit more time than **NND**. By increasing α, the execution time for the 3 methods do not change much. In Fig. 3(h), we vary the value of n, where $\alpha = 0.4$. Still, **MCD** runs much longer than the other two heuristics. When $n = 80$, **MCD** takes nearly 20 ms while the other two only takes about 0.2 ms. Comparing with Fig. 3(c), **MCD** is slower than cheapest insertion.

The insertion order: SF (Algorithm 2) finds the sketch for Q, and inserts the remaining points following the reverse order of the points being deleted. Alternatively, we can insert the remaining points following **FI**, since our algorithm processes points in a manner from far to near. We compare **NND, MCD**, and **2NND**, which use the reverse order the the points being deleted, with the **NND+FI, MCD+FI**, and **2NND+FI**, which use **FI** as the method to insert the remaining points back. Figure 3(i) shows the accuracy by varying α. In general, the reverse order of deletion is better than the order of furthest insertion, which catches the global sketch. **2NND** outperforms the others.

7 Related Works

Since 1930s, numerous works have been proposed to accelerate the processing of TSP, or improve the accuracy of approximate answers [16].

In operation research, TSP is modeled as integer programming, which can be further relaxed as linear programming. The state-of-the-art exact solution, Concorde, is based on linear programming. By using randomization, [17] finds $(1 + \frac{1}{c})$-approximate answer in $O(n(\log n)^{O(c)})$ time, for every fixed $c > 1$. It is known as the best theoretical result. Ant colony algorithm and genetic algorithm are also adopted to improve accuracy.

In database area, researchers prefer to combine TSP with practical applications. [5–7] propose 3 different planning problems which can be reduced to generalized TSP, a variant of normal TSP. There are also queries that aim at minimizing overall distance, with different objective functions. [2] surveys the works of finding the shortest path in static networks. Finding the optimal meeting point is also an interesting topic [4].

8 Conclusion

In this paper, we revisited the traveling salesman problem, and focus on TSP queries with a given set of points Q. We propose a new Sketch-First heuristics by Sketch-First, we select a subset of Q, denoted as Q_g, forming the sketch of Q. For the points in Q_l ($= Q \setminus Q_g$), we add them back to the sketch by local refinement. For selecting Q_g, we discussed three deletion-based heuristics. For inserting points in Q_l back to the sketch, we discussed two algorithms. Both algorithms adopt the reverse order of the points being removed from the consideration of the sketch. The idea is to start from the optimal tour for a small set of points, and to insert an additional point to the previous obtained tour, which is supposed to be better, one by one, iteratively. The reversed order of inserting is shown to be effective to improve the accuracy. In our extensive experimental studies, we confirm that our approach is effective and efficient.

Acknowledgment. This work was supported by grant of the Research Grants Council of the Hong Kong SAR, China 14209314.

References

1. Zhu, A.D., Ma, H., Xiao, X., Luo, S., Tang, Y., Zhou, S.: Shortest path, distance queries on road networks: towards bridging theory and practice. In: SIGMOD, pp. 857–868 (2013)
2. Sommer, C.: Shortest-path queries in static networks. ACM Comput. Surv. **46**(4), 1–31 (2014)
3. Xu, Z., Jacobsen, H.: Processing proximity relations in road networks. In: SIGMOD, pp. 243–254 (2010)
4. Yan, D., Zhao, Z., Ng, W.: Efficient algorithms for finding optimal meeting point on road networks. PVLDB **4**(11), 968–979 (2011)
5. Li, F., Cheng, D., Hadjieleftheriou, M., Kollios, G., Teng, S.-H.: On trip planning queries in spatial databases. In: Medeiros, C.B., Egenhofer, M., Bertino, E. (eds.) SSTD 2005. LNCS, vol. 3633, pp. 273–290. Springer, Heidelberg (2005)
6. Sharifzadeh, M., Kolahdouzan, M.R., Shahabi, C.: The optimal sequenced route query. VLDB J. **17**(4), 765–787 (2008)
7. Cao, X., Chen, L., Cong, G., Xiao, X.: Keyword-aware optimal route search. PVLDB **5**(11), 1136–1147 (2012)
8. Rosenkrantz, D.J., Stearns, R.E., Lewis II, P.M.: An analysis of several heuristics for the traveling salesman problem. SIAM J. Comput. **6**(3), 563–581 (1977)
9. Deineko, V.G., van Dal, R., Rote, G.: The convex-hull-and-line traveling salesman problem: a solvable case. Inf. Process. Lett. **51**(3), 141–148 (1994)
10. Böckenhauer, H.-J., Hromkovič, J., Sprock, A.: Knowing all optimal solutions does not help for TSP reoptimization. In: Kelemen, J., Kelemenová, A. (eds.) Computation, Cooperation, and Life. LNCS, vol. 6610, pp. 7–15. Springer, Heidelberg (2011)
11. Bellmore, M., Nemhauser, G.L.: The traveling salesman problem: a survey. Oper. Res. **16**(3), 538–558 (1968)
12. Nicholson, T.: A sequential method for discrete optimization problems and its application to the assignment, travelling salesman, and three machine scheduling problems. IMA J. Appl. Math. **3**(4), 362–375 (1967)
13. Bentley, J.L.: Fast algorithms for geometric traveling salesman problems. Informs J. Comput. **4**(4), 387–411 (1992)
14. MacGregor, J.N., Ormerod, T.: Human performance on the traveling salesman problem. Percept. Psychophysics **58**(4), 527–539 (1996)
15. Macgregor, J.N., Ormerod, T.C.: Evaluating the importance of the convex hull in solving the euclidean version of the traveling salesperson problem: reply to lee and vickers. Percept. Psychophysics **62**(7), 1501–1503 (2000)
16. Gutin, G., Punnen, A.P.: The Traveling Salesman Problem and Its Variations. Springer Science & Business Media, New York (2002)
17. Arora, S.: Polynomial time approximation schemes for euclidean traveling salesman and other geometric problems. J. ACM **45**(5), 753–782 (1998)

Finding Least On-Road
Travel Time on Road Network

Lei Li, Xiaofang Zhou$^{(\boxtimes)}$, and Kevin Zheng

School of ITEE, The University of Queensland,
St. Lucia, Brisbane, QLD 4072, Australia
l.li3@uq.edu.au, {zxf,kevinz}@itee.uq.edu.au

Abstract. Shortest path and fastest path query on time-dependent road network are widely used nowadays, but none of them can answer a least on-road travel time query, which aims to find a path between two vertices on a time-dependent road network that has the minimum on-road travel time(waiting on any vertex is allowed). In this paper, we propose a cheapest path algorithm which expands *Dijkstra's Algorithm* to solve this problem. The time complexity of it is $O(|V|\log|V| + |V|T + |E|T^2)$, where T is the number of the involving time unit, $|V|$ is the number of vertices and $|E|$ is the number of edges. Extensive experiments are conducted on the two different speed profiles to test the performance of our cheapest path algorithm. The results validate the effectiveness of our work.

1 Introduction

Thanks to the fast development of the mobile network, free map services and the widespread using of GPS devices, traveling on earth has never been easier. A variety of navigation applications can direct people to travel by bike, car or public transportation system to wherever they want based on their current locations. What behind these navigation applications are the path algorithms. In the past decades, people have developed several path algorithms aiming at different using scenarios. For example, shortest path algorithms [1,2] can find a path with the smallest geographical distance between two places with no consideration of travel time, while the fastest path algorithm can return a path that has the least travel time if we depart instantly. If the departure time is an interval(i.e. departing is allowed in the coming time period), the interval fastest path algorithms [3–12] can return an optimal path that has the shortest total travel time. As for the travels by transportation system, earliest arrival path and latest departure path [13,14] are two important paths that bound the total travel time.

However, none of the above path algorithms can support the following application: From a logistics company's perspective, the distance and the travel time between two courier locations are less important than the fuel consumption on the road, which is a critical operating cost. And it is a common sense that the time spent on road is dominating the fuel consumption. So reducing the on-road travel time is more important in this scenario. To reduce the on-road travel time,

© Springer International Publishing AG 2016
M.A. Cheema et al. (Eds.): ADC 2016, LNCS 9877, pp. 137–149, 2016.
DOI: 10.1007/978-3-319-46922-5_11

the driver can pull up his car at the depot area, roadside resting area or public parking lot to wait for the traffic condition to become better. For example, as shown in Fig. 1, the travel cost between the vertices are functions of time. Suppose the starting time is 2:00, then the fastest path is $v_1 \rightarrow v_2$ at 2:00 and $v_2 \rightarrow v_3$ at 2:03, and its total time is 6 min. However, if we wait on v_2 until 2:05, although we will arrive v_3 later on 2:07, which has a longer total travel time, this path has a shorter on-road travel time, which is 5 min. So by taking advantages of these parking possible places, the total on-road travel time can be reduced dramatically.

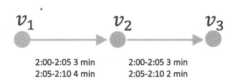

Fig. 1. Least on road travel time example

Assume a road network is organized as a graph together with the time-dependent traffic condition information, a query aiming to find a least on-road travel time path can be generalized as follows: Given a source vertex v_s with a departing time interval and a destination vertex v_d with a latest arrival time, a path $p_{s,d}$ can travel through a series of road each with time c_{t_i} at time t_i and waiting on vertex is allowed. Thus, the total on-road travel time is $T_{on-road} = \Sigma_{i=s}^{d} c_{t_i}$. Our query aims to find a path with the minimum $T_{on-road}$. We call this it the *least-on-road-travel-time(LRTT) problem*. It is different from the *store and forward* related problems [15] which focus on optimizing the flow rather than the individual vehicle. And it is different from all the other existing path problems due to the different minimizing objects and they all have longer on road travel time than a *LRRT* path.

The challenge of solving this problem is that we do not know if waiting on the current vertex will result in a shorter on road travel time or not. And if it does, how long should we wait? The shortest path algorithms cannot solve this problem because their cost functions on roads are static. As for the fastest path algorithms, waiting on vertices, as proved in [4], is insufficient, they will never return the least on road travel time path even if it returns the mth fastest path, where m can be as large as possible. In fact, it aims to find the optimal departure time on the starting vertex rather than on all the vertices. One possible naive approach using the current fastest path algorithm is to select the waiting time on each vertex greedily. First it selects the optimal departure time of the starting vertex using the fastest path algorithm, and the second vertex on this path along its arrival time are determined. Then it finds the optimal departure time from the second vertex, which may generate a different path from the previous one. So a third vertex with its arrival time is determined. This approach runs on and on repeatedly until the destination vertex is reached. Obviously it uses the fastest path algorithm multiple times, so its running time could be much longer. Although it may find a smaller on road travel time than just calling the fastest path only on the starting vertex, there is no guarantee that it can find the optimal solution since it is totally heuristic. Figure 2 shows an example of the comparison of our algorithm, the fastest path [4] and the naive approach. The example computes the paths from location {31.2414, 121.304} to {31.2559, 121.386} in Shanghai, whose shortest distance is 10 km. The starting time interval is set from 10:00 to 16:00 and the latest arrival time is 19:00. The fastest path can find a path with on-road travel time of 1385 s in 5 s. The repeated approach can find a path of 1130 s in 164 s, while our algorithm find a path of 986 s in 11 s. Although it takes longer time than the fastest path to find a path, the on-road travel time of our algorithm is the smallest.

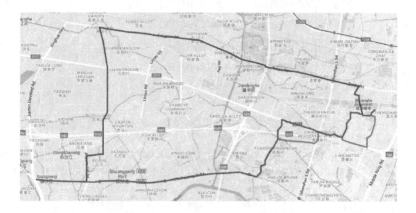

Fig. 2. Least on road travel time of our proposal, fastest path and recursive fastest path. Red line: Our algorithm; Blue line: Fastest path; Pink line: Naive repeated baseline (Color figure online)

Our contributions are listed as follows:

- We propose a cheapest path algorithm to solve the least-on-road-travel-time problem in $O(|V| \log |V| + |V|T + |E|T^2)$ time.
- We evaluate the effectiveness and efficiency of our cheapest path algorithm with extensive experiments.

The rest of the paper is organized as follows. Section 2 defines and analyzes the least on-road traffic time problem formally. Section 3 describes the cheapest path algorithm and proves its correctness and complexity. Experiment results are shown in Sect. 4. Section 5 reviews the related works and we conclude our paper in Sect. 6.

2   Problem Definition and Analysis

Any road network can be represented as a graph where the vertices represent intersections and the edges represent road segments. If we attach the time-dependent traffic condition information on each vertex, this graph is generalized to a time-dependent graph defined below:

Definition 1. *A (**Time-Dependent Graph**) is defined as $G_T(V, E, W)$, where $V = \{v_i\}$ is the vertex set, $E \subseteq V \times V$ is the directed edge set and W is a set of functions. For each edge $(v_i, v_j) \in E$, there is a function $W_{ij}(t) \in W$, where t is a time in time domain $\hat{T}$, that tells how much time it costs to travel from v_i to v_j at time t.*

A path from v_1 to v_k $p = ((v_1, v_2), (v_2, v_3)..., (v_{k-1}, v_k))$ can be decomposed into a series of path segments as shown below:

$$p_{1,2} : \begin{cases} Arrival(v_1) = ts_1, \\ Depart(v_1) = Arrival(v_1) + w_1, \\ Cost_{1,2} = W_{1,2}(Depart(v_1)) \end{cases}$$

$$p_{2,3} : \begin{cases} Arrival(v_2) = Depart(v_1) + Cost_{1,2}, \\ Depart(v_2) = Arrival(v_2) + w_2, \\ Cost_{2,3} = W_{2,3}(Depart(v_2)) \end{cases}$$

$$...$$

$$p_{k-1,k} : Arrival(v_k) = Depart(v_{k-1}) + Cost_{k-1,k}$$

where $Arrival(v)$ is the arrival time at v and $Depart(v)$ is the departure time from v. w_i is the waiting time spent on v_i. The on-road travel time $R_{p_{1,k}}$ is:

$$R_{p_{1,k}} = \Sigma_{i=1}^{k-1} Cost_{i,i+1} = \Sigma_{i=1}^{k-1} W_{i,i+1}(Depart(v_i))$$

Thus, among all the possible paths between v_1 and v_k starting from time interval $[t_{s_1}, t_{s_2}]$ and ending by t_k, there exists a path $p_{i,k}^*$ whose $R_{p_{1,k}^*}$ is no bigger than others'. We call the time on this path the *least on-road travel time*.

Definition 2. *(**Least On Road Travel Time (LRTT) Problem**). Given a Time-Dependent Graph G_T and $Q_{LRTT}(v_s, v_d, t_{s_1}, t_{s_2}, t_d)$, where v_s is the starting vertex, $[t_{s_1}, t_{s_2}]$ is the departing time interval on v_s, v_d is the destination vertex and t_d is the latest arrival time, the LRTT problem is to minimize the on road travel time*

$$R_{p_{s,d}^*} = min\{R_{P_{s,d}}\}$$

The difference between the *LRTT* problem and fastest path problem [3,4] is that *LRTT* allows waiting on any vertex and aims to minimize $\Sigma_{i=1}^{k-1} Cost_{i,i+1}$ which finds the optimal departure time of all the vertices in the time interval $[t_{s_1}, t_d]$, while the fastest path only allows waiting on the starting vertex and aims to minimize $Arrival(v_d) - Departure(v_s)$ which only has to find the optimal departure time of v_s in a much shorter time interval $[t_{s_1}, t_{s_2}]$.

3   Cheapest Path Algorithm

3.1   Overview

In this section, we propose a cheapest path algorithm to solve the $LRTT$ problem. Our algorithm is made up of three parts: *Active Time Interval Computation*, *Minimum Cost Function Update* and *Path Expansion*. The *Active Time Interval Computation* determines the sufficient time interval for each vertex. The time outside the active time interval is out of consideration. *Minimum Cost Function* records a minimum cost from v_s to other vertices at different arrival time within their active time intervals. From now on, it should be noted that the term *cost* means the travel time from one vertex to another on the edge level, and it also means the total on-road travel time on the path level. The update algorithm provides a fast way to update each vertex's minimum cost function. The *Path Expansion* is a *Dijkstra*-based algorithm that can find the minimum on-road travel time of all vertices, including the destination vertex.

Algorithm 1. Cheapest Path Algorithm

    Input: Time-dependent graph $G_T(V, E, C)$, LRTT query $Q(v_s, v_d, t_{s_1}, t_{s_2}, t_d)$
    Output: The least on road travel time $R_{p^*_{s,d}}$ and its corresponding path $p^*_{s,d}$

1 **begin**
2   //Active Time Interval Computation
      $EA(v_i) \leftarrow TemporalDijkstra(G_T, v_s, v_d, t_{s_1}, t_{s_2}, t_d))$ //earliest arrival time
3     $LD(v_i) \leftarrow ReverseTemporalDijkstra(G_T, v_d, v_s, t_d, t_{s1})$ //latest departure time
4
5     $R_{p^*_{s,d}} \leftarrow PathExpansion(G_T, v_s, v_d, t_{s_1}, t_{s_2}, t_d)$

3.2   Active Time Interval Computation

The active time interval computation is a *Dijkstra*-based single start-time fastest path algorithms. But the *Dijkstra algorithm* cannot be applied directly on a time-dependent graph since the costs on edges are dynamic so that a trip departing later on a road may arrive earlier than another earlier departing trip. However, unlike the general graphs, road network has an inherent *FIFO property*, which claims that a vehicle enters a road later than another earlier vehicle cannot leave this road earlier, if the take over is not allowed. Thus, the single start-time fastest path can be computed using *Dijkstra* directly [4].

    The active time interval is computed by two single start-time fastest paths. The lower boundary, which is actually the earliest arrival time from v_s, can be computed by a single start-time fastest path from v_s at earliest departure time t_{s_1}. The upper boundary, which is actually the latest departure time for each vertex, can be computed reversely from v_d at the latest arrival time t_d(the time travels backward). If the latest departure time is later than t_{s2}, we could just simply assign t_{s_2} to it. Since only two *Dijkstra* computations are needed, the time complexity is $O(|V| \log |V| + |E|)$.

3.3   Minimum Cost Function Update

Within each vertex v_i's active time interval, we define a *minimum cost function* $C_i(t)$ to record the minimum on-road time traveling from v_s to v_i that arrives v_i on time t. Such a cost function has a nice *non-increasing* property that can be used to convert the optimal waiting time selection problem to segment intersection problem. By non-increasing, we mean $\forall EA(v_i) \leq t_1 < t_2 \leq LD(v_i) \Rightarrow C_i(t_1) \geq C_i(t_2)$. The proof of this property is quite natural: suppose it takes larger cost if we arrive v_i on t_2 than another path that arrives on t_1, then we can simply take advantage of the latter path and wait on v_i from t_1 to t_2, which will reduce cost of the whole time interval $[t_1, t_2]$ to $C_i(t_1)$. The reason behind this is that waiting on a vertex will just increase the total travel time but not the on-road travel time. Then, our goal is to compute the final minimum cost function $C_i(t)$ for each vertex v_i and find the minimum cost of v_d.

Although all the functions we use here are a set of discrete linear functions, the consecutive functions share the same end points. So we only have to store all the end points in ascending order of time in their active time interval. Thus, the cost function of a vertex is organized as a point set $\{(t_0, C_i(t_0)), (t_1, C_i(t_1)), ..., (t_k, C_i(t_k))\}$. Initially, $C_i(t)$ is a function with a single value $C_i(EA(v_i))$ which is its earliest arrival cost, and it is stored in a set with two end points: $\{(EA(v_i), C_i(EA(v_i)), (LD(v_i), C_i(EA(v_i)))\}$. Then it would be updated by its in-neighbors. Suppose v_j is v_i's in-neighbor whose minimum cost function is $C_j(t)$ and the travel cost from v_j to v_i is $W_{j,i}(t)$. So the cost that travels from v_s to v_i via v_j is $g_{j,i}(t) = C_j(t) + W_{j,i}(t)$. Then, $C_i(t)$ can be updated by $g_{j,i}(t)$ if it has some smaller values than the original $C_i(t)$. Since our minimum cost function is non-increasing, it will always use the smaller value of the two functions as the new function. So the v_i's new minimum cost function is $C_i'(t) = min(C_i(t), g_{j,i}(t)), \forall t \in [EA(v_i), LD(v_i)]$.

However, since $g_{j,i}(t)$ is not necessarily non-increasing, it cannot be used directly. Thus, we refine $g_{j,i}(t)$ first to make it follows the non-increasing property. Suppose $(t_p, g_{j,i}(t_p))$ is the current visiting end point in $g_{j,i}(t)$. For the next point $(t_{p+1}, g_{j,i}(t_{p+1}))$, if $g_{j,i}(t_p) \geq g_{j,i}(t_{p+1})$, $(t_{p+1}, g_{j,i}(t_{p+1}))$ should be inserted into $g_{j,i}'(t)$ and acts as the current visiting point. If $g_{j,i}(t_p) < g_{j,i}(t_{p+1})$, then $(t_{p+1}, g_{j,i}(t_{p+1}))$ is discarded and the algorithm visits the next points until for some point $(t_q, g_{j,i}(t_q))$ that $g_{j,i}(t_p) \geq g_{j,i}(t_q)$. Then, we have to find the intersection point $(t_{q'}, g_{j,i}(t_p))$ of line $g_{j,i}(t) = g_{j,i}(t_p)$ and the line segment determined by $(t_{q-1}, g_{j,i}(t_{q-1}))$ and $(t_q, g_{j,i}(t_q))$. Both of points $(t_{q'}, g_{j,i}(t_p))$ and $(t_q, g_{j,i}(t_q))$ will be inserted into $g_{j,i}'(t)$. After all the points in $g_{j,i}(t)$ has been visited, the new $g_{j,i}'(t)$ is guaranteed to be non-increasing.

Before we can derive the new $C_i'(t)$ by finding the smaller value of two functions $min(C_i(t), g_{j,i}'(t))$, we introduce a technique to test if two line segments intersect. Since these two discrete functions are both made up of a series of consecutive line segments and non-increasing, the original minimum cost function update problem is converted to a line segment intersection problem, which keeps finding the intersections and uses the lower one as the result. Suppose there are two vectors denoted by points $p_1(x_1, y_1), p_2(x_2, y_2)$, and the original point of the vector is $(0,0)$, the cross product $p_1 \times p_2$ is $x_1 y_2 - x_2 y_1$. If $p_1 \times p_2 > 0$, then p_1 is clockwise from p_2. Otherwise, p_1 is counterclockwise from p_2. For a line segment (p_i, p_j) and point p_k, we define a function $Direction(p_i, p_j, p_k) = (p_k - p_i) \times (p_j - p_i)$ to tell which side of (p_i, p_j) that p_k is on. If $Direction(p_i, p_j, p_k) > 0$, then p_k is on the clockwise side of line segment (p_i, p_j), otherwise p_k is on the counter-clockwise side. For two line segments (p_1, p_2) and (p_1', p_2'), we can use the $Direction$ function to tell if they intersect each other or not. First we compute the following four direction indication values: $d_1 = Direction(p_1', p_2', p_1)$, $d_2 = Direction(p_1', p_2', p_2)$, $d_3 = Direction(p_1, p_2, p_1')$ and $d_4 = Direction(p_1, p_2, p_2')$. Suppose $p_1.x < p_2.x$ and $p_1'.x < p_2'.x$, then there are only 2 cases that these two line segments are intersected as shown in Fig. 3. For example, in case (a), if $d_1 > 0, d_2 < 0, d_3 < 0$ and $d_4 > 0$, (p_1, p') is the desired line segment. This approach can prune the situations that two segments do not intersect quickly. For example, if $d_1 > 0$ and $d_2 > 0$ or $d_1 > 0$ and $d_3 > 0$, we can safely draw the conclusion that these two line segments will not intersect.

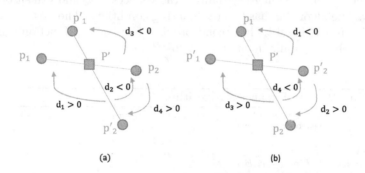

Fig. 3. Line segment intersection

With the technique introduced above, we can describe the update algorithm now. The algorithm works in a *sweeping-line* way. It visits the line segments in the $C_i(t)$ and $g'_{j,i}(t)$ together one by one to derive the new cost function $C_i'(t)$. First it retrieves the first line segments in $C_i(t)$ and $g'_{j,i}(t)$, and their corresponding end points (p_1, p_2) and (p_1', p_2'). Then it computes the corresponding d_1, d_2, d_3 and d_4. If $d_1 > 0, d_2 < 0, d_3 < 0$ and $d_4 > 0$, it is guaranteed that the line segments has a intersection point p' and line segment (p_1, p') should appear in $C_i'(t)$. If $d_1 < 0, d_2 > 0, d_3 > 0$ and $d_4 < 0$, the line segment (p_1', p') should appear in $C_i'(t)$. Such procedure runs on until all the end points are visited. Suppose the active time interval has T time units. In the worst case, there are T end points in the cost function. Within the update of each line segment, it only costs constant time. So the time complexity of the minimum cost function update algorithm is $O(T)$.

3.4   Path Expansion Algorithm

The path expansion algorithm uses a *Dijkstra* way to find the minimum on-road travel time from the source vertex to all the other vertices. Instead of using the "shortest distance" as the sorting key in the original *Dijkstra*, we use the minimum value of the each vertex's minimum cost function as the sorting key. The details are described in Algorithm 2.

Line 2 to line 4 initialize the cost functions of each vertex by adding the two end points $(v_i.EATime, v_i.EATime - t_{s_1})$ and $(v_i.LDTime, v_i.EATime - t_{s_1})$. This initial line is a horizontal line with the single value of the earliest arrival cost, and is bounded by the active time interval. Obviously the source vertex's cost is alway 0. Then these minimum cost functions are organized into a priority queue Q ordered by their minimum values. Each time we retrieve the cost function with the smallest minimum value in the queue, we use it to update the minimum cost functions of its out-neighbors using the update algorithm. If the out-neighbor's minimum cost function is changed, we visit this vertex again. Although it is written as inserting back to Q in line 13, it always takes only $O(1)$ time to visit it again since it has a smaller key than all the vertices in Q and will sit on top of Q directly. The algorithm has two terminating conditions. The first one is when Q becomes empty(line 6). The second one is when the top function's smallest value larger than v_d's minimum on road cost(line 8).

Algorithm 2. Path Expansion Algorithm

Input: Time-dependent graph $G_T(V, E, C)$, query $Q_{LRTT}(v_s, v_d, t_{s_1}, t_{s_2}, t_d)$
Output: The least on road travel time $R_{p^*_{s,d}}$

1 **begin**
2   **for** $v_i \in V$ **do**
3          $C_i(v_i.EATime) = v_i.EATime - t_{s_1}$
4          $C_i(v_i.LDTime) = v_i.EATime - t_{s_1}$

5   Let Q be a priority queue initially containing pairs $(min(C_i t), v_i)$, ordered by $min(C_i t)$ in ascending order
6   **while** $!Q.empty()$ **do**
7          $v_i \leftarrow Q.pop()$
8   **if** $min(C_i(t)) \geq min(C_d(t))$ **then**
9              $break$
10   **for** $v_u \in v_i$'s out-neighbors **do**
11             $C_u = Update(C_i, C_u)$
12   **if** C_u changed **then**
13                 $Q.insert(min(C_u(t)), v_u)$

14   **return** $min(C_d(t))$

Theorem 1. *Algorithm 2 can find the least on road travel time.*

Proof. Initially, the top of Q is v_s's smallest cost, which is 0 because it is the starting vertex. Then, its out-neighbors can all get their minimum on-road travel time after updated from v_s. Suppose v_i is the current top of Q and v_u is v_i's out-neighbor. If $min(v_u) < min(v_i)$, v_i cannot update v_u. In fact, v_u has already find its least on-road travel time that no vertex in Q can update it. If $min(v_i) < min(v_u)$, v_u can find a better path via v_i and gets updated. And since $min(_vi) < min(v_k), \forall v_k \in Q$, v_i has found its least on road travel time. Finally, after a $min(v_i) > min(v_d)$ pops out from Q, it is guaranteed that no vertex in Q can update $min(v_d)$. Thus, v_d had found its least on road travel time.   □

3.5   Complexity Analysis

As mentioned previously, the time complexity of the *active time interval* algorithm is $O(|V| \log |V| + |E|)$. As for the *Path Expansion* algorithm, we use T to denote the maximum number of time unit among all the active time intervals. The priority queue takes $O(|V| \log |V| + |E|)$ time to visit all the vertices and edges once. But in the worst case of our algorithm, each vertex may be revisited T times when its minimum cost function is updated and each time it takes $O(|E|T)$ to update. So the total vertex visit takes $O(|V|T)$ time and the update takes $O(|E|T^2)$ time. To sum up, the total time complexity is $O((|V| \log |V| + |V|T + |E|T^2))$.

As for the space complexity. The time-dependent information takes $O(T|E|)$ space. The minimum cost function takes $O(T|V|)$ space. The graph itself takes $O(|V| + |E|)$ space. So the total space complexity is $O(T(|V| + |E|))$.

4   Experiments

4.1   Experiment Setup

Datasets. We test our algorithm on two maps: Beijing map which contains 302,364 intersection and 387,588 roads, and Shanghai map which has 243,842 intersections and 310,058 roads. The time-dependent information we use is derived from the trajectories collected from these two cities.

4.2   Evaluation on Least On-Road Travel Time

We test our cheapest path algorithm on the two road networks. In each road network, we run tests on 3 different sets, each contains 100 vertex pairs. The distance of between pairs within each set is 5 km, 10 km and 20 km. The intersection pairs are selected randomly. The starting time interval size is set from 10:00 to 15:00 and the latest arrival time is set to 19:00, which is a common business closing time. We compare the on-road travel time computed by our algorithm(CP) with the results of earliest arrival path(EAP), latest departure path(LDP), fastest path(FP) [4], shortest path(SP) and the recursive version of fastest path(RE). The RE serves as our baseline since it can compute the approximate least on-road time by calling FP recursively. At last, we add the total travel time of our algorithm(CPT) into the comparison. It should be noted that only FP and RE allow waiting on vertex. The first row of Fig. 4 shows the

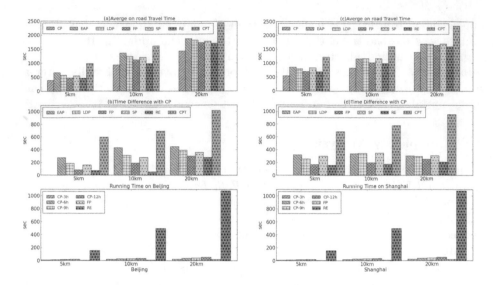

Fig. 4. Experiment result

comparison results. It is obvious that *CF* always has the least on-road travel time followed by *RE*. *FP* always has a smaller value that the remaining algorithms since it can find the fastest path on the no-waiting setting. *CPT* always has the largest time since it also counts the waiting time on vertices. The second row further clarifies the actual difference between *CP* with other algorithms.

Apart from the distance, we also test the effect of the time interval size between the starting time and latest arrival time, i.e. $t_d - t_{s_1}$, which actually determines the T in our time complexity analysis. We set t_{s_1} to 10:00, and set t_d to 13:00, 16:00, 19:00 and 22:00. We test the algorithm running time on these four settings. We also add the algorithm running of *FP* and *RE* in the previous experiment. The results are shown in the third line of Fig. 4. It is obvious that although *RE* can calculate the on road travel time most close to the result of *CP*, its running time is much longer than the others. As for the different *CP* settings, the algorithm running time grows nearly linear to the time interval size and are all merely several seconds. Although our time complexity analysis says the running time is bounded by T^2, it is an extreme worst case. So it still agrees with our analysis. As for the *FP*, it always has a smaller running time since it is only affected by the small starting time interval size rather than the total traveling time interval size.

5   Related Work

Although there were no least on-road travel time algorithm studied before, several similar path problems on different types of time-dependent graphs have been studied throughout the years.

The first kind of the time-dependent graph is the discrete time graph, or "timetable" graph, whose edges' existence is time-dependent like a transportation system. The algorithms in this field aims to compute *earliest arrival time, latest departure time, shortest path* and *shortest duration time* on such graphs. Cooke et al. [16] proved that these queries can be solved with a modified version of the Dijkstra's algorithm. However, it is not sufficient in large network people are facing nowadays, so several techniques are proposed to improve the efficiency [17–20]. The state-of-art works in this field are [13,14]. However, they only work on timetable graphs. Although some of their concepts are used in this work and the *shortest duration time* looks similar, they cannot solve the *LRTT* problem directly. Furthermore, their timetables is easy to acquire from different sources and are static throughout the time, while the time-dependent information in this work is continuous and changes randomly throughout the time.

The second kind of time-dependent graph has a continuous cost function over time. It is a theoretic model to describe a road network. *Dreyfus* [21] first showed the time-dependent fastest path problem was solvable in polynomial time if the graph is restricted to have *FIFO* property in 1969. Other early theoretical works are from *Halpern* [22] in 1997 and *Orda* [23] in 1991. However, these algorithms are so complicated to implement that no works have reported any computational evaluation of them. *DOT* [5] runs for k times to get an approximate result, which is both unbounded and unreliable. [3] uses an extension of A^* algorithm to find the fastest path, but since it is hard to prune their searching space especially when the graph is large, and its efficiency is really poor. [4] proposed a Dijkstra based algorithm to solve the problem faster than the previous methods. From then on, some works [17,18] try to augment pre-computation method like *SHARC* and *CH* to time-dependent graph, but still they are not practical due the long pre-computation time and unsuitable for dynamic environment. What's worse, they are all theoretic works that all of their experiments are conducted on synthetic graphs, which means they are impracticable in real life environment. [7] is the only work that derives their speed profile from real world sensor data and computes the fastest path on road. But still, none of the existing approaches related to the time-dependent graph can give the correct answer of a *LRTT* query.

6   Conclusion

In this paper, we first introduce a path problem on road network or general time-dependent graphs called *least on-road travel time(LRTT) problem* that was considered too hard to cope with previously. By figuring out the *non-increasing* property of the *minimum cost function*, we propose a *cheapest path* algorithm that can solve this problem in $O((|V|\log |V|+|V|T+|E|T^2))$ time. After proving the correctness, we compare the actual on road travel time computed by our *cheap path* algorithm with the existing path algorithms. The result shows that our algorithm can find a path with the least on-road travel time among all the compared path algorithms.

References

1. Dijkstra, E.W.: A note on two problems in connexion with graphs. Numer. Math. **1**(1), 269–271 (1959)
2. Goldberg, A.V., Harrelson, C.: Computing the shortest path: a search meets graph theory. In: Proceedings of the Sixteenth Annual ACM-SIAM Symposium on Discrete Algorithms, pp. 156–165. Society for Industrial and Applied Mathematics (2005)
3. Kanoulas, E., Du, Y., Xia, T., Zhang, D.: Finding fastest paths on a road network with speed patterns. In: Proceedings of the 22nd International Conference on Data Engineering, ICDE 2006, p. 10. IEEE (2006)
4. Ding, B., Yu, J.X., Qin, L.: Finding time-dependent shortest paths over large graphs. In: Proceedings of the 11th International Conference on Extending Database Technology: Advancesin Database Technology, pp. 205–216. ACM (2008)
5. Chabini, I.: Discrete dynamic shortest path problems in transportation applications: Complexity and algorithms with optimal run time. Trans. Res. Record: J. Transp. Res. Board **1645**, 170–175 (1998)
6. Orda, A., Rom, R.: Shortest-path and minimum-delay algorithms in networks with time dependent edge-length. J. ACM (JACM) **37**(3), 607–625 (1990)
7. Demiryurek, U., Banaei-Kashani, F., Shahabi, C., Ranganathan, A.: Online Computation of fastest path in time-dependent spatial networks. In: Pfoser, D., Tao, Y., Mouratidis, K., Nascimento, M.A., Mokbel, M., Shekhar, S., Huang, Y. (eds.) SSTD 2011. LNCS, vol. 6849, pp. 92–111. Springer, Heidelberg (2011). doi:10.1007/978-3-642-22922-0_7
8. Lu, E.H.-C., Lee, W.-C., Tseng, V.S.: Mining fastest path from trajectories with multiple destinations in road networks. Knowl. Inf. Syst. **29**(1), 25–53 (2011)
9. Cai, X., Kloks, T., Wong, C.: Time-varying shortest path problems with constraints. Networks **29**(3), 141–150 (1997)
10. Wang, X., Zhou, X., Lu, S.: Spatiotemporal data modelling, management: a survey. In: Proceedings of the 36th International Conference on Technology of Object-Oriented Languages and Systems, TOOLS-Asia 2000, pp. 202–211. IEEE (2000)
11. Deng, K., Zhou, X., Shen, H.T., Sadiq, S., Li, X.: Instance optimal query processing in spatial networks. VLDB J. **18**(3), 675–693 (2009)
12. Zheng, K., Fung, P.C., Zhou, X.: K-nearest neighbor search for fuzzy objects. In: Proceedings of the 2010 ACM SIGMOD International Conference on Management of Data, pp. 699–710. ACM (2010)
13. Wu, H., Cheng, J., Huang, S., Ke, Y., Lu, Y., Xu, Y.: Path problems in temporal graphs. Proc. VLDB Endowment **7**(9), 721–732 (2014)
14. Wang, S., Lin, W., Yang, Y., Xiao, X., Zhou, S.: Efficient route planning on public transportation networks: a labelling approach. In: Proceedings of the 2015 ACM SIGMOD International Conference on Management of Data, pp. 967–982. ACM (2015)
15. Fratta, L., Gerla, M., Kleinrock, L.: The flow deviation method: An approach to store-and-forward communication network design. Networks **3**(2), 97–133 (1973)
16. Cooke, K.L., Halsey, E.: The shortest route through a network with time-dependent internodal transit times. J. Math. Anal. Appl. **14**(3), 493–498 (1966)
17. Batz, G.V., Delling, D., Sanders, P., Vetter, C.: Time-dependent contraction hierarchies. In: Proceedings of the Meeting on Algorithm Engineering and Experiments, pp. 97–105. Society for Industrial and Applied Mathematics (2009)
18. Delling, D.: Time-dependent sharc-routing. Algorithmica **60**(1), 60–94 (2011)

19. Dibbelt, J., Pajor, T., Strasser, B., Wagner, D.: Intriguingly simple and fast transit routing. In: Bonifaci, V., Demetrescu, C., Marchetti-Spaccamela, A. (eds.) SEA 2013. LNCS, vol. 7933, pp. 43–54. Springer, Heidelberg (2013)
20. Geisberger, R.: Contraction of timetable networks with realistic transfers. In: Festa, P. (ed.) SEA 2010. LNCS, vol. 6049, pp. 71–82. Springer, Heidelberg (2010)
21. Dreyfus, S.E.: An appraisal of some shortest-path algorithms. Oper. Res. **17**(3), 395–412 (1969)
22. Halpern, J.: Shortest route with time dependent length of edges and limited delay possibilities in nodes. Z. fuer Oper. Res. **21**(3), 117–124 (1977)
23. Orda, A., Rom, R.: Minimum weight paths in time-dependent networks. Networks **21**(3), 295–319 (1991)

A Weighted K-AP Query Method for RSSI Based Indoor Positioning

Huan Huo[1](✉), Xiufeng Liu[2], Jifeng Li[3], Huhu Yang[1],
Dunlu Peng[1], and Qingkui Chen[1]

[1] School of Optical-Electrical and Computer Engineering,
University of Shanghai for Science and Technology, Shanghai 200093, China
huo_huan@yahoo.com
[2] Technical University of Denmark, 2800 Kongens Lyngby, Denmark
[3] University of Oulu, 90014 Oulu, Finland

Abstract. The paper studies the establishment of offline fingerprint library based on RSSI (Received Signal Strength Indication), and proposes WF-SKL algorithm by introducing the correlation between RSSIs. The correlations can be transformed as AP fingerprint sequence to build the offline fingerprint library. To eliminate the positioning error caused by instable RSSI value, WF-SKL can filter the noise AP via online *AP* selection, meanwhile it also reduces the computation load. WF-SKL utilizes LCS algorithm to find out the measurement between the nearest neighbors, and it proposes K-AP (P,Q) nearest neighbor queries between two sets based on Map-Reduce framework. The algorithm can find out K nearest positions and weighted them for re-positioning to accelerate the matching speed between online data and offline data, and also improve the efficiency of positioning. According to a large scale positioning experiments, WF-SKL algorithm proves its high accuracy and positioning speed comparing with KNN indoor positioning.

1 Introduction

As the rapid growth of wireless networking and the development of portable devices, a new upsurge of LBS (Location Based Service) applications are in making [1].

Nowadays the LBS has several ways to deal with the Indoor Positioning Service(IPS), such as Bluetooth [2], IR (Infrared Radiation) [3], RF (Radiation Frequency)[3], ultrasound [4] and UWB (Ultra-Wide Bandwidth) [5]. Several of them rely on electromagnetic wave, nevertheless, the electromagnetic wave spreading can be influenced by multipath effect. As the reflection and diffraction caused by the materials and the shapes of interior architecture affect the wave spreading, the signal error would be very high [6]. The Ultrasound and UWB could reach the accuracy within 1 cm [7], however, the high demand of hardware needs a lot of equipments to support, which brings extremely high cost, hence they are not suitable for commercial use. WIFI infrastructures are widely deployed in most shopping centers and supermarkets, and also have relative high

© Springer International Publishing AG 2016
M.A. Cheema et al. (Eds.): ADC 2016, LNCS 9877, pp. 150–163, 2016.
DOI: 10.1007/978-3-319-46922-5_12

accuracy and low cost, therefore WIFI is a popular solution for both academic research and industrial application.

In reality, IPS could be based on either Bluetooth or WIFI signal strength. Currently, WIFI-based IPS could utilize trilateration-centroid algorithm [8], geographic information probability [1] or RSSI-based algorithm [9]. Furthermore, the researchers propose two matching algorithms based on uncertainty and probability being built on RSSI-based algorithm. The uncertainty-based matching algorithm adopts KNN algorithm deriving from Kalman filtering, and the probability-based matching algorithm adopts Bayes probability [10].

Due to the unstable RSSI, WIFI-based indoor positioning still has a great gap of accuracy with costly UWB and ultrasonic wave positioning. The complex indoor architecture causes multipath effect, reflection and diffraction, hence the key point is to improve the stability of signal. For this issue, a general solution is to build a fingerprint library.

According to [5], the researcher proposes a method of building the fingerprint library depending on RSSI values as the clustering pattern (AP signal strength). Although the fingerprint library increases the efficiency of the algorithm via clustering, the data error derived from instable RSSI value, which is called jitter, generates sampling error and reduces the accuracy of position estimation. Also, the calculation of RSSI value is too complex, and it consumes huge amount of store space.

The researchers deal with the uncertainty in [11]. The signal strength sampled at one place will change and fluctuate in a small range with the time pass by. An experiment proves that the sequences of a pair of AP would be stable if the RSSI value exceeds 10 dBm between them [12].

In this paper, we propose an improved indoor WIFI positioning algorithm based on fingerprint library and weighted KNN. Compared with traditional methods which only utilize the uncertain RSSI value, the new method collects fingerprints from APs signal and introduces the AP correlation into the algorithm. Moreover, to form a stable sequence-pair, the new method sorts the signals by the strengths from strong to weak, and utilizes the sequence-pair to solve the issue caused by RSSI fluctuation. Considering the features of APs sequence, we introduce the concept of AP-based reference point(RP) to improve the efficiency of the algorithm. Finally we propose the weighted k-AP(P,Q) query algorithm. The experiment proves that the positioning accuracy is improved after introducing the correlation of the APs RSSI into the algorithm, meanwhile the positioning speed is faster with the fingerprint library clustering.

2   Preliminaries

Before introducing the IPN algorithm, we outline our framework by introducing the following concepts and definitions.

Definition 1 Wireless Access Point (AP). *An AP is a networking hardware device that allows a Wi-Fi compliant device to connect to a wired network. In* *this paper, AP refers to the physical location where Wi-Fi access to a WLAN is available.*

Definition 2 Reference Point (RP). *In the stage of building offline fingerprint library, we need to collect the signal strength value from a series of APs as the control data. These points are considered reference points.*

Definition 3 Longest Common Subsequence (LCS). *LCS is used to measure the similarity between two sequences, which is represented as the length of non-continuous longest common subsequence between two ordered sequences. In this paper, LCS refers to the number of overlapping APs between online AP sequences and offline RP fingerprint of AP sequences. The more overlapping APs, the higher matching degree between two points.*

Definition 4 Received Signal Strength Indication (RSSI). *Received Signal Strength Indication is the criteria to evaluate the quality of connection and signal strength in sending process. It estimates the distance between the sender and the receiver depending on the strength of the received signal, furthermore to position the location based on relevant data.*

2.1   Online AP Selection

Due to the difference and variations between the environments of online and offline data sampling, we need to filter out the APs which are affected by environment significantly, to improve the accuracy of matching. Assuming only a few APs will be influenced by the environment at any certain time point, in other words, part of data is expired in the online testing, hence we need to introduce AP selection mechanism. For example, in a big scale indoor environment, some APs are operated by ISP as public WIFI APs, and some others are temporary individual APs. We can define the APs in this area as a set $R = R_1, R_2, \ldots, R_m, |R| = M$. Via the AP selection algorithm such as MaxMean [6], InfoGain, and Deccorelated space, we can select a new AP set for positioning and delete the invalid APs. The principle of AP selection is to find a subset A which cannot influence the outcome of positioning, meanwhile adding new APs into subset A would not improve the outcome of positioning. We define the subset $A = A_1, A_2, \ldots, A_L, |A| = L, L < M$. This paper adopts RANSAC [5] as the AP selection algorithm. It filters out noise AP being influenced by the environment via probability distribution histogram testing. It improves the accuracy and reduce the calculation based on large scale offline data of APs.

2.2   RSSI for Indoor Positioning

Briefly, the fingerprint library indoor positioning algorithm is a type of RSSI-based KNN algorithm. The principle of the algorithm is to utilize the value of signal strength as the attribute of RP after the RW device receives the wireless signal strength being sent from the AP, then we can use KNN to find out the number K of the reference points, and adopt the locations of reference points to estimate the query point position.

RSSI based indoor positioning has two steps. The first one is to build the offline fingerprint library, and the second step is to apply the KNN matching algorithm on online data. Building offline fingerprint library is the stage of APs signal sampling via selecting huge amount of RPs, as shown in Fig. 1.

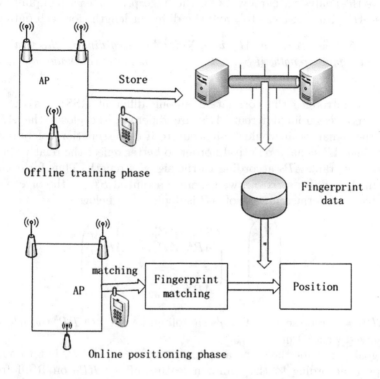

Fig. 1. RSSI based indoor positioning

3   A Weighted K-AP(P,Q) Query Method

3.1   Fingerprint Library Construction

Assume the numbers of APs and RPs are M and N separately, noted as $m = 1, 2, 3, \ldots, M$ and $n = 1, 2, 3, \ldots, N$. For each RP, we sample Q times and define the ith RP sampling from the APs as $S_n^q = (S_{n1}^q, S_{n2}^q, S_{nM}^q)$. So the RSSI matrix for the RPs after Q times sampling is defined as S_n.

$$S_n = \begin{pmatrix} S_n^1 \\ S_n^2 \\ \vdots \\ S_n^Q \end{pmatrix} = \begin{pmatrix} S_{n1}^1 & S_{n2}^1 & \cdots & S_{nM}^1 \\ S_{n2}^2 & S_{n2}^2 & \cdots & S_{nM}^2 \\ \vdots & \vdots & \ddots & \vdots \\ S_{n1}^Q & S_{n2}^Q & \cdots & S_{nM}^Q \end{pmatrix} \qquad (1)$$

In the original RSSI data, the accuracy for weak signals is very low due to the device error, or operative error. For different RPs, some are always receiving weak signal from certain AP, and their RSSI accuracy can not be guaranteed by the device if the RSSI is below the lower bound. Therefore, RSSI matrix can not indicate the signal strength among different APs. In order to reduce the data storage and unnecessary computing, we introduce $NoiseAP$ for offline processing to improve the analysis accuracy. As for the fingerprint of each sampling, we set the $NoiseAP$ with $-\infty$ so as to avoid the different lengths for each data set.

Definition 5 Noise AP. *An AP_i is a $NoiseAP$ according to the weakness of the received signal strength, if $S_{ni} \leq \theta$, where θ is the lower bound of the signal strength received by the device.*

Then we introduce the correlation among different RSSI. As the signal strengths received from different APs are different, the closer the AP, the stronger the signal. So it would be inaccurate if the estimation solely relies on the individual AP signal strength. In order to better reflect the relative position of the RPs, we rank APs according to the signal strength. Thus for each data set S_{nm}^q in the RSSI matrix S_n, we generate a ranked S_{nm}^q with corresponding AP ID. The transformed matrix of AP is depicted as below:

$$
\overline{S}_n = \begin{pmatrix} \overline{S}_n^1 \\ \overline{S}_n^2 \\ \vdots \\ \overline{S}_n^Q \end{pmatrix} = \begin{pmatrix} AP_{n1}^1 & AP_{n2}^1 & \cdots & AP_{nM}^1 \\ AP_{n2}^2 & AP_{n2}^2 & \cdots & AP_{nM}^2 \\ \vdots & \vdots & \ddots & \vdots \\ AP_{n1}^Q & AP_{n2}^Q & \cdots & AP_{nM}^Q \end{pmatrix}
\tag{2}
$$

where $AP_{ni}^q = m$ represents in the $q's$ sampling, for the nth RP, the ith ranked AP is marked with ID m.

Our goal is not for the Qth sampling data for single RP, but to generate a fingerprint according to the common feature of the RPs on RSSI from Q times of sampling. Therefore, we set the ID with the highest frequency on each column as the fingerprint of AP for Q times sampling on the RP. For example, after Q times sampling and transformation, the AP fingerprint of the $nthRP$ is $\bar{S}_n = (M, 20, 18, \ldots, 1)$, assuming that the signal strength from AP M is the strongest one, then the AP with ID 20 follows, and so on.

However, there are some special circumstances. For instance, assuming the AP fingerprint for certain RP is $\bar{S}_i = (38, 20, 20, 25, 9, 10, \ldots, 3)$, there are two identical IDs. This is caused by the interference during the signal transmission, or severe error from measurement. The phenomenon will be eliminated along with the increment of sampling times. The smaller the Q, the higher possibility the repeated AP appears. In theory, the phenomenon can be fully avoided given unlimited Q. But in reality, it is impossible to sample infinite times for large amount of AP.

So we propose a solution for calculating the representative AP. Firstly, we locate the columns with the same repeated AP, compare the repeated frequency of the AP on each column, and set the AP as the representative AP if its frequency is greater than the frequency on others column. Then we exclude the AP with the second largest frequency from the representative AP on the next column, and iterate the same process until there is no repeated AP on each column.

For instance, the AP with ID 20 apears m times on the second column, while the AP with ID 20 apears n times on the third column. If $m > n$, we keep the AP with ID 20 as the representative for the second column, and substitute the second repeated AP with the representative AP for the third column, i.e. 18, and so on. Finally, we can get the AP fingerprint of the RP as $\bar{S}_i = (38, 20, 18, 25, 9, 10, \ldots, 3)$.

Based on the above approach, the RSSI matrix for all the RPs can be transformed to the AP fingerprint library for all the RPs. We name the offline fingerprint library as S'.

$$S' = \begin{pmatrix} \overline{S}_1 \\ \overline{S}_2 \\ \vdots \\ \overline{S}_N \end{pmatrix} = \begin{pmatrix} AP_{11} & AP_{12} & \cdots & AP_{AM} \\ AP_{21} & AP_{22} & \cdots & AP_{2M} \\ \vdots & \vdots & \ddots & \vdots \\ AP_{N1} & AP_{N2} & \cdots & AP_{NM} \end{pmatrix} \tag{3}$$

At last, we analyse the AP fingerprint matrix and perform our clustering methods according to the first column fingerprint, i.e. the strongest signal AP for each RP. For example, if RP_i and RP_j share the same strongest AP from M, we cluster RP_i and RP_j together and marked the APs ID as C_M. Then we get the fingerprint cluster S'', as shown below.

$$S'' = \begin{pmatrix} C_1 \\ C_2 \\ \vdots \\ C_M \end{pmatrix} \tag{4}$$

The fingerprint cluster can speed up the positioning efficiency by fast locating the similar RPs when matching the online data and the offline data.

3.2   A Weighted K-AP(P,Q) Query Method Based on MapReduce

In traditional K-NN algorithm, we can calculate the top K nearest RPs by computing the Euclidean distance between the query point and the offline points, and then estimate the location by the means of their coordinates. With the AP fingerprint library, we do not need to calculate the Euclidean distance to estimate the similarity between the query point and the offline points. Instead, we estimate the similarity by simply counting the number of the overlapped data between the two vectors. The similarity is higher if the two vectors share more common data. Therefore, we calculate the longest common subsequence by the LCS equation below:

$$LCS(s'_m, \bar{S}_n) = \begin{cases} \emptyset & if\, m = 0\, orn = 0 \\ LCS(s'_{m-1}, \bar{S}_{n-1}) & if\, ap_m = AP_n \\ max(LCS(s'_m, \bar{S}_{n-1}), LCS(s'_{m-1}, \bar{S}_n)) & if\, ap_m = AP_n \end{cases}$$
(5)

In Eq. 5, $LCS(s'_m, \bar{S}_n)$ is the common AP length between the online s'_{m_i} and offline $\bar{S}_n$. s'_{m-1} and $\bar{S}_{n-1}$ are the remained parts after removing the first common AP. $ap_m = AP_n$ represents the overlapped AP in the two sequences. When one of the sequences contains no AP, $LCS(s'_m, \bar{S}_n))$ is set to 0. If both sequences contain at least one AP, we recursively compute the maximum LCS. In the algorithm, similar subsequences are not mandatory continuous.

For instance, assume the fingerprint of the query RP is $s = (s_1, s_2, ..., s_M)$, the AP fingerprint is $s' = (ap_1, ap_2, , ap_M)$ after ranking, and the fingerprint vector for ith RP is $\bar{S}_i = (AP_{i1}, AP_{i2}, , AP_{iM})$. Then $LCS(s'_m, \bar{S}_n)$ is computed as the similarity, noted as l.

When estimating the location, we propose a K-AP(P,Q) query algorithm for finding the top k nearest AP pair between two data sets P and Q, based on the $K - NN$ query on MapReduce platform. The main idea is to build an R*-tree index for data set P and Q, then perform K-AP(P,Q) query according to the similarity between APs on MapReduce. As MapRedcuce parallels the original K-NN algorithm, the K-AP(P,Q) algorithm can efficiently find the best matching out of the offline fingerprint library.

The data partition strategy is to locate the strongest AP of the online data in the cluster C_i, and construct set P for all the AP fingerprints in C_i. While set Q contains only the online data. Then we construct the parallel R*-tree index based on the sampling procedure and Hilbert curve construction [13]. The details of the index construction can be referred to [14].

According to the K-AP(P,Q) algorithm, we get the top K nearest RPs, based on which we can calculate the query position with the Eq. 6.

$$\begin{cases} (X, Y) = \dfrac{1}{K} \sum_{i=1}^{K} w_i(x_i, y_i) \\ \sum_{i=1}^{k} w_i = 1 \end{cases}$$
(6)

In Eq. 6, (x_i, y_i) is the real coordinates of the ith RP, and w_i is the weight of the ith RP. We redefine the weight to improve the estimation accuracy as shown in Eq. 7.

$$w_i = \frac{l_j}{\sum_{j=1}^{k} l_j}$$
(7)

Equation 7 indicates that the location similarity will increase along with the increment on the overlapping parts between the query point and the reference points. Based on the above analysis, we propose the WF-SKL algorithm on fingerprint library for the indoor positioning, as shown in Algorithm 1.

Algorithm 1. WF-SKL

Input: (1) N RPs, M RP coordinates, Sampling time Q;
    (2) Original RSSI fingerprint set S;
    (3) Online RSSI fingerprint s;
    (4) Noise AP threshold θ, rank k.
Output: Current point position coordinates: (x, y).
1: Generate M APs from AP selection
2: **for** each RP **do**
3:    collect RSSI from each AP
4:    **if** RSSI $< \theta$ **then**
5:      set $-\infty$
6:    **end if**
7:    get S
8: **end for**
9: **for** each RP in S **do**
10:    **for** each time in S_i **do**
11:      rank (S_i^q)
12:      return $\bar{S}_i$
13:      **while** There exist at least one pair of the same AP in the $\bar{S}_i$ **do**
14:        reallocateAP();
15:      **end while**
16:      return S'
17:    **end for**
18: **end for**
19: cluster (S')
20: return S''
21: // K-AP Query
22: K-AP(P, Q)
23: select C_i as P, s as Q
24: return k APs
25: // Positioning Estimation
26: do $\begin{cases} (X, Y) = \frac{1}{K} \sum_{i=1}^{k} w_i (x_i, y_i) \\ \sum_{i=1}^{k} w_i = 1 \end{cases}$.
27: return (x, y)

4   Experiment

We evaluate the performance of the indoor positioning methods based on AP fingerprint library and the weighted $K - AP(P, Q)$ query algorithm through the experiments taken place on the sixth floor of the University of Shanghai for Science and Technology. The experiment area is $566\ m^2 (35.6\ m \times 15.9\ m)$, located with 16 TP-LINK TL-WDR4300 N750 double band wireless routers functioning as AP. The RSSI values are detected through a Sumsang Note 4 smart phone based on Android 5.0 and collected by the application ADAwifi. We randomly select 50 offline RPs and 20 online testing points. Each RP is sampled 30 times. Figure 2 shows the distribution of all the RPs, testing points and the APs.

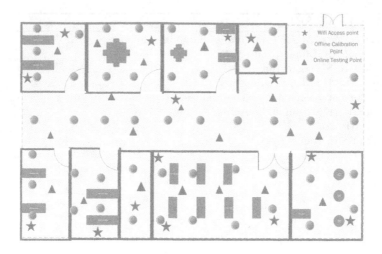

Fig. 2. Distributions of APs, RPs and testing points

The PC configuration for data analysis is Intel(R) Core(TM) i3 M370 with 2.4 GHz CPU, 500 Gbyte hard disk, and Windows8 system. MySQL is the offline storage platform.

4.1   AP Selection Performance

First, in order to verify the influence of the online AP selection, we simulate two groups of testing point positioning experiments in the same environment. We applied WF-SKL algorithm to two groups of experiments, one of which directly queries the position while the other performs online AP selection first. Figure 3 shows the outcome comparison of the positioning error.

In Fig. 3, the positioning error is smaller if the method performs AP selection first. This is because the AP selection filters out the individual AP deviation

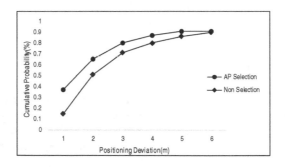

Fig. 3. AP selection performance

caused by the environment vibration. Thus the AP selection helps to improve the positioning accuracy.

4.2 Accuracy Comparison

Then we perform the experiments for the offline AP fingerprint library construction. In the same simulation environment, we apply AP selection first and then estimate the position by the weighted $K - AP(P, Q)$ algorithm, and compare our method with the traditional KNN based fingerprint construction and the KNN based on clusters fingerprint construction [5]. The accuracy comparison outcome is shown in Fig. 4 and the performance comparison outcome is shown in Fig. 5:

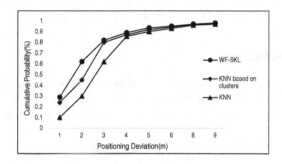

Fig. 4. Accuracy comparison

As seen from Fig. 4, the accuracy of WF-SKL algorithm is much better than the traditional KNN algorithm. The reason why our fingerprint computation method is more accurate than the traditional KNN method is because ours reduces the errors generated by rough estimates rather than simply computing the average. We also find out that our algorithm is more accurate than the KNN based on clusters algorithm. This is because KNN based on clusters algorithm selects the strongest RSSI cluster, but its data is based on single RSSI value, thus can not prevent the errors caused by the unstable RSSI. Our method is a relatively stable AP pairs positioning estimation, which can avoid the errors caused by using individual RSSI values.

Figure 5 indicates the positioning speed of different algorithms. In the real environment, the scale of fingerprint library increases with the declining distance between the testing points, which means the positioning accuracy based on fingerprint library would be higher. Figure 5 shows the KNN algorithm has the lowest processing speed. It does not cluster the data in the offline stage, hence the calculation in the stage of online matching slows down the speed. WF-SKL algorithm has a faster speed because it adopts AP pair as matching data, which is simpler than RSSI being used by KNN based on clusters as matching data. Not like KNN based on clusters algorithm calculating RSSI as the matching

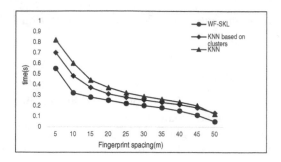

Fig. 5. Performance comparison

data, WF-SKL algorithm calculates the common sequence via LCS to process the matching query, therefore the matching speed is faster.

4.3   Weight Evaluation

We test the performance of weighting in this experiment. The testing dataset adopts weighted WF-SKL algorithm, while the other same dataset is processed with the algorithms without weighting. The results of the experiments are shown in Fig. 6:

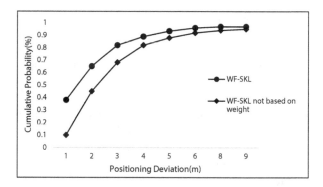

Fig. 6. Weight influence

In Fig. 6, the accuracy of non-weighted WF-SKL algorithm is much lower to estimate the position, however, the accuracy is improved a lot after introducing the weighting. Weighted value can reflect the distance of the matching points, however the average value cannot reflect the distance, therefore the accuracy of non-weighted algorithm is lower.

4.4 Scalability

In the last experiment, we compare different KNN query algorithms in the same environment. The KNN based on clusters algorithm is an improved KNN algorithm by introducing clustering and weighting, in other words, KNN based on clusters algorithm is a kind of KNN query algorithm using Euclidean distance. The performance is shown in Fig. 7.

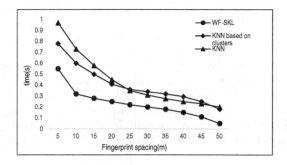

Fig. 7. Parallel performance

As shown in Fig. 7, the speed of the WF-SKL algorithm is faster than others. K-AP(P,Q) algorithm is the R*-tree based KNN query method, which could be accelerated to process big data query based on MapReduce framework. Therefore the processing speed of the paralled algorithm is much faster than the other two algorithms.

Finally, we compare the performances of the three positioning methods in Table 1. The parameters include mean error, error variance, and the error within $2M$. The accuracy of WF-SKL algorithm is in 2.35 m. It is increased 8.2 % compared with KNN algorithm, and also increased 16.96 % compared with cluster-based KNN algorithm. For the WF-SKL algorithm, the rate of the error within 2 m reaches 47.28 %, meanwhile it has the lowest error variance, which means the data positioning of our method is much more reliable.

Table 1. Comparison among different methods

Methods	Mean Error(m)	Error Variance(m)	Error within 2 m
WF-SKL	2.35	1.48	47.28 %
KNN based on clusters	2.56	1.62	38.74 %
KNN	2.83	1.89	32.56 %

5 Conclusion and Future Work

Currently the wireless indoor positioning is widely implemented and deployed. In this paper, we mainly focus on introducing the method of building AP fingerprint library based on the correlations of RSSI. Furthermore, we introduce an AP fingerprint based WF-SKL algorithm, which adapts the KNN algorithm to a weighted K-AP query method. The algorithm eliminates several weaknesses of the old methods and improves the performance and accuracy of IPS.

References

1. Aboodi, A., Wan, T.-C.: Evaluation of wifi-based indoor (wbi) positioning algorithm. In: 2012 Third FTRA International Conference on Mobile, Ubiquitous, and Intelligent Computing (MUSIC), pp. 260–264. IEEE (2012)
2. Rida, M.E., Liu, F., Jadi, Y., Algawhari, A.A.A., Askourih, A.: Indoor location position based on bluetooth signal strength. In: 2015 2nd International Conference on Information Science and Control Engineering (ICISCE), pp. 769–773. IEEE (2015)
3. Want, R., Hopper, A., Falcao, V., Gibbons, J.: The active badge location system. ACM Trans. Inf. Syst. (TOIS) **10**(1), 91–102 (1992)
4. Priyantha, N.B., Chakraborty, A., Balakrishnan, H.: The cricket location-support system. In: Proceedings of the 6th Annual International Conference on Mobile Computing and Networking, pp. 32–43. ACM (2000)
5. Hou, Y., Sum, G., Fan, B.: The indoor wireless location technology research based on Wi-Fi. In: 2014 10th International Conference on Natural Computation (ICNC), pp. 1044–1049. IEEE (2014)
6. Zou, H., Luo, Y., Lu, X., Jiang, H., Xie, L.: A mutual information based online access point selection strategy for wifi indoor localization. In: 2015 IEEE International Conference on Automation Science and Engineering (CASE), pp. 180–185. IEEE (2015)
7. Liu, H., Darabi, H., Banerjee, P., Liu, J.: Survey of wireless indoor positioning techniques and systems. Syst. Man Cybern. Part C: Appl. Rev. IEEE Trans. **37**(6), 1067–1080 (2007)
8. Yang, C., Shao, H.-R.: Wifi-based indoor positioning. Commun. Mag. IEEE **53**(3), 150–157 (2015)
9. Sen, S., Choudhury, R.R., Radunovic, B., Minka, T.: Precise indoor localization using phy layer information. In: Proceedings of the 10th ACM Workshop on Hot Topics in Networks, p. 18. ACM (2011)
10. Fang, Y., Deng, Z., Xue, C., Jiao, J., Zeng, H., Zheng, R., Lu, S.: Application of an improved k nearest neighbor algorithm in wifi indoor positioning. In: China Satellite Navigation Conference (CSNC): Volume III, pp. 517–524. Springer (2015)
11. Jun, J., Chakraborty, S., He, L., Gu, Y., Agrawal, D.P.: Robust and undemanding wifi-fingerprint based indoor localization with independent access points (2015)
12. Yang, S., Dessai, P., Verma, M., Gerla, M.: Freeloc: Calibration-free crowdsourced indoor localization. In: 2013 Proceedings IEEE INFOCOM, pp. 2481–2489. IEEE (2013)

13. Liu, Y., Jing, N., Chen, L., Chen, H.: Parallel bulk-loading of spatial data with mapreduce: An r-tree case. Wuhan Univ. J. Nat. Sci. **16**(6), 513–519 (2011)
14. Dun-lu Peng, J.X.-L.: Algorithm for k-closest pair query based on two sets on mapreduce framework. J. Chin. Comput. Syst. **37**(3), 483 (2016). http://xwxt.sict. ac.cn/EN/abstract/article3302.shtml

AutoCadroid: An Automated Tool for Building Smartphone Indoor Navigation System from AutoCad Files

Shamim Hasnath[✉], Mohammed Eunus Ali, M. Kaysar Abdullah, and S.M. Farhad

Department of CSE, Bangladesh University of Engineering and Technology (BUET), Dhaka 1000, Bangladesh
shamim@hasnath.net

Abstract. With the recent development of smartphone technologies and indoor positioning systems, it is now possible to provide a large variety of location-based services (LBSs) in indoor spaces. Due to the complex nature of indoor space modeling (i.e. re-creating floor plans), the applicability of indoor LBS has been limited to few big installations such as airports, big shopping centers, etc. To facilitate the growth of indoor LBSs to end users, we overcome this major limitation of indoor space recreation by automatically converting an AutoCad DXF file, the most widely used design format, to a *Spatialite* database for smartphones. We develop algorithms that first pre-process the data from a DXF file and then effectively recognize rooms, doors, corridors and obstacles, and finally convert these objects for indoor space navigation through smartphones. We develop a working prototype, AutoCadroid, in Android that enables a user to build her own indoor navigation system of her own office or home from an AutoCad DXF file.

1 Introduction

The recent development of smartphone technologies and indoor positioning systems enable us to render a wide variety of location-based services (LBSs) in indoor spaces [3,4]. An important class of indoor space LBSs is the searching and navigation system where a user searches for a desired point of interest (POI) and asks for a route to the POI. For example, one may want to find the location of an appropriate shop to buy the desired cloth in a large shopping center, or a tourist may want to find a route that guides a tourist to different POIs in a museum. Due to the complex nature of indoor space modeling (i.e., re-creating floor plan designs), the applicability of indoor LBS has been limited to few big installations such as airports, big shopping centers, etc. To facilitate the growth of indoor LBSs to end-user level, we build a system, named AutoCadroid, that automatically converts an AutoCad DXF file [1], the most widely used design

© Springer International Publishing AG 2016
M.A. Cheema et al. (Eds.): ADC 2016, LNCS 9877, pp. 164–176, 2016.
DOI: 10.1007/978-3-319-46922-5_13

format, to a *Spatialite* [5] [1] database for smartphones, which ultimately helps a user build her own indoor navigation system for her own office or home.

AutoCAD is a widely used design tool for making floor plans. As a result the designs of floor plans of almost all the modern buildings are available in AutoCAD Drawing Exchange Format (DXF). The DXF file format facilitates the inter-operability between AutoCAD software and any third party reader to view the file. A DXF file uses simple geometries such as Lines and Polylines to represent indoor objects like rooms, doors, corridors, and obstacles. However, since a designer has the flexibility to use any of these basic geometries to construct different types of indoor objects, it is not trivial to automatically identify these objects from a DXF file.

A *spatialite_gui* [8] is available for importing a DXF file into a database management system. However, this tool directly imports all objects of basic spatial data types, e.g., Lines or Polylines, directly into the database. Thus, semantic structures of objects of interest, e.g., rooms, doors, corridors, etc., are not preserved in this transformation. Moreover, since such an object is made of by joining a large number of basic spatial objects, an automatic conversion using *spatialite_gui* generates a complex database with lots of links among different tables, which makes it unusable for navigation applications.

In [4], Boysen et al. proposed a software system to export a design file (i.e., Digital Building Information (DBI) file) to PostgreSQL with PostGIS, and then use a mobile application to search and navigate. They proposed a manual system, which allows a user to identify doors and rooms manually, and then store these objects into database to facilitate faster search. Manually creating indoor maps especially for bigger venues is a time consuming and expensive process. Thus, recently Xu et al. [9] present an approach to automate the room identification from basic geometries of an AutoCAD DXF file, and then store rooms and doors in an Oracle database to facilitate faster search. However, the system proposed in [9] may not be able to identify rooms that are drawn in Auto-CAD using combination of lines (i.e., $LINE$ object in DXF) and polylines (i.e., $(LW)POLYLINE$ object in DXF). They also eliminate a room if it completely contains by another room, which also generates the incompleteness in the navigation system.

We are the first to propose a tool for Android, namely AutoCadroid, that automatically converts an AutoCad DXF file, the most widely used design format, to a *Spatialite* database for smartphone based indoor navigation. We first develop algorithms that correctly recognize rooms, doors, and corridors of an AutoCad DXF file. Then, we build a path structure by connecting identified objects for indoor navigation. Finally, we export these identified objects of interest along with the connectivity to a *Spatialite* database, which ultimately allows an efficient query processing in an indoor navigation system. AutoCadroid also enables a user to interactively tag and modify these objects for indoor space navigation through smartphones. Hence, AutoCadroid facilitates a user to build

---

[1] A library extension of SQLite database.

her own personalized indoor navigation system by using her smartphone from the already available AutoCad design file of a building.

Note the outcome of our technique for identifying rooms, doors and corridors can be used by any existing indoor navigation system (e.g., [4]). One of the aims of this paper is to utilize Spatialite databases for indoor navigation. Hence, DXF files are converted in a format that the Spatialite database can use for query processing.

In summary, the contribution of this paper is threefold:

- We develop algorithms to automatically identify rooms, doors, and corridors from the floor plans of an AutoCad DXF file. We also identify necessary points to construct a path through the corridors for complex scenarios.
- We import identified indoor space objects into a Spatialite database system for Smartphones, which enable one to exploit the strength of built-in spatial database management system for querying the database.
- We develop a system, AutoCadroid in Android platform that enables a user to build her personalized indoor space navigation system.

2   AutoCadroid Architecture

2.1   Overview

A DXF file has six sections: HEADER, CLASSES, BLOCKS, TABLES, OBJECT and ENTITIES [1]. ENTITIES are used to store all the geometric information about the floor plan. These geometries mostly consist of LINE, POLYLINE, LWPOLYLINE and ARC types of objects. Walls of the rooms in the floor plan can be drawn using either (LW)POLYLINE or LINEs. Moreover, multiple LINEs or POLYLINEs can be used while representing a wall in the design. ARCs are used to represent the door objects. One of our objectives is to automatically identify doors, rooms, corridors, and obstacles (or closed spaces) from these basic geometries.

First, we remove unnecessary (or redundant) lines that are used to make the walls thick in the design, which we call pre-processing step. Then we identify the door ARCs, and convert these ARCs to LINEs so that we can consider both wall LINEs and door LINEs in the same way in subsequent steps. In the next step, we identify unique rooms from door lines and wall lines. Then we also present a model to connect rooms through doors that facilitates navigation from any source location to a destination location. Finally, we store all identified objects in *Spatialite* database so that personalized navigation queries can be executed on this data [2].

2.2   Pre-processing

One of the main challenges in identifying rooms, doors, corridors, and other obstacles (e.g., a closed space) from a DXF file is the redundant geometric shapes (e.g., LINES, POLYLINES) that are included in the design phase. For example, a wall of a room is designed using multiple parallel lines to make it look nicer or thicker. [9] proposed an algorithm that removes line if it is within a threshold distance (vertical) from a base line. It does not consider the lengths and relative positions of two comparing lines and thus may end up with removing important information that are necessary for identifying rooms. For example, Fig. 1(a) shows two parallel lines AB and CD, where AB contains CD. AB and CD may represent a single wall of a room or two walls in two different rooms. In either case we can safely remove CD. This is because, if we assume AB and CD be the walls of two separate rooms, vertical lines at two end points of CD (i.e., walls perpendicular to CD) that form a room, can be extended to connect with AB to form the same room. On the other hand, in the right figure (Fig. 1(b)), we see two parallel lines AB and CD, where AB does not contain CD. In this case if we remove CD based on its small vertical distance from AB, we may miss a room as the two vertical lines at the two ends of CD of the room cannot be extended to connect with AB.

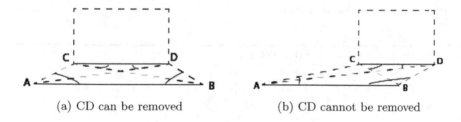

(a) CD can be removed   (b) CD cannot be removed

Fig. 1. Parallel redundant lines

Based on the above observation, we take the following strategy. A line can be removed if (i) the line is parallel to the base line and within a certain vertical distance d from the base line, and (ii) the base line contains the other line completely. In this case AB and CD forms acute angles (which is shown using dotted lines in Fig. 1(a)). We can identify these parallel lines by calculating the dot product, $\overrightarrow{BA}.\overrightarrow{BD} \geq 0$. On the other hand, if two lines form an obtuse angle, e.g., $\angle ABD$, none of thesw lines can be removed (Fig. 1(b)). Thus to remove redundant lines, we first sort all lines in decreasing order of their lengths. Then we compare each line with other smaller lines, first by their slopes and then by their lengths. If a line meets the above criteria, we remove it from the line list. In this step, we get a reduced set of lines that form the boundaries of rooms, corridors, and obstacles. The thickness threshold d is empirically decided by examining the thickness of lines in the design.

2.3 Identifying Doors

In the next step, we identify doors from a DXF file, and then connect those doors with identified rooms for navigation. A door in a DXF file is represented as an

ARC (*start_point, end_point, radius, start_angle, end_angle*). We convert an *ARC* to a *LINE*(u_d, v_d). From the two end points and the radius of an *ARC* we first find the center of the *ARC*. Then, we get two edges by connecting two ARC end points with the center. If those two edges have the same slope, both combinedly form the door LINE (This is the case where the door is on the middle of a room wall and is represented as a full arc.). Otherwise, the edge that does not overlap with any of the wall line, is the door LINE (This is the case if the door is adjacent to two walls of a room, and is represented as an half arc.). We term these identified door lines as door edges.

2.4   Identifying Rooms

In this step, we identify rooms, corridors, or closed spaces from the wall lines and door lines obtained in the previous two steps. These lines are referred as edges. We first merge the edge set of walls with the edge set of doors to form a planar graph. Then we find all cycles (faces) of the planar graph that finds all rooms, corridors or closed spaces.

Algorithm 1. FindRooms(Edge Set E)

INPUT: Edge Set $E\{e_1, e_2, ..., e_n\}$, $|E| = n$;
OUTPUT: Room Set $R\{r_1, r_2, ..., r_m\}$; , $|R| = m$; $r_i = \{e_{i1}, e_{i2}, ..., e_{ik}\}$; $|r_i| = k$; $e_{ij} \in E$.
1: $R \leftarrow \emptyset$
2: $E' \leftarrow \emptyset$
3: **for** each edge $e_i(u_i, v_i) \in E$ **do**
4:     $E' \leftarrow E' \cup e_i(v_i, u_i)$
5: **end for**
6: $E \leftarrow E \cup E'$
7: **while** $E \neq \emptyset$ **do**
8:     $curRoom \leftarrow \emptyset$
9:     $eBegin \leftarrow$ any edge from E
10:    $curEdge \leftarrow eBegin$
11:    **do**
12:        $nextEdge \leftarrow$ edge that produces maximum clock-wise angle with $curEdge$
13:        $curEdge \leftarrow nextEdge$
14:        $curRoom \leftarrow curRoom \cup curEdge$
15:        remove edge $curEdge$ from E
16:    **while** $curEdge \neq eBegin$
17:    $R \leftarrow R \cup curRoom$
18: **end while**
19: **return** R

Algorithm 1 shows the pseudocode of our room finding algorithm. The algorithm takes an edge set E as in input and returns a room set R as output. For each edge $e_i \in E$ of the edge set, we make a duplicate edge e'_i in the opposite

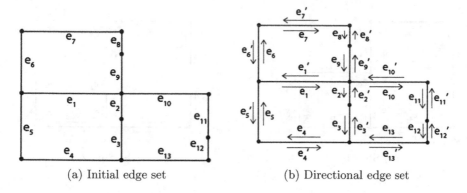

(a) Initial edge set   (b) Directional edge set

Fig. 2. Door and wall edges

direction. Intuitively, a non-directional edge is replaced by two directional edges, e_i (forward direction) and e_i' (backward direction). Thus the input edge set E is updated with directional edges (Lines 3–6).

Figure 2(a) shows the planar graph consisting thirteen edges, $\{e_1, e_2...e_{13}\}$, of rooms and the doors. The right figure shows the bi-directional edge set $\{e_1, e_1', e_2, e_2', ..., e_{13}, e_{13}'\}$.

In the next step, we randomly choose an edge $curEdge$ from the edge set and find the corresponding room of that edge (Lines 7-18). For this, we first find the incident edge, $nextEdge$, that produces maximum clock-wise angle with respect to $curEdge$. $nextEdge$ is then added as the next wall of the current room, and removed from the edge set. Then we update $curEdge$ with $nextEdge$, and repeat the above procedure.

Figure 3 shows the steps of finding a room. First, an edge e_1 is selected as a random edge. Then we find all of its incident edges that start from the end of e_1. These are e_2, e_{10}, e_9' (as shown in Step 2). Then we pick the edge e_2 that produces the maximum clock-wise angle with respect to the edge e_1 (Step 3). Thus e_2 is added as the next wall edge to the currently exploring room. Then we move to edge e_2 and repeat this process again. This way, edges e_3, e_4, e_5 are chosen as next walls. When a cycle is complete, we find a room $\{e_1, e_2, e_3, e_4, e_5\}$ (Step 6).

The above process runs for each remaining edge (i.e., after removing the edges for which rooms are already identified from the original edge list) in the edge list to identify all polygons, which represent doors, corridors, or closed spaces.

2.5   Finding Path Vertices for Indoor Navigation

After finding all the doors, rooms, corridors, and closed spaces in the above steps, we need to build a navigation structure that shows us the path from a source to a destination. For this, we first compute path vertices, and then connect those path vertices to create a navigation structure.

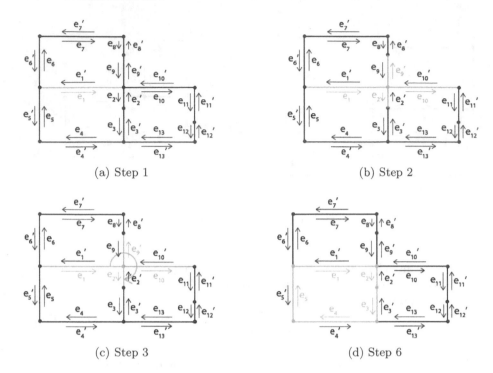

(a) Step 1   (b) Step 2

(c) Step 3   (d) Step 6

Fig. 3. Steps for finding a room

Algorithm 2 generates necessary vertices for path finding. For every door we take two points. One point inside the room and another point outside the room. If we connect these two points, the connecting line becomes perpendicular to the door edge, and distance of any of these point from from the mid-point of the door equals to the half of the door edge's length. Thus for a door edge (u, v), we compute two path vertices as follows. For the vertex inside the room:

$$v_1 = (\frac{u.x + v.x}{2} + \frac{u.y - v.y}{2}, \frac{u.y + v.y}{2} - \frac{u.x - v.x}{2})$$

and for the vertex outside the room

$$v_2 = (\frac{u.x + v.x}{2} - \frac{u.y - v.y}{2}, \frac{u.y + v.y}{2} + \frac{u.x - v.x}{2}).$$

To navigate through corridors and along the convex paths of the rooms, we need to add more path vertices.

For this, we select the corner points that comply with the following properties. We consider the meeting point of a pair of edges as a corner point, if no other edge exists inside the reflex angle (i.e., larger than 180°) between these two edges. Function $reflex\_edge(e_i, e_j, E)$ returns true if e_i and e_j intersect and no edge in E exists between the reflex angle formed by e_i and e_j (Line 10). Figure 4(a) shows two edges that meet the above criteria.

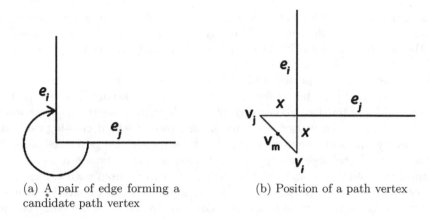

(a) A pair of edge forming a
candidate path vertex

(b) Position of a path vertex

Fig. 4. Finding corner point and the corresponding path vertex. (a) A pair of edge
forming a candidate path vertex (b) Position of a path vertex

---

Algorithm 2. FindPathVertices(Edge Set E, Door Edge Set E_d)

INPUT: Edge Set E, Door Edge Set E_d

OUTPUT: A list of path vertices

1: $path\_vertices \leftarrow \emptyset$
2: **for** $e_i \in E_d$ **do**
3:      $v_{mi} = $ mid-point of e_i
4:      $v_1 = $ vertex at distance $length(e_i)/2$ from v_{mi} left side and $\perp$ to e_i
5:      $v_2 = $ vertex at distance $length(e_i)/2$ from v_{mi} right side and $\perp$ to e_i
6:      add v_1 and v_2 to $path\_vertices$
7: **end for**
8: $x = $ average length of the doors
9: **for** each pair $(e_i, e_j) \in E$ **do**
10:      **if** $reflex\_edge(e_i, e_j, E)$ **then**
11:          $d_i = length(e_i)$
12:          $d_j = length(e_j)$
13:          $o = $ intersecting point of e_i and e_j
14:          $v_i = o + \hat{e}_i * (d_i + x)/d_i$
15:          $v_j = o + \hat{e}_j * (d_j + x)/d_j$
16:          $v_m = $ mid point of (v_i, v_j)
17:          add v_m to $path\_vertices$
18:      **end if**
19: **end for**
20: add to $path\_vertices$ the vertices around the bounding box of the obstacles
21: **return** $path\_vertices$

To compute the corresponding path vertex of a corner point, we extend two
edges by length x (we take x as the half of the average length of all doors) from
the corner point. Then we connect the end points of these two extended edges
and take the middle point of this connecting line as a path vertex (Lines 11-16).

The intuition behind selecting such a point as a path vertex is to establish a navigation point that do not falls into any other room and can be connected from the corridors of both side of the room. Figure 4(b) shows the path vertex for the corresponding corner point A.

We repeat the above procedure for all pair of reflex edges.

Now, we have all the necessary points for path navigations. The two points on the two sides of the door allow us to enter and exit from the room. The corner points are necessary for going a room to another room without crossing the wall edges. Finally, the four points around each obstacles are also generated as path vertices to avoid the obstacles during path navigation. To create a navigation structure, we need to know, to which vertices we can go from each vertex, i.e., neighbor vertices of each vertex. We can do this simply checking if an edge formed by any two path points intersects any wall edges or any obstacle bounding edges. We store such edges in the *Spatialite* database which is described in the next section.

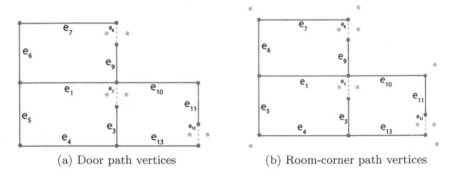

(a) Door path vertices   (b) Room-corner path vertices

Fig. 5. Steps for finding a path vertices

2.6   Autocadroid Spatialite Database

We store all the rooms, corridors and doors along with their connectivity in the *Spatialite* database [7]. We use Spatialite's *VirtualNetwork* module, which can connect the path vertices and generate the path automatically, for finding the path from one room to another room. Figure 6 shows the entity relationship diagram of our database (Fig. 5).

Room polygons are stored in **Room** table. The polygon is stored in *Polygon_Geom* column which is of *POLYGON* type. Typically most of the floors of a building have the same floor plan. So the same polygon can be used for different floors. We store the actual floor wise room entry in **Floor_Room** table. **Room** table has one-to-many relation with **Floor_Room** table. Thus, a room polygon does not need to be duplicated in every floor. A room can have multiple doors.

So the **Room** table has one-to-many relation with the **Door** table. The door edge is stored in *Door_Geom* column which is of *LINESTRING* type.

For each floor plan we make one entry in **Road_Network** table and all the related path edges are stored in **Path_Edge** table. Each floor has a reference to corresponding **Rood_Network**. *Start_Node* and *End_Node* columns in *Path_Edge* table represent the id of the start vertex and the end vertex of the path edge, respectively. *Edge_Geom* is of *LINESTRING* type. It stores the geometry of the path edge. The length of the path edge is stored as *Cost*.

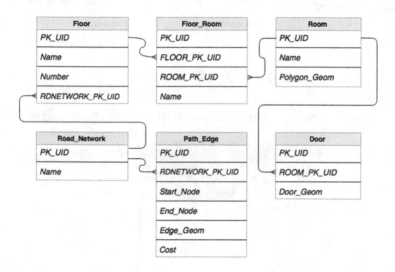

Fig. 6. Entity relationship diagram of indoor spaces

3   A Working Prototype System

We have developed the system, named AutoCadroid, that runs on Android platform. AutoCadroid imports a DXF file into the system and presents every layers of lines, polylines and arcs from the DXF file. If there are more than one floor plans, we can select a floor by drawing a boundary around that floor plan (Fig. 7(a)). In the next step, our system removes redundant lines, where a user can visualize the reduction of unnecessary lines (the screenshot is not shown). After that doors (Fig. 7(b)) and rooms (Fig. 7(c), where the red polygon indicates the corridor) are identified by the system. At the final stage of conversion, our system finds find path vertices for navigation (Fig. 7(d)). Identified data structure of all previous steps are stored in spatialite database as described in the previous section. AutoCadroid also allows to add description to a identified room or any object as shown in Fig. 7(e), and find a path to a destination as shown in Fig. 7(f).

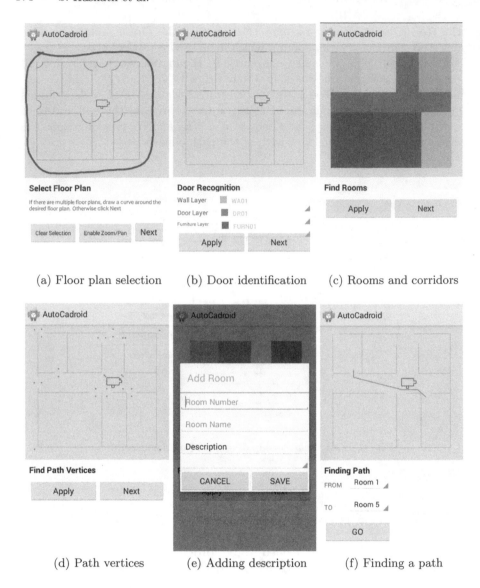

(a) Floor plan selection   (b) Door identification   (c) Rooms and corridors

(d) Path vertices   (e) Adding description   (f) Finding a path

Fig. 7. Screenshots of AutoCadroid

4   Experimental Evaluation

We have run a set of experiments to test the effectiveness and efficiency of our approach. We have used a Mac PC with processor: 2.4 GHz Intel Core i5 and memory: 8 GB 1600 MHz DDR3 running OS X 10.11.1. We test our algorithm using five DXF files. Table 1 shows the accuracy and efficiency of our door and room finding algorithms. Our algorithm identify all doors and rooms correctly. In one case (sample4.dxf), our algorithm cannot identify one room due to the missing line in the design file. We also see that in all cases, our system takes less than 100 ms to process all the steps from the pre-processing step to AutoCadroid database generation.

Table 1. Experiment results

File Name	Edges		Rooms		Doors		Path Vertices	Time (ms)
	Initial	Reduced	Actual	Identified	Actual	Identified		
floor_plan1.dxf	42	40	8	8	8	8	44	70
floor_plan2.dxf	91	57	6	6	9	9	36	54
yoffice3.dxf	65	33	4	4	2	2	44	45
sample4.dxf	114	58	11	10	14	14	59	77
sample5.dxf	139	75	11	11	17	17	67	92

5   Related Works

DXF files are widely used for designing most of the design applications like AutoCad, SolidWorks, and CATIA. To visualize DXF files in smartphones, *GeospatialDataAbstractionLibrary(GDAL)* [6] and *SpatialiteGUI* [8] are the two most popular software systems. They are not suitable for navigation as they reads DXF data types and stores these basic types into Spatialite database.

In [4], Boysen et al. proposed a software system to export a design file (i.e., Digital Building Information (DBI) file) to PostgreSQL with PostGIS, and then use a mobile application to search and navigate. They allow a user to identify doors and rooms manually, which is a time consuming and expensive process especially for bigger venues. Thus, recently Xu et al. [9] present an approach to automate the room identification from basic geometries of an AutoCAD DXF file, and then store rooms and doors in an Oracle database to facilitate faster search. The major limitations of [9] are (i) since they use cluster based technique for removing duplicate lines, this approach may remove lines which are essential for room identification, (ii) they remove all rooms that are completely inside other rooms, which may result in in-accuracy of the navigation, and (iii) this approach also fails to construct paths for complex scenarios, such as a shopping mall having multiple corridors and back doors.

6   Conclusion

In this paper, we have presented a system, AutoCadroid, that automatically identifies indoor space objects from the most widely used floor plan design file - an AutoCad DXF file. We have developed algorithms that can correctly construct rooms, doors, and corridors from a large number of lines, polylines etc. AutoCadroid enables an end-user to build an indoor space navigation system of her own office/home building. To facilitate faster query processing, we have exploited the strength of in-built spatial data management tool, Spatialite, to store the data. In summary, AutoCadroid removes a major barrier, i.e., the necessity of re-creating floor plan designs, towards the growth of the indoor LBSs, and hence AutoCadroid is a major step forward towards the development of indoor LBSs. We have built the whole system for an Android smartphone to facilitate an end-user to process the AutoCad design to build her own navigation system in one single device. In the future, we would like to implement path directions from one floor to another through elevator and stairs.

Acknowledgments. This research is supported by the ICT Division - Government of the People's Republic of Bangladesh.

References

1. A.H.D. (2015). http://help.autodesk.com/view/ACD/2015/ENU/
2. Afyouni, I., Ray, C., Claramunt, C.: Spatial models for context-aware indoor navigation systems: A survey. J. Spatial Inf. Sci. **4**, 85–123 (2012)
3. Ali, M.E., Tanin, E., Zhang, R., Kulik, L.: A motion-aware approach for efficient evaluation of continuous queries on 3d object databases. VLDB J. **19**(5), 603–632 (2010)
4. Boysen, M., de Haas, C., Lu, H., Xie, X., Pilvinyte, A.: Constructing indoor navigation systems from digital building information. In: ICDE, pp. 1194–1197 (2014)
5. S. database for Android with VirtualNetwork module. https://www.gaia-gis.it/fossil/libspatialite/wiki?name=splite-android
6. GDAL. http://gdal.org/drv_sqlite.html
7. Gotlib, D., Gnat, M.: Spatial database modeling for indoor navigation systems. Reports Geodesy Geoinformatics **95**, 49–63 (2013)
8. GUI, S.: https://www.gaia-gis.it/fossil/spatialite_gui/index
9. Xu, D., Jin, P., Zhang, X., Du, J., Yue, L.: Extracting indoor spatial objects from CAD models: a database approach. In: Liu, A., Ishikawa, Y., Qian, T., Nutanong, S., Cheema, M.A. (eds.) DASFAA 2015 Workshops. LNCS, vol. 9052, pp. 273–279. Springer, Heidelberg (2015)

Uncertain Data and Trajectory

Integration of Probabilistic Information

Fereidoon Sadri[(✉)] and Gayatri Tallur

Department of Computer Science, University of North Carolina,
Greensboro, NC, USA
f_sadri@uncg.edu

Abstract. We study the problem of data integration from sources
that contain probabilistic uncertain information. Data is modeled by
possible-worlds with probability distribution, compactly represented in
the probabilistic relation model. Integration is achieved efficiently using
the extended probabilistic relation model. We study the problem of determining the probability distribution of the integration result. It has been
shown that, in general, only probability ranges can be determined for
the result of integration. We show that under intuitive and reasonable
assumptions we can determine the exact probability distribution of the
result of integration. Our methodologies are presented in possible-worlds
as well as probabilistic-relation frameworks.

Keywords: Data integration · Probabilistic data · Uncertain data ·
Probabilistic relation model

1 Introduction

Information integration and modeling and management of uncertain information
have been active research areas for decades, with both areas receiving significant
renewed interest in recent years (*e.g.*, [8,11]). The importance of information
integration *with uncertainty*, on the other hand, has been realized more recently
(*e.g.*, [2,12]). In a world with ever increasing data generated by both humans
and machines alike, the field of computer science has seen a transition from
computation-intensive applications to data-intensive ones. Most of the data,
in particular data discovered through data mining and knowledge discovery, is
uncertain. Hence, integration of *uncertain* data has become a necessity for many
modern applications. It has been observed that "While in traditional database
management managing uncertainty and lineage seems like a nice feature, in data
integration it becomes a necessity" [12].

The widely accepted conceptual model for uncertain data is the possible-
worlds model [1]. For practical applications, a representation of choice is the
probabilistic relation model [10], which provides a compact and efficient representation for uncertain data. It has been shown that integration of uncertain
data represented in the probabilistic relation model can be achieved efficiently
using the extended probabilistic relation model [6].

© Springer International Publishing AG 2016
M.A. Cheema et al. (Eds.): ADC 2016, LNCS 9877, pp. 179–190, 2016.
DOI: 10.1007/978-3-319-46922-5_14

In this paper we concentrate on the integration of *probabilistic* uncertain data. Integration in the probabilistic relation framework is the most efficient approach but this approach faces challenges when probabilities are included. There is no clear way to associate probabilities with extended probabilistic relations (unlike probabilistic relations). Further, it has been shown that, even in the possible-worlds model, it is only possible to obtain probability *ranges* for the result of data integration [14]. In this paper we study the problem of determining the probability distribution of the integration result in the two main frameworks: The probabilistic possible-worlds model, and the probabilistic relation model. We show that, under intuitive and reasonable assumptions, we can determine the exact probability distribution of integration in either of the frameworks. Further, we show that the two approaches are equivalent while the probabilistic relation approach provides a significantly more efficient method in practice.

2 Preliminaries

Foundations of uncertain information integration were discussed in the seminal work of Agrawal *et al.* [2]. The goal of integration is to obtain the best possible uncertain database that contains all the information implied by sources, and nothing more. We presented an alternative integration approach in [14]. These approaches are based on the well-known *possible-worlds* model of uncertain information [1]. The possible-worlds model is widely accepted as the conceptual model for uncertain information, and is used as the theoretical basis for operations and algorithms on uncertain data. But it is not a suitable representation for the *implementation* of uncertain information systems due to lack of efficiency. Instead, compact representations, such as the *probabilistic relation model* [9,10], are more appropriate for the implementation. We also studied the problem of integration of information represented by probabilistic relations in [6], and presented efficient algorithms for the integration. In this section, we will review some of the observations and results from these works.

2.1 Integration Algorithm for Uncertain Data Represented in the Possible-Worlds Model

We begin with the following definition of *uncertain database* from [2].

Definition 1. (UNCERTAIN DATABASE) *An uncertain database U consists of a finite set of tuples $T(U)$ and a nonempty set of possible worlds $PW(U) = \{D_1, \ldots, D_n\}$, where each $D_i \subseteq T(U)$ is a certain database.*

We presented a logic-based approach to the representation and integration of uncertain data in the possible-world model in [14], and showed it was equivalent to the integration approach of [2]. Algorithm 1 below is an alternative integration algorithm. It is easy to show it is equivalent to the aforementioned algorithms. First, we need the following definition:

Definition 2. (COMPATIBLE POSSIBLE-WORLD RELATIONS). *Let S and S' be information sources with possible worlds $\{D_1, \ldots, D_n\}$ and $\{D'_1, \ldots, D'_{n'}\}$, respectively. Let T and T' be the tuple-sets of S and S'. A pair of possible-world relations D_i and D'_j are compatible if for each tuple $t \in T \cap T'$ either both D_i and D_j contain t, (i.e., $t \in D_i$ and $t \in D'_j$), or neither D_i nor D'_j contain t (i.e., $t \notin D_i$ and $t \notin D'_j$). Otherwise D_i and D'_j are not compatible.*

Given information sources S and S', the integration algorithm (Algorithm 1) considers all possible-world pairs from the two sources. If they are compatible, their union forms a possible-world of the integration.

Algorithm 1. Integration of uncertain data represented in the possible-worlds model

---

Given information sources S and S' with possible worlds $\{D_1, \ldots, D_n\}$ and $\{D'_1, \ldots, D'_{n'}\}$ and tuple sets T and T'

For every pair of possible-world relations $D_i \in S, D'_j \in S'$

  if D_i and D'_j are compatible then let $Q_{ij} = D_i \cup D'_j$

The possible-worlds model of the result of integrating S and S' has the set of possible-world relations Q_{ij} for every compatible pair D_i and D'_j, and the tuple set $T \cup T'$.

---

3   Integration of Probabilistic Uncertain Data

3.1   Probabilistic Possible-Worlds Model

Definition 3. *A probabilistic uncertain database U consists of a finite set of tuples $T(U)$ and a nonempty set of possible worlds $PW(U) = \{D_1, \ldots, D_n\}$, where each $D_i \subseteq T(U)$ is a certain database. Each possible world D_i has a probability $0 < P(D_i) \leq 1$ associated with it, such that $\sum_{i=1}^{n} P(D_i) = 1$.*

Our goal is to integrate information from sources containing probabilistic uncertain data, and to compute the probability distribution of the possible-worlds of the result of the integration. It has been shown that, in general, exact probabilities of the result of integration can not be obtained [14]. Rather, only a *range* of probabilities can be computed for each possible world in the integration. In this paper, we show that, under intuitive and reasonable assumptions, it is possible to obtain exact probabilities for the result of integration.

3.2   Integration in the Probabilistic Possible-Worlds Framework

Let S and S' be sources with possible worlds $\{D_1, \ldots, D_n\}$ and $\{D'_1, \ldots, D'_{n'}\}$, respectively. Consider the bi-partite graph G defined by the relation (D_i, D'_j): D_i and D'_j are compatible (See Definition 2). The graph G is called the *compatibility graph* for sources S and S': There is an edge between D_i and D'_j if they are compatible. It has been shown that [14]

– Each connected component of G is a complete bipartite graph.
– Let H be a connected component of G. Then $\sum_{D_i \in H} P(D_i) = \sum_{D'_j \in H} P(D'_j)$.
  These conditions have been called *probabilistic constraints* in [14].

Probabilistic constraints are imposed by the semantics of probabilistic integration. But it is unlikely that they hold in practice. We regard these constraints as important means to adjust (or revise) the original probabilities of the sources when the constraints are violated [15]. Henthforth we assume the probabilities have been adjusted and probabilistic constraints hold.

Example 1. Consider the possible worlds of information sources S and S' shown in Figs. 1 and 2.

The compatibility bipartite graph G for the possible-world relations of these sources is shown in Fig. 3. Note that we have $P(D_1) + P(D_2) = P(D'_1) + P(D'_2)$ and $P(D_3) = P(D'_3) + P(D'_4)$ by the probabilistic constraints.   □

student	course
Bob	CS100

D1

student	course
Bob	CS100
Bob	CS101

D2

student	course
Bob	CS101

D3

Fig. 1. Possible Worlds of source S

student	course
Bob	CS100

D'1

student	course
Bob	CS100
Bob	CS201

D'2

student	course
Bob	CS201

D'3

student	course
Bob	CS201
Bob	CS202

D'4

Fig. 2. Possible Worlds of source S'

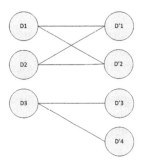

Fig. 3. Compatibility Graph for Example 1

There are 6 possible-world relations in the result of integration, corresponding to the two connected components of the compatibility graph. These 6 relations are shown in Fig. 6. Let us concentrate on the top connected component portion of the compatibility bipartite graph G. This connected component gives rise to 4 possible-world relations corresponding to $D_1 \wedge D_1'$, $D_1 \wedge D_2'$, $D_2 \wedge D_1'$, and $D_2 \wedge D_2'$. We want to compute the probabilities of these possible-world relations, $P(D_1 \wedge D_1')$, $P(D_1 \wedge D_2')$, $P(D_2 \wedge D_1')$, and $P(D_2 \wedge D_2')$, given the probability distribution of the possible worlds of the sources, $P(D_1), P(D_2), P(D_1'), p(D_2')$.

We have four unknowns. We can write the following four equations:

$$P(D_1 \wedge D_1') + P(D_1 \wedge D_2') = P(D_1),$$
$$P(D_2 \wedge D_1') + P(D_2 \wedge D_2') = P(D_2),$$
$$P(D_1 \wedge D_1') + P(D_2 \wedge D_1') = P(D_1'),$$
$$P(D_1 \wedge D_2') + P(D_2 \wedge D_2') = P(D_2').$$

But, unfortunately, these equations are not independent. Note that the probabilistic constraint requires that $P(D_1) + P(D_2) = P(D_1') + P(D_2')$. Hence, any one of the 4 equations can be obtained from the other 3 using the probabilistic constraint. Hence we can only compute a probability range for each of these four possible-world relation.

So, how can we obtain exact probabilities for the possible-world relations of an integration? We make the following *partial independence assumption*.

Partial Independence Assumption: *The only dependencies among the probabilities of possible-world relations are those induced by probabilistic constraints.*

Armed with this intuitive and reasonable assumption, we are able to compute exact probabilities for the result of an integration. We use our example to explain the approach, then present the solution for the general case.

Example 2. Consider again the top connected component in the compatibility graph of Example 1. The structure of the graph tells us that if we have the evidence that the correct database of the first source S is D_1, then we know the correct database of the second source S' is either D_1' or D_2'. Similarly, if we have the evidence that the correct database of the first source S is D_2, then we know the correct database of the second source S' is either D_1' or D_2'. But, by the partial independence assumption, the knowledge of D_1 or D_2 does not influence the probability of D_1'. In other words, $P(D_1' \mid D_1)$ is equal to $P(D_1' \mid D_2)$. Since $P(D_1' \wedge D_1) = P(D_1' \mid D_1)P(D_1)$ and $P(D_1' \wedge D_2) = P(D_1' \mid D_2)P(D_2)$ we get

$$\frac{P(D_1 \wedge D_1')}{P(D_2 \wedge D_1')} = \frac{P(D_1' \wedge D_1)}{P(D_1' \wedge D_2)} = \frac{P(D_1)}{P(D_2)}$$

This serves as an additional equation that enables us to solve for the 4 unknowns. We get:

$$P(D_1 \wedge D_1') = P(D_1)P(D_1')/(P(D_1) + P(D_2))$$
$$P(D_2 \wedge D_1') = P(D_2)P(D_1')/(P(D_1) + P(D_2))$$
$$P(D_1 \wedge D_2') = P(D_1)P(D_2')/(P(D_1) + P(D_2))$$
$$P(D_2 \wedge D_2') = P(D_2)P(D_2')/(P(D_1) + P(D_2))$$

The observations of the above example can be generalized. Let S_1 and S_2 contain information in probabilistic possible-worlds model. Consider a connected component G_1 of the compatibility bipartite graph G of S_1 and S_2. Let

$D_1, \ldots, D_m$ and $D'_1, \ldots, D'_{m'}$ be the nodes of G_1 corresponding to possible worlds of S_1 and S_2, respectively. We can write the following $m + m'$ equations:

$$\sum_{j=1}^{m'} P(D_i \wedge D'_j) = P(D_i), i = 1, \ldots, m; \text{ and } \sum_{i=1}^{m} P(D_i \wedge D'_j) = P(D'_j), j = 1, \ldots, m'$$

But $m + m' - 1$ of these equations are independent. Any one can be obtained from the rest using the probabilistic constraint $\sum_{i=1}^{m} P(D_i) = \sum_{j=1}^{m'} P(D'_j)$. On the other hand, we have $m \times m'$ unknowns $P(D_i \wedge D'_j), i = 1, \ldots, m, j = 1, \ldots, m'$. Additional equations are obtained from the independence assumption $\frac{P(D_1 \wedge D'_j)}{P(D_i \wedge D'_j)} = \frac{P(D_1)}{P(D_i)}$. It can be shown that $(m - 1) \times (m' - 1)$ of these equations are independent. Together with the $m + m' - 1$ equations of the first group we have the needed $m \times m'$ equations to solve for the unknowns. The solutions are,

$$P(D_i \wedge D'_j) = \frac{P(D_i)P(D'_j)}{P}$$

where P is the probabilistic constraint constant $P = \sum_{i=1}^{m} P(D_i) = \sum_{j=1}^{m'} P(D'_j)$.

4   Integration in the Probabilistic Relation Framework

A number of models have been proposed for the representation of uncertain information such as the "maybe" tuples model [7,13], set of alternatives or block-independent disjoint model (BID) [4,5], the probabilistic relation model [9,10], and the U-relational database model [3]. The probabilistic relation model has been widely accepted for compact representation of uncertain and probabilistic data. It is a *complete* model: Any uncertain data in the (probabilistic) possible-worlds model can be represented in an equivalent probabilistic relation [10]. Intuitively, this representation is based on the relational model where each tuple t is associated with a propositional logic formula $f(t)$ (called an *event* in [9].) The Boolean variables in the formulas are called *event variables*. A probabilistic relation r represents the set of possible-world relations corresponding to truth assignments to the set of event variables. A truth assignment μ defines a possible-world relation $r_\mu = \{t \mid t \in r \text{ and } f(t) = true \text{ under } \mu\}$.

In the previous section we presented an approach for the integration of probabilistic uncertain data in the probabilistic possible-worlds framework. As mentioned earlier, the possible-worlds framework is not suitable for practical applications. The size of the input, namely the possible-worlds relations, can be exponential in the size of the equivalent representation in the probabilistic relation framework. Further, we have a very efficient integration algorithm in the probabilistic relation framework. In this section we concentrate on the problem of determining the probability distribution for the integration result in the probabilistic relation framework.

4.1 Probabilistic Data

A probabilistic relation can represent *probabilistic* possible-worlds data by associating probabilities with event variables. Let r be a probabilistic relation. We can compute the probabilities associated with possible-world relations represented by r as follows. Let $V = \{a_1, a_2, \ldots, a_k\}$ be the set of event variables of r. Let μ be a truth assignment to event variables. μ defines a relation instance $r_\mu = \{t \mid t \in r \text{ and } f(t) = true \text{ under } \mu\}$. The probability associated with r_μ is $\prod_{\mu(a_j)=true} P(a_j) \prod_{\mu(a_j)=false} (1 - P(a_j))$. Note that this formula is based on the assumption that event variables are independent of each other. A possible-world relation r_i of r can result from multiple truth assignments to event variables, in which case the probability of r_i, $P(r_i)$ is the sum of probabilities of r_μ for all truth assignments μ that generate r_i.

Example 3. Consider the possible worlds of information sources S and S' from Example 1, shown in Figs. 1 and 2. Assume the probability distributions are $P(D_1) = 0.3$, $P(D_2) = 0.5$, $P(D_3) = 0.2$, $P(D_1') = 0.35$, $P(D_2') = 0.45$, $P(D_3') = 0.05$, and $P(D_4') = 0.15$. Algorithms for producing probabilistic relations for uncertain probabilistic databases have been presented in [6,10]. We have used the algorithm of [6] to obtain the probabilistic relations r_1 and r_2 of Fig. 4 for the uncertain probabilistic databases of Figs. 1 and 2. In Fig. 4, b_1, b_2, b_3, c_1 and c_2 are event variables, and column E records the event formulas associated with each tuple. Probabilities of the event variables are also computed by the algorithm and are: $P(b_1) = 0.35$, $P(b_2) = \frac{9}{13}$, $P(b_3) = 0.25$, $P(c_1) = 0.2$, and $P(c_2) = 0.625$. □

student	course	E
Bob	CS100	$\neg c_1$
Bob	CS101	$c_1 \vee c_2$

pr-relation r

student	course	E
Bob	CS100	$b_1 \vee b_2$
Bob	CS201	$\neg b_1$
Bob	CS202	$\neg b_1 \wedge \neg b_2 \wedge \neg b_3$

pr-relation r'

Fig. 4. Probabilistic relations for sources S and S' of Example 1

4.2 Integration of Uncertain Data Represented in the Probabilistic Relation Model

As mentioned earlier, for efficiency reasons a compact representation of uncertain data is utilized in practice. We will summarize an algorithm for the integration of uncertain data represented in the probabilistic relation model from [6]. First we need the following definition.

Definition 4. *An extended probabilistic relation (epr-relation, for short) is a probabilistic relation with a set of event constraints. Each event constraint is a propositional formula in event variables.*

Semantics of an extended probabilistic relation is similar to that of probabilistic relation, with the exception that only truth assignments that satisfy event constraints are considered. More specifically, A truth assignment μ to event variables is *valid* if it satisfies all event constraints. A valid truth assignment μ defines a relation instance $r_\mu = \{t \mid t \in r$ and $f(t) = true$ under $\mu\}$, where $f(t)$ is the event formula associated with tuple t in r. The extended probabilistic relation r represents the set of relations, called its possible-world set, defined by the set of all valid truth assignments to the event variables. We will use the abbreviation *epr-relation* for extended probabilistic relation henceforth.

Given information sources S and S', let r and r' be the probabilistic relations that represent the data in S and S', respectively. We represent a tuple in a probabilistic relation as $t@f$, where t is the pure tuple, and f is the propositional event formula associated with t. Let $r = \{t_1@f_1, \ldots, t_n@f_n\}$, where f_i is the event formula associated with the tuple t_i. Similarly, let $r' = \{u_1@g_1 \ldots, u_m@g_m\}$. We assume the set of event variables of r (*i.e.*, event variables appearing in formulas $f_1, \ldots, f_n$) and those of r' (*i.e.*, event variables appearing in formulas $g_1, \ldots, g_m$) to be disjoint. If not, a simple renaming can be used to make the two sets disjoint. r and r' can have zero or more common tuples. Assume, without loss of generality, that r and r' have p tuples in common, $0 \le p \le min(n, m)$, $t_1 = u_1, \ldots, t_p = u_p$. The integration algorithm is represented in Algorithm 2. In Algorithm 2, $f_i \equiv g_i$ is equivalent to the logical formula $(f_i \rightarrow g_i) \wedge (g_i \rightarrow f_i)$. We will use the notation $q = r \uplus r'$ to mean that q is the epr-relation that is the result of integration of probabilistic relations r and r'.

Algorithm 2. Integration of uncertain data represented by probabilistic relations

Given information sources S and S', let r and r' be the probabilistic relations that represent the data in S and S'. The result of integration of S and S' is represented by an epr-relation q obtained as follows:

- Copy to q the tuples in r that are not in common with r'
- Copy to q the tuples in r' that are not in common with r For each of the common tuples, copy to q the tuple either from r or from r'.
- For each common tuple t_i, add a constraint $f_i \equiv g_i$, to the set of event constraints of q, where f_i and g_i are the event formulas associated with t_i in r and r', respectively.

It has been shown in [6] that Algorithm 2 is correct. That is, when $q = r \uplus r'$ is obtained by this algorithm, then the possible-worlds of q coincide with the possible-worlds obtained by integrating possible-worlds of r and r' by Algorithm 1.

The complexity of Algorithm 2 is $O(n \log n)$, where n is the size of input (pr-relations of the sources). While the complexity of the possible-worlds integration algorithm (Algorithm 1) is quadratic in the size of its input (possible-world relations of the sources) which itself can be exponential in the size of the input of Algorithm 2.

student	course	E
Bob	CS100	$\neg c_1$
Bob	CS101	$c_1 \vee c_2$
Bob	CS201	$\neg b_1$
Bob	CS202	$\neg b_1 \wedge \neg b_2 \wedge \neg b_3$
$\neg c_1 \equiv b_1 \vee b_2$		

epr-relation $q = r \uplus r'$

Fig. 5. Extended Probabilistic relation for the integration of sources S and S'

Example 4. The result of integration of probabilistic relations of Example 3 (which themselves are equivalent to the possible-worlds relations of Example 1) is the epr-relation of Fig. 5, obtained using Algorithm 2.   ∎

4.3   Determining Probabilities for Extended Probabilistic Relations

While probability computation is straightforward for probabilistic relations as discussed in Sect. 4.1, we do not have a general approach for probability computation for extended probabilistic relations. The reason is that we can no longer assume event variables are independent. Event constraints impose certain dependencies among event variables. Indeed, if we assume event variables are independent, the sum of the probabilities calculated for the possible-worlds of an epr-relation is not equal to 1. This is due to the fact that only a subset of all possible-world relations, those that correspond to valid truth assignments to event variables, are taken into account. We need an approach for probability computation for extended probabilistic relations. If not, we are forced to use the highly inefficient probabilistic possible-worlds approach for the integration of probabilistic data. Further, our probability computation approach for epr-relations must be *equivalent* to the possible-worlds approach. In other words, we have two conceptually equivalent methodologies for probabilistic data: the possible-worlds (highly inefficient) and probabilistic relation (efficient). But whatever we achieve in the probabilistic relation domain must coincide with the possible-worlds domain.

We show that under an intuitive and reasonable assumption regarding the correlation of event variables of epr-relations we are able to compute the probabilities of the result of integration. Further, this assumption is closely related to the *Partial Independence Assumption* in the possible-worlds domain (discussed in Sect. 3.1).

Partial Independence Assumption for Extended Probabilistic Relations: *Event variables are independent except for the relationships induced by the event constraints.*

The following example shows how this assumption allows us to compute probabilities for the extended probabilistic relation framework. We will discuss the correctness of this approach in the next section (Sect. 5).

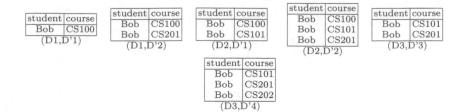

Fig. 6. Possible-world relations of the result of integration of sources S and S'

Example 5. Let's go back to Example 4. The result of integration is the epr-relation shown in Fig. 5. The possible-worlds relations of this epr-relation are shown in Fig. 6.

How can we calculate the probability distribution of the result of integration (possible-world relations of Fig. 6)? The event-variable formulas for the 6 possible-world relations of the integration in this case are $\neg c_1 \wedge \neg c_2 \wedge b_1$, $\neg c_1 \wedge \neg c_2 \wedge \neg b_1 \wedge b_2$, $\neg c_1 \wedge c_2 \wedge b_1$, $\neg c_1 \wedge c_2 \wedge \neg b_1 \wedge b_2$, $c_1 \wedge \neg b_1 \wedge \neg b_2 \wedge b_3$, and $c_1 \wedge \neg b_1 \wedge \neg b_2 \wedge b_3$.

By the partial independence assumption event variables are independent except for the relationships induced by the event constraints. The constraint $\neg c_1 \equiv b_1 \vee b_2$ induces a relationship between c_1 on one hand, and b_1 and b_2 on the other. The rest are still independent. So, for example, c_1 and c_2 are independent, and so are c_2 and b_1; etc. In particular, b_1 and b_2 are also independent. To compute the probability associated with an event-variable formula, we rewrite the formula so that it only contains mutually independent event variables. For example, $\neg c_1 \wedge \neg c_2 \wedge b_1$ is simplified to $\neg c_2 \wedge b_1$ using the equivalence $\neg c_1 \equiv b_1 \vee b_2$. Then we are able to compute the probabilities. In this example, we obtain the following probabilities for the 6 possible-world relations: 0.13125, 0.16875, 0.21875, 0.28125, 0.05, and 0.15.

Let us compare this approach with the integration in the probabilistic possible-worlds framework (Sect. 3.1). It is easy to verify that the probabilistic distribution of the result of the integration computed by the formula $P(D_i \wedge D'_j) = P(D_i)P(D_j)/P$ is exactly the same as the distribution obtained above. For example, the probability of the possible world corresponding to (D_1, D'_1) is $0.3 \times 0.35/(0.3 + 0.5) = 0.13125$.   □

5   Correctness of Probability Computation

In this section we present an overview of theoretical issues that form the basis of probability computation algorithms presented in previous sections. Interested readers are referred to the full paper which contains detailed discussions and proofs [16].

An important issue that we must address is the following:

Let a source S contain uncertain data in the probabilistic possible-worlds model, and r be an equivalent representation in the probabilistic relation model for data contained in S. It is well known that r is not unique. There can be many probabilistic relations representing the data in S.

This means that the result of integration of two sources S and S' in the possible-worlds domain is unique, but it is not unique in the probabilistic relation domain. For example, let r_1 and r_2 be alternative probabilistic relation representations for S, and r'_1 and r'_2 be alternative representations for S'. The result of integration of S and S' in the possible-worlds domain is unique. But in the probabilistic relation domain we can get (extended probabilistic relations) $r_1 \uplus r'_1$, $r_1 \uplus r'_2$, $r_2 \uplus r'_1$, and $r_2 \uplus r'_2$, where $\uplus$ is the integration operator (*e.g.*, using Algorithm 2). We must prove that, by our probability computation algorithm for epr-relations, all these epr-representations are indeed equivalent in the sense that they correspond to the probabilistic possible-worlds representation of the integration of S and S' (e.g., using Algorithm 1) where the probability distribution is obtained according to the approach of Sect. 3.2.

To address the above issues, we have shown the following [16].

- Given a probabilistic (or extended probabilistic) relation r, we associate a logical formula in terms of the event variables of r with each possible-world relation represented by r.
- We considered a subclass of extended probabilistic relations, namely, those that can be obtained through integration of sources represented by probabilistic relations.
- We prove that when the result of integration can be obtained by multiple epr-relations, these relations are equivalent in the following sense:

  • All epr-relations have exactly the same set of possible-worlds relations.
  • The logical formulas associated with each possible-world relation in different epr-relations are (logically) equivalent when event constraints are taken into account.

- Since the probability computation for epr-relations uses the logical formulas plus event constraints, different epr-relations for the integration of the same two sources have exactly the same probability distribution.
- Further, we also show that in the possible-worlds integration approach, the probabilities computed for the possible-world relations are exactly the same as those computed in the equivalent probabilistic relations framework.

Interested readers are referred to [16] for detailed discussions.

6   Conclusion

We focused on data integration from sources containing probabilistic uncertain information, in particular, on computing the probability distribution of the result of integration. We presented integration algorithms for data represented in two frameworks: The probabilistic possible-worlds model and the probabilistic relation model. In the latter case the result of integration is represented by an extended probabilistic relation. We showed that under intuitive and reasonable assumptions the exact probability distribution of the result of integration can be computed in the two frameworks. Alternative approaches to the computation of the probability distribution were presented in the two frameworks.

References

1. Abiteboul, S., Kanellakis, P.C., Grahne, G.: On the representation and querying of sets of possible worlds. In: Proceedings of ACM SIGMOD International Conference on Management of Data, pp. 34–48 (1987)
2. Agrawal, P., Sarma, A.D., Ullman, J.D., Widom, J.: Foundations of uncertain-data integration. Proc. VLDB Endowment **3**(1), 1080–1090 (2010)
3. Antova, L., Jansen, T., Koch, C., Olteanu, D.: Fast and simple relational processing of uncertain data. In: Proceedings of IEEE International Conference on Data Engineering, pp. 983–992 (2008)
4. Barbará, D., Garcia-Molina, H., Porter, D.: The management of probabilistic data. IEEE Trans. Knowl. Data Eng. **4**(5), 487–502 (1992)
5. Benjelloun, O., Sarma, A.D., Halevy, A.Y., Theobald, M., Widom, J.: Databases with uncertainty and lineage. VLDB J. **17**(2), 243–264 (2008)
6. Dayyan Borhanian, A., Sadri, F.: A compact representation for efficient uncertain-information integration. In: Proceedings of International Database Engineering and Applications IDEAS, pp. 122–131 (2013)
7. Codd, E.F.: Extending the database relational model to capture more meaning. ACM Trans. Database Syst. **4**(4), 397–434 (1979)
8. Dalvi, N.N., Ré, C., Suciu, D.: Probabilistic databases: diamonds in the dirt. Commun. ACM **52**(7), 86–94 (2009)
9. Dalvi, N.N., Suciu, D.: Efficient query evaluation on probabilistic databases. In: Proceedings of International Conference on Very Large Databases, pp. 864–875 (2004)
10. Dalvi, N.N., Suciu, D.: Efficient query evaluation on probabilistic databases. VLDB J. **16**(4), 523–544 (2007)
11. Haas, L.: Beauty and the beast: the theory and practice of information integration. In: Schwentick, T., Suciu, D. (eds.) ICDT 2007. LNCS, vol. 4353, pp. 28–43. Springer, Heidelberg (2006). doi:10.1007/11965893_3
12. Halevy, A.Y., Rajaraman, A., Ordille, J.J.: Data integration: The teenage years. In: Proceedings of International Conference on Very Large Databases, pp. 9–16 (2006)
13. Liu, K.C., Sunderraman, R.: On representing indefinite and maybe information in relational databases. In: Proceedings of IEEE International Conference on Data Engineering, pp. 250–257 (1988)
14. Sadri, F.: On the foundations of probabilistic information integration. In: Proceedings of International Conference on Information and Knowledge Management, pp. 882–891 (2012)
15. Sadri, F.: Belief revision in uncertain data integration. In: Sharaf, M.A., Cheema, M.A., Qi, J. (eds.) ADC 2015. LNCS, vol. 9093, pp. 78–90. Springer, Heidelberg (2015). doi:10.1007/978-3-319-19548-3_7
16. Sadri, F., Tallur, G.: Integration of probabilistic uncertain information (2016). CoRR, abs/1607.05702

Top-k Dominance Range-Based Uncertain Queries

Ha Thanh Huynh Nguyen$^{(\boxtimes)}$ and Jinli Cao

La Trobe University, Melbourne, Australia
ht34nguyen@students.latrobe.edu.au,
J.Cao@latrobe.edu.au

Abstract. Most of the existing efforts for probabilistic skyline queries have used data modeling where the appearance of the object is uncertain while the attribute values of objects are certain. In many real-life applications, the values of an uncertain object can be in a continuous range that a probability density function is employed to describe the distribution of the values. In addition, the "interest-ingness" of the objects as a single criterion for measuring skyline probability may result in missing some desirable data objects. In this paper, we introduce a new operator, namely, the Top-k Dominating Range (TkDR) query, to identify the subset of truly interesting objects by considering objects' dominance scores. We devise the ranking criterion to formalize the TkDR query and propose three algorithms for processing the TkDR query. Performance evaluations are conducted on both real-life and synthetic datasets to demonstrate the efficiency, effectiveness and scalability of our proposed approach.

1 Introduction

Data values are inherently uncertain in several important applications, such as data integration, sensor data analysis, moving object detection, market surveillance, trend prediction, economic decision making and mobile data management. The uncertainty which exists in these applications is generally caused by device failure, the limitations of the measuring equipment, a delay in data transfer or updates, etc. [1, 2]. With the rapid increase in data volume and the importance of these applications, the need to develop advanced analytical tools to explore a large amount of uncertain data has become an important task. Unlike query processing in deterministic databases, providing meaningful query answers to end users from such imprecise databases is more complex as the uncertainty in data affects the accuracy of the answers to queries. For these reasons, modelling, processing and analyzing uncertain data have received considerable attention from the database community in recent years.

1.1 Motivation

Skyline queries have been proven to be powerful analytical techniques in a wide spectrum of multi-criteria decision-making applications by helping users identify a subset of desirable objects from the huge amount of available data. In addition, the

© Springer International Publishing AG 2016
M.A. Cheema et al. (Eds.): ADC 2016, LNCS 9877, pp. 191–203, 2016.
DOI: 10.1007/978-3-319-46922-5_15

number of applications involved with uncertain data has significantly increased in recent years. Incorporating uncertainty into the process of managing and querying databases introduces many complex challenges, including query semantics, evaluation, and efficiency [3]. As a result, there have been many attempts to propose semantics and algorithms for skyline queries on uncertain data recently [4–10].

In the literature, uncertain databases can be categorized as tuple-level uncertainty and attribute-level uncertainty. However, the majority of existing work on the development of algorithms is based on the model of tuple-level uncertainty or the model of the discrete case of attribute-level uncertainty. These algorithms cannot be used to handle continuous distributions in attribute values directly.

Due to the uncertain nature or actual system limitation in the data collection phase, continuous attribute uncertainty arises naturally in many real-world applications. Scant work addresses the probabilistic skyline query where data objects are represented continuously within their uncertain regions. In [11], the authors proposed the notion of p-skyline which returns all objects whose skyline probabilities are at least p while in [2], Zhang et al. introduce the top-k SOUND query which retrieves k uncertain objects having the largest skyline probabilities. However, the limitation of these two approaches is that they measure the "interestingness" of an uncertain object merely based on the objects' skyline probabilities which may result in missing some desirable data objects [12]. In a continuous model, every data object is associated with a skyline probability, and as the number of dimensions increases, these skyline probabilities are very small, approaching zero, which makes it difficult to provide a subset of truly important and desirable objects to the users [13]. In addition, in [2, 11], each uncertain object is represented by an uncertain region in a multi-dimensional space associated with a multivariate probability density function (pdf). Once an instance of an object is generated by its pdf, the value of this instance on every dimension is certain, whereas in our data model, the value of an object on each dimension is modelled as an uncertain interval $[lo, hi]$ and is associated with a pdf to describe the distribution of values within the range. Therefore, the algorithms proposed in [2, 11] cannot be directly applied to solve our problem.

Inspired by these limitations, we introduce a new operator, named the Top-k Dominance Range (TkDR) query to extract a subset of truly interesting data objects, which processes probabilistic databases where the attribute values of objects are within a continuous range. We discover the "interestingness" of a data object not merely based on skyline probabilities but also using one extra criterion. We use the intuition that an object is more interesting if it can dominate more other objects and has been dominated by fewer objects.

1.2 Contributions

To the best of our knowledge, this is the first work to use the dominance range concept for uncertain queries. In brief, the key contributions of this paper are summarized as follows:

- Modeling uncertain objects' values as continuous ranges, providing a new ranking criterion and formalizing the Top-k Dominance Range (TkDR) query to return a subset of truly interesting data objects based on their dominance score.
- Proposing two efficient algorithms based on the Aggregate R-Tree (aR-Tree) indexing structure to answer the TkDR query.
- Conducting extensive experiments with both real and synthetic datasets to demonstrate the performance of our proposed algorithms.

The rest of paper is organized as follows. Section 2 surveys the related research work for the background. Section 3 discusses the new ranking criteria and formalizes our problem. Section 4 describes the two proposed algorithms using aR-tree index structure to efficiently process the TkDR query. Extensive experimental results and our findings are reported in Sect. 5. Finally, Sect. 6 concludes our paper with some suggestions for future work.

2 Related Work

The skyline operator was first introduced in [14] and has attracted considerable attention in the database community in recent years. Since then, a number of algorithms for skyline queries have been proposed in the literature, including, to name a few, Bitmap [15], Nearest Neighbor (NN) [16], Branch and Bound (BBS) [10], Lattice [17], Block-Nested-Loop (BNL) [14], Divide and Conquer (D&C) [14] and Sort-First-Skyline (SFS) [18]. In the literature, an uncertain object can be defined either discretely or continuously. Although probabilistic skyline and top-k queries have been well studied in uncertain databases, the majority of existing work on probabilistic skyline queries was limited only to the discrete model where an object is presented as a finite set of instances associated with its existence probability [19–22]. Scant work [2, 11] addresses probabilistic skyline queries where data objects are represented continuously in their uncertain regions. As mentioned in Sect. 1.1, their measures are limited in effectiveness, hence they may miss out some desirable data objects.

Our data model used in this research differs from the data models in [2, 11] where each uncertain object is represented by an uncertain region in multi-dimensional space. The values of an object on each dimension are modelled as an interval with equal bounds (certain value) or different bounds (uncertain value). As a result, the algorithms in [2, 11] cannot handle our Top-k Dominance Range (TkDR) query efficiently.

3 Preliminaries

In this section, we formally define the ranking criterion and the Top-k Dominance Range (TkDR) query in Subsects. 3.1 and 3.2, respectively.

3.1   Ranking Criterion

Let $\mathcal{P}^n$ be a set of uncertain data objects and $n = |\mathcal{P}|$ be the number of uncertain objects in $\mathcal{P}$. The value of object $X \in \mathcal{P}^n$ is modeled by a probability density function, denoted by $f_X(.)$, which describes the distribution of the value of O within the interval $[lo_X, hi_X]$. Generally, $\int_{lo_X}^{hi_X} f_X(x)dx = 1$ and $f_X(x) = 0$ if $x < lo_X$ or $x > hi_X$. Without loss of generality, in the rest of the paper, we assume that bigger values are preferred and all values are in $\mathbb{R}^+$ (i.e., values are non-negative). To keep our model simple, we assume that all uncertain objects in the databases are independent of each other and the probability density function within the uncertain interval of value is uniformly distributed.

Definition 1 (Uncertainty Value Interval). An uncertainty value interval of object X is an interval $[lo_X, hi_X]$ where $lo_X, hi_X \in \mathbb{R}^+$ and $hi_X > lo_X$.

Definition 2 (Dominate Probability). Given two uncertain objects X and Y, where the range of the value of X and Y is uniformly defined in the interval $[lo_X, hi_X]$ and $[lo_Y, hi_Y]$ respectively, assume that X and Y are independent, the probability that X dominates Y in one dimension, denoted by $Pr[X \prec Y]$, is a joint probability distribution and is defined as follows:

$$Pr[X \prec Y] = \int_{lo_X}^{hi_X} \int_{lo_Y}^{x} f_X(x).f_Y(y)dydx \tag{1}$$

Trivial cases.

When $hi_Y \leq lo_X, Pr[X \prec Y] = 1$
When $lo_Y \geq hi_X, Pr[X \prec Y] = 0$

Non-trivial cases. As we assume that the probability density functions within the uncertain intervals of value are uniformly distributed, (1) can be re-written as follows

$$Pr[X \prec Y] = \int_{lo_X}^{hi_X} Pr[Y < x] \frac{1}{hi_X - lo_X} dx \tag{2}$$

where $Pr[Y < x] = \int_{lo_Y}^{x} f_Y(y)dy$ is the cumulative distribution function and is defined as follows:

$$Pr[Y < x] = \left\{ \begin{array}{ccc} 0 & if & x \leq lo_Y \\ \frac{x - lo_Y}{hi_Y - lo_Y} & if & lo_Y < x < hi_Y \\ 1 & if & x \geq hi_Y \end{array} \right\}$$

Then, we have two sub-cases:

- When $lo_X < hi_Y < hi_X$, then

$$\Pr[X \prec Y] = \frac{1}{(hi_X - lo_X)(hi_Y - lo_Y)} \int_{\max(lo_X, lo_Y)}^{hi_Y} (x - lo_Y)dx + \frac{1}{(hi_X - lo_X)} \int_{hi_Y}^{hi_X} dx$$

$$= \frac{1}{(hi_X - lo_X)(hi_Y - lo_Y)} \left(\frac{hi_Y^2}{2} - lo_Y.hi_Y - \frac{\max(lo_X, lo_Y)^2}{2} + lo_Y.\max(lo_X, lo_Y) \right) + \frac{hi_X - hi_Y}{hi_X - lo_X}$$

- When $lo_Y < hi_X < hi_Y$, then

$$\Pr[X \prec Y] = \frac{1}{(hi_X - lo_X)(hi_Y - lo_Y)} \int_{\max(lo_X, lo_Y)}^{hi_X} (x - lo_Y)dx$$

$$= \frac{1}{(hi_X - lo_X)(hi_Y - lo_Y)} \left(\frac{hi_X^2}{2} - lo_Y.hi_X - \frac{\max(lo_X, lo_Y)^2}{2} + lo_Y.\max(lo_X, lo_Y) \right)$$

In the multi-dimensional case, we can calculate the dominate probability between two objects independently in each dimension and then combine them to an overall dominate probability.

Definition 3 (Dominating Score). Given an uncertain object X, the dominating score of X, denoted by $dgs(X)$, is the summation of the probability that X dominates other uncertain objects in the data set.

$$dgs(X) = \sum_{\substack{Y \neq X \\ Y \in \mathcal{P}^n}} \Pr[X \prec Y]$$

Definition 4 (Dominated Score). Given an uncertain object X, the dominated score of X, denoted by $dds(X)$, is the summation of probability that other uncertain objects in the dataset dominate it.

$$dds(X) = \sum_{\substack{Y \neq X \\ Y \in \mathcal{P}^n}} \Pr[Y \prec X]$$

Definition 5 (Dominance Score). Given an uncertain object X, the dominance score of X, denoted by $dns(X)$ is defined as:

$$dns(X) = dgs(X) - dds(X)$$

3.2   Problem Formulation

We now present the formal definition of the Top-k Dominance Range Query based on the ranking criterion defined in Sect. 3.1.

Definition 6 (Top-k Dominance Range Query). Given a set $\mathcal{P}^n$ of uncertain data and a user-defined integer k, the top-k dominance range query retrieves the k uncertain objects that have the largest dominance score.

We use the notion of dominance score to measure the "interestingness" of an uncertain data object. A higher dominance score of an object indicates that it is a more interesting answer.

4   Computing Top-k Dominance Range Query

In this section, we introduce our algorithms to answer the Top-k Dominance Range query which can identify the most interesting uncertain objects, according to the aforementioned ranking criterion. A straightforward algorithm, the Progressive Algorithm, is used to answer the TkDR query following the framework: it iterates through each uncertain object in the dataset; computes its dominating (resp., dominated, and dominance) scores by performing pairwise dominance checks against all other objects; then sorts the list of dominance scores in descending order and returns k objects having the largest scores. However, the computational cost of this method is extremely high since there is such a huge number of pairwise checks to be performed in order to obtain the final answer, regardless of the value of k. Motivated by this problem, we analyzed some properties of the dominance relationship which essentially allow us to disqualify objects without computing their exact probabilities by exploiting the lower and upper bounds of the dominated/dominating scores. As a result, the number of pairwise comparisons is significantly reduced and hence, the efficiency of our algorithms can be greatly improved.

Let X and Y be two uncertain objects where the range of values of X and Y is uniformly defined in the interval $[lo_X, hi_X]$ and $[lo_Y, hi_Y]$, respectively.

Definition 7 (Domination relationship). If $lo_X \geq hi_Y$ then X fully dominates Y, and we denote the relationship as $X \preccurlyeq Y$. We have $\Pr[X \prec Y] = 1$

Property 1. If $X \preccurlyeq Y$, then the dominated score of Y is at least as X, i.e. $dds(Y) \geq dds(X)$, and the dominating score of Y is at most as X, i.e. $dgs(Y) \leq dgs(X)$.

Exploiting the above property allows us to (i) reduce unnecessary dominance checks and (ii) design a stopping condition so that the search operation can be terminated earlier without exhaustively scanning the whole dataset. In order to avoid the computation exact dominance score of each examined object where the expensive cost occurs, we next present two algorithms, namely, the One Scan Algorithm and the Indexed Based Algorithm, which are more efficient and effective in terms of I/O cost and number of dominance checks.

4.1   One Scan Algorithm

The goal of this algorithm is to efficiently answer the TkDR query without exhaustively iterating the whole dataset by establishing lower and upper bounds for the dominated &

dominating scores to discard disqualified objects. In general, an uncertain object U is qualified to be in the result set if it is better than the k-th object in the answer set (i.e. lower dds or higher dgs). To facilitate the algorithm, we assume that all uncertain objects are sorted in descending order of the upper values of the intervals (i.e. hi values) in the input list L_{hi}. We denote DL_X as the dependent list of U which contains all objects having intervals which overlap with X. It is clear that processing object $U \in L_{hi}$ may affect the dominance score of the objects before U in L_{hi} and those in its dependent list.

We also maintain another list $\mathcal{E}$ to store the objects that (i) already examined and (ii) fully dominate as least one other examined object; and a max heap $\mathcal{Q}$ will contain objects that have been temporarily dis-qualified from the result set R. Let X be the currently examined object. We maintain a lower bound for $dds^-(X)$ and upper bound for $dgs^+(X)$ in order to check whether the search process should be terminated. In particular, we define $dds^-(X)$ and $dgs^+(X)$ as follows:

$$dds^-(X) = |\mathcal{E}| \text{ and } dgs^+(X) = dgs(S_k)$$

where S_k is the k-th object in R if $|R| = k$. The termination conditions guarantee that X or any other objects after X in L_{hi} have no chance of being better than S_k. If the result set is empty or less than k objects, X will be inserted to R (line 4). If there exists another object in $\mathcal{Q} \cap S_k$ fully dominates X, then X will be ignored (line 6), otherwise we compare X with every object Y in $R \cap \mathcal{Q}$ to check if X is qualified for the final answer set or if it can turn a loser (objects temporarily dis-qualify) to a winner (objects qualify for the result set) or vice versa (line 10–17).

One Scan Algorithm
Input: the list L_{hi} and the user-defined integer k
Output: the result set of k most interesting data objects

```
1 begin
2 initialize R= ∅, ε = ∅ and Q = ∅ (max heap based on dns)
3 for each object X ∈ L_hi do
4     if 0 ≤ |R| < k then insert X to R
5     else //|R| = k
6          if ∃V ∈ Q∩S_k such that V ≼ X then skip X
7          let S_k is the k-th object in R
8          if |ε| > dds(S_k) and hi_X < lo_{S_k} then return R
9       compute dgs(X), dds(X) and dns(X)
10      for each object Y ∈ R∪Q  do
11           update dgs(Y), dds(Y) and dns(Y)
12           if Y ∉ ε  and Y fully dominate X then
13                insert Y to ε
14           re-sort R and Q
15           let S_k is the current k-th object in R
16           Z:= deheap(Q)
17           Compare X, Z and S_k to update R, ε and Q
18 return R
19 end
```

4.2 Bounding Dominance Score

The One Scan algorithm can efficiently reduce the search space and provide a list of the most interesting candidates. However, there is a limitation in that the One Scan algorithm requires uncertain data objects in the input list to be sorted in descending order of upper values of intervals and the cost for sorting is considerably expensive. To further improve the One Scan algorithm, we propose an enhanced algorithm to process the TkDR query based on the indexing structure aR-tree, an Indexed Based algorithm, which does not require the objects to be pre-sorted. In addition, our experiments show that the efficiency of the Indexed Based algorithm has been significantly improved, compared with the One Scan algorithm, as the number of objects to be accessed is greatly reduced.

Index Based Algorithm
Input: the aR-tree R and the user-defined integer k
Output: the result set of k most interesting data objects

```
1  begin
2  R_min, H_max : new min heap, max heap
3  compute dgs⁺ and dds⁻ of all entries in the root node
4  for all entries e ∈ aR.Root do
5      enheap (H_max, ⟨e, dgs⁺(e), dds⁻(e)⟩)
6      while |H| > 0 and (the top entry's dgs > dds(S_k) or the top
   entry's dds < dds(S_k)) do
7      e := deheap(H)
8      read the child node N pointed by e
9      if N is a non-leaf node then
10         compute dgs⁺ and dds⁻ for all entries in N
11         for all entries e_N in N do
12            enheap (H_max, ⟨e_N, dgs⁺(e_N), dds⁻(e_N)⟩)
13     else // N is a leaf
14         compute dgs⁺ and dns⁻ for all objects O in N
15     update R_min and S_k based on dgs⁺(O) and dds⁻(O), ∀O ∈ N
16 return R_min
17 end
```

Given a subset of objects where the uncertainty in their values is presented by intervals (i.e. $[lo, hi]$), let *MBI* denote the minimum bounding interval (MBI for short) which contains all the objects in the subset. Each non-leaf node e in aR-tree presents a *MBI*, whose interval is $[\mu^-, \mu^+]$ where μ^- and μ^+ denote the lowest and highest point of the *MBI*, respectively. The main idea of the algorithm is to make a decision ahead by comparing the lower/upper bound of the dominating/dominated scores of MBIs with the current k-th object in the result set to decide whether to visit the child node of entry e or not. It is worth noting that there are four cases of a dominance relationship between an object O and a *MBI* (an non-leaf entry in aR-tree): (i) if $lo_O > \mu^+$ (i.e. p fully dominates *MBI*), then O fully dominates all data objects in *MBI*; (ii) if $\mu^- < lo_O < \mu^+$ (i.e. partially dominates), then O may fully dominate some, but not all objects in *MBI*; (iii) if $hi_O < \mu^-$ (i.e. fully dominated), then O is fully dominated by all objects in *MBI*; (iv) if $\mu^- < hi_O < \mu^+$ (i.e. partially dominated), then O may be fully dominated by some, but not all objects in *MBI*. The cases are similar to the dominance relationship between an entry and another entry.

Recall that an uncertain object O qualifies for the result set if it is better than the k-th object S_k in the answer set, implying that its dominating score is greater than S_k or its dominated score is lower than S_k. We next define the upper bound of the dominating score and the lower bound of the dominated score of an entry (i.e. a *MBI*).

Lemma 1. The upper bound of the dominating score of an entry e in aR-tree (i.e. a *MBI*), denoted by $dgs^+(e)$, is the dominating score of the point μ^+ of the *MBI* and the lower bound of the dominated score, denoted by $dds^-(e)$, is the dominated score of the point μ^- of the *MBI*.

The pseudo code of the Indexed-Based algorithm shows that a max-heap H_{max} is employed for organizing the entries to be visited in descending order of their dominating scores; and a min-heap R_{min} is used to manage the current top-k objects having the largest dominance scores as the algorithm progresses, while S_k is the k-th object in R_{min} used for pruning purposes. First, the upper bound of the dominating scores and the lower bound of the dominated scores of the aR-tree root entries are computed (the details of the function have been omitted due to the space limit) for the max-heap H_{max}, while the dgs^+ of H's top entry e is higher than that of S_k or the dds^- is lower than that of S_k, e is deheaped, and the node N pointed by e is visited. If N is a non-leaf node, its entries are enheaped, after computing their upper and lower score bounds. If N is a leaf node, the scores of the points in it are computed and the top-k result set R_{min} with S_k may be updated, if applicable.

5 Experiments

The experimental setup is discussed in Sect. 5.1 and the results are reported in Sect. 5.2. All the experiments are conducted using the Java programming language on a PC computer with a Core-2 Duo CPU running Windows 7 with 4 GB of main memory.

5.1 Experimental Settings

We conduct the experiments to evaluate the efficiency and scalability of our methods by using two real datasets and three synthetic datasets under different configurations (varying k, dimensionality and cardinality of the datasets).

Real datasets. We employ two real datasets to evaluate the performance of our proposed algorithms: (1) APT: which contains 30,000 apartment records obtained from the website apartments.com, as in [23], and has 65 % of records which have uncertain rent values; and (2) NBA: which contains 19,000 records of various technical statistics (i.e. games played, points, rebounds, assists,) of NBA players. Since the records in the NBA dataset originally only contain certain values, we explicitly replace some statistics randomly to achieve the rate of 50 % of the records containing uncertain values.

Synthetic datasets. We employ three synthetic datasets to model different scenarios: (1) UID: a dataset where object attributes are uniformly and independently generated for different dimensions; (2) COR: a dataset where object attributes are correlated (i.e. objects which are good in one dimension are also good in the other dimensions) and (3) AN-COR: a dataset where object attributes are anti-correlated (i.e. an object that is good in one dimension is bad in one or another dimension). In each dataset, the uncertainty intervals are uniformly distributed in [0,1000] and the length of each interval is normally distributed with mean $\mu = 5$ and deviation $\sigma = 1$.

5.2 Results

We first investigate the effect of parameter k on the efficiency of our proposed algorithms (1) PRO: Progressive Algorithm, (2) OS: One Scan Algorithm and (3) IND: Index Based Algorithm. Figures 1 and 2 depict the results on APT, NBA and the three synthetic datasets, respectively, in terms of the number of object accesses and processing time in seconds (i.e., the sum of I/O cost and CPU time). The value of k varies in the range [10,40] and the default value is 20. From the statistics shown in the figures, we can observe that OS and IND outperform PRO. The differences between OS and IND are not remarkable in APT data due to the fact that there is only one uncertain dimension in APT, therefore, the reduction of the number of node accesses using IND is not significant.

Next, we investigate the impact of varying dimensionality $d = [2, 5]$ on the efficiency of our algorithms with k = 20 and cardinality N = 500 K. Figure 3 shows the

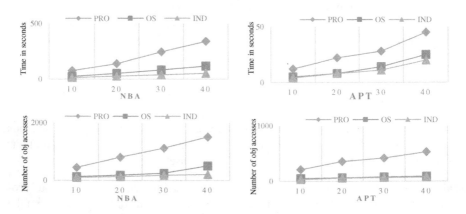

Fig. 1. Query cost vs. varying k on real datasets

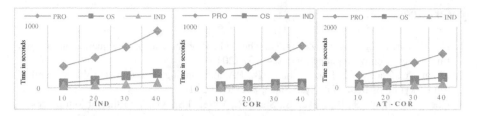

Fig. 2. Query cost vs. varying k on synthetic datasets

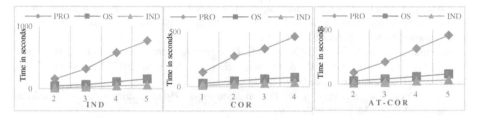

Fig. 3. Query cost vs. varying dimension d on synthetic datasets

experimental results on the three synthetic datasets. As expected, the performance of PRO and OS degrades as the number of dimensions grows. This is due to the fact that the number of dominance checks is significantly increased when the number of dimension increases. The results show that the IND performance is much better than PRO and OS under different settings. Due to the space limitation, we omit the diagrams showing the influence of varying dataset cardinality which exactly follows the same trend of the dimensionality's impact.

6 Conclusion

Probabilistic skyline queries have been well explored in the literature. However, discovering the subset of interesting data objects based merely on skyline probabilities may result in missing desirable objects. In this paper, we are the first to study the problem of the Top-k Dominance Range Query on uncertain databases, which measures the "interestingness" of a data object, based on a newly defined criterion. In addition, we also consider the data model where the value of each object on each dimension is represented as a possible range which exists in many real-world applications. An interesting direction for future work is to consider the problem in a stream setting where each object is associated with an interval time of validity, or offering additional scalability, especially for massively big data.

References

1. He, G., Chen, L., Zeng, C., Zheng, Q., Zhou, G.: Probabilistic skyline queries on uncertain time series. Neurocomputing **191**, 224–237 (2016)
2. Zhang, Y., Zhang, W., Lin, X., Jiang, B., Pei, J.: Ranking uncertain sky: the probabilistic top-k skyline operator. Inf. Syst. **36**(5), 898–915 (2011)
3. Cheng, R., Singh, S., Prabhakar, S., Shah, R., Vitter, J.S., Xia, Y.: Efficient join processing over uncertain data. In: Proceedings of the 15th ACM International Conference on Information and Knowledge Management 2006, pp. 738–747. ACM (2006)
4. Yiu, M.L., Mamoulis, N.: Multi-dimensional top-k dominating queries. VLDB J. **18**(3), 695–718 (2009)
5. Lian, X., Chen, L.: Top-k dominating queries in uncertain databases. In: Proceedings of the 12th International Conference on Extending Database Technology: Advances in Database Technology 2009, pp. 660–671. ACM (2009)
6. Vlachou, A., Doulkeridis, C., Halkidi, M.: Discovering representative skyline points over distributed data. In: Ailamaki, A., Bowers, S. (eds.) SSDBM 2012. LNCS, vol. 7338, pp. 141–158. Springer, Heidelberg (2012)
7. Lee, J., You, G.-W.: Hwang, S.-w.: Personalized top-k skyline queries in high-dimensional space. Inf. Syst. **34**(1), 45–61 (2009)
8. Zhou, B., Yao, Y.: Evaluating information retrieval system performance based on user preference. J. Intell. Inf. Syst. **34**(3), 227–248 (2010)
9. Nanongkai, D., Sarma, A.D., Lall, A., Lipton, R.J., Xu, J.: Regret-minimizing representative databases. Proc. VLDB Endowment **3**(1–2), 1114–1124 (2010)
10. Papadias, D., Tao, Y., Fu, G., Seeger, B.: An optimal and progressive algorithm for skyline queries. In: Proceedings of the 2003 ACM SIGMOD International Conference on Management of Data 2003, pp. 467–478. ACM (2003)
11. Pei, J., Jiang, B., Lin, X., Yuan, Y.: Probabilistic skylines on uncertain data. In: Proceedings of the 33rd international conference on Very large data bases 2007, pp. 15–26. VLDB Endowment
12. Gao, Y., Miao, X., Cui, H., Chen, G., Li, Q.: Processing k-skyband, constrained skyline, and group-by skyline queries on incomplete data. Expert Syst. Appl. **41**(10), 4959–4974 (2014)
13. Gao, Y., Liu, Q., Chen, L., Chen, G., Li, Q.: Efficient algorithms for finding the most desirable skyline objects. Knowl. Based Syst. **89**, 250–264 (2015)
14. Borzsony, S., Kossmann, D., Stocker, K.: The skyline operator. In: Proceedings of the 17th International Conference on Data Engineering, 2001, pp. 421–430. IEEE (2001)
15. Tan, K.-L., Eng, P.-K., Ooi, B.C.: Efficient progressive skyline computation. In: VLDB 2001, pp. 301–310 (2001)
16. Kossmann, D., Ramsak, F., Rost, S.: Shooting stars in the sky: an online algorithm for skyline queries. In: Proceedings of the 28th International Conference on Very Large Data Bases 2002, pp. 275–286. VLDB Endowment
17. Morse, M., Patel, J.M., Jagadish, H.: Efficient skyline computation over low-cardinality domains. In: Proceedings of the 33rd International Conference on Very Large Data Bases 2007, pp. 267–278. VLDB Endowment (2007)
18. Chomicki, J., Godfrey, P., Gryz, J., Liang, D.: Skyline with presorting. In: ICDE 2003, pp. 717–719 (2003)
19. Cormode, G., Li, F., Yi, K.: Semantics of ranking queries for probabilistic data and expected ranks. Paper presented at the Proceedings of the 2009 IEEE International Conference on Data Engineering (2009)

20. Soliman, M.A., Ilyas, I.F., Chang, K.C.-C.: Top-k query processing in uncertain databases. Paper presented at the Proceedings of the 23rd International Conference on Data Engineering
21. Hua, M., Pei, J., Zhang, W., Lin, X.: Efficiently answering probabilistic threshold top-k queries on uncertain data. Paper presented at the Proceedings of the 2008 IEEE 24th International Conference on Data Engineering (2008)
22. Zhang, X., Chomicki, J.: Semantics and evaluation of top-k queries in probabilistic databases. Distrib. Parallel Databases **26**(1), 67–126 (2009)
23. Soliman, M.A., Ilyas, I.F.: Ranking with uncertain scores. Paper presented at the Proceedings of the 25th International Conference on Data Engineering

Classification Based on Compressive Multivariate Time Series

Chandra Utomo[(✉)], Xue Li, and Sen Wang

School of Information Technology and Electrical Engineering,
The University of Queensland, Brisbane, Australia
{c.utomo,xueli,sen.wang}@uq.edu.au
http://www.itee.uq.edu.au

Abstract. Prediction of critical condition in intensive care unit (ICU) becomes one of the current major focuses in hospital healthcare delivery. Most of existing data mining methods only considered single time series signal and worked in original dimension. Consequently, they performed poorly for extended dataset of patient records. The main challenge of ICU prediction is the data too big to be stored and processed in timely manner. The problem in this study is how to compressed the original data into as small as possible while preserved prediction performance. In this paper, we propose multivariate compressed representation (MultiCoRe). Each recorded vital signal is transformed in frequency domain then reduced in low dimensional space. Multivariate distance measurement (MultiDist) is introduced to compute similarity between two patient records directly in MultiCoRe. Experimental results using MIMIC-II dataset show that our proposed method improved prediction accuracy and run hundreds times more efficient than other baseline methods.

Keywords: Time series representation · Machine learning · Medical prediction

1 Introduction

Medical sensors in ICU generate high volume, velocity, variety and value of data that cannot be stored and processed by conventional methods [1]. Research in septic patients revealed that accurate prediction of critical condition during ICU followed by appropriate timely treatment can significantly improve survival rate. Survival rate was 82.7 % if effective treatments were given during first 30 min. It decreased 7.6 % in average for each hours of delay [2]. Consequently, developing state of the art big data frameworks with real-time analytical capabilities is becoming important to save patient life.

The National Institutes of Health (NIH) funded a development of Multi-parameter Intelligent Monitoring in Intensive Care II (MIMIC-II) database. It is remarkably to four aspects that are freely accessible, varied population, de-identified, and high temporal resolution [3]. An extensive variety of studies has

© Springer International Publishing AG 2016
M.A. Cheema et al. (Eds.): ADC 2016, LNCS 9877, pp. 204–214, 2016.
DOI: 10.1007/978-3-319-46922-5_16

been made on MIMIC-II. However, slight improvement has been accomplished in clinical decision support for supervision of critical condition patient.

PhysioNet and Computer in Cardiology created a competition to predict acute hypotensive episode (AHE) of ICU patients as critical condition parameter [4]. The aim is to develop intelligent method to forecast AHE in the next one hour for each patient record by given previous 10-h multivariate time series patient data. Several noteworthy methods [5–7] has been proposed to tackle that challenge that delivered good prediction results in given dataset. However, most of them only used single vital signal, named Mean Arterial Blood Pressure (MABP), to predict AHE. They did not utilized other vital signals that are also important based on domain knowledge. As a results, their performances were dropped significantly when scale to extended patients records [8] (Fig. 1).

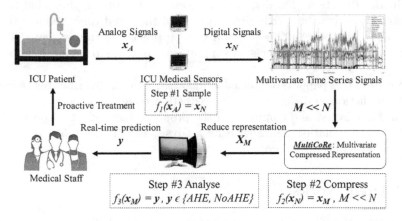

Fig. 1. Time series reduced-dimension representation using MultiCoRe to enable real-time critical condition prediction for proactive treatment

There are two main challenges in this real time ICU prediction based on multivariate time series data. The first challenge is representation methods. In MIMIC II dataset, each patient might have 12 to 66 recorded vital signals spanned in several days or weeks during ICU stay. Making prediction directly with such raw data is impractical in timely manner due to storage and processing limitations. Representation methods for multivariate time series data that can significantly reduce raw data size while still keep principal information in the data is highly needed. The second challenge is similarity measures. The distance between time series need to be defined more carefully than variable with conventional data type such as nominal, categorical, and ordinal. Trends and semantics have to be acquired in order to reveal the essential similarity. In case of multivariate time series, it becomes more challenging since combination of variables need to be considered for computing overall distance.

In this paper, we proposed Multivariate Compressed-dimension Representation (MultiCoRe) based on spectral transformation. Each i-th original time series signal $X_i \in \mathbb{R}^N$ was transformed into $X_i \in \mathbb{R}^M$ where M is significantly smaller than N, then the results were aggregated to form MultiCoRe. Additionally, we also proposed Multivariate Distance (MultiDist) measurement as combination of euclidean distance to calculate two patient records directly in MultiCoRe representation. We tested our proposed framework using MIMIC-II dataset then compare the performance with baseline methods. Experimental results showed that MultiCoRe can run hundreds times faster and require thousand times less storage while maintaining prediction power of original dataset.

There are three main contributions of this paper. Firstly, we showed that integrating multivariate time series signals can improve prediction performances. It is worthy to consider combination of vital signals. Secondly, we proved that it is possible to do prediction directly in the reduced dimension while preserving accuracy. The curse of dimensionality as common in many applications can be solve in this direction. Thirdly, the proposed MultiCoRe is promising for real time prediction. It is hundreds times faster and thousand times reduces storage requirement. We believe this solution can be practically implemented in real medical setting as clinical decision support system to help saving ICU patient life.

The rest of this paper is organized as follows. Section 2 describes related works in ICU prediction and dimensionality reduction techniques. Section 3 explains problem formulation and summary of the proposed methods. Section 4 delivers experimental evaluation along with its analysis. Finally, Sect. 5 concludes the paper and provides future works.

2   Related Works

In 2009, PysioNet/Computers in Cardiology created a challenge to predict critical condition in ICU [4]. Given prediction starting time T_0, the aim is to automatically predict whether there is acute hypotensive episode (AHE) in one hour prediction window based on 10 h historical window. AHE is defined as a period where minimum 90 % of Mean Arterial Blood Pressure (MABP) waveform at least 60 mmHg during 30-min window. Figure 2 illustrates this challenge.

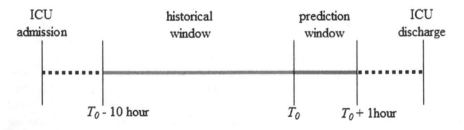

Fig. 2. Historical and prediction window as a 11 h subsequence multivariate time series of ICU patient record during ICU stay.

Henriques and Rocha [7] proposed neural network multi-models to predict evolution of blood pressure signals. The approach produced best results of the challenge. Sun et al. [8] proposed locally supervised metric learning (LSML). This approach delivered better results than multi-model neural network. However, the distance measurement was sensitive to temporal alignment. Ghosh et al. [9] proposed gap constrained sequential contrast patterns using symbolic approximation (SAX). The extracted patterns can be further studied for knowledge discovery in AHE prediction. Although the previous methods delivered good prediction results within given competition dataset, most of them only used single time series signals. This made prediction results performed poorly when dataset were extended into hundreds or thousands patients records.

ICU patient records are essentially high dimension. Working directly in original dataset (time domain) can be time and space consuming. Numerous methods have been proposed for dimensionality reduction. Several widely used techniques are Discrete Fourier Transformation (DFT) [10], Singular Value Decomposition (SVD) [11], Discrete Wavelet Transformation (DWT) [12], Piecewise Aggregate Approximation (PAA) [13], Adaptive Piecewise Constant Approximation (APCA) [14], Chebyshev Polynomial (CHEB) [15], Symbolic Aggregate approXimation (SAX) [16] and Indexable Piecewise Linear Approximation (IPLA) [17]. However, those methods are used to transform single time series into low dimension. Besides high dimension, ICU patient records are multivariate. Several vital signals are recorded simultaneously. Careful novel method is still needed in order to compress multivariate time series data as a feature for prediction.

3 The Proposed Methods

3.1 Problem Formulation

Definition 1. *Given P as a patient record with n time series signals during ICU stay and T_0 as prediction starting time, subsequence records $R = [X, Y]$ is $11 h$ of P consists of $10 h$ historical window X*

$$X = \begin{bmatrix} P_1(T_0 - 10h) \dots P_1(T_0 - 2) \ P_1(T_0 - 1) \\ P_2(T_0 - 10h) \dots P_2(T_0 - 2) \ P_2(T_0 - 1) \\ \vdots \qquad \vdots \qquad \vdots \\ P_n(T_0 - 10h) \dots P_n(T_0 - 2) \ P_n(T_0 - 1) \end{bmatrix} \qquad (1)$$

followed by 1 h prediction window Y

$$Y = \begin{bmatrix} P_1(T_0) \ P_1(T_0 + 1) \dots P_1(T_0 + 1h) \\ P_2(T_0) \ P_1(T_0 + 1) \dots P_2(T_0 + 1h) \\ \vdots \qquad \vdots \qquad \vdots \\ P_n(T_0) \ P_1(T_0 + 1) \dots P_n(T_0 + 1h) \end{bmatrix} \qquad (2)$$

where there is a function $f : Y \to y \in \{AHE, NoAHE\}$ that labels subsequence record R into class AHE or NoAHE.

Given historical window patient records $X \in \mathbb{R}^{n \times N}$ with n signals each in N dimension, prediction algorithm $f(X) = y$, the problem is how to develop compressed model $\overline{X} \in \mathbb{R}^{n \times M}$ with $M \ll N$, such that prediction accuracy based on original historical window is preserved. More formally, it is defined as

$$\underset{\arg M}{\text{minimize}} \quad f(\overline{X}) \geq f(X). \tag{3}$$

3.2   MultiCoRe: Multivariate Compressed Representation

There are three main steps in building MultiCoRe for classification in reduced dimension. The first step is each signal in time domain $X(N)$ is transformed into frequency domain $X(F)$ using Discrete Fourier Transform (DFT)

$$X(f) = \sum_{t=1}^{N} X(t)\omega_N^{(t-1)(f-1)} \tag{4}$$

where $\omega_N = e^{(-2\Pi i)/N}$. We can see that this linear domain transformation is not changing the dimension. Both are still in $\mathbb{R}^N$. In the second step, M first coefficient in $X(F)$ are selected. Now, the reduced representation is in $\mathbb{R}^M$. We aggregated all reduced representation of all signals to form MultiCoRe. Figure 3 illustrates these first two steps.

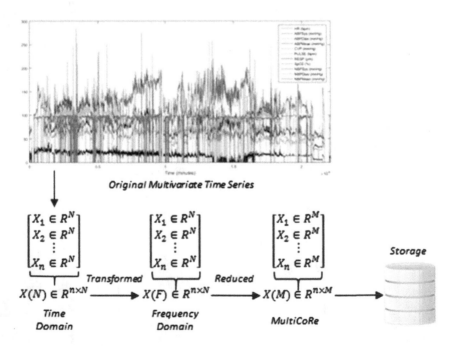

Fig. 3. Transformation from original multivariate time series into MultiCoRe format.

The third step is defining distance measurement between two records directly in MultiCoRe. Suppose two records $X^{[A]}$ and $X^{[B]}$ are in MultiCoRe, multivariate distance $MultiDist(X^{[A]}, X^{[B]})$ is defined as summation of distance combination from all associated signals.

$$MultiDist(X^{[A]}, X^{[B]}) = Dist(X_1^{[A]}, X_1^{[B]}) + ... + Dist(X_n^{[A]}, X_n^{[B]})$$
$$= \sum_{i=1}^{n} Dist(X_i^{[A]}, X_i^{[B]}) \tag{5}$$

where $Dist(X_i^{[A]}, X_i^{[B]})$ is defined as euclidean distance between two time series signals i in records $X^{[A]}, X^{[B]}$.

$$Dist(X_i^{[A]}, X_i^{[B]}) = ED(X_i^{[A]}, X_i^{[B]})$$
$$= \sqrt{\sum_{f=1}^{M} (X_i^{[A]}(f) - X_i^{[B]}(f))^2} \tag{6}$$

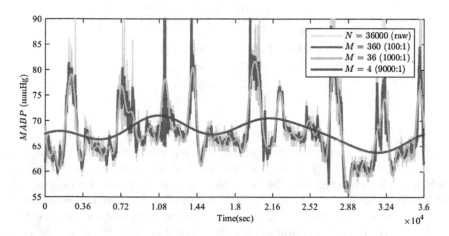

Fig. 4. 10 h Mean Arterial Blood Pressure (MABP) signal sampled per second ($N = 36000$) and its reconstructions by using DFT with some coefficients M.

We decided to use DFT because it is non-data adaptive representation. It means that the transformation matrix can be fixed. Data adaptive representation such as APCA, SVD, and SAX need to find best parameter value based on current dataset. Their transformation matrix require to be redefined to manage performance accuracy if the dataset changing significantly. Other important reasons to use DFT are euclidean distance in time domain is preserved in frequency domain and most time series energy can be concentrated in first few coefficients. Figure 4 shows that reconstruction using small number of coefficients, such as $M = 360$ (99 % compression), is able to capture fundamental characteristics of original time series.

4   Experimental Evaluation

4.1   Dataset

Experimental dataset were taken from MIMIC-II database [3]. There were data collection and preprocessing procedures to transform raw patients records data into well-form data for experimental purposes. First of all, each numerical patient record available in MIMIC-II ATM[1] was annotated by AHE Episodes Annotation Software[2] developed by PhysioNet. Although it was possible to create our own annotation procedure, we decided to use the software directly written by them to make sure validity of annotation results. There are three possibilities for each patient record: $AHE(s)$, $NoAHE$, or $undefined$ due to no MABP signal recorded. We disregarded records with $undefined$ status because they can not be labelled. This initial step produced hundreds number of patient records.

Table 1. Statistical summary and AHE prediction accuracy from each signal.

Signals	Mean		STD		ACC
	AHE	NoAHE	AHE	NoAHE	
HR	86.92	84.74	9.44	6.35	53.94 %
SABP	101.51	124.77	12.15	13.51	72.14 %
DABP	48.16	65.87	7.68	9.37	88.01 %
MABP	64.94	86.12	8.69	10.41	88.79 %
PVC	3.42	4.75	5.96	18.24	50.94 %

At the second step, we extracted all 10 h records before every AHE event and labelled them as AHE or positive samples. Then, we extracted all non-overlapping 10 h records from patients records who had not experienced AHE and labelled them as $NoAHE$ or negative samples. Finally, we chose records that has at least 90 % available MABP for sufficient representation and removed signals that completely missing in that 10 h period because it can not be recovered by linear interpolation. The final experimental dataset consists of 389 AHE records, 512 $NoAHE$ records, and 5 vital signals for every records that are Heart Rate (HR), Systolic Arterial Blood Pressure (SABP), Diastolic Arterial Blood Pressure (DABP), Mean Arterial Blood Pressure (MABP) and Premature Ventricular Contraction (PVC). Statistical summary of these signals in experimental dataset can be seen in Table 1.

4.2   Results and Analysis

In this study, we used k-nearest neighbour (kNN) as classifier. The reason is that it has been reported outperformed other classifiers in time series classification.

---

[1] http://www.physionet.org/cgi-bin/atm/ATM.
[2] https://www.physionet.org/challenge/2009/ahe-detect-1.0/.

Additionally, experimental results is not hard to reproduce because of easy implementation and parameter free property [18]. We used 10-fold cross-validation using kNN classifier with MultiDist similarity measure to compare our proposed representation method MultiCoRe with baseline original dataset. These experiments were performed on Intel (R) Core (TM) i7-4770 CPU @3.40 GHz (8 CPUs) and 16 GB of RAM.

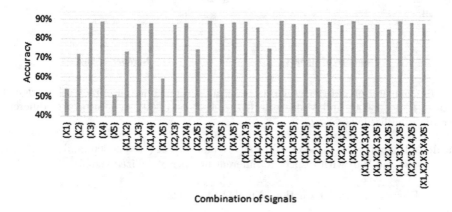

Fig. 5. Accuracy of all 31 possible combinations from 5 signals using original dataset.

Initially, we tested our hypothesis that considering multivariate time series signals will improve prediction performances. We measured accuracy of all possible combinations. Since the experimental dataset has 5 signals, there are 31 combinations. As can be seen in Fig. 5, prediction accuracy using combination of signals may be higher than accuracy using single signal. For example, accuracy using X1 (HR) is 53.94 %, using X2 (PVC) is 50.94 % and using both X1 and X2 is 59.60 %. Another example is accuracy using X3 (DABP) is 88.01 %, using X4 (MABP) is 88.79 %, and using both X1 and X2 is 89.46 %. This showed that using multivariate time series for better prediction is a right direction.

We used prediction using original dataset ($N = 36000$) as a baseline to compare with our proposed MultiCoRe. We tested three different number of coefficients M in MultiCoRe that are 4, 36 and 360. For each number of signals combination, we measured average and maximum accuracy for all data representation. We found that generally accuracy using MultiCoRe is not lower than using original dataset. Interestingly, MultiCoRe outperformed baseline method in most cases as can be seen in Fig. 6. We discussed in the previous section that euclidean distance in time domain (original time series) is preserved in frequency domain. This results confirmed that the first M Fourier coefficients are better features than the whole original series in term of prediction accuracy. We analysed the reason is most energy in time series is represented in first few coefficients.

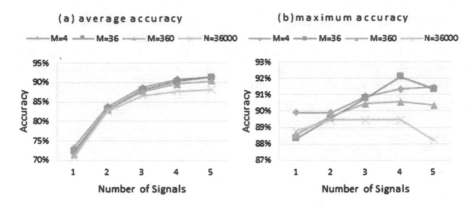

Fig. 6. Comparation of (a) average accuracy and (b) maximum accuracy using original dataset ($N = 36000$) and MultiCoRe ($M = 4$, 36 and 360) with respect to different number of utilized signals.

Table 2. Running time (in seconds) using original dataset ($N = 36000$) and MultiCoRe ($M = 4$, 36 and 360) with respect to different number of utilized signals.

Dimension	Data reduction	Number of signals				
		1	2	3	4	5
$N = 36000$	00.00 %	415.36	938.25	2064.80	2640.50	3542.70
$M = 360$	99.00 %	11.46	23.78	34.23	44.36	54.78
$M = 36$	99.90 %	5.50	8.48	11.42	14.46	17.88
$M = 4$	99.99 %	5.12	7.65	10.17	12.70	15.17

We measured efficiency of the proposed MultiCoRe in term of space and running time. If we used $M = 360$, it means that compression rate is 1:100 (360:36000) and data reduction is 99 %. Even if we used lower number of M such as 36 (99.9 % reduction), the accuracy is still better than original dataset. In term of time, we can see in Table 2 that running time of prediction using original signals are growing significantly higher than MultiCoRe. For example, original data with 1 signal took 415.36 s and with 5 signals took 3542.70 s while MultiCoRe with $M = 36$ only need 5.5 s for 1 signal and 17.88 s for 5 signals. Based on this results, the proposed MultiCoRe is thousand time more efficient in space and hundred time more efficient in running time than original dataset. This kind of efficiency may enable real time critical condition prediction in ICU.

5   Conclusion

In this paper, we developed Multivariate Compressed Representation (Multi-CoRe) framework to save patient records in reduced-dimension using DFT. We also introduced Multivariate Distance (MultiDist) to measure the dis(similarity) between two records as combination of distance measurements directly in reduced dimension. Based on extensive experimental results, we found that the performance of our proposed method outperformed baseline method. These findings confirmed that considering multivariate vital signals is better than single one. Additionally, the data can be significantly reduced up to 99 % of original data size while keep fundamental information in time series and prediction power.

As mostly ICU data are discarded in hospital due to storage limitation, this study therefore indicates that the benefit of MultiCoRe may address big medical data problem with respect to data size. To the best of our knowledge, this is the first study to consider ICU multivariate vital sensor data in reduced-dimension. Although our methods have accommodated all medical sensors data in ICU, it is important to search for best combination of multivariate signals to achieve optimal accuracy. Future work should therefore include follow-up study designed to do feature selection in MultiCoRe.

Acknowledgments. This paper is partially supported by ARC Discovery Project, with Grant ID: DP140100104, Effective Recommendations based on Multi-Source Data.

References

1. Andreu-perez, J., Poon, C.C.Y., Merrifield, R.D., Wong, S.T.C.: Big data for health. IEEE J. Biomed. Heal. Inform. **19**, 1193–1208 (2015)
2. Kumar, A., Roberts, D., Wood, K.E., Light, B., Parrillo, J.E., Sharma, S., Suppes, R., Feinstein, D., Zanotti, S., Taiberg, L., Gurka, D., Kumar, A., Cheang, M.: Duration of hypotension before initiation of effective antimicrobial therapy is the critical determinant of survival in human septic shock. Crit. Care Med. **34**, 1589–1596 (2006)
3. Saeed, M., Villarroel, M., Reisner, A.T., Clifford, G., Lehman, L.-W., Moody, G., Heldt, T., Kyaw, T.II., Moody, B., Mark, R.G.: Multiparameter Intelligent monitoring in intensive care II: a public-access intensive care unit database. Crit. Care Med. **39**, 952–960 (2011)
4. Moody, G., Lehman, L.: Predicting acute hypotensive episodes: the 10th annual physionet/computers in cardiology challenge. Comput. Cardiol. **36**, 541–544 (2009)
5. Chen, X., Xu, D., Zhang, G., Mukkamala, R.: Forecasting acute hypotensive episodes in intensive care patients based on a peripheral arterial blood pressure waveform. In: Computers in Cardiology 2009 (2009)
6. Mneimneh, M.A., Povinelli, R.J.: A rule-based approach for the prediction of acute hypotensive episodes. In: Computers in Cardiology, pp. 557–560 (2009)
7. Henriques, J., Rocha, T.: Prediction of acute hypotensive episodes using neural network multi-models. Comput. Cardiol. **2009**, 549–552 (2009)
8. Sun, J., Sow, D., Hu, J., Ebadollahi, S.: A system for mining temporal physiological data streams for advanced prognostic decision support. In: Proceedings - IEEE International Conference on Data Mining, ICDM, pp. 1061–1066 (2010)
9. Ghosh, S., Feng, M., Nguyen, H., Li, J.: Hypotension risk prediction via sequential contrast patterns of ICU blood pressure. IEEE J. Biomed. Heal. Inform. **PP**, 1 (2015)

10. Agrawal, R., Faloutsos, C., Swami, A.: Efficient similarity search in sequence databases. In: Lomet, D.B. (ed.) FODO 1993. LNCS, vol. 730, pp. 69–84. Springer, Heidelberg (1993). doi:10.1007/3-540-57301-1_5

11. Korn, F., Jagadish, H.V., Faloutsos, C.: Efficiently supporting ad hoc queries in large datasets of time sequences. In: Proceedings of the 1997 ACM SIGMOD International Conference on Management of Data - SIGMOD 1997, pp. 289–300. ACM, New York (1997)

12. Chan, K.-P., Ada Wai-Chee, F.: Efficient time series matching by wavelets. In: Proceedings 15th International Conference on Data Engineering (Cat. No.99CB36337), pp. 126–133. IEEE (1999)

13. Keogh, E., Chakrabarti, K., Pazzani, M., Mehrotra, S.: Dimensionality reduction for fast similarity search in large time series databases. Knowl. Inf. Syst. **3**, 263–286 (2001)

14. Keogh, E., Chakrabarti, K., Pazzani, M., Mehrotra, S.: Locally adaptive dimensionality reduction for indexing large time series databases. In: Proceedings of the 2001 ACM SIGMOD International Conference on Management of Data - SIGMOD 01, pp. 151–162. ACM, New York (2001)

15. Cai, Y., Ng, R.: Indexing spatio-temporal trajectories with Chebyshev polynomials. In: Proceedings of the 2004 ACM SIGMOD International Conference on Management of Data - SIGMOD 04, p. 599. ACM, New York (2004)

16. Lin, J., Keogh, E., Wei, L., Lonardi, S.: Experiencing SAX: a novel symbolic representation of time series. Data Min. Knowl. Discov. **15**, 107–144 (2007)

17. Chen, Q., Chen, L., Lian, X., Liu, Y., Yu, J.X.: Indexable PLA for efficient similarity search. In: VLDB, pp. 435–446. VLDB Endowment, Vienna (2007)

18. Wang, X., Mueen, A., Ding, H., Trajcevski, G., Scheuermann, P., Keogh, E.: Experimental comparison of representation methods and distance measures for time series data. Data Min. Knowl. Discov. **26**, 275–309 (2012)

Adapting ELM to Time Series Classification: A Novel Diversified Top-k Shapelets Extraction Method

Qiuyan Yan[1,2(✉)], Qifa Sun[1], and Xinming Yan[1]

[1] School of Computer Science and Technology,
China University of Mining Technology, Xuzhou 221116, China
{yanqy,yanxm}@cumt.edu.cn, sunqifa@live.com
[2] School of Safety Engineering,
China University of Mining Technology, Xuzhou 221116, China

Abstract. Extreme Learning Machine (ELM for shot) is a single hidden layer feed-forward network, where the weights between input and hidden layer are initialized randomly. ELM is efficient due to its utilization of the analytical approach to compute weights between hidden and output layer. However, ELM still fails to output the semantic classification outcome. To address such limitation, in this paper, we propose a diversified top-k shapelets transform framework to improve representative and interpretative ability of ELM. Specifically, we first define the similar shapelets and diversified top-k shapelets to construct diversity shapelets graph. Then, a novel diversity graph based top-k shapelets extraction algorithm to search diversified top-k shapelets. Finally, we propose a shapelets transformed ELM algorithm named as **DivShapELM** to automatically determine the k value, which is further utilized for time series classification. The experimental results demonstrate that the proposed approach significantly outperforms traditional ELM algorithm in terms of effectiveness and efficiency.

Keywords: Extreme learning machine · Shapelets transformed classification · Diversified query · Feature extraction

1 Introduction

Extreme Learning Machine (ELM for short) [1] was originally developed based on single hidden-layer feed forward neural networks (SLFNs). Compared with the conventional learning machines, it is of extremely fast learning capacity and good generalization capability. Thus, ELM, with its variants [2], has been widely applied in many fields [3]. The results indicate that ELM produces comparable or better classification accuracies with reduced training time and implementation complexity compared to artificial neural networks methods and support vector machine methods.

Unfortunately, as a black-box method, ELM fails to measure up to the task of time series data classification by itself. A possible solution to this issue is

© Springer International Publishing AG 2016
M.A. Cheema et al. (Eds.): ADC 2016, LNCS 9877, pp. 215–227, 2016.
DOI: 10.1007/978-3-319-46922-5_17

to improve the interpretability of ELM by feature selection. If a set of selected features improve the classification accuracy much more than original feature sets, it is reasonable to interpret the results by them. Nevertheless, selection the most representative and interpretative feature can improve the interpretability of ELM and make ELM more adapt to time series classification. In the context of feature selection in time series data analysis, most of the current methods adopt such a framework that ranks subsequences according to their individual discriminative power to the target class and then selects top-k ranked subsequences [4]. These methods have some common drawbacks: (1) the selected features are not the most representative and interpretative, (2) many redundant features are selected, and (3) the number of selected feature is arbitrarily specified by a parameter k.

In this paper, a novel method is proposed to improve ELM representative and interpretative ability by extraction diversified top-k shapelets features.Shapelets was introduced as a primitive for time series data mining [5] and was utilized in classifying time series data [6]. Shapelets transformed classification methods [7,8] were proposed to separate the processing of shapelets selection and classification. The k shapelets are selected in an offline manner, and which can not only improve the affectivity and efficiency of classification, but also introduce a common feature attraction method which can be used in all typical time series classification algorithms. Nevertheless, shapelets based classification methods have been widely discussed and used in many real applications [9,10].

The most challenge is that there are large quantities of redundant shapelets in candidates which decreasing the accuracy of classification and the parameter k is hard to determine. Some works [8,11] detect this problem and use clustering or pruning methods to remove the redundant, but still exist redundant shapelets, also, the k value is determined from experiments. This paper make the following contributions: First, two conceptions including similar shapelets and diversified top-k shapelets are presented. Based on these conceptions, a method of construction diversify shapelets graph is proposed. Second, a diversified top-k shapelets query method is presented to find top-k representative shapeletes of each class. Third, we propose an diversified top-k shapelets transformed ELM algorithm which can automatically determine the parameter k and transform data using the determined k shapelets. The experimental results show that the proposed approach significantly improves the interpretability and performance of ELM.

2   Preliminary

Extreme Learning Machine (ELM) is a generalized single hidden-layer feedforward network. In ELM, the hidden layer node parameters are mathematically calculated instead of being iteratively tuned; thus, it provides good generalization performance at thousands of times faster speed than traditional popular learning algorithms for feedforward neural networks.

Suppose there are N arbitrary distinct training instances (x_i, t_i), where $x_i = [x_{i1}, x_{i2}, ..., x_{in}]^T \in \mathrm{R}^n$, and $t_i = [t_{i1}, t_{i2}, ..., t_{in}]^T \in \mathrm{R}^m$, standard SLFNs with

$\tilde{N}$ hidden nodes and activation function $g(x)$ are mathematically modeled as

$$\sum_{i=1}^{\tilde{N}} \beta_i g_i(\mathbf{x}_j) = \sum_{i=1}^{\tilde{N}} \beta_i g_i(\mathbf{w}_i \cdot \mathbf{x}_j + b_i) = \mathbf{o}_j, j = 1, ..., N, \qquad (1)$$

where $\mathbf{w}_i = [w_{i1}, w_{i2}, ..., w_{in}]^T$ is the weight vector connecting the ith hidden node and the input nodes, $\beta_i = [\beta_{i1}, \beta_{i2}, ..., \beta_{im}]^T$ is the weight vector connecting the ith hidden nodes and the output nodes, and b_i is the threshold of the ith hidden node.If a SLFN with $\tilde{N}$ hidden nodes with activation function $g(x)$ can approximate these N samples with zero error, it then implies that there exist β_i, w_i and b_i, such that:

$$\sum_{i=1}^{\tilde{N}} \beta_i G(\mathbf{w}_i \cdot \mathbf{x}_j + b_i) = \mathbf{t}_j, j = 1, \ldots, N, \qquad (2)$$

The above N equations can be written compactly as

$$H\beta = T, \qquad (3)$$

where

$$H(\mathbf{w}_1, \ldots, \mathbf{w}_{\tilde{N}}, b_1, \ldots, b_{\tilde{N}}, \mathbf{x}_1, \ldots, \mathbf{x}_N)$$
$$= \begin{pmatrix} g(\mathbf{w}_1 \cdot \mathbf{x}_1 + b_1) & \cdots & G(\mathbf{w}_{\tilde{N}} \cdot \mathbf{x}_1 + b_{\tilde{N}}) \\ \vdots & \ddots & \vdots \\ g(\mathbf{w}_1 \cdot \mathbf{x}_N + b_1) & \cdots & G(\mathbf{w}_{\tilde{N}} \cdot \mathbf{x}_N + b_{\tilde{N}}) \end{pmatrix}_{N \times \tilde{N}}$$

$$\beta = \begin{bmatrix} \beta_1^T \\ \cdots \\ \cdots \\ \beta_{\tilde{N}}^T \end{bmatrix}_{\tilde{N} \times m} \quad \text{and} \quad T = \begin{bmatrix} t_1^T \\ \cdots \\ \cdots \\ t_N^T \end{bmatrix}_{N \times m}$$

H is named as hidden layer output matrix of the network, where with respect to inputs $x_1, x_2, \ldots, x_N$ and its jth row represents the output vector of the hidden layer with respect to input x_j.

ELM differs from other training algorithms in that the hidden node parameters w_i and b_i are not tuned during training, but are instead assigned with random values according to any continuous samplings distribution. Equation (3) then becomes a linear system and the output weight β are estimates as Eq. (4).

$$\widehat{\beta} = H^\dagger T, \qquad (4)$$

where $H^\dagger$ is the Moore-Penrose generalized inverse of the hidden layer output matrix H.

3   Diversified Top-k Shapelets Transformed ELM

In this section, we discuss three parts of our work (1) construction the diversity graph of shapelets candidates, (2) querying diversified top-k shapelets, and (3) transforming the data based on diversified top-k shapelets and applying in ELM. The following contents will discuss above three contribution separately.

Before removing the redundant shapelets, we firstly need to get the shapelets candidates set. The original shapelets extraction algorithm is time consuming and complexity is $O(n^2m^4)$, n is the number of time series in the data set, m is the length of each time series. In order to improve the efficiency of shapelets based classification method, we follow the method proposed in [6], which transformed the data sets through SAX method and decreased the time complexity to $O(nm^2)$.

3.1   Construction the Diversity Graph of Shapelets Candidates

Considerable works have focused on the diversified top-k query, but they almost applying on a typical circumstance. In our work, we use the diversity graph [12] to find a general method to extract diversified top-k shapelets.

Given I is a shapelets candidate sets, $I = \{s_1, \ldots s_n\}$, and n is the number of I. The question is how to measure the similarity of two shapelets and how to define the diversified top-k shapelets. So we first give the two definitions.

Definition 1: Similar shapelets. Given two shapelets s_i and s_j , $1 \leq i, j \leq n, i \neq j$ and n is the number of shapelets candidates. The optimal split point of s_i and s_j are $< s_i, d_i >$ and $< s_j, d_j >$, the split threshold are d_i and d_j. We say s_i and s_j are similar shapelets when they satisfy $dis(s_i, s_j) \leq \min(d_i, d_j)$. We denote the similar shapelets as $s_i \approx s_j$.

Definition 2: Diversified top-k Shapelets. Given a shapelets candidates set $I = \{s_1, \ldots s_n\}$, and an integer k where $1 \leq k \leq |I|$. The diversified top-k shapelets query results of I, denoted as $\mathrm{DivTopk}(I)$, is a list of results that satisfy the following three conditions.

(1) $\mathrm{DivTopk}(I) \subseteq (I), |\mathrm{DivTopk}(I)| \leq k$
(2) For any two results $s_i \in I$ and $s_j \in I$ and $s_i \neq s_j$, if $s_i \approx s_j$, then $\{s_i, s_j\} \not\subset \mathrm{DivTopk}(I)$.
(3) $\sum\limits_{s_i \in I} \mathrm{score}(s_i)$ is maximized.

We give a diversity shapelet graph example of ChlorineConcentration dataset as in Fig. 1. There are ten shapelets candidates as shown in Fig. 1-a and the diversity graph of these ten candidates from Algorithm 1 are shown in Fig. 1-b. The black, red and green subsequence are the top-3 shapelets and can get the best classification accuracy. Next section we will explain how to get the diversified top-k shapelets on the diversify graph.

Algorithm 1. conShapeletGraph(allShapelets)

input: shapelets candidates allShapelets
output: diverisity shapelets Graph
 1: Graph = ϕ
 2: sort(allShapelets)
 3: **for** i=1 to |allShapelets| **do**
 4:     Graph.add(allShapelets[i])
 5: **end for**
 6: **for** j=1 to |allShapelets| **do**
 7:     **for** k=1 to |allShapelets| **do**
 8:         **if** (allShapelets[j] $\approx$ allShapelets[k]) **then**
 9:             Graph[j].add(Graph[k])
10:             Graph[k].add(Graph[j])
11:         **end if**
12:     **end for**
13: **end for**
14: **return** Graph

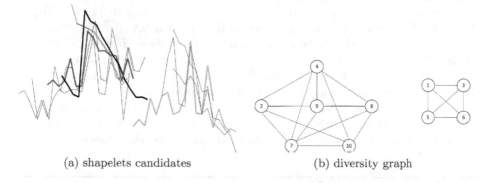

(a) shapelets candidates   (b) diversity graph

Fig. 1. Example of diversity shapelets graph (Color figure online)

3.2 Diversified Top-k Shapelets Extraction

Traditional top-k query only returns the objects with largest k score, however, diversified top-k query concerns not only the score value but also the similarity of each object and remove all the redundant objects from results. According to [12], find top-k results falling into two categories: incremental manner and bounding manner. We noticed that the bounding manner first satisfied the k value, but in each step it may not add the largest score vertex. In our problem, we must maintain the largest information gain shapelets in order to have the best classification accuracy. So we calculate the diversified top-k shapelets via an incremental manner. The detailed search procedure is shown as in Algorithm 2.

Algorithm 2. DivTopkShapelets (Graph, k)

---
input: diverisity Graph, k value
output: k number of shapelets
 1: kShapelets = $\varnothing$, n = |V(Graph)|
 2: kShapelets.add(v1)
 3: while(| kShapelets |< k)
 4: **for** i=2 to n **do**
 5:     **if** (Graph[i] $\cap$ kShapelets = $\varnothing$) **then**
 6:         kShapelets.add(vi)
 7:     **end if**
 8: **end for**
 9: **return** kShapelets

3.3 Diversified Top-k Shapelets Transform for ELM

After getting diversified top-k shapelets, we can use these shapelets to transform data before ELM classification. For each instance T_i in dataset Dt, the subsequence distance is computed between T_i and S_j, S_j is a shapelet in Top-k shapelets. The resulting k distances are used to form a new instance of transformed data, where each attribute corresponds to the distance from each shapelet to the original time series. In order to get the best classification accuracy and also to get rid of the independence on the parameter k, we set k in an interval of $[1, \kappa]$ where κ is an empirical optimal value, according to our experiments (see Sect. 4.1), which is set to 9, then we use the ELM to learning training data and evaluate each diversified top-k shapelets candidate. The k value with the largest prediction accuracy is selected. The details are as in following algorithm 3.

Algorithm 3. DivShapELM(Graph, κ)

---
input: κ value
output: ELM classification results
 1: **for** i=1 to κ **do**
 2:     kShapelets = DivTopkShapelet(Graph,k)
 3:     output = $\varnothing$
 4:     **for** ts= 1 to |Dt| **do**
 5:         transformed = $\varnothing$
 6:         **for** s = 1 to |kShapelets| **do**
 7:             dist = subsequenceDist(ts,s)
 8:             transformed.add(dist)
 9:         **end for**
10:         output.add(transformed)
11:     **end for**
12:     using ELM evaluate output
13:     kShapelets = the output with highest accuracy on ELM
14: **end for**
15: using kShapelets transform testing data
16: **return** ELM classification results

---

4   Experiments

To evaluate our proposed methods, we selected 15 data sets from the UCR time series repository. We use a simple train/test split and all reported results are testing accuracy. All shapelets candidates selection, top-k diversified shapelets extraction and classifier construction is done on the training set. All experiments are implemented in Java within the Weka framework.

4.1   Determination of Shapelets Length and κ

There are two parameters min and max in the procedure of shapelets candidates generation. The two parameters determine the length of shapelets candidates which can influence finding the best representative shapelets. Followed [8], we set min-length and max-length of subsequences to generate shapeless are m/11 and m/2 separately, m is the length of each time series.

First, in order to explain how the k value influences accuracy of classification, we test the average accuracy of six classifier on fifteen data sets with the varying k value. As shown in Fig. 2, with the increasing of k value, average classification accuracy first increases and then becomes stable when k is 9. Accordingly, we set the κ value as 9 and use this value in the following experiments.

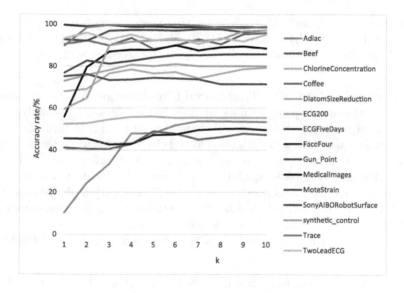

Fig. 2. Accuracy varying with k

4.2   Representation of Optimal Shapelets Sets

In this section, we want to get a visual overlook at what was the optimal shapelets indeed. Because reference [11] has verified that ShapeletSelection can remove the

(a) optimal 14 shapelets of   (b) optimal 2 shapelets of
ShapeletSelection with the best accuracy   DivTopkShapelet with the best accuracy

Fig. 3. Optimal shapelets sets

most redundant shapelets, we only compared the optimal shapelets sets between
DivTopkShapelets and ShapeletSelection on TwoLeadECG dataset when the
two algorithms all have the best classification accuracy, as shown in Fig. 3.
The optimal shapelets sets were acquired from ShapeletSelection when k = 14
(Fig. 3-a), and from DivTopkShapelet when k = 2(Fig. 3-b).

4.3 Accuracy Comparison

In this section, we select six traditional time series classification algorithms
including C4.5, 1NN, Naive Bays(NaB), BayesianNetwork(BaN), RandomFor-
est(RaF) and RotationForest(RoF) to compare the accuracy with our proposed
methods.

Accuracy comparison with Traditional Classification. Firstly, we directly
use selected classification algorithms to classify the datasets. Secondly,we use
DivTopkShapeletsto extract optimal shapeletes sets and transform data, then
classify transformed data sets with six selected classification algorithms. Results
are presented in Table 1, the column captions with classifier name plus S(see
C4.5(S)) means **DivTopkShapelets** transformed classification results. From
Table 1 we can see that compared to traditional classification algorithms, **Div-
TopkShapelets** transformed classification methods can improve accuracy of 9
out of 15 datasets. For all six classification algorithms, average accuracy are
improved. Especially for NaiveBays, **DivTopkShapelets** improves 13 data sets
accuracy.

Accuracy comparison with ShapeletSelection Algorithm. We compared
the relative accuracy of **DivTopkShapelets** with ShapeletSelection method as
shown in Table 2. From results we can draw the conclusion that compared with
ShapeletSelection, **DivTopkShapelets** transformed classification method can
improve accuracy of 9 out of 15 datasets. On Adiac dataset, **DivTopkShapelets**
has the best performance, the average accuracy improved 20.08 %. For classifiers,

Table 1. Accuracy comparison with traditional classification algorithm (The method/s with the highest accuracy in each database are shown in bold)

(a)

Data	C4.5	C4.5(S)	1NN	1NN(S)	NaB	NaB(S)
Adiac	53.19	49.36	59.34	56.27	56.52	57.54
Beef	56.67	40	60	53.33	50	60
Chlorine	64.3	56.82	68.52	58.59	34.61	45.52
Coffee	57.14	92.86	75	**100**	67.86	92.86
Diatom	71.24	67.65	93.46	**94.44**	87.91	78.76
ECG200	72	79	**89**	78	77	80
ECGFiveDays	72.12	98.61	80.6	**99.77**	79.67	96.4
FaceFour	71.59	78.41	87.5	**97.73**	84.09	82.95
Gun_Point	77.33	92	92	96.67	78.67	95.33
MedicalImages	62.5	49.08	67.89	46.45	44.87	51.84
MoteStrain	78.67	78.67	85.78	89.62	84.19	85.7
SonyAIBORobot	65.56	92.51	67.72	95.51	92.85	95.51
Synthetic_control	81	95	88	**98**	96	96.33
Trace	74	**100**	82	98	80	95
TwoLeadECG	71.82	91.22	72.52	98.95	69.8	**99.65**
Average	68.61	77.41	77.95	**84.09**	72.28	80.89
Improve datasets		10		10		13

(b)

Data	BaN	BaN(S)	RaF	RaF(S)	RoF	RoF(S)
Adiac	50.9	38.87	62.15	58.82	**62.29**	60.1
Beef	60	33.33	53.33	46.67	**80**	53.33
Chlorine	59.9	56.09	70.76	57.68	**81.93**	57.5
Coffee	64.29	96.4	64.29	92.86	94.12	96.43
Diatom	94.12	77.45	88.89	75.16	86.93	78.1
ECG200	75	81	82	82	83	80
ECGFiveDays	78.05	98.26	68.99	99.19	90.71	99.65
FaceFour	89.37	88.64	87.5	92.05	75	96.59
Gun_Point	85.33	**99.33**	96	96	86	98
MedicalImages	40.13	47.37	71.71	53.03	**72.37**	53.16
MoteStrain	85.62	85.78	85.98	84.5	84.82	**89.94**
SonyAIBORobot	74.04	**95.84**	71.38	95.67	72.88	95.51
Synthetic_control	92.67	96.67	93.67	96.33	92.67	97.67
Trace	82	**100**	80	**100**	91	96
TwoLeadECG	73.22	97.98	71.73	93.42	91.66	94.73
Average	73.64	79.54	76.56	81.56	83.03	83.11
Improve datasets		10		9		9

Table 2. Relative accuracy between **DivTopkShapelets** and ShapeletSelection (The highest average relative accuracy in each database and each classifier are shown in bold. The negative value means accuracy are not improved.)

Data	C4.5(S)	1NN(S)	NaB(S)	BaN(S)	RaF(S)	RoF(S)	Average
Adiac	16.62	21.23	17.14	16.37	26.09	23.02	**20.08**
Beef	−3.33	3.33	16.67	0.00	3.33	10.00	5.00
Chlorine	0.10	14.24	−10.70	−0.65	13.54	0.78	2.89
Coffee	−3.57	7.14	0.00	0.00	−7.14	7.14	0.59
Diatom	5.56	9.15	−6.86	−4.90	−16.34	−5.89	−3.21
ECG200	4.00	−6.00	2.00	3.00	5.00	0.00	1.33
ECGFiveDays	−0.46	−0.23	−1.28	−0.58	−0.35	0.70	−0.37
FaceFour	3.41	0.00	1.13	−7.95	−1.13	9.09	0.76
Gun_Point	0.67	−2.00	−1.34	−0.67	−3.33	−0.67	−1.22
MedicalImages	1.84	8.55	0.53	−4.08	15.00	5.92	4.63
MoteStrain	−4.63	1.77	−5.11	−5.59	−6.15	−1.03	−3.46
SonyAIBORobot	16.97	32.61	5.99	8.49	12.64	11.32	14.67
Synthetic_control	3.00	1.67	−0.67	0.67	−1.00	0.00	0.61
Trace	6.00	0.00	−4.00	2.00	0.00	−2.00	0.33
TwoLeadECG	−3.07	−0.52	0.88	−0.35	−6.05	−2.46	−1.93
Average	2.87	**6.06**	0.96	0.38	2.27	3.73	2.71
Data sets improved	10	11	8	7	7	10	10

DivTopkShapelets increase 1NN classifier most with the accuracy improved 6.06 %.

Accuracy Comparison with ELM. In this section, we compare the classification accuracy between **DivShapELM** and traditional ELM and average accuracy of six traditional classification algorithms(column named as avg_T). As shown in Table 3, **DivShapELM** can obvious improve traditional ELM accuracy. In 13 out of 15 datasets, the accuracy of ELM is improved and is close to or even better than average accuracy of traditional classification algorithms. Especially, in Trace dataset, **DivShapELM** has the accuracy of 99.20 %, better then ELM 39.40 %.

4.4   Runtime Comparison

DivShapELM has three extra pre-procedures: shapelets candidate selection, diversified shapelets selection and data transform. Once the transformed data are prepared, the rest procedure is a usual classification process. Table 4 gives the extra time and classification time of **DivShapELM** and ELM. The time cost of diversified shapelets selection are varied with data sets, but which can be conducted in an offline manner. Because **DivShapELM** can transform a dataset

Table 3. Accuracy comparison between **DivShapELM** and rival methods (The method/s with the highest accuracy in each database are shown in bold)

Data	ELM	DivShapELM	avg_T
Adiac	30.44	43.52	**57.40**
Beef	**70.67**	44.00	60.00
Chlorine	54.95	57.35	**63.34**
Coffee	61.79	**92.15**	70.45
Diatom	70.43	64.31	**87.09**
ECG200	78.60	79.60	**79.67**
ECGFiveDays	70.12	**97.45**	78.36
FaceFour	43.18	**89.54**	82.51
Gun_Point	83.60	**95.80**	85.89
MedicalImages	54.42	56.72	**59.91**
MoteStrain	58.33	69.08	**84.18**
SonyAIBORobot	54.71	71.68	**74.07**
Synthetic_control	65.23	**97.23**	90.67
Trace	59.80	**99.20**	81.50
TwoLeadECG	67.69	**78.68**	75.13

Table 4. Runtime of **DivShapELM** and ELM (The method with minimum runtime in each database are shown in bold and time in seconds)

Data	Candidate selection	Diversified top-kshapelets	Data transform	DivShapELM	ELM
Adiac	1277	18.81	0.811	**0.04992**	0.13728
Beef	1026	2267.046	0.702	0.0156	0.0156
Chlorine	2636	28.224	7.363	**0.01716**	0.05148
Coffee	337	2109.798	0.436	0.04368	0.00936
Diatom	95	4286.979	2.294	**0.02652**	0.0312
ECG200	216	8.145	0.14	**0.0156**	0.02496
ECGFiveDays	84	7.722	0.717	**0.02184**	0.02652
FaceFour	634	602.46	0.671	**0.02184**	0.06864
Gun_Point	151	28.503	0.218	**0.0312**	0.09048
MedicalImages	732	1.971	0.514	**0.00468**	0.05304
MoteStrain	30	6.183	0.483	**0.03276**	0.04368
SonyAIBORobot	28	3.51	0.249	0.01404	0.00624
Synthetic_control	340	3.087	0.25	**0.04368**	0.07488
Trace	1168	2413.485	1.201	**0.01716**	0.1638
TwoLeadECG	25	2.385	0.546	**0.039**	0.1017

from $n \times m$ length to a $\mathbf{R}^{n \times k}$ matrix(k ≪ m), the runtime of **DivShapELM** can be reduced. As shown in Table 4, **DivShapELM** has the less classification time on 12 out of 15 datasets.

5   Conclusion and Future Work

In this paper, we proposed a novel method to adapt ELM to time series classification by extraction diversified top-k shapelets. Our work includes three parts:

(1) we introduce two conceptions of similar shapelets and diversified top-k shapelets, based on these conceptions, a method of construction diversity shapelets graph is presented, (2) we propose a diversified top-k shapelets extraction method, named as **DivTopkShaplete**, to find out all of the most representative and interpretative features of each class, and (3) we put forward a shapelets transformed ELM algorithm, named as **DivShapELM**, which automatically determine k value and get the diversified top-k shapelets to improve performance of ELM. The experiments results show that **DivShapELM** can improve the efficiency and interpretative of ELM.

For future work, we plan to leverage multi-view feature representations [13–20] to achieve the performance improvement.

Acknowledgments. Supported by the Youth Science Foundation of China University of Mining and Technology under Grant No. (2013QNB16), Natural Science Foundation of Jiangsu Province of China (BK20140192).

References

1. Huang, G.-B., Zhu, Q.-Y., Siew, C.-K.: Extreme learning machine: a new learning scheme of feedforward neural networks. In: Proceedings of the IJCNN2004, pp. 985–990 D(2004)
2. Zong, W., Huang, G.-B., Chen, Y.: Weighted extreme learning machine for imbalance learning. J. Neurocomputing **101**, 229–242 (2013)
3. Savojardo, C., Fariselli, P., Casadio, R.: BETAWARE: a machine-learning tool to detect and predict transmembrane beta barrel proteins in prokaryotes. J. Bioinf. **29**(4), 504–505 (2013)
4. Zhao, Y., Wang, G., Yin, Y., Li, Y., Wang, Z.: Improving ELM-based microarray data classification by diversified sequence features selection. J. Neural Comput. Applic. **27**(1), 155–166 (2016)
5. Ye, L., Keogh, E.: Time Series Shapelets: A New Primitive for Data Mining. In: Proceedings of the 15th ACM SIGKD, pp. 947–956 (2009)
6. Rakthanmanon, T., Keogh, E.: Fast shapelets: a scalable algorithm for discovering time series shapelets. In: Proceedings of the 13th SDM, pp. 668–676 (2013)
7. Lines, J., Davis, L.M., Hills, J., et al.: A shapelet transform for time series classification. In: Proceedings of the 18th ACM SIGKDD, pp. 289–297 (2012)
8. Hills, J., Lines, J., Baranauskas, E., et al.: Classification of time series by shapelet transformation. J. Data Min. Knowl. Discovery **28**(4), 851–881 (2014)
9. Zakaria, J., Mueen, A., Keogh, E.: Clustering time series using unsupervised-shapelets. In: Proceedings of the 12th ICDM, pp. 785–794 (2012)
10. Xing, Z., Pei, J., Philip, S.Y., et al.: Extracting interpretable features for early classification on time series. In: Proceedings of the 11th SDM, pp. 247–258 (2011)
11. Yuan, J.D., Wang, Z.H., Han, M.: Shapelet pruning and shapelet coverage for time series classification. J. Softw. **26**(9), 2311–2325 (2015)
12. Qin, L., Yu, J.X., Chang, L.: Diversifying top-k results. Proc. VLDB Endowment **5**(11), 1124–1135 (2012)
13. Wang, Y., Lin, X., Wu, L., et al.: Robust subspace clustering for multi-view data by exploiting correlation consensus. IEEE Trans. Image Process. **24**(11), 3939–3949 (2015)

14. Wang, Y., Zhang, W., Wu, L., Lin, X., Zhao, X.: Unsupervised metric fusion over multiview data by graph random walk-based cross-view diffusion. IEEE Trans. Neural Netw. Learn. Syst. 1–14 (2015)
15. Wang, Y., Lin, X., Wu, L., Zhang, W.: Effective multi-query expansions: robust landmark retrieval. In: ACM Multimedia, pp. 79–88 (2015)
16. Wang, Y., Lin, X., Wu, L., Zhang, W., Zhang, Q.: LBMCH: learning bridging mapping for cross-modal hashing. In: ACM SIGIR, pp. 999–1002 (2015)
17. Wang, Y., Lin, X., Zhang, Q.: Towards metric fusion on multi-view data: a cross-view based graph random walk approach. In: ACM CIKM, pp. 805–810 (2013)
18. Wu, L., Wang, Y., Shepherd, J.: Efficient image and tag co-ranking: a bregman divergence optimization method. In: ACM Multimedia, pp. 593–596 (2013)
19. Wang, Y., Lin, X., Wu, L., Zhang, W., Zhang, Q.: Exploiting correlation consensus: towards subspace clustering for multi-modal data. In: ACM Multimedia, pp. 981–984 (2014)
20. Yang, W., Wenjie, Z., Lin, W., Xuemin, L., Meng, F., Shirui, P.: Iterative views agreement: an iterative low-rank based structured optimization method to multi-view spectral clustering. In: IJCAI, pp. 2153–2159 (2016)

Scalable and Fast Top-k Most Similar Trajectories Search Using MapReduce In-Memory

Douglas Alves Peixoto$^{(\boxtimes)}$ and Nguyen Quoc Viet Hung

The University of Queensland, Brisbane, Australia
{d.alvespeixoto,q.nguyen}@uq.edu.au

Abstract. Top-k most similar trajectories search (k-NN) is frequently used as classification algorithm and recommendation systems in spatial-temporal trajectory databases. However, k-NN trajectories is a complex operation, and a multi-user application should be able to process multiple k-NN trajectories search concurrently in large-scale data in an efficient manner. The k-NN trajectories problem has received plenty of attention, however, state-of-the-art works neither consider in-memory parallel processing of k-NN trajectories nor concurrent queries in distributed environments, or consider parallelization of k-NN search for simpler spatial objects (i.e. 2D points) using MapReduce, but ignore the temporal dimension of spatial-temporal trajectories. In this work we propose a distributed parallel approach for k-NN trajectories search in a multi-user environment using MapReduce in-memory. We propose a space/time data partitioning based on Voronoi diagrams and time pages, named Voronoi Pages, in order to provide both spatial-temporal data organization and process decentralization. In addition, we propose a spatial-temporal index for our partitions to efficiently prune the search space, improve system throughput and scalability. We implemented our solution on top of Spark's RDD data structure, which provides a thread-safe environment for concurrent MapReduce tasks in main-memory. We perform extensive experiments to demonstrate the performance and scalability of our approach.

1 Introduction

GPS trajectory data carry rich information about moving objects, and have been extensively used for a great number of real-world applications, such as city traffic planing, alternative routes suggestion, trip recommendation, drivers pastern analysis, dynamic event identification, and so on [7,15,32].

Given a query trajectory T, a constant k, a time interval $[t_0, t_1]$, and a trajectory dataset S, the top-k nearest neighbor trajectories problem (k-NN), is to find in S the k closest (or most similar) trajectories from T active during $[t_0, t_1]$. k-NN trajectories is one of the most traditional query operations in trajectory databases, and has received plenty of attention, e.g [6,20,23,25]. Applications include, for example, to identify the top-k vehicle's trajectories in a frequent

© Springer International Publishing AG 2016
M.A. Cheema et al. (Eds.): ADC 2016, LNCS 9877, pp. 228–241, 2016.
DOI: 10.1007/978-3-319-46922-5_18

path in order to calculate their average fuel consumption during a certain period of time, for logistics optimization. However, processing k-NN trajectories in a multi-user environment is challenging; the application may be serving hundreds of requests over the network, and k-NN search in general demands extensive use of computational resources. Furthermore, k-NN search for trajectories is a complex operation, unlike other simpler spatial objects, trajectories are essentially non-uniform sequential data with variable length, attached with both spatial and temporal attributes.

Besides, the massive amount of GPS data available, as well as the increasing number of trajectory data application users, demands more robust, fast and scalable solutions, since location-based service should be able to serve multiple requests over large-scale datasets. Therefore, a typical solution is to consider distributed parallel computation with frameworks such as MapReduce (MR) [8], which provides an abstraction for parallel computation and efficient resources allocation of concurrent threads. Frameworks like Spark [29], on the other hand, provides a MR solution for faster data processing using in-memory data storage.

Current state-of-the-art for k-NN trajectories, however, mainly focus on single-thread/single-user paradigm, and cannot be easily tailored to the MR model [5,6,20,23]. Existing research to support spatial queries using MR, e.g. [1,3,10,14], utilize either a multi-core *divide-and-conquer* strategy, where each *mapper* is responsible to process a sub-query over a subset of the dataset, while the intermediate results from the *map* are refined by the *reducers*; or utilize spatially-aware partitioning in order to organize the space into disjoint groups of spatially close objects. Spatial-aware partitioning strategies in MR can achieve up to 10x faster performance than *divide-and-conquer* by maintaining data locality [10,33], since only a smaller number of partitions containing query candidates are selected for processing, reducing query latency and avoiding unnecessary I/O.

Contribution. The current MR works on k-NN, however, either apply for the spatial dimension only, ignoring the sequential nature and temporal dimension of trajectories, i.e. [2,12,14,31]; or only supports range selection for trajectories [16,27]. To overcome this limitation, we propose a bulk-loading in-memory partitioning strategy based on Voronoi diagrams and time pages, named **Voronoi Pages**, to support multiple k-NN trajectories query in MR, and a spatial-temporal composite index, named **VSI** (Voronoi Spatial Index) and **TPI** (Time Page Index), to prune the search space and speed up trajectory similarity search. Voronoi-based partitioning have been successfully used for distance-based search in MR [2,12,14]. For the best of our knowledge, this is the first work to address similarity-based search for trajectory data in MR.

2 Problem Statement and Overall Approach

2.1 Top-k Trajectories Problem

A trajectory T of a moving object is a multidimensional sequence of spatial-temporal points, where each point is described as a triple (x, y, t), where (x, y)

are the spatial coordinates of the moving object at a time t. We need a distance function $d(T_a, T_b)$ to calculate the distance between two trajectories [24]. In this work, we adopt the Edit Distance with Projections (EDwP) [20], as trajectory similarity function. EDwP is threshold-free and can cope with local time shifts and non-uniform sampling rates, which are essential in real-world trajectory datasets.

Problem Statement (k-NN). Given an input trajectory dataset $\mathbb{S}$, a trajectory distance function $d(T_a, T_b)$, and a batch of queries $\mathbb{Q} = [(T_1, k_1, t_{1a}, t_{1b}), ..., (T_n, k_n, t_{na}, t_{nb})]$ from application users. For each input query $(T_i, k_i, t_{ia}, t_{ib})$ we want to find in $\mathbb{S}$ the k_i closest trajectories from T_i w.r.t. $d(T_a, T_b)$, and active in $[t_{ia}, t_{ib}]$, named k_i-NN(T_i, t_{ia}, t_{ib}).

Our goal is to improve performance and throughput of k-NN trajectory search using MR in-memory, and allow concurrent queries in multi-user servers. The cost of executing a k-NN query can be measured by the number of input records it has to read and process [3]. Following we describe the challenges of processing k-NN search in multi-user environments over large-scale trajectory datasets.

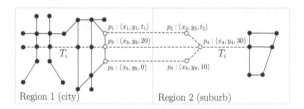

Fig. 1. Road network example.

Temporal dimension: Since trajectory data are queried by both spatial and temporal attributes, temporal dimension must be taken into account [26]; for instance, consider the road network in Fig. 1 connecting two spatial regions, there may be thousands of trajectories passing through the road T_i, however the client may be interested in retrieving similar trajectories within a specific time period. For example, imagine three trajectories passing through $T_1 : \{p_1, p_2\}$, $T_2 : \{p_3, p_4\}$ and $T_3 : \{p_5, p_6\}$, if we want the closest trajectory to T_1 within time $t = [0, 10]$, the application should return T_3 instead of T_2. Only grouping trajectories by spatial region in that case is not strict enough.

Skewness: In Fig. 1 the density of moving object's passing through Region 1 (city region) is much larger than in Region 2 (suburb region). Therefore, we must provide a partitioning strategy as uniform as possible to avoid load imbalance, yet keeping spatial proximity, which is a key factor in spatial data processing in MR [9]. If many concurrent queries are accessing spatial-temporal regions with high density of trajectory points, a poor partitioning may cause I/O bottleneck and impair the system throughput.

System efficiency and reliability: The number of spatial-temporal regions containing the answer of a k-NN query highly depends on the value of k and the time interval $[t_0, t_1]$, hence an iterative neighborhood search may be necessary to select candidate trajectories. In-memory based structures are more suitable for iterative MR jobs, where it is necessary to apply a function repeatedly on the working set of data [29]. Furthermore, we must take concurrency control into account, since different tasks might be working over the same data partitions. Therefore, we provide an off-line data partitioning on top of Spark's in-memory data structure (RDD) [28]. Spark's scheduler is fully thread-safe and supports multiple on-line requests over its RDD.

2.2 Scalable k-NN query using Partitioning

An efficient way to answer k-NN trajectories search is to partition the dataset in a spatial-temporal aware manner, such that the amount of data processed by each query is minimized. Spatial-aware partitioning strategies, such as grid cells and Voronoi diagrams (VD), aim to organize the data into smaller partitions of spatially close objects to reduce the number of query candidates, hence reducing network and I/O costs [30,34]. In this work we extend a VD data partitioning for spatial-temporal trajectories in MR, for it maintains data proximity and provides uniform distribution for skewed datasets. VD is particularly suitable for distance-based search, where grid partitioning suffer from a significant loss of pruning power [2,11,12,14].

Overall process. We uniformly partition the space into Voronoi cells using k-Means clustering, and each Voronoi cell into static temporal partitions (i.e. pages). Trajectories are split into sub-trajectories according to their spatial-temporal extent, such that each sub-trajectory is mapped to one Voronoi Page. We build our Voronoi Pages partitions on top of RDDs to speed up query processing. We process a k-NN query in parallel in a *filter-and-refinement* fashion, first filtering candidate pages, and then running a precise check on the candidate pages. Each process unit can manage a number of pages within a RDD in parallel, and concurrent queries can be served by Spark over its RDD.

3 Voronoi-Based Partitioning with Temporal Awareness

Given an input trajectory dataset, we read and split each trajectory into a set of sub-trajectories, according to its spatial and temporal extent, such that each sub-trajectory is assigned to only one spatial-temporal partition.

Space Partitioning. Given a set of n generator pivots in the dataset space, $PV = \{p_1, ..., p_n\}$, where $p_i = (x_i, y_i)$, we partition the dataset space into n disjoint spatial partitions, where each trajectory sample point is assigned to its closest pivot (i.e. Voronoi cell). Figure 2 illustrates eight trajectories, T_1 to T_8, partitioned across seven Voronoi cells, P_1 to P_7. Boundary trajectories, e.g. T_1 and T_4, are split into sub-trajectories, where each sub-trajectory is assigned to its overlapping cell.

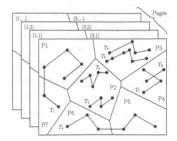

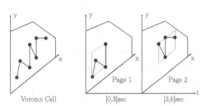

Fig. 2. Trajectories partitioned across Voronoi cells.

Fig. 3. Sub-trajectory partitioning into Voronoi Pages, $TW = 3$ s. Each page contains sub-trajectories that overlap with both the Voronoi polygon area and time window.

Time Partitioning. Given a time window size TW we split the time space of each Voronoi cell into static time pages of size $= TW$, and assign each sub-trajectory sample point inside a polygon to a time page according to its time-stamp. Figure 3 illustrates a sub-trajectory in a given Voronoi cell split into time pages. For the sake of simplicity we assume each sub-trajectory sample point in Fig. 3 was uniformly collected every one second, that is $t_i = [0, 6]$ s; however, this approach is for both uniform and non-uniform samples.

Voronoi Page. Each time page for a given Voronoi cell is called a Voronoi Page (VPage), identified by a spatial-temporal index $\langle VSI, TPI \rangle$, where **VSI** (Voronoi Spatial Index) is the page's polygon identifier, and **TPI** (Time Page Index) is the page's time window identifier. Each VPage is composed of two structures: (1) a local R-Tree of sub-trajectories in the page, and (2) a list of the trajectories' IDs in the page. In our implementation, we use simple R-Tree of sub-trajectory bounding boxes. However, any other index access method for sub-trajectories can be used within a VPage.

Handling Boundary Trajectories. While partitioning both space and time we expect some trajectories to intersect more than one VPage. To minimize replication, we split boundary trajectories and replicate only the boundary segments, that is, if a trajectory segment $\overline{p_i p_{i+1}}$ crosses any polygon boundaries, we split the trajectory and assign the boundary segment to both sub-trajectories; each sub-trajectory is assigned to its overlapping polygon. For the temporal dimension the situation is likewise.

Generator Pivots. We choose the number of generator pivots n based on the size of the dataset and the default RDD block size (i.e. 64 MB), so that each task can process data blocks with roughly the same number of polygonal partitions. We study the effect of n for the system performance in Sect. 5.2. In addition, we must choose the pivots in order to break the space into uniform clusters to avoid load imbalance. Therefore, we use the parallel k-Means++ heuristic [4], provided in the Spark machine learning library (MLlib) [17], which provides a fair approximation of the deterministic k-Means.

3.1   MapReduce Implementation

We assume each input file contains one trajectory per line, as a sequence of spatial-temporal points; the data is initially in the HDFS. We build our VPages structure as an RDD with a `map()` and `reduce()` functions as follows. The partitioning process returns an RDD of VPages (i.e. $\mathbb{P}_{RDD}^{pages}$) cached in-memory. Further details can be found in the technical report [21];

Map: The *mapper* reads and splits a trajectory T into m sub-trajectories, according to its spatial-temporal dimension, and emits a list of $\langle (VSI, TPI), T_i^{sub} \rangle$ with m pairs, $i \in [1, .., m]$, consisting of a sub-trajectory T_i^{sub} as *value*, and the spatial-temporal index of the VPage containing T_i^{sub} as *key*.

Reduce: The *reducer* receives a list of sub-trajectories (*values*), and groups them by index (*key*), adding each sub-trajectory to the VPage R-Tree. At the end of the parallel process, the *reduce* returns an RDD of $\langle (VSI, TPI), VPage \rangle$ pairs, consisting of the spatial-temporal VPage index, and the final VPage.

3.2   Trajectory Track Table (TTT)

We must keep track of sub-trajectories across VPages, so that we can retrieve and rebuild a trajectory when processing a k-NN query. For this purpose, we propose a table-like structure, named **Trajectory Track Table** (TTT). The TTT is a in-memory structure, where each tuple of the table is a pair composed of a trajectory ID and a set of references to VPages (page index hash) containing the pages a trajectory intersects with. The TTT is constructed as an RDD (i.e. $\mathbb{T}_{RDD}^{table}$) so that all nodes have access to it without the need of replication. We build the $\mathbb{T}_{RDD}^{table}$ with MR as follows.

Map: The *mapper* reads and map each input trajectory to a list of $\langle T_{id}, (VSI, TPI) \rangle$ pairs, containing the trajectory identifier for each VPage index T_{id} overlaps with.

Reduce: The *reducer* groups indexes by trajectory T_{id} into a set of VPage indexes $\langle T_{id}, \{(VSI, TPI)\} \rangle$. Each pair $\langle T_{id}, \{(VSI, TPI)\} \rangle$ is henceforth called a table tuple.

4   k-NN Trajectories Query Answering

Given a query trajectory Q, and a time interval $[t_0, t_1]$, we want to retrieve the k-NN of Q within the time interval $[t_0, t_1]$. By using a VD-based approach we focus on the spatial proximity to the specified query location. Let $VP(Q)$ be the set of Voronoi polygons covered by Q, and $VP_N(Q)$ be the set of neighbor polygons of $VP(Q)$. To process k-NN trajectory queries we take advantage of the neighborhood properties of VDs as follows. The proof of these properties can be found at [18].

1. The nearest generator pivot p_j from another generator pivot p_i is among the pivots whose Voronoi polygons share edges with $VP(p_i)$ (locality preserving property).

2. Let n and n_e be respectively the number of pivots and the number of edges in a VD, then $n_e \leq 3n - 6$. And given that every edge in a VD is shared by exactly two polygons, then the average number of edges per Voronoi polygon is less equal than six, i.e., $2(3n - 6)/n = 6 - 12/n \leq 6$.

NN Trajectory Search. From property 1, the nearest neighbor $NN(Q)$ of a query object Q is either in $VP(p_i)$, where p_i is the nearest pivot from object Q, or among the neighbor cells of $VP(p_i)$, for Q might be a boundary object. However, because our query object Q is a trajectory, we must check all polygons intersecting with Q and their neighbors. Moreover, we are interested in a spatial-temporal k-NN, thus we have to look in the specific time pages inside each partition. More precisely, assuming our query object is T_4, and we are interested in a time interval $[t_0, t_1]$, we search for the $NN(T_4, t_0, t_1)$ inside the VPages set $\mathbb{F} = \{(2, [t_0, t_1]), (4, [t_0, t_1]), (5, [t_0, t_1]), (6, [t_0, t_1]), (7, [t_0, t_1])\}$. Nevertheless, trajectories in $\mathbb{F}$ may span to other VPages depending on their spatial and temporal extent, for instance, T_1 in P_7 also spans to P_1. We must ensure the whole trajectories are returned from the previous step in order to evaluate their distances. Thus, from this point we visit the TTT to retrieve the index of other VPages containing the trajectories in $\mathbb{F}$ (if there is any new). We filter from the $\mathbb{P}_{RDD}^{pages}$ the sub-trajectories in the VPages returned from the $\mathbb{T}_{RDD}^{table}$ and append the remainder sub-trajectories to $\mathbb{F}$. A post-processing step is done to merge sub-trajectories in $\mathbb{F}$, and compute the $NN(Q, t_0, t_1)$.

k-NN Trajectories Search. Similar to [2], to calculate the remainder (k-1)-NN of Q, suppose both Q and $NN(Q)$ are in P_3, $Q = T_5$ and $NN(T_5) = T_6$, thus we also look for the second NN of Q in pages inside the neighborhood of P_3, that is P_1, P_2 and P_4. The remainder NNs are retrieved in the same recursive process; the search stops at the k_{th} iteration if the number of candidates c is $c \geq k$, or continues the search until $c \geq k$. From property 2, the number of neighbor partitions we have to look for time pages in every iteration is at most six for every partition containing the current candidate.

4.1   K-NN Search Using RDD

The VPages containing the k-NN result are unknown until the query is executed, thus, we calculate k-$NN(Q, t_0, t_1)$ with k iterative *filter-and-refinement* MR jobs, so that in every i_{th} iteration we have the i_{th}-$NN(Q, t_0, t_1)$ result. More technical details and implementation can be found at [21].

First Filter-Refinement: We first select from $\mathbb{P}_{RDD}^{pages}$ all pages in the interval $[t_0, t_1]$, for every polygon $P_i \in (VP(Q) \cup VP_N(Q))$. Finally, we collect all trajectories inside the filtered VPages, and active during $[t_0, t_1]$ – we use the $\mathbb{T}_{RDD}^{table}$ to track other VPages containing trajectories in $(VP(Q) \cup VP_N(Q))$. This step returns a RDD of candidate trajectories $\mathbb{T}_{RDD}^{cddt}$. The *refinement* receives

$\mathbb{T}_{RDD}^{cddt}$ from the *filter* step, and returns a list of trajectories from $\mathbb{T}_{RDD}^{cddt}$ sorted by distance to Q. If one is interested in the 1-NN(Q, t_0, t_1) only, we return the first element in the sorted list as the 1-NN(Q, t_0, t_1) result.

Iterative Search: For every i_{th}-NN of Q remaining, we perform a *filter-refinement* process in a fashion as `similar as before, using an iterative neighborhood search over the $\mathbb{P}_{RDD}^{pages}$ as stated in Sect. 4. At the end of each i_{th} stage the intermediate results are collected and the candidates list is updated in the application master.

5   Experiments

We present a set of highlighted experiments on a real trajectory dataset to evaluate the performance and scalability of our approach.

Experimental Setup. We use a 16 GB trajectory dataset collected from the southern region of China. The dataset contains 4 million heterogeneous trajectories from taxis and personal vehicles in a period of five days. The data is initially stored in HDFS. All algorithms are implemented in the Spark Java library version 1.5.1. Experiments are conducted on a cluster with 30 nodes. Each node is a Ubuntu 14.04 LTS with dual-core processor and 3 GB of RAM. We employ Spark-JobServer [22] to support multiple concurrent jobs in our application. We evaluate our method for both NN and k-NN trajectory queries. Due to space limitations we set $k = 10$ by default. We use $1,000$ as the default number of cells, and fixed the time window at $1,200$ s (based on the mean $\mu = 543$ s and standard deviation $\sigma = 700$ s of trajectories duration), so most trajectories fit into one time page. As query input, we randomly selected 100 trajectories from the dataset, the query time was set as the beginning and ending time of each query trajectory; we perform the queries in batches of 5 concurrent threads by default.

We compare our approach against a Grid-cell based approach, also commonly used in spatial MR works, e.g. [10,27]. The grid-based approach is similar to the VD one, except the space is partitioned into a uniform grid. Throughout this section we refer to the VD approach and Grid approach as *VPages* and *GPages* respectively. To process k-NN queries in *GPages* we employ a technique similar to that in SpatialHadoop [10] to prune the search space, except we use the trajectories' centroid distance to select candidate trajectories. To a fair comparison, we apply the same splitting strategy and spatial-temporal granularity on both implementations.

5.1   VPages Construction Evaluation

Index Construction Scalability. Figure 4(a) demonstrates the execution time for reading the data from HDFS and building both *VPages* and *GPages* RDDs for different dataset sizes, i.e. from $1/4$x to 1x the original dataset. *GPages*

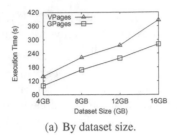

(a) By dataset size.

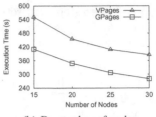

(b) By number of nodes.

Fig. 4. Index construction time evaluation.

Table 1. Trajectories distribution across *VPages* by number of pivots.

#Pivots	#VPages	#Sub-Trajectories	#Splits (Avg)	Latency (s)
250	78,715	6,045,863	1.51	247.5
500	146,479	6,276,712	1.57	301.5
1,000	265,700	6,538,746	1.63	385.0
2,000	464,912	6,949,443	1.74	497.0

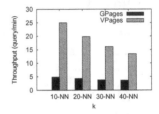

Fig. 5. Query Number of k.

outperformed *VPages* on index construction time on all scenarios due to the one-to-one complexity of parsing trajectory data points to a uniform grid, against the $O(n * k)$ complexity of Voronoi diagram construction. This is also true for different numbers of computing nodes as shown in Fig. 4(b). However, *VPages* outperformed *GPages* in query latency and throughput as we discuss further on this section.

Effect of the Number of Pivots. Table 1 gives statistical information about trajectories distribution across *VPages* and the execution time on building the *VPages* RDD for different numbers of Voronoi cells. As expected, the execution time increase with the number of cells, this is due the increasing number of comparisons during the map phase. The number of trajectories' splits increase with the spatial partitioning granularity, this is due to increasing number of boundary trajectories in more tight partitions. However, system throughput improves for larger numbers of cells as we further discuss.

5.2 System Performance and Scalability

Scalability Evaluation. Figure 6(a) shows the system throughput for NN and 10-NN queries on both *VPages* and *GPages* RDDs. We measure system throughput by the number of queries completed per minute. Overall, *VPages* performed up to 10x better than *GPages* for both NN and k-NN queries as the dataset grows. This is mainly due to two reasons: first the filter step of *VPages* is more accurate than its *GPages* counterpart on filtering candidate trajectories; secondly, *VPages* presented a more uniform data distribution across partition using

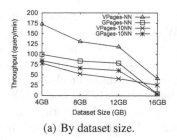

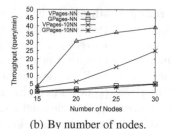

(a) By dataset size.

(b) By number of nodes.

Fig. 6. System throughput evaluation.

k-Means clustering than the grid-based approach, which caused the load imbalance in *GPages*. However, for dataset smaller than 16 GB, *GPages* performed *k*-NN search slightly better than *VPages*; this is due to the iterative neighborhood search on *VPages*, which seeks for the query result on neighbor cells even for small input datasets. This difference, however, disappears as the dataset grows due to the most homogeneous data distribution of *VPages*. Near 16 GB for *GPages*, however, the cluster resources utilization reaches its limits for the default parameters, once each concurrent query needs to cache and process its own copy of the filtered RDD partitions, which causes Spark to shuffle more data and spill some data to disk for larger input datasets, thus the performance deterioration on *GPages*. The situation is likewise with number of nodes smaller than 20 nodes, as shown in Fig. 4(b), where *VPages* outperformed *GPages* in all scenarios up to 25x in NN search and up to 10x in *k*-NN.

Effect of the Number of Pivots. Figure 7(a) gives the system throughput using *VPages* for different numbers of Voronoi cells. Overall, finer-grained partitions tends to positively affect query latency and throughput, this is due to the filter step to be more precise when retrieving candidate trajectories. This improvement in query latency increases the resources availability, hence increasing parallelism and system throughput.

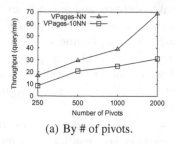

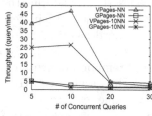

(a) By # of pivots.

(b) By # of concurrent queries.

Fig. 7. System throughput by number of pivots and by number of concurrent queries.

Concurrency Evaluation. Here we evaluate the effect of the number of concurrent queries to the system throughput. We submit queries in batches of 5 to 30, and start one thread per query job using the Spark-JobServer [22] framework. Queries are executed in a "round-robin" fashion using Spark's *FAIR* job scheduling, so that all queries get a roughly equal share of cluster resources. Figure 7(b) gives the overall results. For *VPages* on both NN and k-NN queries the system throughput increased from 5 to 10 concurrent queries, this is due to the best use of our cluster resources. Near 10 concurrent queries, however, the resources utilization reaches its peak, hence its maximum throughput. Furthermore, even with dataset in main-memory the overhead of managing large numbers of concurrent jobs can lead to more contentions and strongly limit the system scalability [19]. For *GPages* the situation was much worse, with its peak near 5 concurrent jobs. In summary, *VPages* demonstrated to be up to 15x better than *GPages* on handling multi-user application and concurrent jobs.

Effect of Number of Neighbors (k). Figure 5 gives a comparative on the system throughput as k grows. The partitions containing the k-NN are unknown until the query is executed; however, spatial locality is not always preserved in grid-based, which means we need to extent the search space in *GPages* further than in *VPages* to retrieve the candidate trajectories, which negatively impacts the performance of *GPages*. Due to the locally preserving property of VDs, most neighbors of a given object are in the nearby polygons, thus are retrieved in the first iterations; and due to the homogeneous data distribution in VPages, the number of trajectories in the neighbor partitions to retrieve are roughly the same, which leads to near linear effect on query latency as k increases; the system throughput, therefore, is directly affected by queries latency. *VPages* is more sensitive to k than *GPages*, however, *VPages* is only as poor as *GPages* for very large values of k, where a great number of partitions need to be track.

6   Related Work

MR-based Solutions for k-NN of spatial points. Lu et al. [14] and Akdogan et al. [2] use a VD-based approach to partition the space and index spatial objects based on its closest pivots during the *map* phase, and processing k-NN and RNN queries [2] and k-NN join [14] in iterative *MR* tasks; and outperforms similar MR works based on grid-based partitioning for k-NN query [33] and k-NN join [31]. Our partitioning method is closely related to that in [2,14], except we extend VD for spatial-temporal dimension of trajectories in order to support trajectory similarity search, we also use RDD to support in-memory based computation and concurrent queries.

Unified Frameworks for Spatial Queries in MR. SpatialHadoop [10] has been developed to support spatial data operations and spatial data indexing using MR. SpatialHadoop outperformed Hadoop and traditional approaches by using an extensive set of spatial partitioning structures to improve spatial queries performance in MR [9]. Similarly, Aly et al. proposed AQWA [3], an adaptive spatial data partitioning based on kd-Trees to support range selection and k-NN search in MR; unlike SpatialHadoop, AQWA provides a dynamic space partitioning which reacts to changes on both the query workload and new incoming data. Similar to SpatialHadoop, Hadoop-GIS [1] and ScalaGiST [13] presented another general purpose solution for spatial queries and cost-efficient spatial data indexing in MR. However, the aforementioned works only support nearest neighbors search for spatial points, and do not provide any support for spatial-temporal trajectories.

Centrally-based Indexing for Similarity Search. This includes LCSS [23], ERP [5], EDR [6] and TrajTree [20]. The goal is to extend tree-based indexing to organize the trajectory dataset space for similarity search. However, all these methods index trajectories based on spatial similarity only, ignoring its temporal dimension. The main drawbacks of these centrally-based structures are that they do not provide fully decentralization for parallel computation, and do not scale for large datasets. Furthermore, all aforementioned works are for disk-based computation, whereas our solution takes advantage of in-memory structure to speed up similarity search. SharkDB [25] is a in-memory storage architecture for trajectories, which partitions the dataset into time frames, in order to support general purpose operations. However, SharkDB uses only temporal partitioning, and focus on column-oriented architectures. By using RDD, however, we can provide a distributed and fault-tolerant solution for large datasets and concurrent tasks.

7   Conclusions

In this work we present a multi-user system to process concurrent k-NN trajectories search using Spark's RDD, a thread-safe and resilient distributed data structured for large-scale data processing in main-memory using MapReduce model. We introduced a novel spatial-temporal data partitioning approach, named Voronoi Pages, built on top of RDD to a scalable and fast processing of concurrent k-NN trajectories search in MR. Voronoi Pages provides both homogeneous data partitioning and spatial-temporal locality preserving, essentials for MR-based systems. Our experimental results based on a real trajectory dataset demonstrates the performance and good scalability of our approach against another common approach used in MR for spatial data.

Acknowledgments. This research is partially supported by the Brazilian National Council for Scientific and Technological Development (CNPq).

References

1. Aji, A., et al.: Hadoop-GIS: a high performance spatial data warehousing system over mapreduce. VLDB **6**(11), 1009–1020 (2013)
2. Akdogan, A.: Voronoi-based geospatial query processing with mapreduce. In Cloud- Com, pp. 9–16. IEEE (2010)

3. Aly, A.M., et al.: AQWA: adaptive query workload aware partitioning of big spatial data. VLDB **8**(13), 2062–2073 (2015)
4. Bahmani, B., et al.: Scalable k-means++. VLDB **5**(7), 622–633 (2012)
5. Chen, L., Ng, R.: On the marriage of lp-norms, edit distance. In: VLDB, pp. 792–803 (2004)
6. Chen, L., Özsu, M.T., Oria, V.: Robust, fast similarity search for moving object trajectories. In: SIGMOD, pp. 491–502 (2005)
7. Dai, J., et al.: Personalized route recommendation using big trajectory data. In: ICDE, pp. 543–554 (2015)
8. Dean, J., Ghemawat, S.: MapReduce: simplified data processing on large clusters. Commun. ACM **51**(1), 107–113 (2008)
9. Eldawy, A., Alarabi, L., Mokbel, M.F.: Spatial partitioning techniques in Spatial-Hadoop. VLDB **8**(12), 1602–1605 (2015)
10. Eldawy, A., Mokbel, M.F.: SpatialHadoop: a MapReduce framework for spatial data. In: ICDE, pp. 1352–1363 (2015)
11. Kolahdouzan, M., Shahabi, C.: Voronoi-based k nearest neighbor search for spatial network databases. In: VLDB, pp. 840–851 (2004)
12. Li, C., et al.: Processing moving k NN queries using influential neighbor sets. VLDB **8**(2), 113–124 (2014)
13. Lu, P., et al.: ScalaGiST: scalable generalized search trees for mapreduce systems [innovative systems paper]. VLDB **7**(14), 1797–1808 (2014)
14. Lu, W., et al.: Effcient processing of k nearest neighbor joins using mapreduce. VLDB **5**(10), 1016–1027 (2012)
15. Luo, W., et al.: Finding time period-based most frequent path in big trajectory data. In: SIGMOD, pp. 713–724 (2013)
16. Ma, Q., et al.: Query processing of massive trajectory data based on mapreduce. In: International Workshop on Cloud Data Management, pp. 9–16. ACM (2009)
17. MLlib: http://spark.apache.org/docs/latest/mllib-guide.html
18. Okabe, A., et al.: Spatial tessellations: concepts and applications of Voronoi diagrams, vol. 501. Wiley, New York (2009)
19. Pandis, I., et al.: Data-oriented transaction execution. Proc. VLDB Endowment **3**(1–2), 928–939 (2010)
20. Ranu, S., et al.: Indexing, matching trajectories under inconsistent sampling rates. In: ICDE, pp. 999–1010 (2015)
21. Scalable and fast top-k most similar trajectories search using MapReduce in-memory. Technical report (2016). https://www.researchgate.net/publication/303487238
22. Spark-JobServer: https://github.com/spark-jobserver/spark-jobserver
23. Vlachos, M., Gunopulos, D., Kollios, G.: Discovering similar multidimensional trajectories. In: Agrawal, R., Dittrich, K.R. (eds.) ICDE, pp. 673–684 (2002)
24. Wang, H., et al.: An effectiveness study on trajectory similarity measures. In: ADC, pp. 13–22 (2013)
25. Wang, H., et al.: SharkDB: an in-memory column-oriented trajectory storage. In: CIKM, pp. 1409–1418 (2014)
26. Wang, X., Zhou, X., Lu, S.: Spatiotemporal data modelling, management: a survey. In: TOOLS-Asia, pp. 202–211. IEEE (2000)
27. Yang, B., Ma, Q., Qian, W., Zhou, A.: TRUSTER: TRajectory data processing on ClUSTERs. In: Zhou, X., Yokota, H., Deng, K., Liu, Q. (eds.) DASFAA 2009. LNCS, vol. 5463, pp. 768–771. Springer, Heidelberg (2009)

28. Zaharia, M., et al.: Resilient distributed datasets: a fault-tolerant abstraction for in-memory cluster computing. In: USENIX Conference on Networked System Design and Implementation, p. 2 (2012)
29. Zaharia, M., et al.: Spark: cluster computing with working sets. In: USENIX Conference on Hot Topics in Cloud Computing, p. 10 (2010)
30. Zamanian, E., Binnig, C., Salama, A.: Locality-aware partitioning in parallel database systems. In: SIGMOD, pp. 17–30 (2015)
31. Zhang, C., Li, F., Jestes, J.: Effcient parallel kNN joins for large data in MapReduce. In: EDBT, pp. 38–49 (2012)
32. Zheng, Y., Zhou, X.: Computing with Spatial Trajectories. Springer, New York (2011)
33. Zhong, Y., et al.: Towards parallel spatial query processing for big spatial data. In: IPDPSW, pp. 2085–2094. IEEE (2012)
34. Zhou, X., Abel, D.J., Truffet, D.: Data partitioning for parallel spatial join processing. In: Scholl, M., Voisard, A. (eds.) SSD 1997. LNCS, vol. 1262, pp. 178–196. Springer, Heidelberg (1997). doi:10.1007/3-540-63238-7_30

Data Mining and Analytics

Prescriptive Analytics for Big Data

Reza Soltanpoor[1(✉)] and Timos Sellis[2]

[1] RMIT University, Melbourne, Australia
reza.soltanpoor@rmit.edu.au
[2] Swinburne University of Technology, Melbourne, Australia
tsellis@swin.edu.au

Abstract. Prescriptive analytics is considered as the next frontier in the area of business analytics. It provides organizations with adaptive, automated, and time-dependent courses of actions to take advantage of likely business opportunities. Given enterprises' objectives, prescriptive analytics assists them maximize their business values and at the same time mitigates their likely risks by recommending optimal sequences of actions. In this work, a federated prescriptive analytics framework comprising descriptive, predictive and prescriptive components is proposed. The framework also links the extracted insight from the data to their pertinent generated actions. Finally, a few indicative use cases are presented to indicate the necessity of this new analytics paradigm.

Keywords: Prescriptive analytics · Descriptive analytics · Predictive analytics · Big data

1 Introduction

Proper data-driven analytics approaches are of significant importance in assisting enterprises improve their business values in the era of big data [8]. Further, to take advantage of their business opportunities, organizations with considerable amounts of data are interested in using optimal data processing and pertinent insight extraction techniques according to their objectives [8]. Therefore, utilization of relevant business analytics methods plays a key role in helping firms boost their success [8]. Analytics, as a multidisciplinary concept, is defined as the means to acquire data from diverse sources, process them to elicit meaningful patterns and insights, and distribute the results to proper stakeholders [8, 18, 22]. A lot of research has been conducted in discovering effective methods to collect, report, process, comprehend and extract insight from big data. These approaches enable enterprises understand what has happened in the past and what is likely to happen in the future [15]. Therefore, two essential types of analytics were introduced to take the responsibility for comprehending past and future events: "descriptive analytics" and "predictive analytics" [20]. *Descriptive analytics* is focused on the past. It summarizes the data and produces information from diverse set of heterogeneous data [11]. *Predictive analytics*, on the other hand, is the forecasting analytics and is concerned with the future [12]. It builds accurate

© Springer International Publishing AG 2016
M.A. Cheema et al. (Eds.): ADC 2016, LNCS 9877, pp. 245–256, 2016.
DOI: 10.1007/978-3-319-46922-5_19

predictive models based on the unified data and answers questions like what and why something is likely to happen in the future. Both descriptive and predictive analytics assist organizations in extracting proper insights from the data they own.

However, having the insight on what has happened in the past and what will happen in the future is not sufficient to take advantage of business opportunities. Enterprises need to utilize analytics to transform information into insights and act upon them to meet their objectives [3,8,15]. Therefore, there is a gap between the generated insights and optimal operational courses of actions based on those insights to be recommended [5]. To address this issue, a new kind of analytics has emerged, *Prescriptive analytics* which is considered as the next frontier in the business analytics [13]. It is concerned with the recommendation and guidance and provides organizations with adaptive, automated and time-dependant sequences of operational actions. It answers questions like "What should be done?" and "Why should it be done?" [11].

In this study, we propose a federated data-driven prescriptive analytics paradigm incorporating descriptive, predictive and prescriptive components. Our generic architecture can be applied to a diverse range of use cases to assist them with their decision making process in an optimal and near real-time manner. Our main contribution is proposing an integrated data-driven architecture which enables collecting heterogeneous data, applying optimal data-processing techniques on them, generating accurate extrapolations based on the extracted patterns from data, and providing the firms with adaptive, actionable recommendations matching their business objectives in-time. The unique way of combining three analytics tasks and their components together is another contribution of this work.

The rest of this article is organized as the following: in Sect. 2, recent studies in descriptive, predictive and prescriptive analytics are reviewed. In Sect. 3, our novel federated prescriptive analytics architecture is proposed. In Sect. 4, a couple of use cases are studied and one particular application is elaborated to demonstrate the significance of the proposed paradigm. In Sect. 5, a prescriptive analytics conceptual framework is introduced to address some major open issues with the prescriptive analytics - Sect. 5.1 - and one abstract framework in the context of learning analytics is described - Sect. 5.2. Finally, in Sect. 6, this work's outcomes and its future research directions are indicated.

2   Related Studies

Big enterprises are striving to utilize analytics in order to transform information into insight, extract value out of it and act upon that [3]. Companies want to know what is happening now, what will happen in the future, and what are the courses of actions they can take to reach an optimal result [15]. Big enterprises hold data and need to extract insight from these data to take advantage of future opportunities and decrease possible risks [9]. Therefore, they need operational courses of actions based on those insights to be recommended [5]. Prescriptive analytics is mentioned as the new business analytics paradigm which helps the firms to fill this gap. According to the literature, business analytics is categorized to three main stages [4,10,21] (Fig. 1 also demonstrates business analytics components):

Descriptive Analytics - which is the "data summarization" phase and reports the past. It answers the question "What has happened?" and extracts information from raw data [11]. Business and management reports including sales, customer and operations are examples of descriptive analytics. There is also an extension to the descriptive analytics named "diagnostic analytics" which reports the past but tries to answer the questions like "Why did it happen?". It helps organizations in grasping the reasons of the events happened in the past. Diagnostic analytics gives the enterprises the cognition to understand relationships among different kinds of data.

Predictive Analytics - which is the "forecasting or extrapolation" phase and incorporates the descriptive analytics output as well as some machine learning (ML) algorithms and simulations techniques to build accurate models that predict the future. It answers the questions "What will happen?" and "Why will it happen?" in the future [11,12]. Predictive analytics assists enterprises in identifying future opportunities and likely risks by distinguishing specific patterns over the historical data. A key challenge in predictive analytics is having as much data as possible. More data means more validated models and therefore more accurate predictions. Some prominent techniques which are utilized in this phase are data mining (DM), text/web/media mining, and forecasting approaches. The output of the predictive analytics is multiple predictions and their equivalent probability scores. Forecasting the demand for goods in a certain region or for a specific group of customers and adjusting the production based on that forecast is an example of predictive analytics.

Prescriptive Analytics - which is the "recommender or guidance [1]" phase provides enterprises with adaptive, automated, time-dependent, and optimal decisions [6]. Its goal is to bring business value through better strategic and operational decisions. In general, prescriptive analytics is a predictive analytics which prescribes one or more courses of actions and shows the likely outcome/influence of each action. It answers the questions "What should I do?" and "Why should I do it?". It is purely built on the "what-if" scenarios [14]. Main elements of the prescriptive analytics are optimization [16,19], simulation, and evaluation methods [7]. Simply put, it provides advice based on predictions and enterprise's constraints. Prescriptive analytics takes into consideration the output of predictive analytics with compliance rules and business constraints to generate best courses of actions as the optimal decision. In other words, it takes a solid, actionable predictive model and the feedback data collected from those actions and recommends optimal decisions to assist decision makers in reaching their desired outcomes [2,17]. Prescriptive analytics systems generally have two important characteristics: (1) They provide the enterprise with the actionable outcomes. These outputs generate comprehensible prescriptions in terms of actions. (2) they support

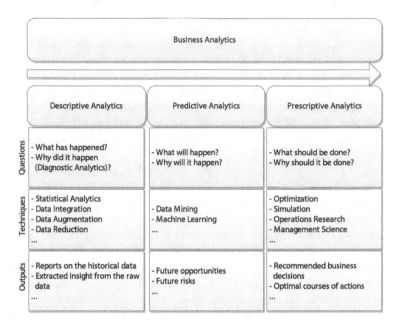

Fig. 1. Business analytics stages

feedback mechanisms in terms of tracking the suggested recommendations as well as unprecedented events occurrence in system's lifetime.

To the best of our knowledge, there has not been any particular study which incorporates these three types of analytics together to provide a single framework. Therefore, we propose a novel integrated prescriptive paradigm to address mentioned requirements in an adaptive and time-dependant way in Sect. 3. Our approach deals with heterogeneous data from different sources, builds dynamic and accurate predictive models based on the collected data, and addresses issues around the decision making in-time with actionable recommendations.

3   Proposed Architecture

To gain the advantages of analytics in achieving optimal and automated decisions in near real-time, a federated architecture comprising descriptive, predictive and prescriptive analytics as well as different data generator models and a holistic data warehouse is proposed in this section. The detailed representation of the architecture is illustrated in Fig. 2. It depicts a prescriptive analytics environment which gets the data from disparate sources with wide range of types as input and generates sequences of actions as output. Furthermore, each prescriptive module is considered a predictive module with aggregated functionalities such as intervention, feedback, assessment, adaptation, recommendation and personalization. Similarly, each predictive module is a descriptive element with augmented

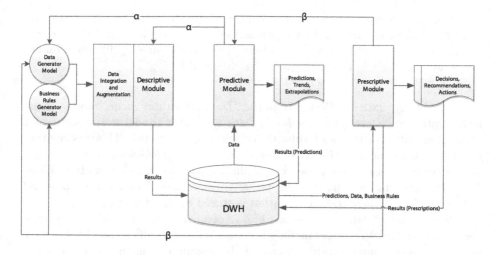

Fig. 2. Proposed federated prescriptive analytics architecture

prediction and data mining capabilities. The key building blocks of the proposed integrated prescriptive analytics architecture are listed below:

Data and Business Rules Generator Models - which are responsible for providing different kinds of data and domain-specific information to the system.

Descriptive Analytics Module - which is the data summarization/reduction unit. It pre-processes the collected data utilizing data cleaning and integration units and generates features for the processed data. Finally, it unifies the processed data into a particular form understandable by other analytical modules. This module also provides a diverse range of reports over the historical data.

Predictive Analytics Module - which is the forecasting unit. This component builds a predictive model over the unified data provided by the descriptive module. To improve the accuracy of the predictive model, this unit sends proper feedback towards the data and the descriptive analytics components and requests for amendments in the generated data or the extracted features. These feedback lines from predictive analytics module to the data model and the descriptive analytics module are labeled as "α" in Fig. 2.

Prescriptive Analytics Module - which is the recommendation unit and comprises simulation, optimization and evaluation/feedback units. To validate the accordance of the prescriptions with the system objectives, this module sends feedback towards the data and business rules models as well as predictive module. These feedback lines from prescriptive unit to data, business rules and predictive units are labeled as "β" in Fig. 2. Each prescriptive analytics module has three core elements: *simulation, optimization,* and *evaluation.* There are also other sub-components such as *decision-making, feedback* and *adaptation* which are taken into consideration inside these core elements.

Data Warehouse - which is the central data storage unit. It stores and retrieves a diverse range of data types as well as intermediate and final results of each analytics module mentioned above. Interactions among different components and the data warehouse unit is also depicted in Fig. 2.

The whole process is described in five stages as follows:

Step1) First, data is collected from the *"Data Generator Model"* and is transferred into the *"Descriptive Module"*. Data generator model is the source for producing data (in terms of volume, velocity, and variety). There is another generator model named *"Business Rules Generator Model"* which gives the system specific business data (rules, constraints, preferences, policies, etc.). These data elements will be utilized by the prescriptive module. The result of data and business rules augmentation/integration is stored in the storage.

Step2) The *"Data Integration and Augmentation Module"* as a sub-component of the descriptive module is concerned with unification of those data streams into one interpretable format. This module cleans the collected data, reduces data dimensions, and transforms/integrates the data into a unified format which is operational to other analytics components.

Step3) The descriptive module then stores the summarized data and business rules in the storage. The descriptive module, also, provides some statistical analysis and reports over the historical data.

Step4) Next, the *"Predictive Module"* queries the updated summarized data from the storage as input and builds an accurate predictive model based on the historical data (incorporating a diverse set of Machine Learning algorithms). The predictive model elicits patterns among the input data and extrapolates different possible futures from the historical data with their probability scores. Given the predictions and future trends, enterprises can distinguish their future opportunities and likely risks. These predictions are also stored in the storage. Under certain circumstances, the system needs to update its data or disseminate some feedback according to the generated outputs. Therefore, the update requests/feedback are redirected to the designated components.

Step5) Finally, the *"Prescriptive Module"* takes the stored updated predictions as well as updated business rules. This module comprises of three key elements: (1) optimization, (2) simulation, and (3) evaluation units. The process starts by feeding the simulation unit with the predicted scenarios from the predictive analytics. It generates multiple scenarios according to the available data. This unit answers the question "What should be done?" by producing a list of possible actionable recommendations. To narrow-down the simulated results and improve their accuracy, the evaluation unit measures their accordance to the defined objectives. Based on the results, the simulation unit stores the proper future scenarios in the storage. Finally, the optimization unit retrieves validated simulated scenarios along with defined objectives and business rules. The best and optimized scenario from the list of scenarios will be selected and will be sent to the evaluation unit for the final validation. The validated scenario will be considered as the system's final result in terms of action sequences. The optimization unit answers the question "Why should we do it?". The output will be actionable recommendations represented in different forms: decisions (like "YES/NO" suggestions), scalar values or vector/matrix of values (such as recommended prices, amount, fairs, etc.), or a full-fledged production plan.

4   Application Scenario

Prescriptive analytics is applicable to a wide range of use cases. For example, in Transportation, prescriptive analytics assists in managing transportation capacities, recommending variant ticket prices at different rates during times of increased or decreased demand to maximize capacity and profitability. In Oil and Gas Industry, based on the collected data from past drilling processes and production history, prescriptive analytics assists in locating and producing oil and gas in a cost effective manner, recommending where and how to drill to maximize production and minimize cost as well as environmental impacts. In Logistics, prescriptive analytics can help the organizations to find the best optimal route for deliveries and freights which can be updated in near real-time given past performances, load information, current conditions, and vehicle specifications. In this paper, we focus on the use of prescriptive analytics in the following area: in education and a particularly emerging field named "learning analytics" in Sect. 4.1.

4.1   Use Case - Learning Analytics in Educational Institutions

Currently, educational institutions possess a great number of data repositories which incorporate heterogeneous educational data (student grades, student/faculty/staff profiles, course information, resource repository information and so forth) from diverse sources (learning management systems, personalized learning environments, social media, etc.). These pedagogical data items have been growing exponentially recently. It is crucial to make proper decisions based on huge volumes of educational data. Traditionally, those institutions have been inefficient in getting adequate insights over their data repositories. Therefore, educational systems require pertinent types of analytics to meet their objectives. Given the above-mentioned facts, educational institutions are relevant use cases to consider because they own big data (in the context of education) and they need particular types of analytics to extract insight from the data, make proper decisions and act upon them to enhance their learning processes.

To address this requirement, "learning analytics" has emerged as a new type of analytics in the context of education. It is concerned with collecting educational data from different sources, building accurate predictive models on those data items, and making decisions based on generated predictions to enhance the quality of learning. Learning analytics is applicable to a wide educational levels from micro-layer which is focused on students and teachers to the macro-layer where cross-institutional processes are taken into consideration. According to the literature, key learning analytics requirements are categorized into: data collection, learner profiling, data interpretation, prediction and decision making, and adaptation and personalization. To satisfy mentioned needs, several models have been proposed among which core learning analytics processes have emerged: monitoring and analysis, prediction and intervention, tutoring, assessment and feedback, adaptation and personalization, and reflection. To effectively utilize learning analytics capacity, it should be formalized in the context of analytics. Our proposed prescriptive analytics paradigm can be utilized to model learning analytics as described in the following which targets most of learning analytics processes:

The descriptive analytics module is focused on acquiring educational data from different sources like students' learning management systems (LMS) or their social activities and other digital foot-prints like their tracks on massive open online courses (MOOCs). These data items need to be cleaned and then transformed into one unified format. Efficient reports and visualizations should also be generated in this stage. These outputs assist educational stakeholders in understanding what/why has happened in their learning processes till now (pedagogic trends, students progress, teachers feedback, resource allocation/utilization trends, etc.).

The predictive module applies accurate predictive models on the unified educational data to forecast future opportunities and risks in the environment. These predictions include recognizing at risk students, indicating students retention/drop-out rates, identifying patterns of success/failure, student performance, courses outcomes/effectiveness, students achievements and overall student experience.

The prescriptive module takes predicted outcomes, particular institution's objectives and business rules to generate optimal sequences of operational actions that help the institution improve student experience/satisfaction, make their pedagogic materials more adapted to students' needs, and accelerate intelligent feedback/intervention toward at risk students. Some actions might affect the general institutional regulations like admission policies, funding/scholarship distributions, resource allocation policies, and staff/faculty contract amendments.

A generic prescriptive analytics-oriented framework which is customized for the context of learning analaytics is elaborated in Sect. 5.

5   A Prescriptive Analytics Conceptual Framework

In this section we first outline a "prescriptive analytics conceptual model" and a class diagram of which is illustrated in Fig. 3. Next, an abstract framework to model key learning analytics processes mentioned in Sect. 4.1 is introduced.

5.1   Prescriptive Analytics Model

Major issues to be addressed through the prescriptive analytics conceptual model include:

Design and implementation of a federated prescriptive analytics architecture. This is a complex task to accomplish due to inherent difficulties in proposing a suitable analytical combination approach. There should be a proper way of incorporating descriptive, predictive and prescriptive analytics modules along with their subsidiary components. The relations among different elements should be devised in a way that helps the whole system to work appropriately. In addition, for the prescriptive analytics module in particular, it is critical to define and combine its key elements properly.

Design of an adaptive prescriptive paradigm. Change is an indispensable aspect of modern real-world applications. Taking the change in different levels of the system into consideration contributes to a more complex and adaptive system design. Dynamic and adaptive response to changes/feedback plays a key role in organizations' success. The response can be defined in terms of amended decisions, changed recommendations or a new course of actions.

Visualization of the outcomes. There has been a large body of research in visualizing descriptive and predictive analytics reports and predictions. However, not a lot of research has been conducted to fill the gap between generated actions based on the insights and illustrating them in a proper way to the stakeholders. Displaying operational actions in a comprehensible and responsive way has an influential effect on making invaluable decisions and taking advantage of future opportunities.

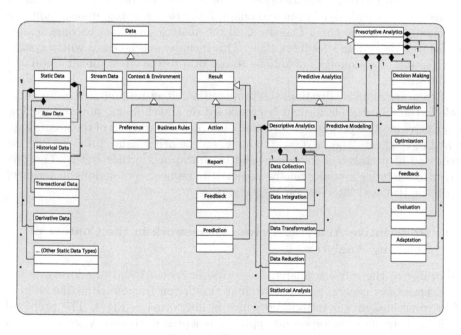

Fig. 3. Prescriptive conceptual model

The proposed prescriptive analytics paradigm in Sect. 3 addresses above-mentioned issues effectively as the following:

The novel way of composing three analytics modules and their sub-components is described in Sect. 3. According to Fig. 2, the descriptive module is responsible for collecting data from different sources and unifying them in a format which is comprehensible to other analytics modules. Its results in terms of report and visualizations are also stored in the holistic storage unit. The predictive analytice module, later, gets the unified data and builds predictive models over it. Its outcomes are also stored in the storage. The prescriptive module, finally, retrieves the predicted scenarios and business rules from the storage and applies particular simulation and optimization techniques on them to come up with optimal recommendations. Further, the class diagram of Fig. 3-the "prescriptive conceptual model" - illustrates key components of the proposed prescriptive paradigm and their relations. It encompasses a wide range of functional elements. For example, the data component is composed of *static* and *stream* data where the static element can include *historical, transactional,* and *derivative* data elements. Another example is the descriptive module which comprises *data collection, data integration, data transformation, data reduction* and *statistical analysis* sub-components. As the final example, the important prescriptive analytics module incorporates *decision making, simulation, optimization, evaluation, adaptation* and *feedback* components.

With regard to being adaptive, the feedback lines illustrated in Fig. 2, α and β labeled lines, provide dynamism and adaptability to the system in responding to the changes near real-time. Therefore, all the analytics modules become flexible to changes via system's feedback lines. This includes special cases where system re-prescribes the sequences of actions to conform to the environment's updated conditions.

Regarding the visualization, the proposed architecture prepares comprehensible outputs in flexible formats/views based on stakeholders' preferences. The descriptive module prepares reports and historical illustrations of the events happened in the past. The predictive module, on the other hand, provides forecast trends to help stakeholders understand organization's future status. The prescriptive analytics module, finally, depicts the sequences of actions in terms of feedback, text and intervention suggestions.

5.2   Prescriptive Analytics-Driven Framework in the Context of Learning Analytics

According to the learning analytics use case in Sect. 4.1 and to address its key requirements, a generic prescriptive analytics-driven framework in the context of learning analytics is introduced which is illustrated in Fig. 4. The proposed framework justifies one important application of prescriptive analytics in the area of education. This framework comprises two modules: conceptual module and logical module. The *conceptual module* is the proposed prescriptive analytics model described in Sects. 3 and 5.1 which is the core analytical engine. The *logical module* is focused on representing major learning analytics processes. All logical module components are related to their corresponding conceptual module

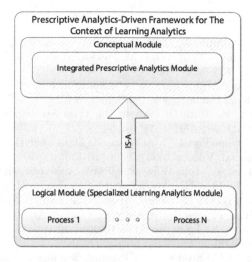

Fig. 4. Specialized prescriptive analytics framework for learning analytics

unit using the "IS-A" relation. Therefore, the logical module is the specialized case of the conceptual module.

6   Conclusions and Future Work

A federated data-driven prescriptive analytics paradigm is proposed in this work. This architecture comprises different analytical components: descriptive, predictive and prescriptive and addresses major issues regarding the linkage between extracted insights and recommended actions to be taken based on those insights. The proposed architecture is also capable of processing heterogeneous data from diverse range of resources in huge amounts - the key characteristics of big data. Further, an application scenario from the area of education - learning analytics - is provided to indicate the importance and advantage of the proposed paradigm and its application. Furthermore, key prescriptive analytics open issues are investigated and we demonstrate how our approach is capable of addressing those concerns.

The future work will focus on applying the proposed architecture in particular analytical domain of education - learning analytics - to assist the educational institutions in improving their learning processes. By formalizing specific educational use cases in the context of prescriptive analytics, those institutions can enhance their learning environments and improve student experience. Our prescriptive analytics architecture will also be implemented in detail. Further, the work's outcomes will be compared with similar conducted research in this field.

References

1. Adomavicius, G., Tuzhilin, A.: Toward the next generation of recommender systems: A survey of the state-of-the-art and possible extensions. Knowl. Data Eng. IEEE Trans. **17**(6), 734–749 (2005)
2. Apte, C.: The role of machine learning in business optimization. In: Proceedings of the 27th International Conference on Machine Learning (ICML-10), pp. 1–2 (2010)
3. Baker, P., Gourley, B.: Data Divination: Big Data Strategies. Delmar Learning (2014). ISBN: 1305115082 9781305115088
4. Banerjee, A., Bandyopadhyay, T., Acharya, P.: Data analytics: Hyped up aspirations or true potential. Vikalpa **38**(4), 1–11 (2013)
5. Barga, R., Fontama, V., Tok, W.H.: Predictive Analytics with Microsoft Azure Machine Learning: Build and Deploy Actionable Solutions in Minutes. Apress (2014). ISBN-13 (pbk): 978-1-4842-1201-1 and ISBN-13 (electronic): 978-1-4842-1200-4
6. Basu, A.: Five pillars of prescriptive analytics success. Anal. Mag. **8**, 8–12 (2013)
7. Bertsimas, D., Kallus, N.: From predictive to prescriptive analytics. arXiv preprint (2014). arXiv:1402.5481
8. Chen, H., Chiang, R.H., Storey, V.C.: Business intelligence and analytics: From big data to big impact. MIS Q. **36**(4), 1165–1188 (2012)
9. Davenport, T.H., Dyché, J.: Big data in big companies (2013)
10. Delen, D.: Real-World Data Mining: Applied Business Analytics and Decision Making. FT Press, New Jersey (2014)
11. Delen, D., Demirkan, H.: Data, information and analytics as services. Decisi. Support Syst. **55**(1), 359–363 (2013)
12. Eckerson, W.W.: Predictive analytics. Extending the value of your data warehousing investment. TDWI Best Pract. Report **1**, 1–36 (2007)
13. Evans, J.R., Lindner, C.H.: Business analytics: the next frontier for decision sciences. Decis. Line **43**(2), 4–6 (2012)
14. Haas, P.J., Maglio, P.P., Selinger, P.G., Tan, W.C.: Data is dead. without what-if models. PVLDB **4**(12), 1486–1489 (2011)
15. Kaisler, S.H., Espinosa, J.A., Armour, F., Money, W.H.: Advanced analytics-issues and challenges in a global environment. In: 2014 47th Hawaii International Conference on System Sciences (HICSS), pp. 729–738. IEEE (2014)
16. Liberatore, M., Luo, W.: Informs and the analytics movement: The view of the membership. Interfaces **41**(6), 578–589 (2011)
17. Marathe, M.V., Mortveit, H.S., Parikh, N., Swarup, S.: Prescriptive analytics using synthetic information. Emerging Methods in Predictive Analytics: Risk Management and Decision-Making: Risk Management and Decision-Making, p. 1 (2014)
18. Power, D.J.: Using 'big data' for analytics and decision support. J. Decis. Syst. **23**(2), 222–228 (2014)
19. Schniederjans, M.J., Schniederjans, D.G., Starkey, C.M.: Business Analytics Principles, Concepts, and Applications: What, Why, and How. Pearson Education, Inc. (2014). ISBN-10: 0-13-355218-7 and ISBN-13: 978-0-13-355218-8
20. Schölkopf, B., Platt, J.C., Shawe-Taylor, J., Smola, A.J., Williamson, R.C.: Estimating the support of a high-dimensional distribution. Neural Comput. **13**(7), 1443–1471 (2001)
21. Sharda, R., Asamoah, D.A., Ponna, N.: Business analytics: Research and teaching perspectives. In: Proceedings of the ITI 2013 35th International Conference on Information Technology Interfaces (ITI), pp. 19–27. IEEE (2013)
22. Van Barneveld, A., Arnold, K.E., Campbell, J.P.: Analytics in higher education: Establishing a common language. EDUCAUSE Learn. Initiative **1**, 1–11 (2012)

Topical Event Detection on Twitter

Lishan Cui$^{(\boxtimes)}$, Xiuzhen Zhang, Xiangmin Zhou, and Flora Salim

School of Science (Computer Science and Information Technology),
RMIT University, Melbourne 3001, Australia
{lishan.cui,xiuzhen.zhang,xiangmin.zhou,flora.salim}@rmit.edu.au

Abstract. Event detection on Twitter has attracted active research. Although existing work considers the semantic topic structure of documents for event detection, the topic dynamics and the semantic consistency are under-investigated. In this paper, we study the problem of topical event detection in tweet streams. We define topical events as the bursty occurrences of semantically consistent topics. We decompose the problem of topical event detection into two components: (1) We address the issue of the semantic incoherence of the evolution of topics. We propose to improve topic modelling to filter out semantically inconsistent dynamic topics. (2) We propose to perform burst detection on the time series of dynamic topics to detect bursty occurrences. We apply our proposed techniques to the real world application by detecting topical events in public transport tweets. Experiments demonstrate that our approach can detect the newsworthy events with high success rate.

Keywords: Dynamic topic modelling · Topic mutation · Event detection · Burst detection

1 Introduction

Recent years have seen an astonishing increase in the usage of the Twitter platform for various applications. On Twitter, users post messages, share information and communicate with friends. Twitter messages are often expressions by people with personal and public activities that occur worldwide, many of which describe real world facts and events. The ubiquitous use of Twitter has proven that the posts are updated more frequently than traditional news channel, and are distributed all over the world [18]. Therefore, detecting events over the Twitter platform is more effective and efficient in time-critical applications.

Existing event detection is mostly focused on detecting breaking news [19], query-based events [13], or monitoring disaster events [18]. A number of studies have been done for event summarisation [7,13], but focusing more on the description of events. Online detection of new events (sometimes called first story) over the tweet stream [4,17] very often results in noisy events like "7th Billionths Child Born", but the retrospective detection of semantically significant events is very important. For example, for public transports, discovering the prominent issue-related events are of utmost importance for the public transport administrative authority.

© Springer International Publishing AG 2016
M.A. Cheema et al. (Eds.): ADC 2016, LNCS 9877, pp. 257–268, 2016.
DOI: 10.1007/978-3-319-46922-5_20

In this paper, we study the problem of retrospective analysis of tweets for discovering significant topics, and the hot periods of events that are associated. We focus on the detection of *topical events* in public transportation domain. Different from breaking events, topical events occur during a certain period and fluctuate over time. For example, the train delay recurrently happens on Monday and Friday. The cancellation always happens when the weather is unusual. Topical event analysis helps city planners to discover these "recurring" situations and provide more reliable public transport services. Frequent occurrences of semantically-related events form a topical event. Announcements on planned service disruptions or special concerns are not a topical event.

It is important to note that our study in this paper is different from TDT (topic detection and tracking) [1], which attempts to cluster documents as events using clustering techniques. Rather than clustering documents into topical events, our purpose is to discover the hidden topics in documents (tweets) based on analysing their dynamic semantic structures over streams of documents, along their time dimension. Topics are generally known, but we want to discover dynamic topics rather than a collection of independent topics. Under the same semantic structure, a dynamic topic evolves over time. The consistence of the topic is to measure the semantic consistency of the topic along the time series. When a topic is semantically consistent along the time series, the topic is defined as semantically consistent. To achieve effective topical event detection on Twitter, this paper is focused on the following research questions:

- How to detect the semantically consistent dynamic topics over time?
- How to detect the bursty periods of topical events?

Unlike the traditional event detection based on a given query topic, we do semantic summarisation and event detection at the same time. Therefore, conventional keywords based detection is not applicable to our work. In this paper, we propose to capture topics that evolve over Twitter content and epochs. We define a topical event as the bursty occurrences of a set of semantically consistent topics.

We decompose the problem of topical event detection into two components: (1) We address the issue of semantic mutation of dynamic topics during evolution over time. We propose to improve topic modelling to filter out semantically inconsistent dynamic topics. (2) We propose to perform state-based burst detection technique [3] to identify bursty occurrences of topical events in a discrete temporal sequence of dynamic topics. Experiments on the dataset of Melbourne public transport tweets show that our approaches can accurately detect and describe significant events over epochs. Our discovered events include persistent issues like "delay" as well hidden recurrent hot topics such as "accident", "cancelled", etc. In addition to the topical labels for semantically summarising events, we also pinpoint the timestamps for events.

2   Related Work

Event detection on Twitter has been discussed in the literature. The assumption is that all relevant documents for a topic contain some old or new events of interest [2]. First story detection (FSD) was introduced to examine the Topic Detection and Tracking (TDT) task that ran as part of the TREC conference [1]. The objective of FSD was to detect the first occurrence of an article that was related to a given topic. The feature-pivot techniques model an event in text streams by grouping words together while rising sharply in frequency [8–10]. An event is represented by a number of keywords showing burst in frequency count [10]. The underlying assumption is that some of the related words about an event would show a spike in the usage when the event is happening.

Using topic distributions rather than the bag of words representing documents reduces the effect of lexical variability while retaining the overall semantic structure of the corpus [22]. Pan *et al.* [16] proposed event detection approaches by combining the LDA model with temporal segmentation and spatial clustering models. The Space-Time LDA is a Spatial Latent Dirichlet allocation (SLDA) [20] adapted from the detection of segments in images to the detection events in the text corpus. The Location-Time constrained Topic (LTT) [23] represents a social message as a probability distribution over a set of topics and captures the unknown composite social events by measuring the similarity between two messages over social media streams. These existing studies focus on discovering the hidden semantic topics for given breaking news events like natural disasters, rather than discovering events from the dynamics of hidden topics.

Our study aims to detect topical events from the tweet stream. The problem can be viewed as a retrospective event detection problem by analysing the Twitter archive. Different from existing event detection research on specific events [4,7,17,21], we aim for detecting significant unspecified events. Rather than relying solely on the word frequency count for event detection [8–10], we focus on the semantic structure of the document collection, which reduces discovering semantically insignificant events. Different from existing work on the topic-related event detection [16,20,23], we model the temporal dynamics when forming topics and our topical events are defined regarding the topic dynamics and the semantic consistence.

3 Problem Formulation

In this section, we formally define the problem of topical event detection on Twitter streams.

Given a text stream $D = \{d_1, d_2, \ldots, d_i\}$ where d_i is a tweet message with time stamp t_i and $t_i \leq t_j$ if $i < j$, and given a fixed time window (for example a week), a Twitter stream D can be divided into epochs of tweets. A *dynamic topic model* comprises a set of dynamic topics that evolve over the epoches under the same term distributional structure. It is well recognised that topic modelling can produce topics lacking semantic coherence [14,15]. The semantics for dynamic topics not only evolves but also can mutate over time, and the mutant topics consist of words that are statistically important for the dynamic topic in epochs but completely lack semantic coherence. Our experiments (Sect. 5.2) show that some topics for public transport can mutate completely, and it is hard for human annotators to discern their semantics for each epoch.

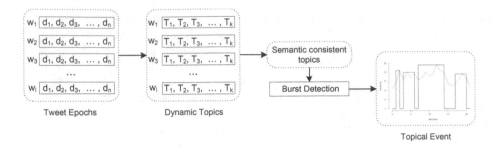

Fig. 1. The framework for topical event detection

The problem of topical event detection is to find a set of semantically consistent topics for the text stream that comprises occurrences of events. Figure 1 shows the framework of the topical event detection. The Twitter stream divided to fixed time window w_i. Let K be the number of topics in each epoch, N_l be the number of documents for each epoch w_i. Applying topic models on tweet epoch, the documents on each epoch modeled as topics under the same semantic structure. The problem of topical event detection can be decomposed into two tasks:

1. Discovery of semantically consistent dynamic topics over the tweet stream along epochs.
2. Detection of bursty occurrences events associated with the topics discovered in the first task.

In the first task of summarising epochs of tweets into dynamic topics, we need to model the temporal dynamics of topics and also address the issue of semantic mutation of dynamic topics. Topic modelling has been widely used to discover hidden topics in a collection of documents. In our problem setting, the discovered topics should be semantically consistent over epochs. In this regard, we propose to measure the semantic coherence of topics across epochs and filter out semantically inconsistent topics. The temporal dynamics for the frequency of topics over epochs form a time series, and our second task of event detection is to detect the bursty occurrences of topics from the time series. To detect significant topical events from tweet posts associated with the topic, it is important to filter out the trivial or gibberish discussions of the topic. We propose to apply a state-based burst detection algorithm to detect topical events in a discrete temporal sequence.

4 Methodology

We will first describe our approach based on non-parametric topic model [6,12] to discover semantically consistent dynamic topics, then demonstrate the detection of bursty occurrences associated topics.

4.1   Discovering Semantically Consistent Dynamic Topics

We aim for discovering hidden topics for a stream of tweets with timestamps, specifically the semantically consistent topics overt time. When the tweet stream is divided into epochs, the semantics of topics summarised for epochs evolve over time. We propose to apply dynamic topic modelling to discover hidden topics from the tweet stream over epochs and then measure the semantic coherence to filter out inconsistent dynamic topics.

We applied the non-parametric dynamic topic model [6,12] that introduced components evolution as a chain, extended the standard topic model LDA [5] to identify semantically consistent latent topics over epochs.

The non-parametric dynamic topic models apply the hierarchical Pitman-Yor process (PYP) to model both the document-topic proportions and the topic-word distributions evolving over epochs. In each epoch, the process is similar to the standard LDA except for the PYP process. The posterior distribution of topics depends on the information from both word and time slice.

The non-parametric dynamic topic models identify and describe the topics over epochs. The topic consistency is ensured by maximizing the estimation. However, the topic sometimes may contain the mutation, the variation that happened during the topic evolution over epochs. The topical event is detected from a semantically consistent dynamic topic.

The Kullback-Leibler divergence [11] is a measure of the difference between two probability distributions. The semantic mutation of dynamic topics during evolution can be measured by KL divergence. Given dynamic topic T and l epochs, let T_x and T_y denote respectively the word probability distribution for topic T with epoch x and y, which $x, y \in l$. The semantic distance of T_x and T_y definded as:

$$s(T_x, T_y) = \sum_i T_x(i) * ln\frac{T_x(i)}{T_y(i)}. \tag{1}$$

where i denotes the words under consideration. For each dynamic topic, we measured the semantic distance between epoch pairs by counting word frequency.

$s(T_x, T_y)$ measures the semantic distance between time slices. To measure the consistency level across epochs for a dynamic topic, we define a threshold θ for the distance metric of a time series against the first epoch to measure the topic semantic consistency. In this paper, we use the mean of the semantic distance over epochs as the threshold θ. When the semantic distances over epochs are less than θ, we consider the topic as a semantically consistent topic over the epochs.

4.2   Topical Event Detection

The composition of topics for epochs changes over time. The frequency for a dynamic topic fluctuates and forms a time series. We aim to detect significant bursts of a dynamic topic as the event associated with the topic. We propose to apply state-based burst detection algorithm to detect the topical events in a discrete temporal sequence.

In the area of Topic Detection and Tracking [1], the sequence of documents for a distinct topic is analysed as time series. The analysis of time series comprises of methods for extracting meaningful dynamic characteristics of the data. Among them, detecting and modelling bursts is a common task. We propose to apply state-based burst detection algorithm proposed by Araujo *et al.* [3] to detect the bursts of topical events in a discrete temporal sequence.

Events can happen at any time instant. Moreover, the timescale of some events is varied. For instance, low-intensity earthquakes have a timescale of days, and car crash accidents have a timescale of hours. Topical events detection need detect long-term vibrating bursts and short-term sharp bursts simultaneously. In the state-based burst detection model, the transit probability refers to the probability of changing the state. Differencing the transit probability can capture differences in the frequencies.

Under the state-based burst detection model, time intervals with different event frequencies are modeled as different states. The whole sequence of the event is regarded as a Hidden Markov Model of the states. The Poisson process is used to model arrivals. The detection of bursts is then achieved by applying the dynamic programming to find sequences of states that best fit the time series. Moreover, the fitness function generalizes this model to time series data that consists of sequences of events obtained from repeated measurements of time.

In our work, we model binary states model, which includes a non-bursty state and a bursty state. The states are pre-defined with different mean values according to the probabilistic distribution. The desired estimation for the sequence of frequencies $\{\lambda_1, \ldots, \lambda_T\}$ can be obtained by maximizing this probability, namely

$$f(\lambda_1, \ldots, \lambda_T) = \sum_{t=1}^{T} (x_t \ln \lambda_t - \lambda_t) + K \sum_{t=2}^{T} \delta_{\lambda_t, \lambda_{t-1}}. \tag{2}$$

where $\delta_{\lambda', \lambda} = 1$ if $\lambda' = \lambda$ and 0 otherwise. The parameter $K = \log [p(E-1)/(1-p)]$, with E is the number of frequencies in a discrete temporal sequence.

5   Experiments

We detected topical events from a collection of Twitter dataset gathered over a period of five months (1st January 2014 to 31st May 2014). The dataset was collected through Twitter's streaming API.

5.1   Datasets

The focus of our study is Melbourne public transport related event detection. To get Melbourne transport related tweets, we investigated three location related attributes, such as GPS coordinates, place name, and author's location indicated as Melbourne. The statistics of the Melbourne-based dataset are shown in Table 1. Then we aggregated the tweet contents by day.

Table 1. Melbourne-based tweets dataset

Month	# GeoTag	# Place	# User	# Combined
January	2,322	298	60,550	61,780
Feburay	1,335	241	41,574	42,338
March	1,918	335	56,281	57,307
April	1,652	687	57,992	58,936
May	1,877	1,375	61,174	62,214
Total	9,104	2,936	277,571	**282,575**

Table 2. The sample summarisations of Week 1 news articles

Week	Date	Summary	Topic Label
1	6 Jan	Delay traffic delays return work	delay
1	6 Jan	Hit killed by train Frankston	accident
1	10 Jan	Travelling tram seat beside syringes	service

The ground truth public transport events. We need to establish a reliable ground truth for public transport events to evaluate the topical events detected by our approach. To this end, we crawled the online news articles from The Age, a well known daily Melbourne newspaper, published for the same period as our Twitter dataset from 1st January 2014 to 31st May 2014. There are 799 pieces of news in five months from The Age for that duration, out of which 50 news articles are related to the public transport according to the keyword search. The fifty news articles have information of the date, snippet and news title. Two annotators summarised the content for each piece of news using five

Table 3. The semantic distance and Jaccard results for ten topics over 22 weeks

Topic	Topic Label	Avg	$s(T_x, T_y) \leq \theta$ (%)	Jaccard Distance
Topic0	late home	2.003	**95.5**	0.673
Topic1	service	2.399	86.4	0.605
Topic2	delay	1.997	**95.5**	0.486
Topic3	accident	2.269	**95.5**	0.555
Topic4	service	2.267	**95.5**	0.586
Topic5	stop	2.017	86.4	0.682
Topic6	train	2.284	90.9	0.691
Topic7	roadwork	2.150	**95.5**	0.564
Topic8	public	2.237	90.9	0.582
Topic9	driver	2.133	90.9	0.527

keywords from the news snippets and titles. The agreement was reached by discussions. The sample summarisations for the first week of news articles are shown in Table 2. The corresponding labels for each news were done manually by annotators based on the topic labels in Table 3 (See Sect. 5.2).

5.2   Extraction of Consistent Dynamic Topics

As described in Sect. 4, we performed two steps to extract semantically consistent dynamic topics over five months. We first semantically summarised tweets topics to discovery semantically consistent dynamic topics. In our experiments, we set the parameter $Epoch = 22$, indicates the summarisation for the weekly in our corpus. We investigated the results that when $K = 10$, where the K is the number of topics. Figure 2 shows the trend of each topic by week for ten topics. The summaries of the meaning of each topic are shown in Table 3 (Topic Label). The labels for topics were manually assigned by annotators.

We can see from Fig. 2, during most of the period, the frequency of Topic6 is less than 15. Topic1 has a peak value of 32 at week 4, and Topic3 has another peak value of 31 at week 16. Topic7 has stable frequencies over all weeks compare to other topics.

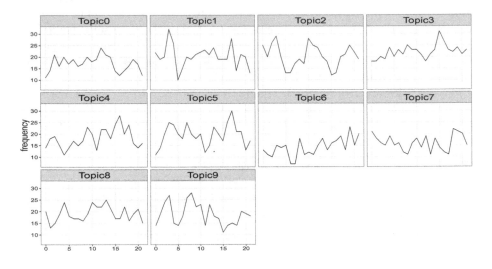

Fig. 2. The dynamic topics by week

The second step is to discover semantically consistent topics. We estimated the semantic consistence for topics over epochs. Table 3 lists the average semantic distance scores for each dynamic topic. Topic2 has the lowest average score while Topic1 has the highest score.

To discover the consistent topics, we applied the threshold θ to filter out the mutant topics. We set θ as the mean of the semantic distances over epochs. For each topic, if there are more than 95 % of semantic distance scores $(s(T_x, T_y))$ less than the threshold, we treat this topic as a semantically consistent dynamic topic. The results are shown in Table 3 as the percentage of the consistency. The topics in bold are semantically consistent dynamic topics.

We evaluated the semantically consistent topics from the view of semantic topic evolution. We aggregated the word frequency for each topic over epochs, then selected the top ten high-frequency words for each topic as the topic representative words. These topic representative words lead the meaning of each topic. During the topic evolution, these words should have high probability occurrence. For each topic, we calculated the occurrence distributions of topic representative words over the epoch.

We applied Jaccard distance to calculate the dissimilarity between each topic over 22 weeks using the top ten high-frequency topic words. The formula is as follow:

$$d_J(E_i, E_j) = 1 - J(E_i, E_j) = \frac{|E_i \bigcup E_j| - |E_i \bigcap E_j|}{|E_i \bigcup E_j|}. \tag{3}$$

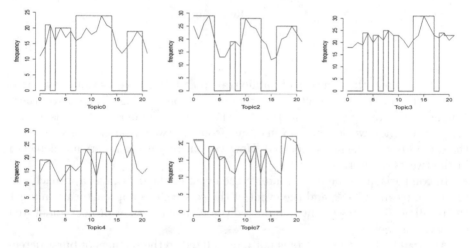

Fig. 3. The topical event detection

The results are shown in Table 3. Topic5 and Topic6 have the highest dissimilarity. In our approach, these two topics were filtered out as well. For Topic2 and Topic3, the Jaccard values are lower, means these two topics are semantically consistent over 22 weeks.

5.3   Detection of Event Occurrences

The state-based burst detection was applied to detect the period of topical events. We treat dynamic topics as temporal series, and the time scale is a week. In total, we ran the algorithm for burst detection on 22 weeks series.

The burst detection results for the five semantically consistent topics are shown in Fig. 3. Topic0 detected four occurrences, including week 2, from week 4 to week 6, from week 8 to week 14 and week 18 to week 20. Topic3 detected six bursts and Topic4 detected five occurrences.

By and large, the news of public transport is only about major accidents that caused very long delays and disruptions. Such accidents usually involved casualties or major injuries, e.g. a person being hit by a train. On the other hand, the tweets have more diverse topics when it comes to public transport. People send updates on the delays or their grievances on trivial matters.

Table 4. Event detection

	News	Event
January	9	7
Feburary	4	2
March	20	17
April	4	3
May	13	6
Sum	50	35

Table 4 shows the numbers of public transport related events from news articles and the number of events correctly detected on Twitter. In January, there is nine public transport news featured in the news. From our method, using five consistent topics, we detected seven events from our tweets dataset correspond to the same week as the news reported. However, there are two bursts not detected in the tweets dataset.

In comparison to the news content, the event of the delay, our approach can reach the recall of 73 %, and the event of the service, our approach can only get the recall of 70 % since the news reported the planned work or project annotated as service as well.

Accurately detecting bursts is not only related to the settings of burst detection parameters, but the numbers of consistent topics are also important. The larger number of consistent topics can increase the accuracy of event detection, but decrease the quality of the event summary.

6   Conclusions

In this paper, we studied the problem of topical event detection in a stream of tweet messages. We decomposed the problem of topical event detection into two components: (1) Semantically consistent dynamic topic discovery: We applied dynamic topic modelling to discover dynamic topics. More importantly, to address the issue of semantic mutation for the evolution of dynamic topics, we proposed to use the KL divergence measure to filter out semantically inconsistent dynamic topics. (2) Detection of events burst occurrences: We applied state-based burst detection on the time series for dynamic topics to detect bursty occurrences of topical events. We applied our proposed technique to the real world by detecting topical events for a public transport Twitter dataset, for five months. Our results demonstrated that our approach can detect the newsworthy events for public transport with high success rate. For future work, we will focus on developing a unified model that combines dynamic topic modelling and mutation topic pruning. We will also investigate how to achieve online topical event detection for live Twitter streams.

References

1. Allan, J.: Introduction to topic detection and tracking. In: Allan, J. (ed.) Topic Detection and Tracking, pp. 1–16. Springer, New York (2002)
2. Allan, J., Lavrenko, V., Jin, H.: First story detection in TDT is hard. In: Proceeding 19th International Conference on Information and Knowledge Management, pp. 374–381. ACM (2000)
3. Araujo, L., Cuesta, J.A., Merelo, J.J.: Genetic algorithm for burst detection and activity tracking in event streams. In: Runarsson, T.P., Beyer, H.-G., Burke, E.K., Merelo-Guervós, J.J., Whitley, L.D., Yao, X. (eds.) PPSN 2006. LNCS, vol. 4193, pp. 302–311. Springer, Heidelberg (2006)
4. Becker, H., Naaman, M., Gravano, L.: Beyond trending topics: real-world event identification on Twitter. ICWSM **11**, 438–441 (2011)
5. Blei, D.M., Ng, A.Y., Jordan, M.I.: Latent Dirichlet allocation. Mach. Learn. Res. **3**, 993–1022 (2003)
6. Buntine, W.L., Mishra, S.: Experiments with non-parametric topic models. In: Proceeding of 20th ACM SIGKDD, pp. 881–890. ACM (2014)
7. Cordeiro, M.: Twitter event detection: combining wavelet analysis and topic inference summarization. In: Proceeding Doctoral Symposium on Informatics Engineering, DSIE, p. 123–138 (2012)
8. Fung, G.P.C., Yu, J.X., Yu, P.S., Lu, H.: Parameter free bursty events detection in text streams. In: Proceeding of 31st International Conference on Very Large Data Bases, pp. 181–192. VLDB Endowment (2005)
9. He, Q., Chang, K., Lim, E.P.: Analyzing feature trajectories for event detection. In: Proceeding of 30th ACM SIGIR, pp. 207–214. ACM (2007)
10. Kleinberg, J.: Bursty and hierarchical structure in streams. Data Min. Knowl. Disc. **7**(4), 373–397 (2003)
11. Kullback, S., Leibler, R.A.: On information and sufficiency. Ann. Math. Stat. **22**(1), 79–86 (1951)
12. Li, J., Buntine, W.: Experiments with dynamic topic models. In: Proceeding of NewsKDD workshop on Data Science for News Publishing. ACM (2014)
13. Metzler, D., Cai, C., Hovy, E.: Structured event retrieval over microblog archives. In: Proceeding of the North American Chapter of the Association for Computational Linguistics, pp. 646–655. Association for Computational Linguistics (2012)
14. Mimno, D., Wallach, H.M., Talley, E., Leenders, M., McCallum, A.: Optimizing semantic coherence in topic models. In: Proc. Empirical Methods in Natural Language Processing. pp. 262–272. Association for Computational Linguistics (2011)

15. Newman, D., Lau, J.H., Grieser, K., Baldwin, T.: Automatic evaluation of topic coherence. In: Proceeding of the North American Chapter of the Association for Computational Linguistics, pp. 100–108. Association for Computational Linguistics (2010)
16. Pan, C.C., Mitra, P.: Event detection with spatial latent dirichlet allocation. In: Proceeding of 11th ACM/IEEE International joint Conference on Digital libraries, pp. 349–358. ACM (2011)
17. Petrović, S., Osborne, M., Lavrenko, V.: Streaming first story detection with application to Twitter. In: Proceeding of the North American Chapter of the Association for Computational Linguistics, pp. 181–189. Association for Computational Linguistics (2010)
18. Sakaki, T., Okazaki, M., Matsuo, Y.: Earthquake shakes Twitter users: real-time event detection by social sensors. In: Proceeding of 19th International Conference WWW, pp. 851–860. ACM (2010)
19. Sankaranarayanan, J., Samet, H., Teitler, B.E., Lieberman, M.D., Sperling, J.: Twitterstand: news in tweets. In: Proceeding 17th ACM SIGSPATIAL International Conference on advances in geographic information systems, pp. 42–51. ACM (2009)
20. Wang, X., Grimson, E.: Spatial latent Dirichlet allocation. In: Proceeding Advances in Neural Information Processing Systems, pp. 1577–1584 (2008)
21. Weng, J., Lee, B.S.: Event detection in Twitter. ICWSM **11**, 401–408 (2011)
22. Yao, L., Mimno, D., McCallum, A.: Efficient methods for topic model inference on streaming document collections. In: Proceeding of 15th ACM SIGKDD International Conference on Knowledge Discovery and Data Mining, pp. 937–946. ACM (2009)
23. Zhou, X., Chen, L.: Event detection over Twitter social media streams. Int. J. VLDB **23**(3), 381–400 (2014)

A Time and Opinion Quality-Weighted Model for Aggregating Online Reviews

Yassien Shaalan$^{(\boxtimes)}$ and Xiuzhen Zhang

RMIT University, Melbourne, Australia
{yassien.shaalan,xiuzhen.zhang}@rmit.edu.au

Abstract. Online reviews are playing important roles for the online shoppers to make buying decisions. However, reading all or most of the reviews is an overwhelming and time consuming task. Many online shopping websites provide aggregate scores for products to help consumers to make decisions. Averaging star ratings from all online reviews is widely used but is hardly effective for ranking products. Recent research proposed weighted aggregation models, where weighting heuristics include opinion polarities from mining review textual contents as well as distribution of star ratings. But the quality of opinions in reviews is largely ignored in existing aggregation models. In this paper we propose a novel review weighting model combining the information on the posting time and opinion quality of reviews. In particular, we make use of helpfulness votes for reviews from the online review communities to measure opinion quality. Our model generates aggregate scores to rank products. Extensive experiments on an Amazon dataset showed that our model ranked products in strong correspondence with customer purchase rank and outperformed several other approaches.

Keywords: Product ranking · Rating aggregation · Online review · Opinion evaluation

1 Introduction

Product reviews can be considered one of the most popular tools to support buying decisions of customers. Also, merchants can get useful feedback on their products. Pioneers of e-commerce like Amazon and eBay have been using product reviews since 1997. In this way they give customers the ability to evaluate products before purchasing by reading previous customer reviews. According to [18], about 57 % of customers considering buying electronics read reviews prior to actual buying and about 40 % of customers even say they wouldn't buy electronics without reading online reviews first.

[12] showed that 63 % of customers are more likely to make a purchase from a site which has user reviews.

Aggregate review scores are widely used by E-commerce sites to rank products and help online consumers to make purchasing decisions. The average rating is the de facto aggregate review score by most E-commerce sites. However the

© Springer International Publishing AG 2016
M.A. Cheema et al. (Eds.): ADC 2016, LNCS 9877, pp. 269–282, 2016.
DOI: 10.1007/978-3-319-46922-5_21

average rating does not always accurately reflect products' overall quality, especially for products with few ratings [15].

Another problem with the simple averaging approach is that it tends to produce high, very close aggregate rating scores and product qualities can not be effectively measured and ranked [4]. The problem is also noted as "the all good reputation problem" [23]. For instance, we computed the average ratings for over 150,000 books and around 10 million reviews in the Books category of our Amazon dataset (see the Experiments section) and found that about 75 % products have an average rating higher than 4. Taking a closer look, we found that the number of products with maximum difference of 0.1, for example between 4.1 and 4.2, is around 15,000 products. This further complicates the ability of ranking these products with very subtle differences in average star ratings.

Online review aggregation models for measuring product qualities have been proposed in recent research, and can be classified into three approaches. The probabilistic and statistical approach [1,7,8,14,16] computes the expected rating for a product, assuming a normal distribution or binomial distribution for all review ratings. Another approach is the weighted aggregation approach [11,17, 21,22] where weighting heuristics like reviewer biases are considered. The third approach [23–25] aggregates opinion polarities extracted from review textual contents.

We propose to measure review quality making use of the helpfulness votes in the online review community to accurately measure the true quality of products. Intuitively the quality of reviews is an important factor that needs to be considered when measuring product quality using reviews. However, in the aggregation models of all existing studies the quality of opinion expressed in reviews is largely ignored. It is well acknowledged that opinions expressed in reviews are of varying qualities [26]. Research has shown that helpfulness votes for reviews in online communities are strong indicators for quality of reviews and are influenced by the posting time of reviews [5,19]. We propose to measure the opinion quality based on user's reviews helpfulness under each product category separately. The reason is that ratings can vary on the same product based on personal biases, due to prior expectations of the product's quality or personal preferences.

The posting time of reviews is also a very important factor we consider when ranking products. It is shown in [2,24] that recent reviews are better than earlier ones because they reflect the latest opinions about the product which is considered the most helpful for forthcoming transactions. However, not all reviewing systems consider the importance of reviews posting time, which can easily lead to overall ratings that are skewed towards high rating as depicted by [2,4,6,9]. On the other hand, it is also shown that reviews posted early, especially when close to product release time, is closely related to spam reviews [13,20].

In this paper we propose a review quality and posting time-weighted model for the aggregation of ratings from online reviews. In terms of aggregation function, rather than the weighted average linear model, we propose an exponential function for the review weight that combines both factors. The reason why a linear model is not preferred is that it gives a regular and gradual increase of weights with time which does not effectively reflect importance of latest product ratings fast enough (see the Experiment section). Our extensive experiments on 96,500 products from 15 categories with 10 million reviews crawled from Amazon.com showed that our model produced an accurate product ranking which in turn can guide users to make better purchasing decisions.

2   Related Work

Research on quantifying product qualities based on customer reviews ranges from statistical and heuristic-based models for aggregating star ratings to models based on the aggregation of opinion polarities mined from review textual contents.

Several statistical models for computing overall rating for a product by aggregating star ratings of online reviews to quantify product quality were introduced in [16], including average rating, median rating, lower bound on normal and binomial confidence. On the other hand, [7] compared aggregation methods including arithmetic mean, weighted mean, median and mode. It is found that median or mode aggregation are better in terms of quality and robustness. On the other hand, probability-based approaches are mostly interested in removing the impact of outlier ratings by employing probability distributions [8,14]. They represent the overall rating as a probability value, but better estimate relies on the availability of many ratings. In [14] a multinomial probability distribution based on the Dirichlet probability distribution is proposed, which makes explicit use of counts of each star rating level. The model computes aggregate scores similar to the average model on datasets with dense ratings however better with sparse ratings.

Heuristic-based models obtain the aggregate score using different weighting heuristics. In [7] weighted average of ratings is based on the heuristics of reviews' influence. [22] presented a model that captures various important factors like rating aggregation rules and rating behavior that influence the product quality assessment under the existence of partial information. Another weighted scheme is based on rating distribution, where ratings are converted to a normalized scale that is assumed to fit into a particular distribution. One popular distribution is the normal distribution proposed by [1], where they give weights to ratings based on normal distribution of ratings. The more frequent the rating level is, the higher the weight the level will get. However, the problem with this approach is that it assumes a large number of reviews per product to provide the best results, which is not always the case for real applications. In general it is shown that the aggregate score by different aggregating strategies differ significantly, especially in the presence of noisy ratings obtained from various users [15].

There has also been research on mining opinions from review textual contents and aggregating their polarities to measure product qualities. This approach is based on the observation that the star rating may be biased and instead textual contents are more reliable source for opinions [23].

In fact, [23,25] presented a feature-based product ranking technique that mine customer review text, by identifying product features and analyzing their frequencies. A model based on weighted directed graph is applied to evaluate products' relative quality. Furthermore, [24] introduced a model that incorporates review text opinion, reviews helpfulness votes and review posting date to compute the products' ranking. They did not use the star rating; instead, they employed a sentiment analysis technique to compute the review polarity score. It simply subtracts the number of negative sentences from the number of positive sentences to generate the overall review polarity.

Reputation systems is another line of related research, where the objective is to rank merchants using reputation from transactions or to rank reviewers using reputation from reviews. In [21] it is assumed that the average rating of an item is the closest measure they can obtain to the true value of that item, thus, any reviewer who consistently ranks items near their ultimate average can be considered a reliable reviewer and its weight get increased. In addition, [11] proposed another weighted average model, they consider reviewers having different standards in assigning scores. They classify users into either lenient or strict users based on a computed leniency value which is then used as a weight for that user's ratings. This model provides accurate reputation scores for datasets of reviewers with sparse ratings. Moreover, fuzzy models were adopted by [17] as reputation scoring mechanism, because fuzzy logic introduces reasoning rules. Their rankings are based on the customers' own preferences and also on the information provided by search engines about the products. However, [3] proposed a fuzzy model for trust reputation of a user based on his accuracy of prediction to certain item in comparison to other users. Nonetheless the objective of these reputation systems is not to measure product quality.

The quality of opinion expressed in reviews has been discussed in the literature [5,19]. It is shown in [19] that reviews' textual properties, author's reputation and product characteristics are important features for determining whether a review is helpful or not, and that review quality is measured in multiple dimensions, including topical relevancy, author reputation, believability and objectivity. In addition, by analyzing Amazon helpfulness votes specifically, [5] showed that the decision on the helpfulness for a review by a user is not only dependent on its content, but also on a number of social factors – theories from sociology and social psychology are applied to analyse these factors. In general, both studies show that helpfulness votes in online review communities are very helpful opinion quality metric. Moreover, [19] further uncovered a strong correlation between reviews chronological order and helpfulness.

3   The Posting Time and Opinion Quality of Reviews

Computing an overall product rating from all available ratings for a product can become a complex matter due to the fluctuation of ratings on the same product which is introduced by individual bias or spam. Customers who actually purchased and used the product in question can give real and more reliable information about the product's real quality. Unfortunately, this valuable piece of information is not easily accessible to all online reviewing systems, which further complicates the problem. Our proposed model rely mainly on two factors to accurately rank products – the helpfulness votes for a review that reflects the quality of opinion expressed in the review, and the review posting time that has been shown closely related to helpfulness vote and opinion spamming behaviours [13, 20].

In order to have a good understanding of the importance of time and opinion quality factors for ranking products, we need to understand the distribution of ratings for products. We divided product categories in our Amazon dataset into 3 groups according to the average number of reviews per product category, ranging from small, medium to large.

We also aim to analyse the distribution of ratings for products in terms of varying qualities. To do this we make use of Amazon Sales Rank [24], which measures how popular a product is considering its sales within a certain time window. In fact, [24] used sales ranks as ground truth as an evaluation metric for ranking products. The sales ranks have different values for each category like books, games, software, etc. A value of 1 refers to top ranked product, 2 is the second ranked, etc. We chose three product categories to represent products with small, medium and large number of reviews, and from each category, we picked the top ranked high for popular products and the bottom ranked least popular products according to Amazon sales rank. We divided the products' life span into 10 equal periods. Table 1 shows a product category sample from groups with small, medium and large average number of reviews showing it within parenthesis. From each category we picked the top ranked and bottom ranked products showing the actual number of reviews of each product. Then, we show the average rating for each product versus each time period of the product life span - time period 1 refers to earliest time period and 10 refers to latest time period - and finally concluding with the overall average rating for these products.

Although the average number of reviews varies across product categories, within each category, the top ranked popular products have more reviews than the bottom ranked less popular products, although not all of them are necessarily good. Moreover, the overall average rating for the bottom ranked product in Arts category in Table 1 is higher than that of the top ranked which does not make sense in this case. This shows how ranking using average rating can easily lead to wrong results. However, according to the statistical law of large numbers [10], which states that the average of the results obtained from a large number of trials should be close to the expected value. We can conclude that when the number of ratings per product is high, the mean can be considered an adequate approximation of the overall rating for a given product. Thus, the overall average rating in case of these top ranked products can be considered correct assuming low spamming effect. In Table 1, we compare the average rating per time period to the overall average as a meaningful rating aggregation. The difference between average rating of the last 5 periods collectively and the overall rating is smaller

Table 1. Average rating vs time periods for top and bottom ranked products from different product groups

Period	Art(58)		Camera(97)		Baby(110)	
	P_Top	P_Bottom	P_Top	P_Bottom	P_Top	P_Bottom
	#Revs(235)	#Revs(21)	#Revs(197)	#Revs(23)	#Revs(625)	#Revs(21)
	Average	Average	Average	Average	Average	Average
1	4.696	5.0	4.421	5.0	4.581	5.0
2	3.957	4.0	4.421	5.0	4.306	5.0
3	4.696	4.0	4.895	4.5	4.435	4.5
4	4.087	4.0	4.526	4.5	4.226	4.0
5	4.391	4.5	4.368	3.0	4.645	3.0
6	3.870	4.5	4.632	5.0	4.435	3.0
7	4.348	5.0	4.684	3.0	4.323	5.0
8	3.957	4.5	4.842	3.5	4.371	5.0
9	4.522	5.0	4.895	1.0	4.645	5.0
10	3.457	4.75	4.609	1.0	4.71	4.75
Overall	4.217	4.524	4.64	3.217	4.443	4.429

than that of the earliest time periods. In fact, this shows that in the latest time periods the average rating is closer to the overall average in cases of top products in comparison to earlier time periods. This suggests that the ratings of latest time periods are more indicative than that of early ones in qualifying a product. Thus, it is favorable to add more weight to ratings in latest time periods. On the other hand, the bottom ranked products has fluctuating and mostly overrated ratings, which complicates the matter of only using time as a feature not talking into account the rating quality weight.

We next analyse the distribution of opinion quality in terms of helpfulness votes with time. According to Fig. 1, we can see that the number of helpfulness votes decrease with time. Note that period time 1 is the earliest and 10 is the latest in product life span. One reason is that the review has been there for a long time which allowed for the accumulation of helpfulness votes. Another reason may be because it is an act of spam, which complies with the findings of

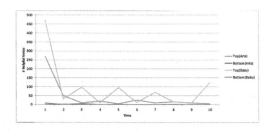

Fig. 1. Number of helpfulness votes with time from earliest to latest time span periods

[13, 20] where spamming effects usually take place in early periods of a product life span. In both cases, the opinion quality measure fades away with time and not necessarily because of bad opinions, but may be it just did not have the time to accumulate more votes as it is still fresh. In order to address these issues, we increase the weight of helpfulness in latest time period through our exponential model.

4 *TQRank*: Time and Quality of Opinion Ranking Model

Based on the previous section analysis, it can be seen that coupling the time and opinion quality factors of reviews can lead to building an accurate product ranking model. In addition, as confirmed with our experiments, a linear weighting function is found to be less effective, because it can not reflect our findings about where in the life time are the important ratings that decide the real quality of a product.

The above analysis leads us to use an exponential function as the weighting model, which simulates the growing importance of ratings over time and in combination with review quality. It can also be considered as if we are applying a forgetting factor for earlier ratings as they were shown to be less effective. On the other hand, later ratings are resonate with the overall average rating. The same principle applies to the opinion quality with an exponential weighting function, each additional helpfulness vote over time makes the rating carries more weight.

The final ranking score for a product p is the weighted aggregation of all ratings S_r, $r = 1..n$, and the weighting function W_r is the weight of review r, as follows:

$$S(p) = \sum_{r=1}^{n} (W_r * S_r)/n \qquad (1)$$

Note that a review r has a posting time t and is for a product in a product category g and by a reviewer u and is computed as follows:

$$W_r = W_r(t, u, g) = e^{\beta(T_t + Q(u,g))} \qquad (2)$$

T_t is the elapsed time from product release, normalised by the whole liftetime for the product. $Q(u, g)$ is an accumulated measure of the opinion quality for user u under product category g with values ranging from 0 to 1 as below:

$$Q(u, g) = \sum_{i=1}^{m} H(i, u, g) / \sum_{i=1}^{m} A(i, u, g) \qquad (3)$$

In the above equation $H(i, u, g)$ and $A(i, u, g)$ are respectively the number of helpfulness votes and the total number of votes, helpful and non-helpful, user u gets for his review i for a product under a category g. $Q(u, g)$ measures the quality of user u for product category g, accumulated over all m reviews that u post for product category g.

Note that our measure of opinion quality is different from the other models which compute this factor per product review [16]. By aggregating all reviews for a user in a product category, our approach reduces the bias due to sparse vote information about reviewer opinion quality at the product level while at the same time captures the closely relevant information at the product category level. To simplify notations, when the product r is clear from the context, $W_r(t, u, g)$ is simply written as $W(t, u, g)$.

5   Experimental Results

Our experiments were all performed on an Amazon dataset[1]. It contains product reviews and meta-data, including 143.7 million reviews spanning from May 1996 - July 2014. This dataset includes product information like ratings, text, and helpfulness votes about reviews, and product meta-data include descriptions, category information, price, brand, sales rank, and image features. We performed some data preprocessing of removing duplicate reviews and grouping all products reviews together according to product categories. The preprocessed dataset contains 33 product categories with 634,586 products and 38,731753 total number of reviews. For presentation purpose, we will show a representative subset of the dataset. We divided product categories into 3 groups according to the average number of reviews per product category. One group with average number of reviews of less than 50, another group with average number of reviews of between 50 and 100 and one last group with an average of more than 100 reviews for a product. In the end our subdataset comprises 96,500 products for 15 categories with 10 million reviews. This way we have representatives for all distributions of reviews with all categories. Table 2 shows statistics about product categories groups mentioned earlier. For each product category, we specify number of products, number of reviews, average number of reviews, total number of votes and helpfulness votes for all products.

We use the Sales Rank for products under a category as the ground truth for evaluation of our ranking model, as in previous studies [24]. Sales rank is a good indicator for products popularity under a category as described earlier. In fact, sales can reflect products' quality. In general, the higher a product's quality, the higher its customer satisfaction, and then the higher it is ranked on the Sales Rank. We apply Spearman correlation to evaluate the correlation of ranking models against sales rank. The correlation values range from -1 for complete non-correlation to 1 for perfect correlation.

5.1   Comparison with Other Baseline Ranking Models

We evaluated our *TQRank* model against several strong baseline models representing research directions on product ranking.

– AverageRank: This approach is still widely used by many online shopping applications.

---

[1] http://jmcauley.ucsd.edu/data/amazon/.

Table 2. Product reviews statistics for product categories groups from our Amazon dataset

Category(# Products)	# Reviews	Avg#Reviews	#Total Votes	#HelpfulVotes
Small Average #Reviews Group				
Appliances(28)	1642	59	10,074	7799
Arts& Crafts(3,107)	179,117	58	521073	449,087
Gift Cards(10)	316	32	4359	2536
Industrial(2,235)	131,961	59	347,684	288,505
Jewelry(4,689)	253,740	54	3,12234	2,60477
Medium Average #Reviews Group				
Camera(2,800)	272,057	97	2,133827	1,696800
Computers(941)	85,428	91	233,488	176,889
Kitchen(19,262)	1,821699	95	5,972960	5,032804
Toys(20,260)	1,277219	63	2,887232	2,310573
Video Games(10,659)	1,029574	97	4,555411	2,701881
Large Average #Reviews Group				
Baby(186)	20,469	110	38811	30,958
Cell Phones(23,355)	2,557402	110	3,109081	2,385656
Electronics(6,666)	678,429	102	2,493038	1,943635
Office Products(162)	16,951	105	29,543	24,436
Software(2,271)	237,709	105	1,224112	968,020

- MiningRank [24]: This is a recent heuristic weighted rank model based on mining review textual content, review helpfulness and time.
- DirichletRank [14]: This is a representative of probabilistic models.

We can see from Table 3 that *TQRank* outperforms all baselines. In fact, 10 out of 15, or 66 %, categories are ranked with high correlation value of greater than or equal to 0.6, which is generally considered strong correlation. Moreover, 6 out of 10 categories got a correlation score of greater than or equal to 0.7, which indicates very strong correlation with the Sales Rank compared to other baselines. We can see that our model works best with medium and dense product categories. This is because with large number of ratings we can easily differentiate between their qualities as our weighing function provides a wider range of values to map their degree of importance. In addition, it is still perform better than other baselines for categories with sparse number of reviews for the same reason of having an effective importance mapping scale.

The second best model is the DirichletRank. However the correlation is still weak. DirichletRank demonstrates consistent performance in ranking of sparse data and get to be similar to AverageRank with dense data. However, the main drawback of such a model is that it only make use of the counts of each star rating level irrespective to the time or the quality of these ratings.

Table 3. Correlation comparison with Sales Rank generated from different ranking methods for product categories with small, medium and large number of reviews per product category

Category(#Products)	AverageRank	MiningRank	DirichletRank	$TQRank$
Small Average #Reviews Group				
Appliances(28)	0.56	−0.136	0.473	**0.661**
Arts & Crafts(3107)	0.061	0.180	0.242	0.481
Gift Cards(10)	−0.042	0.151	0.33	0.120
Industrial Scientific(2235)	0.131	0.1899	0.295	0.483
Jewelry(4689)	−0.02	0.132	0.223	0.581
Medium Average #Reviews Group				
Camera(2800)	0.162	0.162	0.204	**0.754**
Computers(941)	−0.002	0.154	0.152	**0.791**
Kitchen(19262)	0.233	0.186	0.363	**0.700**
Toys & Games(20260)	0.198	0.1561	0.287	**0.707**
Video Games(10659)	0.220	0.227	0.360	**0.741**
Large Average #Reviews Group				
Baby(186)	0.077	0.290	0.257	**0.609**
Cell Phones(23355)	0.083	0.239	0.192	**0.631**
Electronics(6666)	0.109	0.032	0.198	**0.751**
Office Products(162)	0.1153	0.404	0.344	0.541
Software(2271)	0.311	0.259	0.372	**0.699**

MiningRank was fluctuating around the correlation value of 0.2 which strange as they use similar features. However, they use them ineffectively. For one reason, we can see that they on review helpfulness totally ignoring those with less than 10 votes irrespective of when these reviews were posted. In fact, this is the case for most recent reviews, they do not get enough time to accumulate reviewers votes. In a large scale real datasets, we can see the number of reviews having number of up to 10 number votes make at least 40 % of the whole space, thus it must be used accurately. One other observation is that, it was tested only on two popular categories of TV and Camera with almost 500 products and around 12,000 reviews in total. Most of the reviews within these two categories received number of votes between 10 and 200. Furthermore, we can see that the camera category in our dataset contains 2,800 products with 272,052 reviews where 88 % of reviews having number of votes ranging from 0 to 10 which shows why it performed poorly. AverageRank also performed poorly even in cases with dense categories, which may be due to the effect of spam plus the individual bias. The highest correlation score AverageRank got is 0.5 for the Appliances category. The reason to perform reasonably in this case, is that these products had an average number of 60 reviews per each product and equal product life cycles.

Also, 70 % of reviews were shown to be helpful which shows the effect less bias. AverageRank is mostly misleading, this is because it puts equal weight to all rates which doesn't reflect the fact that different reviewers yield different levels of expertise. Moreover, AverageRank suffers from the problem of all good ratings mentioned earlier and with large number of products having slight differences in values it fails to rank products appropriately.

5.2   Evaluation of the *TQrank* Weighting Function

We further evaluated the weight function of *TQrank* against other weighing functions.

– $QRank^l$: Linear weighting function using only the opinion quality $Q(u, g)$
– $TRank^l$: Linear weighting function using only the posting time T_t
– $TQRank^l$: Linear weight function combining opinion quality and posting time

$$W'(u, g, t) = 0.5 * W'(u, g) + 0.5 * W'(t)$$

– $QRank$: Exponential weighing function using only the opinion quality $Q(u, g)$
– $TRank$: Exponential weighing function using only the posting time T_t

Table 4. Correlation with Sales Rank generated using different functions

Category(#Products)	$TQRank^l$	$QRank$	$TRank$	$TQRank$
Small Average #Reviews Group				
Appliances(28)	0.455	0.236	0.450	0.661
Arts &Crafts(3107)	0.050	0.210	0.288	0.481
Gift Cards(10)	−0.111	0.101	0.113	0.120
Industrial Scientific(2235)	0.089	0.295	0.371	0.483
Jewelry(4689)	0.09	0.232	0.427	0.581
Medium Average #Reviews Group				
Camera(2800)	0.23	0.321	0.537	0.754
Computers(941)	0.172	0.464	0.452	0.791
Kitchen(19262)	0.277	0.361	0.489	0.700
Toys & Games(20260)	0.212	0.406	0.509	0.707
Video Games(10659)	0.230	0.397	0.401	0.741
Large Average #Reviews Group				
Baby(186)	0.038	0.290	0.449	0.609
Cell Phones(23355)	0.082	0.339	0.464	0.631
Electronics(6666)	0.182	0.321	0.5067	0.751
Office Products(162)	0.155	0.201	0.330	0.541
Software(2271)	0.278	0.259	0.498	0.699

Table 4 shows the ranking correlation between different weight functions. The ranking correlation of models $QRank^l$ and $TRank^l$ are so weak, so we will only show the their linear combination represented by $TQRank^l$. However, the correlation of $TQRank^l$ with Sales Rank is still very weak, which confirms our hypothesis that linear models does not effectively reflect product ratings correctly. On the other hand, $QRank$ and $TRank$ using only one feature but using exponential increasing weights show better correlation than the linear model. This is because it can better reflect the ratings importance to the weight function. However, combing both feature using $TQRank$ yields the best correlation results across all categories.

6 Conclusion and Future Work

The importance of a fast and accurate product ranking model motivated our research. In this paper, we presented a novel approach to ranking products by weighting the given star ratings of online reviews with review quality in the context of the product category and the review posting time. Our experiments on the Amazon dataset showed that our model's correlation with Amazon Sales rank which indicates product quality measured by sales transactions. Our extensive experiments on 15 product categories from Amazon showed that our model achieved a high correlation with Amazon Sales Rank for most product categories. Our model achieves significantly more effective ranking of products than other approaches. We believe there can be improvements to our model as future work. First, we intend to apply our model to data from other online shopping websites other than Amazon. This will allow us to test our model's performance with varied patterns. It will also allow us to generate an adaptive model based on various reviewers' behaviors. Second, we intend to devise a spam detection module to remove spam's negative effect. In fact, we believe that spam is responsible for most of the biases in the data set in comparison to individual biases.

References

1. Abdel-Hafez, A., Xu, Y., Josang, A.: A normal-distribution based rating aggregation method for generating product reputations. Web Intell. **13**(1), 43–51 (2015)
2. Wang, B.C., Zhu, W.Y., Chen, L.: Improving the Amazon review system by exploiting the credibility and time-decay of public reviews. In: International Conference on Web Intelligence and Intelligent Agent Technology, pp. 123–126. IEEE/WIC/ACM (2008)
3. Bharadwaj, K., Al-Shamri, M.: Fuzzy computational models for trust and reputation systems. Electron. Commer. Res. Appl. **8**(1), 37–47 (2009)
4. Chevalier, J.A., Mayzlin, D.: The effect of word of mouth on sales: online book reviews. J. Mark. Res. **43**(3), 345–354 (2006)
5. Danescu-Niculescu-Mizil, C., Kossinets, G., Kleinberg, J., Lee, L.: How opinions are received by online communities: a case study on Amazon.com helpfulness votes. In: 18th international conference on World wide web WWW 2009, pp. 141–150 (2009)

6. Leberknight, C.S., Sen, S., Chiang, M.: On the volatility of online ratings: an empirical study. In: Shaw, M.J., Zhang, D., Yue, W.T. (eds.) WEB 2011. LNBIP, vol. 108, pp. 77–86. Springer, Heidelberg (2012). doi:10.1007/978-3-642-29873-8_8

7. Garcin, F., Flaing, B., Jurca, R.: Aggregating reputation feedback. In: 1st International Conference on Reputation: Theory and Technology, pp. 62–74 (2009)

8. Fernández, Miriam, Vallet, David, Castells, Pablo: Probabilistic score normalization for rank aggregation. In: Lalmas, Mounia, MacFarlane, Andy, Rüger, Stefan, Tombros, Anastasios, Tsikrika, Theodora, Yavlinsky, Alexei (eds.) ECIR 2006. LNCS, vol. 3936, pp. 553–556. Springer, Heidelberg (2006). doi:10.1007/11735106_63

9. Zacharia, G., Moukas, A., Maes, P.: Collaborative reputation mechanisms for electronic marketplaces. Decis. Support Syst. 4(29), 371–388 (2000)

10. Grimmett, G.R., Stirzaker, D.R.: Probability and Random Processes. Oxford University Press, Oxford (2001)

11. Lauw, H.W., Lim, E.P., Wang, K.: Quality and leniency in online collaborative rating systems. ACM Trans. Web (TWEB) 6(1), 1–27 (2012)

12. iPerceptions Releases Retail E-Commerce Industry Report Q3, December 2011. http://finance.yahoo.com/news/iPerceptions-Releases-Retail-iw-1564944333.html

13. Jindal, N., Liu, B.: Review spam detection. In: 16th International Conference on World Wide Web, pp. 1189–1190, May 2007

14. Josang, A., Haller, J.: Dirichlet reputation systems. In: 2nd International Conference on Availability, Reliability and Security, IEEE, pp. 112–119. IEEE (2007)

15. Mao, A., Procaccia, A.D., Chen, Y.: Better human computation through principled voting. In: 27th AAAI Conference on Artificial Intelligence, pp. 1142–1148, July 2013

16. McGlohon, M., Glance, N., Reiter, Z.: Star Quality: aggregating reviews to rank products and merchants. In: 4th International Conference on Weblogs and Social Media (ICWSM), pp. 1844–1851. AAAI (2010)

17. Mohanty, B.K., Passi, K.: Web based information for product ranking in e-business - a fuzzy approach. In: 8th International Conference on Electronic Commerce, pp. 558–563. ICEC (2006)

18. Nielsen: Global Online Shopping Report, June 2010. http://www.nielsen.com/us/en/insights/news/2010/global-online-shopping-report.html

19. Otterbacher, J.: Helpfulness in online communities: a measure of message quality. In: SIGCHI Conference on Human Factors in Computing Systems CHI 2009, pp. 955–964 (2009)

20. Rayana, S., Akoglu, L.: Collective opinion spam detection: bridging review networks and metadata. In: ACM SIGKDD Conference on Knowledge Discovery and Data Mining, pp. 985–994 (2015)

21. Riggs, T., Wilensky, R.: An algorithm for automated rating of reviewers. In: 1st ACM/IEEE-CS Joint Conference on Digital libraries, pp. 381–387. ACM/IEEE (2001)

22. Xie, H., Lui, J.C.S.: Mathematical modeling and analysis of product rating with partial information. ACM Trans. Knowl. Disc. Data 9(4), 26 (2015)

23. Zhang, X., Cui, L., Wang, Y.: Commtrust: computing multi-dimensional trust by mining E-commerce feedback comments. IEEE Trans. Knowl. Data Eng. 26(7), 1631–1643 (2014)

24. Zhang, K., Cheng, Y., Liao, W.k., Choudhary, A.: Mining millions of reviews: a technique to rank products based on importance of reviews. In: 13th International Conference on Electronic Commerce ICEC 2011 (2011)

25. Zhang, K., Narayanan, R., Choudhary, A.: Voice of the customers: mining online customer reviews for product feature-based ranking. In: 3rd conference on Online Social Networks WOSN 2010, p. 11, June 2010
26. Liu, J., Cao, Y., Lin, C., Huang, Y., Zhou, M., Detection, low-quality product review in opinion summarization. In: EMNLP-CoNLL, pp. 334–342 (2007)

An Effective Spatio-Temporal Approach for Predicting Future Semantic Locations

Hamidu Abdel-Fatao$^{(\boxtimes)}$, Jiuyong Li, Jixue Liu, and Rahman Ashfaqur

School of Information Technology and Mathematical Sciences,
University of South Australia, Adelaide, Australia
hamidu.abdel-fatao@mymail.unisa.edu.au

Abstract. Human mobility prediction in ubiquitous computing is the ability of a system to forecast the anticipated movement of an individual or a group of persons. This interdisciplinary problem has gained traction in fields of academic and industrial research mainly because it is fundamental to achieving system efficiency and marketing efficacy in many applications. This study seeks to develop a novel heuristic technique that predicts the actual geo-spatial locations associated with the most probable semantic tags of locations (e.g. restaurant) that individuals are likely to visit. The intuition of this work lies in the fact that, for any given probable future semantic tag there exists multiple geo-spatial locations associated with it, hence the need to disambiguate the actual destination location. We develop an algorithm *STS_predict*, that exploits the spatio-temporal relationships between the current location of a target individual and candidate geo-spatial locations associated with future semantic tags to predict the actual destination location. We evaluate our approach on a real world GPS trajectory dataset.

1 Introduction

Human mobility is a phenomenon characterised by changing positions of an individual or a group of persons with respect to time. Current proliferation of mobile devices equipped with location sensing technologies (e.g. GPS) has immensely contributed to the generation of massive amounts of historical human mobility data. The wide availability of these data presents an unprecedented opportunity for research into understanding the dynamics of human mobility. Such knowledge constitutes the basis upon which many applications in ubiquitous computing are built. One such application which is also the primary focus of this study, is human mobility prediction based on GPS trajectory data.

Human mobility prediction, typically built on some probabilistic/statistical models [1,2,4,7,8,11] or pattern mining approaches [5,14] essentially aims at forecasting anticipated movement of an individual or a group of persons. Accurately predicting human mobility is the bedrock of many applications which are useful in deploying services that can enable proactive experiences and trigger prompt actions. These applications include traffic control and navigational services, control of disease spread during an outbreak, targeted advertisement,

© Springer International Publishing AG 2016
M.A. Cheema et al. (Eds.): ADC 2016, LNCS 9877, pp. 283–294, 2016.
DOI: 10.1007/978-3-319-46922-5_22

surveillance in counter-terrorism etc. [1,8,11]. For instance, digital assistants developed recently by Microsoft Cortana[1] and Google Now[2] mainly rely on prediction of human mobility to furnish users with relevant information, or assist them to accomplish tasks without their intervention e.g. pre-heating (or cooling) the house when the user is on the way home [12].

Evidence in the existing body of literature suggests that, except for the probabilistic works Abdel-Fatao et al. [1], Gambs et al. [4] and sequential pattern based approach Ying et al. [14], many studies on human mobility prediction based GPS trajectories [2,5,13] only focus on predicting the geo-spatial and/or temporal context of human mobility without considering semantic dimension. Pure geo-spatial predictive algorithms typically utilise raw trajectory data directly. These approaches face two critical issues. Firstly, they impose on users, the burden of making sense of their prediction outcomes which are usually in the form of geographic coordinates. Besides that, their approaches are often computationally expensive [5] due to massive amounts of data involved in dealing directly with high resolution raw GPS trajectory data.

To overcome the drawbacks with these existing works, [1,4] develop Markov chain models for predicting location types (herein referred to as semantic tags) likely to be visited by users. Similar to Gambs et al. [4], Abdel-Fatao et al. [1] firstly transform raw historical trajectories into semantically annotated sequences of significant locations. Subsequently, they build Markov chain models for predicting future semantic tags. However, different from [4], Abdel-Fatao et al. [1] take into account temporal information in developing a timeslot-based Markov chain model for predicting future semantic tags. This makes the model by Abdel-Fatao et al. [1] more effective in comparison to Gambs et al. [4]. The semantic transformation above has two benefits - (i) It drastically reduces the amounts of trajectory data, and by extension resolves the computational problems associated with dealing directly with it; (ii) It precisely expresses semantic meanings of users' mobility behaviours, hence lifting the burden of interpreting prediction outcomes from users.

In spite of the fact that Abdel-Fatao et al. [1] provide more realistic outcome to human mobility prediction, a crucial issue remains unresolved. Their model only seeks to reveal knowledge of semantic tags of future locations a target individual is likely to visit, without indicating the exact geo-spatial context of the semantic tags. For example, they are only able to predict the most probable destination of a target individual to be a restaurant without indicating exactly which restaurant out the repertoire of possibilities that may exist. Consequently, where there exists a multiple geo-spatial locations associated with a semantic tag, the prediction details solely based on such approach will fall short of completeness.

In this study, we address the shortcomings of [1] by developing an effective heuristic technique that disambiguates and predicts the actual geo-spatial context associated with future semantic tags. More specifically, our algorithm exploits temporal relationships between the current and candidate geo-spatial

---

[1] http://www.windowsphone.com/en-us/how-to/wp8/cortana/meet-cortana.

[2] http://www.google.com/landing/now/.

locations by evaluating temporal a score for each candidate location that minimises arrival time and maximises stay duration. Further, our algorithm exploits distributions of candidate locations about the current one in form of radii of gyration, as well as historical transitions between the current and candidate locations in estimating a spatial score for each candidate location. To make prediction, a ranking score that is a weighted total of temporal and spatial scores is evaluated, and the location with the highest score predicted to be the most likely future location of a target user.

We summarize the contribution of our work as follows.

- We develop a novel heuristic algorithm, *STS_predict* for predicting the actual geo-spatial locations associated with future the semantic tag a target individual is likely to visit.
- We also show through empirical analysis that temporal information dominates spatial information in determining the accuracy of predictions.

We evaluate our approach on a real world GPS trajectory dataset.

2 Preliminaries

In this section, we give an overview of the steps involved in pre-processing of raw trajectory data, as well as clarify relevant concepts used in this study.

2.1 Data Pre-processing and Mobility Modelling

Human mobility behaviour can be reconstructed from streams of spatio-temporal points known as trajectory points. Trajectory points organized sequentially in ascending of timestamps represent mobility traces known simply as *trajectory*.

Definition 1. *A trajectory denoted by* $\mathcal{P} = \langle p_1, p_2, ..., p_z \rangle$ *is a sequence of trajectory points organised in ascending order of timestamps, where* $\{p_i \in \mathcal{P} : p_i = (x_i, y_i, t_i)\}$ *is a trajectory point and* $t_i < t_{i+1}$, $\forall i \in [1, z - 1]$.

Typically, trajectory points are sampled at high frequencies by devices, resulting in large volumes of trajectory data. Analysing such raw trajectory data directly will involve significant computational overheads. To overcome this challenge, we adopt the data processing technique by Abdel-Fatao et al. [1], described below.

Firstly, given historical trajectory data of an individual, significant locations visited by the person are identified. These locations known as *stay points*, defined by maximum distance and minimum stay time thresholds, are characterised by clusters of trajectory points. Stay points extraction is in line with the intuition that, when people visit places of interest they stay within nearby areas for considerable periods of time. For example, when people visit the cinema they usually stay within the Cinema Hall for considerable periods of time watching movies.

Next, a density-based clustering algorithm, *OPTICS*, is performed to group stay points representing the same location but having different coordinates into non-overlapping clusters. This step is necessary because GPS positioning error of devices makes it almost impossible to record identical coordinates for the same location even though a user may visit the same place at different times. We therefore group stay points according to geographical proximity and represent each discovered cluster with a single point called *reference point* defined below.

Finally, semantically meaningful representations are assigned to locations by annotating reference points discovered with their corresponding semantic tags. Each semantically annotated reference point is known as *semantic location*. The semantic annotations of reference points is achieved using of a PoI database. Specifically, Foursquare category database is employed, and proximity of PoIs to each reference point is utilized to assign appropriate semantic tags. For more detailed description of the data preprocessing technique please refer to [1].

2.2 Problem Statement

Let $L_i = [(x_i, y_i) : l_i, t_i.a, t_i.s]$ denote a location context where L_i is a semantic location defined by the attributes: (x_i, y_i) − a (latitude, longitude) coordinate pair; l_i − semantic tag of (x_i, y_i); $t_i.a$ and $t_i.s$ − arrival and stay times respectively. Also, let L_c denote the current semantic location; l_d, the semantic tag of a future location; and $L_D = \{L_1, L_2, \ldots, L_n\}$, a finite set of candidate destination geo-spatial locations, where the semantic tag of each $L_i \in L_D$ is l_d. Note that unless otherwise specified, the above notations hold for the rest of this paper.

The problem addressed in this work is to evaluate a ranking score $\hat{r}(L_i)$ for each $L_i \in L_D$ such that the most likely geo-spatial location associated with l_d to be visited is ranked the highest. We develop a heuristic algorithm, *STS_predict* which is essentially a ranking technique to solve the problem. The main challenge posed is how to rank candidate geo-spatial locations so as to assign the actual destination location with the highest rank. We elaborate on how we address this challenge in solving the problem in the section below.

3 Methodology

This section explains our ranking technique for solving the problem herein.

The ranking technique developed in this study is based on the assumption that for any semantic tag, there exists more than one associated candidate geo-spatial locations. We therefore train a model on a list of candidate geo-spatial locations based on their spatio-temporal relationships with the current location. The solution to this problem involves two complementary sub-ranking stages

- temporal mobility prediction and
- geo-spatial mobility prediction.

We present the details for tackling each stage below.

3.1   Temporal Mobility Prediction

The aim of the temporal mobility prediction stage is to rank candidate geo-spatial locations based on their temporal relationships with the current location. We believe that human mobility is strongly influenced by different facets of temporal information such as transition time, time of the day etc. For instance, if an individual intends to visit the supermarket for shopping after leaving the office, he/she will likely prefer a supermarket that minimises arrival time. To this end, we propose a ranking function referred to as *temporal score* denoted by $\hat{r}_t(L_i)$, to rank geo-spatial locations based on their temporal relationships with the current location. We formulate $\hat{r}_t(L_i)$ as follows.

Given L_c during a timeslot t_j and the set $L_D = \{L_1, L_2, \ldots, L_n\}$ associated with l_d during timeslot t_{j+1}, we wish to find the best estimate of $\hat{r}_t(L_i)$ such that the arrival time of the most likely destination geo-spatial location is minimal. For any $L_i \in L_D$, we let $L_i.arrT$ and $L_i.depT$ represent arrival and departure times respectively. We define the transition time between L_c and each $L_i \in L_D$ denoted by $transT(L_c, L_i)$ using the expression $transT(L_c, L_i) = |L_i.arrT - L_c.depT|$. For each $L_i \in L_D$, we compute the mean transition time denoted by μ_{L_i}, based on historical transitions between L_c and L_i using the expression $\mu_{L_i} = \sum_{i=1}^{m} \frac{|L_i.arrT - L_c.depT|}{m}$, where m is the number of historical transitions.

Based exclusively on the current instance of time denoted by $L_c.curT$, we also estimate the anticipated transition time between L_c and each L_i denoted by $transT(L_c, L_i)^{curT}$, using the expression

$$transT(L_c, L_i)^{curT} = \frac{|L_c.curT - L_i.EarrT| + |L_c.curT - L_i.LarrT|}{2} \qquad (1)$$

where $L_i.EarrT$ and $L_i.LarrT$ are the earliest and latest arrival time of L_i.

Next, we compute the overall mean transition time between L_c and all candidate geo-spatial locations $L_i \in L_D$ using $\mu(L_c, L_D) = \sum_{i=1}^{n} \frac{\mu_{L_i}}{n}$, where n is the number of candidate geo-spatial locations. To determine how close the arrival times at candidate locations match the average transition time, we measure the standard deviation σ of transition times for each candidate location $L_i \in L_D$ using the expression.

$$\sigma = \sqrt{\frac{1}{n} \sum_{i=0}^{n} (L_i.arrT - \mu(L_c, L_i))^2} \qquad (2)$$

Finally, we evaluate the temporal score $\hat{r}_t(L_i)$, for each L_i in terms of standard score to reflect how well its transition time matches the average value using Eq. 3.

$$\hat{r}_t(L_i) = w_t(L_i) \times \frac{|transT(L_c, L_i)^{curT} - \mu_{L_i}|}{\sigma} \qquad (3)$$

where $w_t(L_i) = \frac{\text{mean stay duration at } L_i}{\sum_{i=1}^{n} \text{mean stay duration at } L_i \in L_D}$ is a weight we employ to give importance to amount of time users like spend in locations they visit.

The Algorithm 1 summarises the complete process of the temporal score estimation.

Algorithm 1. Temporal Score Estimation

Require: L_c, $L_D = \{L_1, L_2, \ldots, L_n\}$
Ensure: $\hat{r}_t(L_i)$
1: **for** $L_i \in L_D$ **do**
2:      Compute the mean transition time μ_{L_i}, between L_i and L_c
3:      Compute the expected transition time $transT(L_c, L_i)^{curT}$, between L_i and L_c
         based on current time
4: **end for**
5: Estimate overall mean transition time between L_c and all $L_i \in L_D$
6: Compute σ using Eq. 2
7: Generate $\hat{r}_t(L_i)$ based on Eq. 3

3.2 Geo-Spatial Mobility Prediction

The geo-spatial mobility prediction stage seeks to exploit dependence of human mobility on geographic distance in estimating a ranking score for each $L_i \in L_D$. Studies [6,8,9,11] have shown that human mobility behaviours follow a power law characterised by heavy tail distributions of travel distances. That is, people mostly prefer to visit nearby locations, except for occasional deviations. In line with this fact, we propose a spatial ranking function $\hat{r}_s(L_i)$, to rank each $L_i \in L_D$ according to its geographic distance from the current location L_c.

Given L_c during timeslot t_j, and the set $L_D = \{L_1, L_2, \ldots, L_n\}$ associated with l_d during t_{j+1}. We wish to find the best estimate of $\hat{r}_s(L_i)$ such that the most likely destination geo-spatial location is geographically as close as possible to L_c, and is among the most visited locations. To this end, we utilize three geo-spatial scale parameters that influence human mobility namely: travel distance, radius of gyration, r_g, and transition support, $Tr_{sup}(L_i)$, in computing $\hat{r}_s(L_i)$.

Firstly, we compute the travel distance between each $L_i \in L_D$ and L_c denoted by $\Delta_d(L_i, L_c)$, as a measure of geographic distance of L_i from L_c. To determine $\Delta_d(L_i, L_c)$, we employ the Haversine formula [10]. We emphasise on the importance of nearness of locations to users and define a *willingness factor* denoted by $\omega_d(L_i)$, to reflect the preference of users to visit nearby locations by taking the inverse logarithms of travel distances. More formally, this is given by

$$\omega_d(L_i) = \frac{1}{log(\Delta_d(L_i, L_c))} \tag{4}$$

We also explore the influence of geo-spatial distribution of candidate locations about the current location on users mobility. To do this, we compute the radii of gyration [8,11] of the candidate locations about the current one based on the geo-spatial distribution of the candidate locations relative to the current location. We express the radius of gyration more formally as

$$r_g = \sqrt{\frac{1}{n} \sum_{i=1}^{n} |dist(L_i - L_c)|^2} \tag{5}$$

where $|dist(L_i - L_c)|$ is the geographic distance between the coordinates of L_i and L_c based on the Haversine formula. The value of r_g gives a holistic view of the ranges of candidate locations from the current location.

Further, we evaluate the transition support $Tr_{sup}(L_i)$, for each $L_i \in L_D$ defined as the ratio of the sum of historical contiguous transitions between L_c during timeslot t_j and L_i during timeslot t_{j+1} to the total sum of contiguous transitions between the L_c and all $L_i \in L_D$. More formally, the transition support is expressed as

$$Tr_{sup}(L_i) = \frac{count[L_c \to L_i]}{\sum_{i=1}^{n} count[L_c \to L_i \in L_D]}. \tag{6}$$

$Tr_{sup}(L_i)$ is used as a measure of how frequent a user visits a candidate location associated with a semantic tag given the current location L_c, relative to other candidate locations associated the same semantic tag. Higher values of $Tr_{sup}(L_i)$ represent an implicit function of the importance of candidate locations to users.

Finally, we compute the spatial ranking score $\hat{r}_s(L_i)$, for each candidate location using Eq. 7.

$$\hat{r}_s(L_i) = Tr_{sup}(L_i)[1 + \omega_d(L_i)r_g] \tag{7}$$

The first part of the Eq. 7 ranks L_i in terms of visiting frequency, while the second part ranks L_i according to its geo-spatial distance from L_c relative to radius of gyration. The higher the value of $\hat{r}_s(L_i)$ the nearer a geo-spatial candidate location to the current one.

The Algorithm 2 summarises the entire procedure of spatial score estimation.

Algorithm 2. Spatial Score Estimation

Require: L_c, $L_D = \{L_1, L_2, \ldots, L_n\}$
Ensure: $\hat{r}_s(L_i)$
 1: **for** $L_i \in L_D$ **do**
 2:     Evaluate travel distance $\Delta_d(L_i, L_c)$ between L_i and L_c
 3:     Evaluate willingness factor $\omega_r(L_i)$
 4: **end for**
 5: Compute the radius of gyration r_g of L_c relative to L_D
 6: Compute the transition support $Tr_{sup}(L_i)$
 7: Generate $\hat{r}_s(L_i)$ based on Eq. 7

3.3   Ranking Score

The ranking score $\hat{r}(L_i)$, for each $L_i \in L_D$ associated with l_d relative to L_c at any given instance of time, is a measure of the likelihood of an individual to visit

L_i next. We express the ranking score as the weighted sum of the temporal and spatial score of L_i. That is

$$\hat{r}(L_i) = \alpha\hat{r}_t(L_i) + (1 - \alpha)\hat{r}_s(L_i) \tag{8}$$

where $\alpha \leftarrow [0, 1]$ is a tuning parameter that controls the influence of spatial and temporal information on the ranking score.

The first term of Eq. 8 is a weighted temporal score which reflects temporal patterns of visits to geo-spatial locations. Our assumption here is that, human mobility between locations is not only dependent on absolute distance between places but also the density of opportunities and resources available in these locations [9]. As a result nearby locations may not necessarily be locations of choice. We therefore utilize temporal score to capture the influence of temporal information on the choice of a potential destination. The second term of Eq. 8 is a weighted spatial score which captures the influence of geo-spatial distances on the choice of destination locations. Here utilize the relative geo-spatial distances between L_c and $L_i \in L_D$ to rank each L_i accordingly. This is analogous to *rank distance* metric proposed by Noulas et al. [9].

We predict the actual destination of an individual to be the geo-spatial location with the highest value for $\hat{r}(L_i)$. Note that, in spite of the fact that we did not consider self-transitions (transitions within a given timeslot) in our ranking scores, our technique can easily be extended to tackle such cases.

4   Experiment

In this section, we validate of our approach using experiments.

4.1   Dataset, Metrics and Baselines

Dataset. In this study, we utilized *GeoLife* 1.3 dataset[3], a well known, real-world GPS trajectory dataset collected from 182 individuals mostly in Shanghai over 5 years (April 2007–August 2012) using GPS devices. From the dataset, we filtered individuals having trajectories spanning a period of at least one week, in order to increase our chances of finding trajectories which exhibit routine mobility behaviours. We found that trajectories of 149 users satisfied this requirement and processed their datasets accordingly.

Performance Metrics. To evaluate the performance of our algorithm, we employ the following metrics: (1) Accuracy@N; (2) Average percentile rank (APR).

Accuracy@N is a measure of the percentage of locations accurately predicted in the list of top N predicted locations [12].

---

[3] http://research.microsoft.com/en-us/downloads/b16d359d-d164-469e-9fd4-daa38f2b2e13/.

A Percentile Rank (PR) [9] is a score based on the ranking assigned a predicted location L_i among a list of predictions such that PR evaluates to 1 when the actual location that will be visited is ranked as the first and linearly decreases to 0 as the actual location is demoted down the list. Mathematically, this is expressed as $PR = \frac{|N|-rank(L_i)+1}{|N|}$. The APR is obtained by finding the average PR across the testing dataset.

We train our model of 80 % dataset and hold the remaining 20 % for testing.

Baselines. We compare our model with two baselines namely (1) *Most Frequent location predictor (MF)* and (2) *Naive Bayes-based location predictor (NB)*.

The MF location predictor coined from [3] computes a probability for each candidate location only based on the frequency of historical visits. Specifically, it assigns the most frequently visited location the highest probability, hence the predicted location during a specified timeslot. The model is simple and yet intuitive enough for prediction of human mobility. For example, if one is asked to guess the most likely location of a friend at 11 am, one will probable suggest the place of work based on historical observations during that hour.

The NB location predictor on the other hand is a probabilistic model that employs the Naive Bayes approach to assign a conditional probability to each candidate location. Specifically, based on temporal information, NB estimates a posterior probability for each candidate location expressed as $P(L_i^{t_{j+1}}) = \frac{P(L_c^{t_j})P(L_c^{t_j}|L_i^{t_{j+1}})}{P(L_i^{t_{j+1}})}$, where L_c and L_i respectively represent the current and candidate locations and t_j and t_{j+1} are the current and next timeslots respectively.

4.2 Experimental Results

Results Based on Prediction Accuracy@N. We aim to investigate the extent to which spatial and temporal information impact upon the accuracy of our model by varying the tuning parameter α, in computing the prediction score based on our Eq. 8. We vary the values of α from 0 to 1.

Figure 1 shows the results obtained average accuracy of five experimental runs for different values of top N predicted locations using different values of α. We observe that the average accuracy generally increases with increase in N, regardless of the value of α. This makes sense and is line with our logical expectation in that, an increase in the value of N is

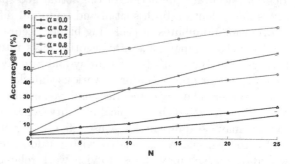

Fig. 1. Accuracy for various values of α

expected to increase the sample size of candidate results, hence enhancing the chances of making more accurate predictions.

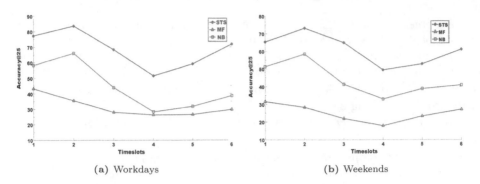

(a) Workdays   (b) Weekends

Fig. 2. Comparison of prediction accuracies for workdays and weekends

Taking a critical view of values of $accuracy@N$ with varying values of α reveals an interesting observation. Setting $\alpha = 0$ suggests that the ranking score is entirely dependent upon spatial information. When this is the case, the value of $accuracy@N$ obtained is very low. Even at the highest value of $N = 25$, $accuracy@N$ is only 14 %. Increasing the value of α indicates increasing the consideration for temporal information in computing the ranking score. We observe that at $\alpha = 0.8$, the highest accuracy was obtained for all values of N, an indication that temporal information is a more dominant factor that impacts on accuracy of human mobility prediction. This makes sense in that, an individual is likely to give prominence to time of the day before deciding to visiting the library in the morning instead of the nightclub, regardless of travel distance.

Finally, we also note that at $\alpha = 1$, indicating prediction of next location is completely dependent upon temporal information, $accuracy@N$ snaps back to new low for all values of N. We reason that, this is possible due to the fact that taking spatial information alone into account in predicting future locations is not enough in achieving reasonably high accuracies of prediction. From these results, we can confidently conclude that, given a known semantic tag of a future location, accurate prediction of the actual geo-spatial location of the semantic tag is dependent on both spatial and temporal information. However, temporal information dominates in achieving higher accuracy of prediction.

To make comparison, we categorise days into *Workdays* (i.e. Mondays to Fridays) and *Weekends* (i.e. Saturday and Sunday), and compare the prediction accuracy of our approach during various timeslots to the baseline methods mentioned earlier. The day categories employed ensure that, user mobility behaviours are well represented since typical mobility behaviours during these day categorisations may be different for a user [9,12]. For instance, a user may typically prefer to visit places associated with outdoor activities on weekend while on workdays, he/she may choose to visit work related locations.

We fix $N = 25$ (and set $\alpha = 0.8$ in the case of our model denoted by STS) and measure prediction accuracy@25 for workdays and weekends. The results are shown in Fig. 2a and b. We observe that, for all the methods prediction accuracy is generally higher on workdays compared with weekends. This is understandable in that workdays are characterised by a more mobility data for prediction in comparison to weekends as people tend to visit more places on workdays. We also notice that, the prediction accuracies are relatively higher during timeslots 1, 2 and 6 (between late evening and early mornings) for both workdays and weekends. This is attributable to the fact that people are more stable during these timeslots, usually in their home locations. The lowest accuracies are recorded during timeslot 4, a period between midday and late afternoon when people sometimes randomly deviate from their routine mobility behaviours.

Finally, in terms of specific models, our model (STS) significantly outperforms MF which is based entirely on frequency of visit to candidate locations and NB, a method that estimates probability of the next location by conditioning on temporal information. Also significant is the fact that the prediction accuracy of NB is higher than that of MF and this can be explained by the fact that NB takes extra temporal information into account in making predictions. Our model outperforms the two baselines because, aside from relying on temporal information, we additionally consider spatial information in making predictions. This does not only give us improved prediction accuracies over methods that may not rely on these features simultaneously, but it also present our model as a more realistic method of predicting future locations.

Results Based on APR. The APR results presented here are used to demonstrate how well our model ranks the next location to be visited amongst candidate locations. In this experiment, we investigate how temporal and spatial information affect our ranking of candidate locations based on APR scores. We measure APR scores for various timeslots on workdays and weekends, and report the results in Table 1.

Table 1. APR for various values of α

α	Workdays						Weekends					
	t_1	t_2	t_3	t_4	t_5	t_6	t_1	t_2	t_3	t_4	t_5	t_6
0.0	0.33	0.31	0.26	0.22	0.30	0.31	0.27	0.26	0.23	0.20	0.27	0.27
0.2	0.49	0.49	0.46	0.44	0.47	0.48	0.46	0.44	0.41	0.39	0.46	0.44
0.5	0.74	0.72	0.69	0.65	0.68	0.72	0.68	0.68	0.64	0.61	0.69	0.67
0.8	0.86	0.86	0.84	0.82	0.85	0.86	0.82	0.81	0.79	0.76	0.84	0.82
1.0	0.68	0.68	0.65	0.60	0.64	0.66	0.62	0.60	0.59	0.54	0.63	0.63

Similar to our observation regarding accuracy measurements, we notice that timeslots 1, 2 and 6 for both workdays and weekends are synonymous with higher APR values. Another noticeable feature is that, there are pronounced similarities between how variations in α values affect accuracy measurements and APR scores. This leads us to conclude that factors that improve the accuracy of next location predictions are also potential drivers of improved APR values.

5 Conclusion

In this work, we develop a novel heuristic technique to predict the actual geo-spatial location associated with the future semantic tags an individual is likely to visit based on historical GPS trajectory data. In particular, we exploit temporal relationships between the current location of an individual and candidate geo-spatial locations in computing ranking scores for candidate geo-spatial locations such that the most probable geo-spatial location is ranked highest.

References

1. Abdel-Fatao, H., Li, J., Liu, J.: STMM: semantic and temporal-aware markov chain model for mobility prediction. In: Zhang, C., Huang, W., Shi, Y., Yu, P.S., Zhu, Y., Tian, Y., Zhang, P., He, J. (eds.) ICDS 2015. LNCS, vol. 9208, pp. 103–111. Springer, Heidelberg (2015). doi:10.1007/978-3-319-24474-7_15
2. Ashbrook, D., Starner, T.: Using GPS to learn significant locations and predict movement across multiple users. Pers. Ubiquit. Comput. **7**(5), 275–286 (2003)
3. Cho, E., Myers, S.A., Leskovec, J.: Friendship and mobility: user movement in location-based social networks. In Proceedings of the 17th ACM SIGKDD International Conference on Knowledge Discovery and Data Mining, KDD 2011, pp. 1082–1090. ACM, New York (2011)
4. Gambs, S., Killijian, M.-O., del Prado Cortez, M.N.: Next place prediction using mobility markov chains. In: MPM 2012, NY, USA (2012)
5. Gidófalvi, G., Dong, F.: When and where next: individual mobility prediction. In: Proceedings of the First ACM SIGSPATIAL International Workshop on Mobile Geographic Information Systems, MobiGIS 2012 (2012)
6. Gonzalez, M.C., Hidalgo, C.A., Barabasi, A.-L.: Understanding individual human mobility patterns. Nature **453**(7196), 779–782 (2008)
7. Lin, M., Hsu, W.-J., Lee, Z.Q.: Predictability of individuals' mobility with high-resolution positioning data. In: Proceedings of the 2012 ACM Conference on Ubiquitous Computing, UbiComp 2012, pp. 381–390. ACM, New York (2012)
8. Lu, X., Wetter, E., Bharti, N., Tatem, A.J., Bengtsson, L.: Approaching the limit of predictability in human mobility. Sci. Rep. **3** (2013)
9. Noulas, A., Scellato, S., Lathia, N., Mascolo, C.: Mining user mobility features for next place prediction in location-based services. In: Proceedings of the 2012 IEEE 12th International Conference on Data Mining, ICDM 2012, pp. 1038–1043. IEEE Computer Society, Washington, DC (2012)
10. Sinnott, R.W.: Virtues of the haversine. Sky and Telescope **68**(2), 158–160 (1984)
11. Song, C., Qu, Z., Blumm, N., Barabási, A.-L.: Limits of predictability in human mobility. Science **327**(5968), 1018–1021 (2010)
12. Wang, Y., Yuan, N.J., Lian, D., Xu, L., Xie, X., Chen, E., Rui, Y.: Regularity and conformity: location prediction using heterogeneous mobility data. In: KDD, pp. 1275–1284. ACM, New York (2015)
13. Yang, N., Kong, X., Wang, F., Yu, P.S.: When and where: predicting human movements based on social spatial-temporal events. In: Proceedings of 2014 SIAM International Conference on Data Mining (2014)
14. Ying, J.J.-C., Lee, W.-C., Tseng, V.S.: Mining geographic-temporal-semantic patterns in trajectories for location prediction. ACM Trans. Intell. Syst. Technol. **5**(1), 1–33 (2013)

Automatic Labelling of Topics via Analysis of User Summaries

Lishan Cui[1], Xiuzhen Zhang[1(✉)], Amanda Kimpton[2], and Daryl D'Souza[1]

[1] School of Science (Computer Science and Information Technology),
RMIT University, Melbourne 3001, Australia
{lishan.cui,xiuzhen.zhang,daryl.dsouza}@rmit.edu.au
[2] School of Health and Biomedical Sciences,
RMIT University, Melbourne 3001, Australia
amanda.kimpton@rmit.edu.au

Abstract. Topic models have been widely used to discover useful structures in large collections of documents. A challenge in applying topic models to any text analysis task is to meaningfully label the discovered topics so that users can interpret them. In existing studies, words and bigram phrases extracted *internally* from documents are used as candidate labels but are not always understandable to humans. In this paper, we propose a novel approach to extracting words and meaningful phrases from *external* user generated summaries as candidate labels and then rank them via the Kullback-Leibler semantic distance metric. We further apply our approach to analyse an Australian healthcare discussion forum. User study results show that our proposed approach produces meaningful labels for topics and outperforms state-of-the-art approaches to labelling topics.

Keywords: Topic model labels · User generated content · Dependency relation parser

1 Introduction

Topic models are useful tools for analysing large collections of documents [13, 17,20,22]. With a topic model, each latent *topic* is a multinomial distribution over words, and each document is described as a mixture of latent topic distributions. Several topic models have been proposed in the literature [1,5], including the popular Latent Dirichlet Allocation (LDA) and other recent developments (e.g., [2]). An important problem is to label the topics produced by topic models so that the topics are understandable and interpretable to humans [7,8,12,15]. The topic below (Topic 0 of Table 4) is generated by the non-parametric LDA [2] from 623 documents (stories) sourced from the Patient Opinion Australia (https://www.patientopinion.org.au/), a web forum for patients to post their stories about healthcare and freely express their opinions. Further details of such use of the LDA model appear in [2].

© Springer International Publishing AG 2016
M.A. Cheema et al. (Eds.): ADC 2016, LNCS 9877, pp. 295–307, 2016.
DOI: 10.1007/978-3-319-46922-5_23

october; appointment; outpatient; referral; letter; answer; confirmed; september; ulcer; telephoned.

The topic is represented by the top ten words as the topic label, ordered by their marginal probability in the topic [1,5]. However such a conventional topic label is difficult for humans to make sense and interpret the meaning [7,8,11,13, 19]. Specifically words *october* and *september* refer to the temporal information whereas *appointment* and *referral* refer to communication; the single words make it hard for people to infer the semantics of the topic "appointment letter". It is therefore critically important to assign semantically meaningful labels to topics so that humans can interpret and understand the topics.

It has been shown in the literature that phrases as labels for topics are more understandable to human beings [11,13,19]. However, existing solutions make the implicit assumption that n-gram phrases are good candidate topic labels. Mei *et al.* [12] used bi-gram phrases from the document collection as candidate labels. Recognising the limitation that words appearing in the document collection cannot represent topics at higher semantic levels, Lau *et al.* [7] proposed to extract words and phrases from external data sources like Wikipedia as candidate topics labels. Still, n-gram phrases extracted from the general knowledge base Wikipedia, are not always meaningful labels for specific domains.

On the other hand, with the development of Web 2.0 technology, user-generated content becomes widely available on the web. User posted documents and comments appear in various discussion forums as well as community question-answer sites like Yahoo!Answers and Stack Overflow. User comments in web forums have been employed for computing seller-buyer trust for E-commerce applications [23], news recommendation [9], and predicting the popularity of online articles [18].

In this paper we propose to utilize user generated summaries for labelling topics for a document collection. The user generated summaries in the same domain as the document collection are ostensibly more relevant to the domain than general external knowledge sources like Wikipedia. Moreover, the *external* user-generated summaries often contain words and phrases at higher semantic levels than words from the *internal* document collection. To the best of our knowledge our work is the first of its kind to take advantage of user-generated content for labelling topic models for document collections.

Our approach, Labelling By Summaries (LBS), can produce meaningful topic labels easy to interpret by humans. Looking again at the topic in the previous example, LBS generates the following words and phrases as its generated label:

`administration; appointment letter`

Note that `administration` is a single-word expression frequently appearing in user summaries but not in the topic word list. Importantly `administration` summarises nicely the administrative aspect details expressed by several topic words including *october, outpatient, referral, answer, confirmed, september* and *telephoned.* On the other hand `appointment letter` is a noun phrase frequently appearing in user summaries generated by the dependency relation parser [4] and it summarises meaningfully the topic words *letter* and *appointment*.

The research questions we address are as follows:

- How to generate meaningful phrases from user summaries as candidate labels for topics?
- For a topic, how to measure the semantic association between candidate labels and the topic to rank the candidate labels?

We make several contributions for automatic labelling of topic models. First we propose to apply the NLP dependency relation parser [4] and frequency-based noise filtering to generate meaningful phrases from user summaries as candidate topic labels. To rank candidate labels for a topic, we further propose a metric to evaluate the semantic association between a topic and the candidate labels based on Kullback-Leibler (KL) divergence [6]. We formulate the problem as minimising the KL divergence between a topic and the candidate labels. We apply LBS with the non-parametric LDA topic model proposed by Buntine *et al.* [2] to summarise documents in the healthcare domain – the Patient Opinion Australia (POA) web forum (details in Sect. 4). We design a user study to evaluate labels for topics. Results show that, compared with state-of-art approaches [8,12], LBS can generate more meaningful labels that can help understand and better interpret topics.

2   Related Work

Existing studies on automatic labelling for topic models can be categorised into internal and external approaches based on the candidate labels used. The internal approaches use topic words that are sourced and generated by topic models from the document collection [1,5,8], whereas external approaches use word sources that are external to the topic words and could be extracted from data sources like Wikipedia or document collection [7,10,12]. The standard approach for labelling topics is to use the top N (typically $N = 10$) topic words ranked by their marginal probability [1,5,8]. However, it is recognised that the list of top words very often do not present coherent semantics [14,16] and are difficult for humans to understand and interpret the meaning of topics. To enhance the topic interpretation, Lau *et al.* [8] proposed to select the best topic words to label topics. Several measures are proposed to re-rank the top ten topic words to select the best label for the topic. Still, the assumption is that the best label for a topic can be found among the internal topic words generated by topic models.

In external approaches there are typically two steps involved: generating candidate labels for topics and ranking candidate labels. Lau *et al.* [7] proposed an approach to automatically label topics from topic models via generation of candidate labels from external data sources (Wikipedia) and supervised learning for ranking the candidate labels. The candidate labels generated can be single terms or phrases from Wikipedia. Mei *et al.* [12] proposed an approach to automatically label topics, comprising generating candidate labels by extracting bi-grams or noun chunks from the document collection. To rank the candidate labels for a topic, topics and labels are represented as distributions of words and the KL divergence is used to measure their semantic distance. Magatti *et al.* [10] proposed a method for labelling topics induced by hierarchical topic models. Their candidate labels are based on the Google Directory hierarchy and the labelling approach relies on a pre-existing ontology and the associated class labels. As a result the approach can only be applied in limited applications.

Although not directly related to topic labelling, Chang *et al.* [3] were one of the first to raise the question of labelling topic models for human understanding. They identified the notion of *intruder* words. Such words (in the top topic words) impede or compromise human understanding of topic meaning and are inconsistent with the semantic meaning in inferred topics.

In the literature, the study closest to our approach is that of Lau *et al.* [7] and Mei *et al.* [12], which are both external approaches. Although our approach in spirit is also an external approach, it has several significant differences from these approaches, in terms of both generating candidate labels and ranking them for labelling: (1) We generate candidate labels from different sources and via a different approach; we employing user summaries and apply dependency relation parsing to generate meaningful candidate labels. (2) We rank candidate labels differently; we represent topics and labels as distributions of documents rather than words to measure the semantic association (distance) between topics and candidate labels.

3   Labelling Topics with User Summaries

The problem of topic model labelling can be decomposed into two sub-problems: generating candidate labels from user summaries and ranking candidate labels. We describe these steps below.

3.1   Generating Candidate Labels by Dependency Relation Parsing

User summaries on web forums are generally short, containing single words and short phrases, but also long phrases and sentences. To label the topics for a document collection, we aim to generate user interpretable words and phrases as candidate labels by analysing collectively all user summaries. Words as well as phrases of two or three words frequently appear in user summaries. Examples of such words and phrases include "hospital", "midwife", "comfort", "aged care" and "first class care". An important observation is that *the user generated succinct expressions of single words or short phrases are generally nouns or noun phrases and express meanings easily understandable by humans*. In our application of this observation, we set short phrases to contain at most three words that appear frequently (frequency of at least four by default) in user summaries as candidate labels.

For longer summaries that are sentences and phrases of more than three words, we apply the typed dependency parser [4] to analyse summaries and generate candidate labels. The typed dependency relation parser has been shown

Table 1. Examples for dependency relation parsing

User summary	POS tag	Typed dependency	Candidate label
Help from nurses on maternity ward	help/VB, from/IN, nurses/NNS, on/IN, maternity/JJ, ward/NN	root(ROOT-0, help-1), prep_from(help-1, nurses-3), amod(ward-6, maternity-5), prep_on(nurses-3, ward-6)	Maternity ward
Medical staff at hospital	medical/JJ, staff/NN, at/IN, hospital/NN	amod(staff-2, medical-1), root(ROOT-0, staff-2), prep_at(staff-2, hospital-4)	Medical staff

to be an effective tool for analyzing short informal text [23]. With typed dependency relation parsing, a sentence is represented as a set of dependency relations between pairs of words in the form of *(head, dependent)*, where content words are chosen as heads, and other related words depend on the heads. Table 1 shows some examples of candidate labels generated from user summaries by the dependency relation parser. The numbers after each word indicate the position of the word in the sentence. The word pairs that form dependency relations are listed in the third column, where *root*, *prep_from*, *amod* and *prep_on* are dependency relation types. For example *amod(ward-6, maternity-5)* represents that the word pair "maternity ward" forms an *adjective modifying* dependency relation. Details of the dependency types are described in [4].

After dependency relation parsing, dependency relations for adjacent words, and where at least one word is a noun, are selected as candidate labels. This strategy lends support to the idea that the noun in a dependency relation generally expresses content and adjacent words indicate that the relation forms noun phrases. For example in Table 1, for the user summary "help from nurses on maternity ward", nouns *nurses* and *ward* are identified, from the POS tag. Among the dependency relations on these nouns, which are *prep_from(help-1, nurses-3)*, *prep_on(nurses-3, ward-6)* and *amod(ward-6, maternity-5)*, *maternity ward* is extracted as a candidate label since *ward* is a noun and the two adjacent words *(maternity-5, ward-6)* form a noun phrase.

Noun phrases generated by the dependency relation parser form candidate labels. These candidate labels may contain errors. We apply a simple yet effective strategy to remove noises—candidate labels occurring in less than a frequency threshold (by default 1%) in the user summaries are considered as noise and filtered out.

3.2 Labelling Topics with Candidate Labels

According to Mei et al. [12], a *topic* θ is a probability distribution of words $p(w \in V|\theta)$ where V is a vocabulary, and $\sum_{w \in V} p(w|\theta) = 1$. A topic label, or simply *label*, l for a topic θ is a sequence of words or phrases that is semantically meaningful and covers the latent meaning of θ. The *relevance score* of a label to a topic $s(\theta, l)$ measures the semantic similarity between θ and l. Note that topics are generated from documents, and the candidate labels are generated from user summaries of the documents.

We present our metric to measure the semantic association between candidate labels generated from user summaries and the topics produced by topic models, in order to rank candidate labels for topics. We employ Kullback Leibler (KL) divergence [6] as the measure of the difference between two probability distributions.

On social media websites like discussion forums, user summaries or comments are for specific documents. As a result phrases generated from user summaries as candidate labels are associated with a multinomial distribution of documents. On the other hand, from the document-topic probability distribution matrix a topic can also be represented as a distribution of documents. Given θ and l, let P_θ and Q_l denote respectively the document probability distribution for θ and l. The relevance score of a candidate label, l, for topic, θ, is defined as the opposite of the KL divergence between the document distributions of l and θ:

$$s(\theta, l) = -D_{KL}(P_\theta || Q_l) = -\sum_i P_\theta(i) * ln \frac{P_\theta(i)}{Q_l(i)}. \tag{1}$$

In the above definition, i denotes the documents under consideration.

Given a topic θ and a set of candidate labels, $s(\theta, l)$ is computed for each candidate label, and the top k labels with the highest relevance scores for topic θ become the labels for the topic. By default $k = 2$, as it is shown in the literature that two phrases are preferred by human annotators as labels and generally have high semantic consistency [12].

4   Analysing the Patient Opinion Australia Stories

The Patient Opinion Australia (POA) is a web forum wherein patients post their stories about health services. In addition to user stories (documents), the POA website allows patients to post summaries (called "Story summary"), that are divided into two types: positive summaries are entered in the field

Table 2. The number of stories with different types of user summaries

Type of user summaries	# of stories (%)
Positive & negative user summaries	135 (21.67 %)
Only positive summaries	333 (53.45 %)
Only negative summaries	125 (20.06 %)
No summaries	30 (4.82 %)
Total	623

"What is Good", and negative summaries are entered in the field "What could be improved". As shown in Table 2, documents may have one or both types of summaries.

We crawled data from the POA website to evaluate our approach, Labelling By Summaries (LBS). We applied the non-parametric LDA topic model [2] to generate 50 topics for collected documents. In later discussions, these 50 topics are numbered as 0..49 as the results of the topic modelling output. Following the topic modelling convention, each topic is represented by the top ten topic words with high marginal probability.

We benchmarked LBS against three other approaches:

- *Top Words*: the conventional approach that labels topics by the top k ($k = 10$) topic words with high marginal probability.
- *Top Word Re-rank*: the straight forward approach Re-rank where top topic words are re-ranked by their frequency in user summaries.
- *PHR*: the approach by Mei *et al.* [12] where the top 1000 bi-gram phrases from documents are candidate labels.

4.1 Data, Topics and Manual Labels

We crawled 623 stories and 593 summaries from the POA website in early August 2014. The number of stories that contained different types of user summaries is shown in Table 2. Stories can have positive ("What is good") or negative ("What could be improved"), or mixed user summaries. A small portion (4.82 %) did not have any user summaries. In terms of the language features, summaries include single words, short or long phrases, and complete sentences. The average length of user summaries was four words, while the longest summary contained 29 words. Five sample user summaries are listed in Table 3, where long phrases of at least four words and sentences are highlighted.

Table 3. Five sample user summaries

	Summary
Doc. 5	GP; bad care; disgusting; emotional wellbeing; fobbed off; lack of compassion; pain; pain relief; rude consultant; understanding
Doc. 24	addressing my concerns; earlier surgical appointment; good care; *private gynaecologist listening to my concerns*
Doc. 36	admission; anaesthetist; good experience; greeted warmly; *I felt I was in good hands*; nurse attitude; nurse care
Doc. 145	initial service; My GP; *staff in surgical ward*; doctor attitude; doctor care; lack of understanding; *treatment in emergency department*
Doc. 567	*communication within RSL Care*; consistancy; *employing people who know how to clean properly*

Table 4. Sample topic labels by different approaches

Topic	Top 10 topic words	LBS	Top Words	Top Word Re-rank	PHR	Human
0	october; appointment; outpatient; referral; letter; answer; confirmed; september; ulcer; telephoned	administration; appointment letter	appointment; letter	appointment; referral	appointment within; first appoint-ment	appointment letter; referral letter
7	neck; wrist; thumb; treatment; brace; report; painkiller; agony; file; skin	medical record; post surgery	neck; wrist	x-ray; neck	neck brace; neck pain	medical record; surgery care
8	healthy; program; kate; lifestyle; men; programme; encouraging; healthier; shed; meet	lifestyle program; continuity of care	healthy; program	program; lifestyle	lifestyle program; healthy lifestyle	healthy lifestyle; lifestyle program
22	wife; hearing; urgency; ENT; surgeries; genetic; westmead; loss; costs; aids	hospital; initial diagnosis	wife; hearing	ENT; loss	hearing loss; hearing aids	ENT emergency; emergency service

It is well recognised that topics produced from topic models may sometimes be statistically important but not convey much semantically consistent information [14,16]. These topics should be discarded rather than labelled. We aim to automatically remove the semantically incoherent topics based on user summaries. Intuitively, individual words in phrases from user summaries are semantically associated. So, topics that break a phrase in user summaries – not every word in a phrase appears as a top topic word for a topic – indicates that the topic lacks semantic coherence. Generally the more phrases broken by a topic the less likely that the topic is semantically coherent. To control noise in user summaries, only frequent phrases (frequency is set to 4) in user summaries are considered. In total we extracted 94 two-word phrases from user summaries.

Applying the frequent user phases to filter incoherent topics, 9 out of 50 topics were filtered out. The remaining 41 topics were then manually labelled by five human annotators as follows: for each topic, the top 10 candidate labels generated by each of the approaches, LBS, Top Words (or Top Word Re-rank), and PHR, were pooled together. As a result approximately 30 (there may be duplicates among the three models) candidate labels (words or phrases), were randomised and suggested to annotators for manual labelling. The full-text stories (documents) were also presented at the side to help understand the context.

Table 5. Ratings by assessors of topic labels by different approaches. Rating 3 indicates "very good label" whereas 0 indicates "completely inappropriate label"

Topic	Approaches	Ratings					Avg rating
0	LBS	3	2	3	1	3	2.4
	Top Words	2	2	2	2	2	2.0
	Top Word Re-rank	2	2	2	2	2	2.0
	PHR	1	1	0	0	1	0.6
7	LBS	3	1	3	3	3	2.6
	Top Words	2	1	1	2	2	1.6
	Top Word Re-rank	2	2	1	2	1	1.6
	PHR	1	2	2	1	1	1.4
8	LBS	3	3	3	3	3	3.0
	Top Words	2	2	2	2	2	2.0
	Top Word Re-rank	2	1	2	2	1	1.6
	PHR	3	3	3	3	3	3.0
22	LBS	2	2	1	1	2	1.6
	Top Words	0	0	0	0	1	0.2
	Top Word Re-rank	2	1	2	2	1	1.6
	PHR	2	3	3	3	1	2.4

It is hard for annotators to unanimously choose the same best label for a topic and so each annotator was asked to choose the top two candidates for each topic and the top two candidates frequently chosen by annotators were deemed the label for the topic. In equal frequency cases, the topic label was resolved by a discussion by the first two authors. Table 4 presents some sample topics and their manual labels (indicated as Human). It is worth noting that all human-suggested labels are phrases, which is consistent with previous finding that phrases are preferred labels for topics and easier for humans to understand [7,12].

4.2 LBS Labels: Quantitative Analysis

We benchmarked LBS against other approaches for automatic labelling. For all approaches, the top two ranked words/phrases were used to label each of the 41 topics. To quantitatively measure the quality of automatic labels, we asked human assessors to rate the labels by each approach. To avoid bias due to presentation, we presented the four types of labels by different methods in random orders. To help a human assessor to interpret the topics, the top three documents with the highest marginal probability for the topic were also presented. Given the labels for a topic, an assessor was asked to rate the labels according to the quality of the labels. The rating levels are: *3: very good label; 2: reasonable label; 1: somewhat related, but bad as a topic label; 0: completely inappropriate topic label.* Each topic was assessed by five assessors who were volunteers, different from the previous assessors for human labelling. For each topic, the ratings of five assessors were averaged to compute the rating for each approach.

We first examined the variance of ratings for topic labels across assessors. Table 5 lists the ratings by assessors of topic labels by different approaches for the four sample topics in Table 4. The table shows that, although assessors rated an approach differently for different topics, they rated an approach surprisingly consistently for a specific topic. For example, PHR received consistent ratings of two 0 s or three 1 s for Topic 0 but consistent ratings of five 3 s for Topic 8. So, we can safely say that the ratings by assessors are reliable. We next discuss the overall performance of all approaches for 41 topics.

Table 6. Quantitative evaluation of LBS labels

	LBS	Top Words	Top Word Re-rank	PHR
Avg rating	1.815	1.322	1.341	1.356
p-value	-	0.0016	0.0009	0.0138
correct labels	34	26	29	24

We compared the ratings for LBS with each of the three other approaches for 41 topics using the Wilcoxon signed-rank test [21]. The overall average ratings for each approach and the Wilcoxon signed-rank test p-value for the other three approaches are shown in the first two rows of Table 6. LBS has the highest average of 1.815, indicating that human assessors judge the labels as in between *somewhat related* to *reasonable label*. All three other approaches have significantly lower average ratings ($p < 0.05$) than LBS. In other words the labels are deemed by assessors as nearly *somewhat related but bad as a topic label*. Note that PHR, using phrases as labels, has better average ratings overall than the other two approaches, which are both based on words. This result confirms that using phrases as topic labels are more understandable to humans [11,13,19]. On the other hand, as PHR generates topic labels from bi-grams in documents rather than from user summaries, the phrases are not always meaningful and this is reflected in their significantly lower ratings by assessors than those for LBS.

The automatic labels of all approaches are also compared with the human labels. The last row of Table 6 shows the number of correct labels generated by automatic approaches compared with the human-generated labels for the 41 topics. We asked human assessors to judge the labels by each approach compared with the human-generated labels. Each label was assessed by five assessors who were volunteers, different from the previous assessors. A label is considered correct if at least three assessors vote the label as correct. Clearly LBS has the largest number of 34 topics with correct labels, indicating a high level of consistency with the human-generated labels. In contrast, top topic words and PHR has the lowest number of 26 and 24 topics with correct labels, respectively, and a low level of consistency with the human-generated labels.

4.3 LBS Labels: Qualitative Analysis

Table 4 lists automatic labels generated by different approaches compared with the human-generated labels for some sample topics. We can see that the automatic labels generated by LBS very often do not appear at all in the word list for topics. Interestingly the labels generated by LBS can very often conceptually generalise the user phrases captured by topics. For example for Topic 0, the LBS label "appointment letter" generalises conceptually the topic words "referral" and "letter". This result can be attributed to the use in LBS of the dependency relation parser of user-generated summaries, which can produce meaningful candidate labels that are more understandable to humans.

Comparing the labels by LBS with the manual labels, it is clear from Table 4 that the LBS generated phrase labels can more accurately capture the meaning of topics. For example for Topic 7, the LBS labels of "medical record" and "post surgery" closely match the human-generated labels of "medical record" and "surgery care". As is also shown in Table 6, LBS has the highest number of labels matching the human-generated labels.

Table 4 also sheds light on the performance of automatic labels by other approaches. Take again Topic 7 as an example. Only LBS correctly labelled it as "medical record; post surgery", while the labels by other approaches are not related to "medical reports; surgery care", but rather only describe detailed aspects of medical reports and surgery care. The word-based Top Words and Top Word Re-rank approaches label the topic as "x-ray", "neck", "wrist", which is difficult for humans to interpret. The PHR phrase approach can generate more meaningful labels like "neck pain", which are phrases extracted from the document collection but are not high level summarisation of the document content.

5 Conclusions

The problem of automatic labelling for topic models with human understandable labels is crucial for the wide applications of topic modelling for many text analysis tasks. Existing work on automatic labelling for topic models either relies on the document collection itself or external web-based resources. In this paper we proposed a novel approach of making use of user summaries to label topic models. We made several contributions: (1) We proposed the novel application of dependency relation parsing to extract meaningful and human understandable phrases from user summaries as candidate labels. (2) We proposed KL divergence to measure the semantic similarity between candidate labels and topics and rank candidate labels. (3) We applied our automatic topic labelling approach to analyse a real-world user discussion forum for healthcare. Results show that our approach can generate meaningful and human interpretable topic labels.

For future work we will design further refinements to our approach in terms of deploying user input. It would also be interesting to extend our approach to other web-based forums for products and services.

Acknowledgements. Thanks go to Mr. Bin Lu for crawling the Patient Opinion Australia website and preliminary analysis of the data. Thanks also go to Associate Professor Michael Greco of Patient Opinion Australia for explanation of the data.

References

1. Blei, D.M., Ng, A.Y., Jordan, M.I.: Latent Dirichlet allocation. J. Mach. Learn. Res. **3**, 993–1022 (2003)
2. Buntine, W.L., Mishra, S.: Experiments with non-parametric topic models. In: Proceeding KDD 2014, pp. 881–890. ACM (2014)
3. Chang, J., Gerrish, S., Wang, C., Boyd-graber, J.L., Blei, D.M.: Reading tea leaves: how humans interpret topic models. In: Proceeding Advances in Neural Information Processing Systems, pp. 288–296 (2009)
4. De Marneffe, M.C., Manning, C.D.: The stanford typed dependencies representation. In: Proceeding of the workshop on Cross-Framework and Cross-Domain Parser Evaluation, pp. 1–8. Association for Computational Linguistics (2008)
5. Griffiths, T.L., Steyvers, M.: Finding scientific topics. Proc. Nat. Acad. Sci. U.S.A. **101**(1), 5228–5235 (2004)
6. Kullback, S., Leibler, R.A.: On information and sufficiency. Ann. Math. Stat. **22**(1), 79–86 (1951)
7. Lau, J.H., Grieser, K., Newman, D., Baldwin, T.: Automatic labelling of topic models. Proc. ACL HLT **2011**, 1536–1545 (2011)
8. Lau, J.H., Newman, D., Karimi, S., Baldwin, T.: Best topic word selection for topic labelling. In: Proceeding COLING 2016: Posters, pp. 605–613 (2010)
9. Li, Q., Wang, J., Chen, Y.P., Lin, Z.: User comments for news recommendation in forum-based social media. Inf. Sci. **180**(24), 4929–4939 (2010)
10. Magatti, D., Calegari, S., Ciucci, D., Stella, F.: Automatic labeling of topics. In: Proceeding 9th International Conference on ISDA, pp. 1227–1232. IEEE (2009)
11. Mei, Q., Liu, C., Su, H., Zhai, C.: A probabilistic approach to spatiotemporal theme pattern mining on weblogs. In: Proceeding of WWW 2006. ACM (2006)
12. Mei, Q., Shen, X., Zhai, C.: Automatic labeling of multinomial topic models. In: Proceeding KDD 2007, pp. 490–499. ACM (2007)
13. Mei, Q., Zhai, C.: A mixture model for contextual text mining. In: Proceeding KDD 2006, pp. 649–655. ACM (2006)
14. Mimno, D., Wallach, H.M., Talley, E., Leenders, M., McCallum, A.: Optimizing semantic coherence in topic models. Proc. EMNLP **2011**, 262–272 (2011)
15. Newman, D., Karimi, S., Cavedon, L., Kay, J., Thomas, P., Trotman, A.: External evaluation of topic models. Proc. ADCS **2009**, 1–8 (2009)
16. Newman, D., Lau, J.H., Grieser, K., Baldwin, T.: Automatic evaluation of topic coherence. Proc. NAACL HLT **2010**, 100–108 (2010)
17. Steyvers, M., Smyth, P., Rosen-Zvi, M., Griffiths, T.: Probabilistic author-topic models for information discovery. In: Proceeding KDD 2004, pp. 306–315. ACM (2004)
18. Tatar, A., Leguay, J., Antoniadis, P., Limbourg, A., de Amorim, M.D., Fdida, S.: Predicting the popularity of online articles based on user comments. In: Proceeding International Conference on Web Intelligence, Mining and Semantics, p. 67. ACM (2011)
19. Wang, X., McCallum, A.: Topics over time: a non-markov continuous-time model of topical trends. In: Proceeding KDD 2006, pp. 424–433. ACM (2006)
20. Wei, X., Croft, W.B.: LDA-based document models for ad-hoc retrieval. In: Proceeding SIGIR 2006, pp. 178–185. ACM (2006)

21. Wilcoxon, F.: Individual comparisons by ranking methods. Biometrics Bull. **1**(6), 80–83 (1945)
22. Zhai, C., Lafferty, J.: Model-based feedback in the language modeling approach to information retrieval. Proc. CIKM **2001**, 403–410 (2001)
23. Zhang, X., Cui, L., Wang, Y.: Commtrust: computing multi-dimensional trust by mining E-commerce feedback comments. Ieee Trans. Knowl. Data Eng. **26**(7), 1631–1643 (2014)

Content-Based Top-N Recommendation Using Heterogeneous Relations

Yifan Chen[1], Xiang Zhao[1,3(✉)], Junjiao Gan[2], Junkai Ren[1], and Yanli Hu[1,3]

[1] National University of Defense Technology, Changsha, China
xiangzhao@nudt.edu.cn
[2] Massachusetts Institute of Technology, Cambridge, USA
[3] Collaborative Innovation Center of Geospatial Technology, Wuhan, China

Abstract. Top-N recommender systems have been extensively studied. However, the sparsity of user-item activities has not been well resolved. While many hybrid systems were proposed to address the *cold-start* problem, the profile information has not been sufficiently leveraged. Furthermore, the heterogeneity of profiles between users and items intensifies the challenge. In this paper, we propose a content-based top-N recommender system by learning the global term weights in profiles. To achieve this, we bring in PathSim, which could well measure the node similarity with heterogeneous relations (between users and items). Starting from the original TF-IDF value, the global term weights gradually converge, and eventually reflect both profile and activity information. To facilitate training, the derivative is reformulated into matrix form, which could easily be paralleled. We conduct extensive experiments, which demonstrate the superiority of the proposed method.

1 Introduction

Recommender systems typically leverage two types of signals to effectively recommend *items* to *users* - user activities, and content matching between user and item profiles. Depending on what to use, the recommendation models in literature are usually categorized into collaborative filtering, content-based and hybrid models [1]. In real-world applications, solely employing collaborative filtering or content-based models can not achieve desirable results, as is often the case that single source of information tends to be incomplete.

To better illustrate, we motivate the following example in architecture. Recently, Vanke, a leading real-estate corporation in China, started a Uber-For-Architects project, namely NOSPPP, which tries to match architects with appropriate projects based on previous project information of architects and firms. During the running of NOSPPP, the data collected are two-fold, with the participated projects and the resumes, respectively. In terms of recommender system, the former is named "feedback" (of users) while the latter is referred as "profiles" (of items). Due to the sparsity of feedback, collaborative filtering based recommenders would face the *cold-start* problem, and hence, we have to resort

© Springer International Publishing AG 2016
M.A. Cheema et al. (Eds.): ADC 2016, LNCS 9877, pp. 308–320, 2016.
DOI: 10.1007/978-3-319-46922-5_24

to content-based or hybrid models [8]. However, unlike the applicant-job scenario therein, where the profiles of users and jobs could well match, the profiles of architects and projects describe things in two different worlds. Specifically, the profiles of architects presents the working experience and skills, while the profiles of projects tells the area, the interior and exterior constructions, etc. This is rational, as designing architectures is in the form of art, where it is hard to specify the conditions or requirements through decomposition. In response to applications like NOSPPP, we explore recommendation utilizing both *sparse feedback* and *heterogeneous profiles*.

In this paper, we exploit a hybrid recommendation method that ensembles both sources of information. For ease of exposition, we consider the case that auxiliary information exists only on the side of items, and propose a item-based top-N recommendation algorithm[1]. Classic item-based collaborative filtering uses the direct link information for recommendation without diffusing the influence of other user-item links. By regarding user-item interactions as a bi-type information network, we observe that such influence can be captured by item node similarity, where PathSim [18] via *meta-path* is served. Moreover, while content matching between heterogeneous profiles of users and items does not produce explicable results, methods including [19–23] suggest high similarity among objects within the same subspace, thus we contend that it can be employed for matching profiles between item-item or user-user, since profiles of same type are naturally homogeneous. A standard way to measure similarity between two profiles is computing the cosine similarity of the two bags of words, and each word is weighted by term frequency tf (within the document) × inverted document frequency idf (of the term within the corpus). While the local term frequency could be computed offline, it has been suggested [8] that the global term weights idf requires further optimization to achieve better precision. Thus, we optimize the global term weights with the guidance of the similarity from PathSim.

In summary, the major contribution of the paper is a novel hybrid recommendation method, the overview of which is outlined as follows:

(1) Derive item similarity measured by meta-paths using PathSim;
(2) Optimize the global term weights guided by PathSim; and
(3) Recommend top-N items based on nearest neighbor collaborative filtering.

Organization. Section 2 discusses related work. We present the method for deriving initial similarity between items in Sect. 3, and then, introduce the learning method for optimizing global term weights in Sect. 4. Experiment results are in Sect. 5, followed by conclusion in Sect. 6.

2   Related Work

Top-N recommender systems have been extensively studied during the last few years, which could be classified into two categories.

---

[1] Without loss of generality, it is straightforward to extend the idea to the case of auxiliary information on both sides of users and items.

The first category is neighborhood-based collaborative filtering, which could be further classified into three classes: item-based and user-based. Given a certain user, *user-based-nearest-neighbor* (userkNN) [7,11,15] first identifies a set of similar users, and then recommends top-N items based on what items those similar users have purchased. Similarly, *item-based-nearest-neighbor* (itemkNN) [16] identifies a set of similar items for each of the items that the user has purchased, and then recommends top-N items based on those similar items. There are plenty of ways to measure user/item similarity, e.g., Pearson correlation, cosine similarity, and so forth.

The second category is model-based collaborative filtering, in which the latent factor models have achieved the state-of-the-art performance. Cremonesi et al. proposed a simple model-based algorithm PureSVD [6], where users' features and items' features are represented by the principle singular vectors of the user-item matrix. Koren proposed the well-known SVD++ model [6]. Wu applied Regularized Matrix Factorization (RMF), Maximum Margin Matrix Factorization (MMMF), and Nonnegative Matrix Factorization (NMF) to recommender systems. Weighted Regularized Matrix Factorization (WRMF) was introduced by Hu et al. [9]. The key idea of these methods is to factorize the user-item matrix to represent the users preferences and items characteristics in a common latent space, and then estimate the user-item matrix by the dot product of user factors and item factors. All these methods assume that only a few variables impact users preference and items features, which means the low-rank structure of user-item matrix.

Another model-based method, SLIM, proposed by Ning et al. [13], predicts the user-item matrix by multiplying the observed user-item matrix by the aggregation coefficient matrix. SLIM estimates the coefficient matrix by learning from the observed user-item matrix with a simultaneous regression model. Specifically, it introduces sparsity with ℓ_1-norm regularizer into the regularized optimization and formed an elastic net problem to benefit from the smoothness of ℓ_2-norm and the sparsity of ℓ_1-norm. Later, plenty of research has been done based on SLIM. SSLIM [14] integrates the side information. LorSLIM [4] involves the nuclear-norm to induce the low-rank property of SLIM. HOSLIM [5] uses the potential higher-order information to generate better recommendation.

3   Initializing Item Similarity

This section presents the method for measuring item similarity.

3.1   Preliminaries

Definition 1 (*User-Item Network*). *User-item network is a bi-type information network, where nodes are of two types and in the form of $G = \langle \{U, I\}, E, W \rangle$. $\{U, I\}$ represents the nodes of the network, where U represents all users and I represents all items. We assume the user-item network is merely derived from rating matrix, that is, E contains only links between user and item. W is the set of weights values on the links.*

We use meta-path [18] to represent the relations in user-item network. Given a meta-path $\mathcal{P}$, the PathSim between two nodes s and t can be calculated as follows:

$$S_{\mathcal{P}}(s,t) = \frac{2S_{\mathcal{P}}(s,t)}{S_P(s,:) + S_{\mathcal{P}^{-1}}(:,t)}. \tag{1}$$

In the above, $S_{\mathcal{P}}(s,t)$ is a Path Count measure and can be calculated as the number of path instances between s and t. $\mathcal{P}^{-1}$ denotes the reverse meta-path of $\mathcal{P}$. $S_{\mathcal{P}}(s,:)$ denotes the path count value following $\mathcal{P}$ and starting with s; and $S_{\mathcal{P}}(s,:)$ denotes the path count value following $\mathcal{P}^{-1}$ ending with t.

We proceed to define the meta path. As in the case of user-item network, towards item-item relation, the meta path can be only in the form of $\mathcal{P}_n = (I(UI)^n)$. For instance, $n = 1$ corresponds to $\mathcal{P}_1 = (IUI)$ and $n = 2$ corresponds to $\mathcal{P}_2 = (IUIUI)$. It is easy to verify the symmetry of $\mathcal{P}_n$, which is required by the original definition of PathSim. Similarly, the meta path for user-user relation could be formulated as $(U(IU)^n)$.

Suppose we define N_p meta paths $\mathcal{P}_1, \mathcal{P}_2, \cdots, \mathcal{P}_{N_p}$, with the corresponding similarities $\boldsymbol{S}_1, \boldsymbol{S}_2, \cdots, \boldsymbol{S}_{N_p}$, the overall similarity should be measured as the weighted aggregation, e.g. $\boldsymbol{S} = \sum_{p=1}^{N_p} \beta_p \boldsymbol{S}_p$, where $\sum_{n=1}^{N_p} \beta_p = 1$. As is suggested in [18] that the meta path with relatively short length is good enough to measure similarity, and a long meta path may even reduce the quality, we set smaller weights for longer meta path.

3.2   Measuring Item Similarity

To measure item similarity through PathSim, we first define the meta-path in the form of $\mathcal{P}_n = (A(BA)^n)$ (the mined frequent patterns [3]). For instance, $n = 1$ corresponds to $\mathcal{P}_1 = (ABA)$ and $n = 2$ corresponds to $\mathcal{P}_2 = (ABABA)$. It is easy to verify the symmetry of $\mathcal{P}_n$ and thus PathSim can be applied. The associated commuting matrix for $\mathcal{P}_n$ is $M = (W_{AB}W_{BA})^n$ and consequently the similarity between item i and item j can be computed by Eq. (1).

Suppose we define N meta-paths $\mathcal{P}_1, \mathcal{P}_2, \ldots, \mathcal{P}_N$, with the corresponding similarities $s_1, s_2, \ldots, s_N$, the overall similarity should be measured as the weighted aggregation, e.g. $s = \sum_{n=1}^{N} \alpha_n s_n$, where $\sum_{n=1}^{N} \alpha_n = 1$. As is suggested in [18] that the meta-path with relatively short length is good enough to measure similarity, and a long meta-path may even reduce the quality, we set smaller weights for longer meta-paths. We naturally set the weights as $\alpha_n = \frac{2^{N-n}}{2^N-1}$. We further denote $\boldsymbol{S}^p$ for the matrix of PathSim, where s_{ij}^p represents the element of $\boldsymbol{S}^p$ in the ith row and j^{th} column.

4   Optimizing Profile Similarity

We prompt to measure the item similarity based on the profiles. Prior to the discussion, we first list the notations used in this section in Table 1. Note that vectors and matrices are made **bold**.

Table 1. Table of notations

N_u	Number of users
N_v	Number of items
N_w	Number of terms
λ	ℓ_2 Norm weight
$\boldsymbol{S}^f, s_{ij}^f$	Similarity derived from item profiles
$\boldsymbol{S}^p, s_{ij}^p$	Similarity derived form pathsim
$\boldsymbol{W}^l, \boldsymbol{w}_i^l, w_{ik}^l$	Local term weights
$\boldsymbol{w}^g, w_{ik}^g$	Global term weights
$\boldsymbol{w}, w_k$	The weights to learn, where $w_k = (w_k^g)^2$
$\boldsymbol{P}, \boldsymbol{p}_k$	The normalized $tf \times idf$ weights

Each profile contains rich text to describe the feature. Thus more effective content analysis methods and text similarity measures are crucial for the recommendation. Most designed recommender systems involving text similarity measure applied cosine similarity of two bags of words, where each word is weighted by $tf \times idf$ [2,17]. Nevertheless, it is possible to go beyond the definition of $tf \times idf$, where tf represents the local term weights and idf the global term weights. While tf could be derived offline with various methods, idf requires further optimization as suggested by [8]. Thus the global term could be optimized with the guidance of PathSim and the similarity derived from profiles could be calculated by the following Equation:

$$s_{ij}^f = \frac{\boldsymbol{d}_i \cdot \boldsymbol{d}_j}{\|\boldsymbol{d}_i\|_2 \|\boldsymbol{d}_j\|_2}, \tag{2}$$

where $\|\cdot\|_2$ is the ℓ_2 norm of a vector, and $\boldsymbol{d}_i$ represents the term vector, each dimension represents a term, and the value in each dimension represents the weight of the term. $\boldsymbol{d}_i$ could be decomposed as $\boldsymbol{w}_i^l \circ \boldsymbol{w}^g$, where $\boldsymbol{w}_i^g$ denotes for the local weights for item i. $\boldsymbol{w}^g$ denotes for the global term weights, which is initially set with the original Inverted Document Frequency, and optimized gradually. $\circ$ is a binary operation, conducting the element-wise product of two vectors, thus the result is also a vector. By letting $w_k = (w_k^g)^2$, we could further formalize Eq. 2 as follows:

$$s_{ij}^f = \frac{\sum_{k=1}^t w_{ik}^l w_{jk}^l w_k}{\left[\sum_{k=1}^t (w_{ik}^l)^2 w_k\right]^{\frac{1}{2}} \left[\sum_{k=1}^t (w_{jk}^l)^2 w_k\right]^{\frac{1}{2}}},$$

and the partial derivative could be derived as:

$$
\begin{aligned}
\frac{\partial s_{ij}^{f}}{\partial w_k} &= \frac{1}{\|\boldsymbol{d}_i\|_2^2 \|\boldsymbol{d}_j\|_2^2} \left\{ w_{ik}^l w_{jk}^l \|\boldsymbol{d}_i\|_2 \|\boldsymbol{d}_j\|_2 - \left[\frac{\|\boldsymbol{d}_j\|_2}{2\|\boldsymbol{d}_i\|_2}(w_{ik}^l)^2 + \frac{\|\boldsymbol{d}_i\|_2}{2\|\boldsymbol{d}_j\|_2}(w_{jk}^l)^2 \right] \boldsymbol{d}_i \cdot \boldsymbol{d}_j \right\} \\
&= \frac{w_{ik}^l w_{jk}^l}{\|\boldsymbol{d}_i\|_2 \|\boldsymbol{d}_j\|_2} - \frac{s_{ij}^f}{2} \left[\frac{(w_{ik}^l)^2}{\|\boldsymbol{d}_i\|_2^2} + \frac{(w_{jk}^l)^2}{\|\boldsymbol{d}_j\|_2^2} \right].
\end{aligned}
$$

To optimize the global term weights, we should define the loss function to measure the difference between s_{ij}^p and s_{ij}^f. We develop the squared loss function and the associated optimization methods.

4.1 Squared Error Loss Function

Due to the sparsity of the user-item information network, we could also expect the sparsity of similarities measured by PathSim. If item i can not reach item j through the bi-type information network, according to Eq. 1, $s_{ij} = 0$.

In this section, the loss function is defined as the squared error, given by

$$
\mathcal{L} = \sum_{i=1}^{N_v} \sum_{j=1}^{N_v} (s_{ij}^f - s_{ij}^p)^2 = \|\boldsymbol{S}^f - \boldsymbol{S}^p\|_F^2,
$$

based on which, we could minimize the following objective function to optimize the global term frequency:

$$
\min_{\boldsymbol{w}} J - \frac{1}{2}\|\boldsymbol{S}^f - \boldsymbol{S}^p\|_F^2 + \frac{\lambda}{2}\|\boldsymbol{w}\|_2^2 \tag{3}
$$
$$
s.t. \ \boldsymbol{w} \geq 0
$$

where $\|\cdot\|_F$ is the Frobenius norm, which is actually the squared sum of all elements of the matrix. $\boldsymbol{w}$ stands for the vector of w_k, and we penalize ℓ_2 norm on the global term weights $\boldsymbol{w}$ to avoid over fitting and sparsity result. $\boldsymbol{S}^p$ is denoted for PathSim matrix, whereas $\boldsymbol{S}^f$ for the profile similarity matrix. We reformulate the problem into the following element-wise form, to facilitate the deduction of partial derivative over w_k, e.g., $\frac{\partial J}{\partial w_k}$.

$$
\min_{w_k} J = \frac{1}{2} \sum_{i=1}^{N_v} \sum_{j=1}^{N_v} (s_{ij}^f - s_{ij}^p)^2 + \frac{\lambda}{2} \sum_{k=1}^{N_w} w_k^2
$$
$$
w_k \geq 0, k = 1, \ldots, N_w
$$

Solution. The partial derivative is given in Eq. 4.

$$
\frac{\partial J}{\partial w_k} = \sum_{i=1}^{N_v} \sum_{j=1}^{N_v} (s_{ij}^f - s_{ij}^p) \left\{ \frac{w_{ik}^l w_{jk}^l}{\|\boldsymbol{d}_i\|_2 \|\boldsymbol{d}_j\|_2} - \frac{\tilde{s}_{ij}}{2} \left[\frac{(w_{ik}^l)^2}{\|\boldsymbol{d}_i\|_2^2} + \frac{(w_{jk}^l)^2}{\|\boldsymbol{d}_j\|_2^2} \right] \right\} + \lambda w_k. \tag{4}
$$

We further define $q_{ij} = s_{ij}^f - s_{ij}^p, p_{ik} = \frac{w_{ij}^l}{\|d_i\|_2}$ and $r_{ij} = (s_{ij}^f - s_{ij}^p)s_{ij}^f$. Thus we have:

$$\sum_{i=1}^{N_v}\sum_{j=1}^{N_v}(s_{ij}^f - s_{ij}^p)\frac{w_{ik}^l w_{jk}^l}{\|d_i\|_2\|d_j\|_2} = \sum_{i=1}^{N_v}\sum_{j=1}^{N_v} q_{ij}p_{ik}p_{jk} = \boldsymbol{p}_k^T \boldsymbol{Q}\boldsymbol{p}_k$$

$$\sum_{i=1}^{N_v}\sum_{j=1}^{N_v}(s_{ij}^f - s_{ij}^p)\frac{s_{ij}^f}{2}\left[\frac{(w_{ik}^l)^2}{\|d_i\|_2^2} + \frac{(w_{jk}^l)^2}{\|d_j\|_2^2}\right] = \sum_{i=1}^{N_v}\sum_{j=1}^{N_v}\frac{1}{2}r_{ij}(p_{ik}^2 + p_{jk}^2)$$

$$= \sum_{i=1}^{N_v}\sum_{j=1}^{N_v} r_{ij}p_{ik}^2 = \boldsymbol{p}_k^T \boldsymbol{R}\boldsymbol{p}_k,$$

where $\boldsymbol{p}_k$ is a vector of p_{ik}, $\boldsymbol{Q}$ is $N_v \times N_v$ matrix of q_{ij} and $\boldsymbol{R}$ is a diagonal matrix with the i-th element of principal diagonal equals $\sum_j r_{ij}$. By defining $\boldsymbol{L} = \boldsymbol{Q} - \boldsymbol{R}$, we find the following close form of derivative:

$$\frac{\partial J}{\partial w_k} = \boldsymbol{p}_k^T \boldsymbol{L}\boldsymbol{p}_k + \lambda w_k.$$

It could be further represented into the matrix form:

$$\frac{\partial J}{\partial \boldsymbol{w}} = \text{diag}(\boldsymbol{P}^T \boldsymbol{L}\boldsymbol{P}) + \lambda\boldsymbol{w}, \tag{5}$$

where $\text{diag}(\cdot)$ extracts the principal diagonal and form as a vector.

Following the common practices for top-N recommendation [10], the loss function is computed over all entries of $\boldsymbol{S}$. As for the non-negative constraint, we apply Projected gradient method [12] to solve it. The summation above contains $n \times n$ terms, namely all pairwise items in the dataset. To ensure good performance while achieve reasonable training time, the algorithm is paralleled by CUDA.

5   Experimental Evaluation

To evaluate our proposed method, extensive experiments have been conducted. However, due to space limitation, we only present part of the results.

5.1   Experiment Setup

The results reported in this section is based on the NIPS dataset[2]. It contains paper-author and paper-word matrices extracted from co-author network at the NIPS conference over 13 volumes. We regard authors as users, papers as items and the contents of papers as the profile of items. Thus the data has 2037 users (authors) and 1740 items (papers), where 13649 words have been extracted from

---

[2] http://www.cs.nyu.edu/~roweis/data.html.

the corpus of item profiles. The content of the papers is preprocessed such that all words are converted to lower case and stemmed and stop-words are removed. One may note that NIPS dataset is very sparse, that is, some author may publish only one or two papers, which shows the importance of properly leveraging side information for recommendation.

We applied 5-time Leave-One-Out cross validation (LOOCV) to evaluate our proposed method. In each run, each of the dataset is split into a training set and a testing set by randomly selecting one of the non-zero entries of each user and placing it into the testing set. The training set is used to train a model, then for each user a size-N ranked list of recommended items is generated by the model. We varies N as $5, 10, 15, 20$ to compare the result difference. Our method has two parameters, n_p and λ. n_p measures the length of meta-path and λ measures the degree of regularization.

The recommendation quality is measured using Hit Rate (HR) and Average Reciprocal Hit Rank (ARHR) [7]. HR is defined as

$$HR = \frac{\#hits}{\#users},$$

where $\#users$ is the total number of users and $\#hits$ is the number of users whose item in the testing set is recommended (i.e., hit) in the size-N recommendation list. A second measure for evaluation is ARHR, which is defined as

$$ARHR = \frac{1}{\#users} \sum_{i=1}^{\#hits} \frac{1}{p_i},$$

where if an item of a user is hit, p is the position of the item in the ranked recommendation list. ARHR is a weighted version of HR and it measures how strongly an item is recommended, in which the weight is the reciprocal of the hit position in the recommendation list.

We implement our algorithm in C++. As our method involves optimizing the global weights over the whole vocabulary of item profiles, to expedite the training efficiency, the training process is paralleled in GPU and implemented by CUDA[3]. All experiments are done on a machine with 4-core Intel i7-4790 processor at 3.60 GHz and Nvidia GeForce GTX TITAN X graphics card.

5.2 Effect of Initial Value

We first evaluate the influence of initial value we set for global weights on the performance. We compare two settings, random and idf. The initial value is randomly set in the first setting while it is set as the value of inverted document frequency in the latter one. Here λ is set as 0.01 and n_p as 1. The result is reported in Fig. 1, which shows the superiority of idf over random. The result demonstrates the usefulness of side information in this dataset and we set the initial value of global term as idf thereafter.

---

[3] http://www.nvidia.cn/object/cuda-cn.html.

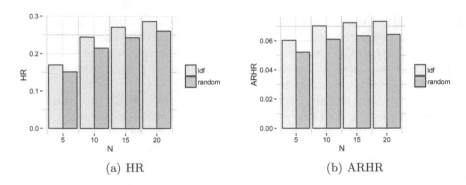

Fig. 1. Effect of initial value

5.3   Effect of Parameters

In this set of experiments, the validation is conducted to select the most suitable parameters. λ is varied from 0 to 0.05 and stepped 0.005 and n_p is set as 1,2,3. We draw the lines in Fig. 2. Three lines are drawn to distinguish $n_p = 1$ (red line), $n_p = 2$ (blue line) and $n_p = 3$ (green line) respectively. The result shows that n_p should be set 1 to achieve better performance. This result is consistent with [18], which suggests shorter length of meta-path is good enough to measure similarity.

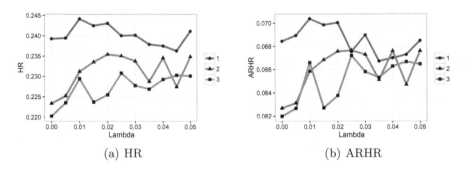

Fig. 2. Effect of parameter λ and n_p (Color figure online)

As Fig. 2 depicts the performance along with λ, we find the best value as 0.01 for $n_p = 1$, 0.02 for $n_p = 1$ and 0.025 for $n_p = 3$. It has also been shown that the method performs more robustly when $n_p = 1$ while it varies dramatically with λ when $n_p > 1$. Based on the observation, we finally pick n_p and λ as 1 and 0.01 for the rest of the experiments.

5.4   Recommendation for Different Top-N

By setting the global term as the inverted document frequency, and letting $\lambda = 0.01$, we evaluate the top-N recommendation performance, the result of which is illustrated in Fig. 3. Obviously, with the increase of N, the performance improves. We also compare the different setting of n_p, which further demonstrates $n_p = 1$ could be a better choice. We also found in this set of experiments that when N increase from 5 to 10, the performance shows relatively higher improvement.

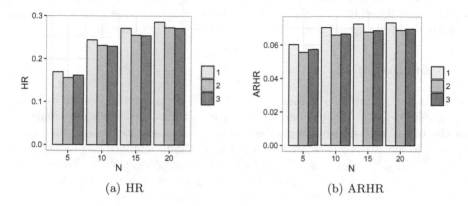

(a) HR   (b) ARHR

Fig. 3. Performance of Proposed Method

5.5   Comparison of Algorithms

We finally compares our method with other algorithms in this set of experiments. As top-N recommendation methods have been extensively studied, we compare only with some state-of-the-art methods, e.g. Slim [13] and LCE [17]. We also incorporate the pure tfidf method to calculate the item similarity for recommendation. To distinguish, we name our proposed method as Mist (Meta

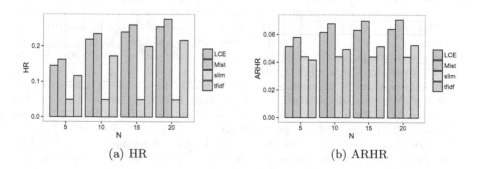

(a) HR   (b) ARHR

Fig. 4. Algorithm comparison

path based item similarity to learn global term weights). We depict the result in Fig. 4, where all the compared algorithms were optimized to the best settings.

Figure 4(a) shows the recommendation of Mist is consistently better than other three methods. Note that Slim has the worst performance, this could be attributed to the sparsity of dataset. LCE also took advantage of item profiles, thus it has achieved good performance. It is also worth noting that the pure tfidf shows relatively acceptable results and Mist could be regarded as the collaborative optimized tfidf.

When it comes to ARHR, showing in Fig. 4(b), Mist also behaves the best, which was followed by LCE, tfidf and Slim. In conclusion, the learned global term weights can well capture both structural and textual information.

6   Conclusion

In this paper, we proposed a content-based top-N recommender system by leveraging item profiles. We first employed PathSim to measure the item similarity on the top of heterogeneous relations between users and items, and then optimized the global term weights towards the PathSim similarities. To facilitate training, the derivation was reformulated into matrix form, which could easily be paralleled. We conducted extensive experiments, and the experimental results demonstrate the superiority of the proposed method.

Acknowledgment. X. Zhao was partially supported by NSFC 61402494 and NSF Hunan 2015JJ4009; Y. Hu was partially supported by NSFC 61302144.

References

1. Balabanović, M., Shoham, Y.: Fab: content-based, collaborative recommendation. Comm. ACM **40**(3), 66–72 (1997)
2. Barjasteh, I., Forsati, R., Masrour, F., Esfahanian, A., Radha, H.: Cold-start item and user recommendation with decoupled completion and transduction. In: Proceedings of the 9th ACM Conference on Recommender Systems, RecSys 2015, Vienna, Austria, 16–20 September 2015, pp. 91–98 (2015)
3. Chen, Y., Zhao, X., Lin, X., Wang, Y.: Towards frequent subgraph mining on single large uncertain graphs. In: 2015 IEEE International Conference on Data Mining, ICDM 2015, Atlantic City, NJ, USA, 14–17 November 2015, pp. 41–50 (2015)
4. Cheng, Y., Yin, L., Yu, Y.: LorSLIM: low rank sparse linear methods for top-n recommendations. In: 2014 IEEE International Conference on Data Mining, ICDM 2014, Shenzhen, China, 14–17 December 2014, pp. 90–99 (2014)
5. Christakopoulou, E., Karypis, G.: HOSLIM: higher-order sparse linear method for top-n recommender systems. In: Proceedings, Part II Advances in Knowledge Discovery and Data Mining - 18th Pacific-Asia Conference, PAKDD 2014, Tainan, Taiwan, 13–16 May 2014, pp. 38–49 (2014)
6. Cremonesi, P., Koren, Y., Turrin, R.: Performance of recommender algorithms on top-n recommendation tasks. In: Proceedings of the 2010 ACM Conference on Recommender Systems, RecSys 2010, Barcelona, Spain, 26–30 September 2010, pp. 39–46 (2010)

7. Deshpande, M., Karypis, G.: Item-based top-N recommendation algorithms. ACM Trans. Inf. Syst. **22**(1), 143–177 (2004)
8. Gu, Y., Zhao, B., Hardtke, D., Sun, Y.: Learning global term weights for content-based recommender systems. In: Proceedings of the 25th International Conference on World Wide Web, WWW 2016, Montreal, Canada, 11–15 April 2016. pp. 391–400 (2016)
9. Hu, Y., Koren, Y., Volinsky, C.: Collaborative filtering for implicit feedback datasets. In: Proceedings of the 8th IEEE International Conference on Data Mining (ICDM 2008), 15–19 December 2008, Pisa, Italy, pp. 263–272 (2008)
10. Kabbur, S., Ning, X., Karypis, G.: FISM: factored item similarity models for top-n recommender systems. In: The 19th ACM SIGKDD International Conference on Knowledge Discovery and Data Mining, KDD 2013, Chicago, IL, USA, 11–14 August 2013, pp. 659–667 (2013)
11. Karypis, G.: Evaluation of item-based top-n recommendation algorithms. In: Proceedings of the 2001 ACM CIKM International Conference on Information and Knowledge Management, Atlanta, Georgia, USA, 5–10 November 2001, pp. 247–254 (2001)
12. Lin, C.J.: Projected gradient methods for nonnegative matrix factorization. Neural Comput. **19**(10), 2756–2779 (2007)
13. Ning, X., Karypis, G.: SLIM: sparse linear methods for top-n recommender systems. In: 11th IEEE International Conference on Data Mining, ICDM 2011, Vancouver, BC, Canada, 11–14 December 2011, pp. 497–506 (2011)
14. Ning, X., Karypis, G.: Sparse linear methods with side information for top-n recommendations. In: Sixth ACM Conference on Recommender Systems, RecSys 2012, Dublin, Ireland, 9–13 September 2012, pp. 155–162 (2012)
15. Papagelis, M., Plexousakis, D.: Qualitative analysis of user-based and item-based prediction algorithms for recommendation agents. Eng. Appl. Artif. Intell. **18**(7), 781–789 (2005)
16. Resnick, P., Iacovou, N., Suchak, M., Bergstrom, P., Riedl, J.: Grouplens: an open architecture for collaborative filtering of netnews. In: CSCW 1994, Proceedings of the Conference on Computer Supported Cooperative Work, Chapel Hill, NC, USA, 22–26 October 1994, pp. 175–186 (1994)
17. Saveski, M., Mantrach, A.: Item cold-start recommendations: learning local collective embeddings. In: Eighth ACM Conference on Recommender Systems, RecSys 2014, Foster City, Silicon Valley, CA, USA, 06-10 October 2014, pp. 89–96 (2014)
18. Sun, Y., Han, J., Yan, X., Yu, P.S., Wu, T.: PathSim: meta path-based top-k similarity search in heterogeneous information networks. PVLDB **4**(11), 992–1003 (2011)
19. Wang, Y., Lin, X., Wu, L., Zhang, W.: Effective multi-query expansions: robust landmark retrieval. In: Proceedings of the 23rd Annual ACM Conference on Multimedia Conference, MM 2015, Brisbane, Australia, 26–30 October 2015, pp. 79–88 (2015)
20. Wang, Y., Lin, X., Wu, L., Zhang, W., Zhang, Q.: Exploiting correlation consensus: towards subspace clustering for multi-modal data. In: Proceedings of the ACM International Conference on Multimedia, MM 2014, Orlando, FL, USA, 03–07 November 2014, pp. 981–984 (2014)
21. Wang, Y., Lin, X., Wu, L., Zhang, W., Zhang, Q.: LBMCH: learning bridging mapping for cross-modal hashing. In: Proceedings of the 38th International ACM SIGIR Conference on Research and Development in Information Retrieval, Santiago, Chile, 9–13 August 2015, pp. 999–1002 (2015)

22. Wang, Y., Lin, X., Wu, L., Zhang, W., Zhang, Q., Huang, X.: Robust subspace clustering for multi-view data by exploiting correlation consensus. IEEE Trans. Image Process. **24**(11), 3939–3949 (2015)
23. Wang, Y., Zhang, W., Wu, L., Lin, X., Zhao, X.: Unsupervised metric fusion over multiview data by graph random walk-based cross-view diffusion. IEEE Trans. Neural Netw. Learn. Syst. **PP**(99), 1–14 (2015). doi:10.1109/TNNLS.2015. 2498149

Exploring Data Mining Techniques in Medical Data Streams

Le Sun$^{(\boxtimes)}$, Jiangang Ma, Yanchun Zhang, and Hua Wang

Centre for Applied Informatics, Victoria University,
Melbourne, VIC 3011, Australia
le.sun1@live.edu.au.vu
{jiangang.ma,yanchun.zhang,Hua.Wang}@vu.edu.au

Abstract. Data stream mining has been studied in diverse application domains. In recent years, a population aging is stressing the national and international health care systems. Anomaly detection is a typical example of a data streams application. It is a dynamic process of finding abnormal behaviours from given data streams. In this paper, we discuss the existing anomaly detection techniques for Medical data streams. In addition, we present a process of using the Autoregressive Integrated Moving Average model (ARIMA) to analyse the ECG data streams.

1 Introduction

The fast development of the sensor, wireless and cloud computing technologies enable a smart healthcare mechanism that supports the consistent remote monitoring on the physical conditions of patients, elderly people or babies, and the efficient processing of the large sensing data sets. Such a smart healthcare mechanism can enhance the quality of life significantly. However, as investigated by the Cloud Standards Customer Council [6], the healthcare institutions are not keen on building smart healthcare systems based on the IT technologies, especially in developing countries. The under-utilization of IT technologies prevents the wide information sharing and the fast information processing in healthcare industry.

Data mining is a process of identifying useful yet previously hidden knowledge from data sets, containing a series of techniques, such as classification, clustering, and association rule mining [10]. Huge volume healthcare data are being produced from various sources like disease monitoring, patient historical records, specialist diagnosis, and hospital devices. These healthcare data are valuable and useful to assist the care givers in making decision efficiently. Data mining techniques have been applied in healthcare industry in terms of a variety of areas [10], for example, tracking the states of patients who have chronic diseases or physical disabilities, analysing the diagnosis results and recommending the treatment programs, identifying the patterns or rules of the 'fraud and abuse' to detect the fraudulent claims or the wrong referrals.

Anomaly detection is a typical example of a data streams application. Here, anomalies or outliers or exceptions often refer to the patterns in data streams that deviate expected normal behaviours. Thus, anomaly detection is a dynamic

© Springer International Publishing AG 2016
M.A. Cheema et al. (Eds.): ADC 2016, LNCS 9877, pp. 321–332, 2016.
DOI: 10.1007/978-3-319-46922-5_25

process of finding abnormal behaviours from given data streams. For example, in medical monitoring applications, a human electrocardiogram (ECG) (vital signs) and other treatments and measurements are typical data streams that appear in a form of periodic patterns.

In this paper, we discuss the state-of-the-art of anomaly detection in Medical data streams, and apply the Autoregressive Integrated Moving Average model (ARIMA) to analyse the ECG data streams. The structure of this paper is: in Sect. 2, we reviewed the anomaly detection techniques in healthcare area; Sect. 3 describes the analysing process of ARIMA on ECG data streams; Sect. 4 concludes this paper.

2   Survey on the Anomaly Detection Techniques in Healthcare Area

In this section, we discuss the techniques of the time series anomaly detection in recent five years. Our discussion focuses on two types of work: online anomaly detection in data streams, and off-line anomaly detection in static time series.

2.1   Previous Survey on Anomaly Detection

Chandola et al. [4] did a comprehensive survey on the techniques (developed in the past two decades, until 2010) of anomaly detection in discrete sequences, and discussed how to apply these techniques to the problem of time series anomaly detection. The authors categorised the existing techniques into three general groups: (1) detect abnormal sequences in a sequence database; (2) detect abnormal subsequences in a long sequence; and (3) identify abnormal patterns in a sequence. As the authors stated, the techniques in these three categories can be easily adapted to the domain of anomaly detection in time series, so I present the main techniques analysed in [4] in this section.

Chandola et al. [4] classified techniques in this category as: Similarity-based anomaly detection, Window-based anomaly detection, and Markov-model-based anomaly detection.

Similarity-Based Techniques. Similarity-based techniques group sequences into a number of clusters based on the similarities (or distances) between these sequences. Different clustering techniques were discussed in [4], such as the k-nearest neighbor(kNN) techniques [24], the k-medoid algorithm [2], the Probabilistic Suffix Trees [29], and Maximum Entropy models [23]. The work of [4] also presented two main distance calculation techniques for discrete sequences: the Simple Matching Coefficient [15], and the length of the longest common subsequence [5].

Sliding-Window-Based Techniques. Window-based techniques use the abnormal degrees of the fixed-length windows applied on a sequence to measure the anomalous degree of this sequence. There are various techniques developed for calculating the anomaly scores of windows and of a sequence. The main techniques of anomaly score calculation of windows include the threshold-based sequence time delay embedding (t-STIDE) [27], anomaly score calculation based on the lookahead pairs [11], anomaly score calculation based on the normal [16] or abnormal dictionaries [8,9], abnormal window identification based on classification techniques [12].

When the anomaly score of each window is obtained, the anomaly score of the entire sequence is calculated by aggregating the anomaly scores of all the windows. Chandola et al. [4] discussed three main aggregation techniques: the average anomaly score [16], the locality frame count [27], and the leaky bucket [13].

2.2 Online Anomaly Detection in Data Streams

Masud et al. [22] focused on developing the data stream classification techniques with the consideration on the concept-drifting and concept-evolution. They designed the ECSMiner algorithm to automatically identify novel classes in a consistent data stream on the basis of the traditional classification techniques. The ECSMiner uses a delayed time period to determine the classification category of a test instance and adopts another time delay and a buffer space to label and store a set of test time points. In addition, to speed up the classification process, the authors defined a 'pseudopoint' that summarizes a cluster of similar time points to reduce both the computation and space complexities.

Tan et al. [26] proposed a one-class anomaly detection algorithm: Streaming Half-Space Trees (HS-Trees). It can dynamically process endless data streams by consuming constant memory volume, and it takes constant time to build and update the tree. The proposed anomaly detection algorithm uses 'Mass' (i.e., the number of data items) to measure the anomaly degree of a streaming data item. The HS-Tree of a data stream can be constructed very fast as the tree construction depends only on the Mass in a data space. The data streams are divided into equal-sized segments (i.e., having similar number of items), and two consecutive windows are defined: the reference window and the latest window. The initial mass profile of the data stream is first learned based on the data items in the reference window. Then the new arriving data are stored in the latest window, and the anomaly scores of these data are calculated based on the initial mass profile. The reference window is updated consistently as long as the latest window is full, which adapts to the evolution of the data streams.

Das et al. [7] studied the anomaly detection in commercial fleet data streams. The authors developed a multiple kernel learning method to detect safety anomalies in both discrete and continuous data. The proposed method applied the one-class support vector machine (SVM) method to detect anomalous, in which the various types of knowledge are incorporated by using multiple kernels. At first, the SVM method solves an optimization problem: minimize

$$Q = \frac{1}{2}\Sigma_{i,j}\alpha_i\alpha_j k(x_i, x_j) \tag{1}$$

subject to $0 \leq \alpha_i \leq \frac{1}{lv}$, $\Sigma_i \alpha_i = 1, v \in [0,1]$;

where α is the Lagrange multiplier, k is a kernel matrix, l is the length of the data sequence, and v is a user-defined parameter for upper-bounding the fraction of the outliers and lower-bounding the fractions of the support vectors.

By solving the optimization problem of 1, I can get a set of support vectors $\{x_i|i \in [0,l], \alpha_i > 0\}$. Then the value of a decision function is calculated to estimate the label of a point x_j by using the formula 2. If the f_{x_j} is negative, x_j is an abnormal point.

$$f(x_j) = sgn(\Sigma_{i \in I_m} \alpha_i k(x_i, x_j) + \Sigma_{i \in I_{nm}} k(x_i, x_j) - \rho) \tag{2}$$

where $I_m = \{i|0 < \alpha_i < 1\}$, and $I_{nm} = \{i|\alpha_i = 1\}$.

I use a kernel function 3 to measure the distance between two discrete points x_i and x_j.

$$K_d(x_i, x_j) = \frac{|LCS(x_i, x_j)|}{\sqrt{l_{x_i}, l_{x_j}}} \tag{3}$$

where $LCS(x_i, x_j)$ represents the longest common sequence of x_i and x_j.

The continuous data sequences, the continuous values will be first transformed to discrete variables by using the SAX technique [19], and then the kernel function 3 is applied. The kernel for the mixture data sets is defined as an integration of the discrete and continuous data sets by using the weight aggregation function 4.

$$k(x_i, x_j) = wK_d(x_i, x_j) + (1 - w)K_c(x_i, x_j) \tag{4}$$

Liu et al. [20] investigated the anomaly detection problem in continuous time series using isolation techniques. They pointed out that the time series item classification methods based on the density and distance based measures are too time and space consuming to be applied in large databases. To implement the isolation-based anomaly detection, they utilized two key features (few and different) of the abnormal data items compared with the normal items. They proposed to use an isolation tree (iTree) to isolate the data items: the abnormal items tend to be closer to the root of the iTree, while the normal items are usually at further locations. Then an ensemble of iTrees is constructed to form an isolation forest (iForest), by sub-sampling the subsets of the time series. The iForest technique can identify the abnormal items by selecting the nodes in the iTrees that are close to the roots in terms of average path lengths. The authors also illustrated how to determine the size of sub-sampling and the number of ensemble iTrees. In particular, a few trees (e.g., 128) and a small subset size for sub-sampling are capable of achieving a high performance on the anomaly detection. The author also indicated that a small sub-sampling size for iTree construction can reduce the swamping and masking degree.

There is various work in anomaly detection in network traffic monitoring. Brauckhoff et al. [1] focused on solving the problem of inaccurate anomaly detection caused by the preprocessing steps for the traffic data streams. The authors analysed the impact of the random packet sampling and temporal aggregation on the signal properties and the correlation structures of the captured traffic streams. They proposed a solution of using a low-pass filter to reduce the aliasing influence. Silveira et al. [25] introduced an traffic anomaly detection approach that is based on a statistical model, namely ASTUTE (A Short-Timescale Uncorrelated-Traffic Equilibrium), to detect strongly correlated flow changes, rather than based on the historical data sets. Khalid et al. [18] developed a motion learning system that models trajectories using DFT-based coefficient feature space representation. They proposed a m-mediods method to represent the class containing n members with m mediods, and provided four anomaly detection algorithms based on the m-mediods method.

Liu et al. [21] explored an approach of automatically identifying outliers in one-class classification problem. The proposed approach performs classification in an unsupervised way to deal with the cases that neither positive nor negative labels is known in advance. The optimization function for this one-class classification is defined as in formula 5.

$$min_{\alpha,\tilde{y}}Q(\alpha,\tilde{y}) = \alpha^T K(I + \gamma_1 L)K\alpha - 2\alpha^T K\tilde{y} \tag{5}$$

s.t. $\|\alpha\| = 1$, $\tilde{y} \in \{c^+ + \frac{\gamma_2}{\|\tilde{y}_+\|}, c^-\}^{n\times 1}$, where $\alpha = [\alpha_1, \cdots, \alpha_n]^T \in R^n$ is the coefficient vector; $K = [k(x_i, x_j)]_{1\leq i,j\leq n} \in R^{n\times n}$ is the kernel matrix; $L = D-W$ is the graph Laplacian matrix; $\tilde{y}$ is the new label assignment; $\|\tilde{y}\|_+$ represents the number of positive items in $\tilde{y}$; and $\gamma_1, \gamma_2 > 0$ are two trade-off parameters.

3   ARIMA Model for ECG Analysis

In this section, we introduce the application of autoregressive integrated moving average (ARIMA) model to ECG analysis. A time series is defined by Definition 1.

Definition 1. *A **time-series** TS is an ordered real sequence: TS = $(v_1, \cdots, v_n)$, where v_i, $i \in [1, n]$, is a point value on the time series at time t_i.*

An ARIMA model is defined by Definition 2 [28].

Definition 2. *An ARIMA model of a time series TS = $(v_1, \cdots, v_n)\}$ is represented by Eq. 6.*

$$(1 - \sum_{i=1:p} \alpha_i L^i)v_t = (1 + \sum_{j=1:q} \theta_j L^j)\varepsilon_t \tag{6}$$

where $\alpha_i, i \in [1, p]$ are the autoregressive parameters, $\theta_j, j \in [1, q]$ are the moving average parameters, $\varepsilon_t, t \in [1, n]$ are the error values, and L is the lag operator.

We analyse a single-variable time series of an ECG record: record n01 of the AF Termination Challenge Database (abbr. AFTDB) [14]. In AFTDB, each record is comprised of two ECG variables, recording one-minute samples of atrial fibrillation with frequency of 128 samples/s. Our objective of this step is to validate the prediction accuracy of the Autoregressive integrated moving average (ARIMA) model for ECG time series. We use R to implement this process.

At first, we plot the single-variable AFTDB time series to identify its trend and seasonality in Fig. 1. The x-axis is the time horizon (each time point corresponds to one signal sample), and y-axis indicates the amplitudes of ECG records. We can see that the time series does not have explicit trend, but seasonality is obvious. Figure 1 too shows the smoothed ECG time series by using the techniques of linear smoothing [3] and LOWESS smoothing [17].

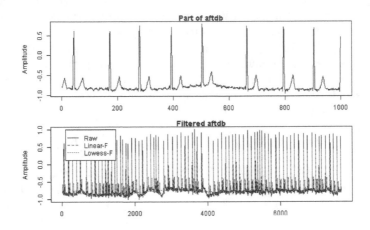

Fig. 1. AFTDB ECG data stream

Figure 2 shows the decomposed AFTDB time series. The raw data (the 1st row) contains three components: seasonal, trend, and irregular signals (i.e. the remainder in Fig. 2). As we observed, the AFTDB-n01 does not have upward or downward trends. From Figs. 1 and 2, there is a strong seasonality with period of 200 time points appoaximately.

In addition, we apply the SI (seasonal*irregular) chart to AFTDB-n01 time series to see the influence degrees of seasonal and irregular components on the change of time series respectively. We show the SI charts of the first four periods (1–200, 200–400, 400–600, 600–800) in Fig. 3. The regular move of seasonal series indicates the dominance role of the seasonal component.

Applying ARIMA fitting model on ECG time series. We show how the ARIMA fits the ECG time series. We continue using AFTDB-n01 as example. From Figs. 1 and 2, we can see the strong seasonality and weak trend in the time series. To determine the parameters of p (Auto-Regressive Parameters), q (Moving Average Parameters), and d (differencing times), we test the autocorrelation fuction (ACF) and the partial autocorrelation function (PACF) of both the raw time series and the first differenced time series. The ACF and PACF of a stationary time series are given by Eqs. 7 and 8 respectively.

$$\rho = Corr(v_t, v_{t-k}) = \frac{Cov(v_t, v_{t-y})}{\sqrt{V(v_t)V(v_{t-k})}} = \frac{Cov(v_t, v_{t-y})}{V}(v_t) = \frac{\gamma_k}{\gamma_0} \quad (7)$$

$$\theta_k = Corr(v_t, v_{t-k}|v_{t-1}, ..., v_{t-k+1}) \quad (8)$$

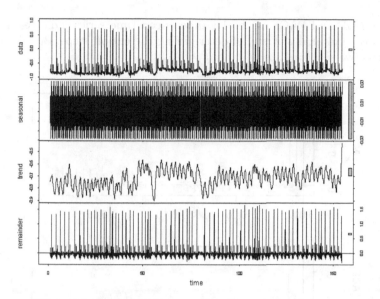

Fig. 2. Decomposition of AFTDB: seasonal component, trend component, and irregular component

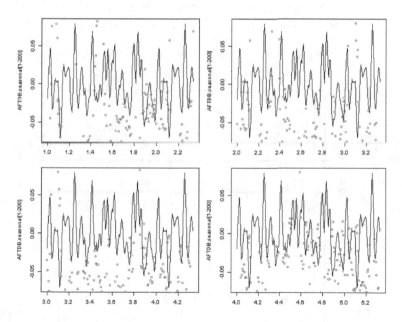

Fig. 3. SI (seasonal*irregular) chart of the first four periods of AFTDB

The ACF and PACF of AFTDB-n01 are shown in Figs. 4 and 5. We can see from Fig. 4 that the ACF of the raw time series shows a periodic pattern, while the ACF and PACF of the 1st differenced time series indicates a better convergence, and an easier way of determining the cut-off steps. Based on these two figures, we set $d = 1$, $p = 6$, and $q = 8$.

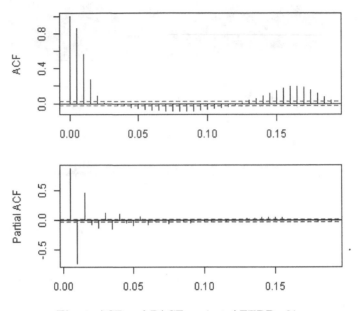

Fig. 4. ACF and PACF against AFTDB-n01

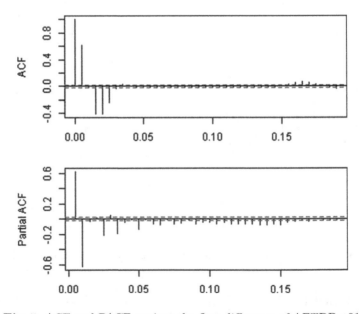

Fig. 5. ACF and PACF against the first difference of AFTDB-n01

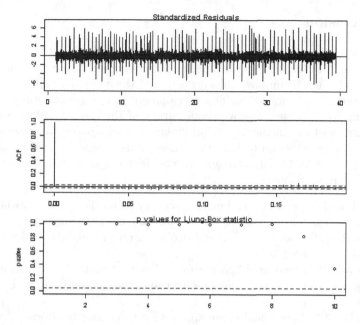

Fig. 6. The standardized residuals, ACF of residuals, and p-values for Ljung-Box statistic of ARIMA fitting for AFTDB-n01

We compare the setting of (6,1,8) with three other settings: (1,1,1), (5,1,7), and (7,1,8). The values of Akaike's Information Criterion (AIC) and Bayesian information criterion (BIC) are shown in Table 1, where the AIC of (6,1,8) is the smallest among these settings, and the BIC of (6,1,8) is almost similar as the BIC of (5,1,7). Therefore, we use the setting of (6,1,8) to fit the ARIMA model of AFTDB-n01.

Table 1. AICs and BICs of four settings for ARIMA fitting of AFTDB-n01

Setting	(6,1,8)	(1,1,1)	(5,1,7)	(15,1,15)
AIC	−17694.2	−12151.16	−17691.21	−17661.7
BIC	−17590.01	−12130.32	−17600.91	−17446.37

We then check the residuals of the fitted ARIMA(6,1,8) and conduct the portmanteau test to see whether or not the residuals are white noise. The checking results are shown in Fig. 6, which shows the residuals have limited correlations and all the p-values are larger than the threshold constraint. The checking results indicate the high quality of the fitting model.

4 Conclusion

Making healthcare decisions is challenging, which is influenced by a variety of factors: the lack of medical knowledge, the subjective mistakes of the healthcare givers, the false and incomplete information, and the misunderstanding and misinterpretation of the knowledge. The development of the medical sensor and the data mining technologies improve the accuracy of the medical information and the correctness of the medical decision making. In this paper, we discussed the existing anomaly detection techniques in healthcare area, and used the ARIMA model to analyse ECG data streams. In the next stage, we are going to make the following innovations:

- develop anomaly detection and motif discovery methods for multi-variate medical time series, considering the correlations among the time series.
- consider the influence of the concept evolution on the medical decisions based on the data stream mining.
- develop a ECG-based healthcare application that consistently monitors the patients, and online detect and predict the diseases based on the ECG data streams.
- extend the ECG-based healthcare application to various healthcare areas, like EEG data stream mining, body temperature monitoring, and speaking voice analysis.

References

1. Brauckhoff, D., Salamatian, K., May, M.: A signal processing view on packet sampling and anomaly detection. In: 2010 Proceedings IEEE INFOCOM, pp. 1–9, March 2010
2. Budalakoti, S., Budalakoti, S., Srivastava, A., Otey, M., Otey, M.: Anomaly detection and diagnosis algorithms for discrete symbol sequences with applications to airline safety. IEEE Trans. Syst. Man Cybern. Part C: Appl. Rev. 39(1), 101–113 (2009)
3. Buja, A., Hastie, T., Tibshirani, R.: Linear smoothers and additive models. Ann. Stat. 17, 453–510 (1989)
4. Chandola, V., Banerjee, A., Kumar, V.: Anomaly detection for discrete sequences: a survey. IEEE Trans. Knowl. Data Eng. 24(5), 823–839 (2012)
5. Chandola, V., Mithal, V., Kumar, V.: Comparative evaluation of anomaly detection techniques for sequence data. In: IEEE International Conference on Data Mining, pp. 743–748 (2008)
6. Council, C.S.C.: Impact of cloud computing on healthcare. Technical report, Cloud Standards Customer Council, November 2012
7. Das, S., Matthews, B.L., Srivastava, A.N., Oza, N.C.: Multiple kernel learning for heterogeneous anomaly detection: algorithm and aviation safety case study. In: Proceedings of the 16th ACM SIGKDD International Conference on Knowledge Discovery and Data Mining, pp. 47–56. ACM (2010)
8. Dasgupta, D., Majumdar, N.: Anomaly detection in multidimensional data using negative selection algorithm. In: Proceedings of the 2002 Congress on Evolutionary Computation, CEC 2002, vol. 2, pp. 1039–1044 (2002)

9. Dasgupta, D., Nino, F.: A comparison of negative and positive selection algorithms in novel pattern detection. In: 2000 IEEE International Conference on Systems, Man, and Cybernetics, vol. 1, pp. 125–130 (2000)
10. Durairaj, M., Ranjani, V.: Data mining applications in healthcare sector a study. Int. J. Sci. Technol. Res. **2**(10) (2013)
11. Forrest, S., Hofmeyr, S.A., Somayaji, A., Longstaff, T.A.: A sense of self for unix processes. In: Proceedings of the 1996 IEEE Symposium on Security and Privacy, SP 1996, p. 120 (1996). http://dl.acm.org/citation.cfm?id=525080.884258
12. Gao, B., Ma, H.Y., Yang, Y.H.: Hmms (hidden markov models) based on anomaly intrusion detection method. In: Proceedings of the 2002 International Conference on Machine Learning and Cybernetics, vol. 1, pp. 381–385 (2002)
13. Ghosh, A.K., Schwartzbard, A., Schatz, M.: Learning program behavior profiles for intrusion detection. In: Proceedings of the 1st Conference on Workshop on Intrusion Detection and Network Monitoring, ID'99, vol. 1, p. 6. USENIX Association, Berkeley (1999). http://dl.acm.org/citation.cfm?id=1267880.1267886
14. Goldberger, A.L., Amaral, L.A., Glass, L., Hausdorff, J.M., Ivanov, P.C., Mark, R.G., Mietus, J.E., Moody, G.B., Peng, C.K., Stanley, H.E.: Physiobank, physiotoolkit, and physionet components of a new research resource for complex physiologic signals. Circulation **101**(23), e215–e220 (2000)
15. Gower, J.C.: A general coefficient of similarity and some of its properties. Biometrics **27**, 857–871 (1971)
16. Hofmeyr, S.A., Forrest, S., Somayaji, A.: Intrusion detection using sequences of system calls. J. Comput. Secur. **6**(3), 151–180. http://dl.acm.org/citation.cfm?id=1298081.1298084
17. Jacoby, W.G.: Loess: a nonparametric, graphical tool for depicting relationships between variables. Electoral. Stud. **19**(4), 577–613 (2000)
18. Khalid, S.: Activity classification and anomaly detection using m-mediods based modelling of motion patterns. Pattern Recogn. **43**(10), 3636–3647 (2010). http://www.sciencedirect.com/science/article/pii/S0031320310002074
19. Liu, J., Keogh, E., Wei, L., Lonardi, S.: Experiencing sax: a novel symbolic representation of time series. Data Min. Knowl. Discov. **15**(2), 107–144 (2007). http://dx.doi.org/10.1007/s10618-007-0064-z
20. Liu, F.T., Ting, K.M., Zhou, Z.H.: Isolation-based anomaly detection. ACM Trans. Knowl. Discov. Data **6**(1), 3:1–3:39. http://doi.acm.org/10.1145/2133360.2133363
21. Liu, W., Hua, G., Smith, J.: Unsupervised one-class learning for automatic outlier removal. In: 2014 IEEE Conference on Computer Vision and Pattern Recognition (CVPR), pp. 3826–3833, June 2014
22. Masud, M., Gao, J., Khan, L., Han, J., Thuraisingham, B.: Classification and novel class detection in concept-drifting data streams under time constraints. IEEE Trans. Knowl. Data Eng. **23**(6), 859–874 (2011)
23. Pavlov, D.: Sequence modeling with mixtures of conditional maximum entropy distributions. In: Third IEEE International Conference on Data Mining, ICDM 2003, pp. 251–258, November 2003
24. Ramaswamy, S., Rastogi, R., Shim, K.: Efficient algorithms for mining outliers from large data sets. In: Proceedings of the 2000 ACM SIGMOD International Conference on Management of Data, SIGMOD 2000, pp. 427–438. ACM, New York (2000). http://doi.acm.org/10.1145/342009.335437
25. Silveira, F., Diot, C., Taft, N., Govindan, R.: Astute: Detecting a different class of traffic anomalies. In: Proceedings of the ACM SIGCOMM 2010 Conference, SIGCOMM 2010, pp. 267–278. ACM, New York (2010). http://doi.acm.org/10.1145/1851182.1851215

26. Tan, S.C., Ting, K.M., Liu, T.F.: Fast anomaly detection for streaming data. In: Proceedings of the Twenty-Second International Joint Conference on Artificial Intelligence, IJCAI 2011, vol. 2, pp. 1511–1516. AAAI Press (2011). http://dx.doi.org/10.5591/978-1-57735-516-8/IJCAI11-254

27. Warrender, C., Forrest, S., Pearlmutter, B.: Detecting intrusions using system calls: alternative data models. In: Proceedings of the 1999 IEEE Symposium on Security and Privacy, pp. 133–145 (1999)

28. Wei, W.W.S.: Time series analysis. Addison-Wesley publ. Reading (1994)

29. Yang, J., Wang, W.: Cluseq: efficient and effective sequence clustering. In: Proceedings. 19th International Conference on Data Engineering, 2003, pp. 101–112, March 2003

Using Detected Visual Objects
to Index Video Database

Xingzhong Du[1], Hongzhi Yin[1], Zi Huang[1], Yi Yang[2], and Xiaofang Zhou[1(✉)]

[1] The University of Queensland, Brisbane, Australia
zxf@itee.uq.edu.au
[2] University of Technology Sydney, Ultimo, Australia

Abstract. In this paper, we focus on how to use visual objects to index the videos. Two tables are constructed for this purpose, namely the unique object table and the occurrence table. The former table stores the unique objects which appear in the videos, while the latter table stores the occurrence information of these unique objects in the videos. In previous works, these two tables are generated manually by a top-down process. That is, the unique object table is given by the experts at first, then the occurrence table is generated by the annotators according to the unique object table. Obviously, such process which heavily depends on human labors limits the scalability especially when the data are dynamic or large-scale. To improve this, we propose to perform a bottom-up process to generate these two tables. The novelties are: we use object detector instead of human annotation to create the occurrence table; we propose a hybrid method which consists of local merge, global merge and propagation to generate the unique object table and fix the occurrence table. In fact, there are another three candidate methods for implementing the bottom-up process, namely, recognizing-based, matching-based and tracking-based methods. Through analyzing their mechanism and evaluating their accuracy, we find that they are not suitable for the bottom-up process. The proposed hybrid method leverages the advantages of the matching-based and tracking-based methods. Our experiments show that the hybrid method is more accurate and efficient than the candidate methods, which indicates that it is more suitable for the proposed bottom-up process.

Keywords: Index · Visual object · Video database

1 Introduction

Videos generate at an unprecedented speed in recent years. Since videos have many differences to texts, video database is specifically proposed to perform data management on the videos [4]. Previous studies on video database mainly focus on how to query the videos. These work equip the video database with not only all the query types that a general database has, but also has the enhanced query types based on the videos, such as low-level query [8,17], spatial-temporal query [5,12] and semantic query [18,25]. Compared to the general query types, these

© Springer International Publishing AG 2016
M.A. Cheema et al. (Eds.): ADC 2016, LNCS 9877, pp. 333–345, 2016.
DOI: 10.1007/978-3-319-46922-5_26

enhanced query types accept visual objects or video shots as input and they are usually able to return results with more spatial and temporal information.

The video database should store location, time and appearance information to support the enhanced query types. This usually requires the video database to maintain two tables simultaneously for index: one is the unique object table which stores the unique objects appear in the videos, the other one is the occurrence table which stores the occurrences of the unique objects in the videos [3,11,14,16]. In previous works, the construction of these two tables is a top-down process: first, hiring experts to generate the unique object table; second, hiring more labors to locate the unique visual objects in the videos frame by frame; third, gathering the location and time information to generate the occurrence table. Since the location, time and appearance information are fully associated with these two tables, they can well support the enhanced query types.

However, there is one major limitation that all the recent works have. That is, the construction process of these two tables heavily depends on human labors [4,12]. It means that, every time a new visual object introduced into the video database, a great deal of human annotations on the existing videos should be conducted to update the tables. Obviously, it is inefficient for the real world large-scale dynamic video database. Recent years' progress on object detector has shown that machine is very possible to locate visual objects as precise as human in the near future [2,10,19]. So one promising way to improve the efficiency is using the detector instead of human labors to locate visual objects as much as possible. However, it is worth noting that precise detector is not enough to relieve labors from annotations. This is because the detector just provides an incomplete occurrence table rather than both tables. To indeed index the videos, we need to innovate existing methods to accomplish two tasks, namely, generating the unique object table and fixing the occurrence table.

There exist three intuitive methods for addressing above problems, namely, recognizing-based, matching-based and tracking-based methods. All of them may link the same visual objects in the occurrence table so as to provide evidences for generating the unique object table or fixing the occurrence table.

- **Recognizing-based** method leverages the recognized information to link the objects. For example, if two car plates are detected and recognized to have the same plate number, they will be linked together. The recognizing-based method is only for generating the unique object table task. Its limitation is that the recognized information is not reliable and applicable always.
- **Matching-based** method leverages low-level visual features to achieve linking. It can be used for both task. It links very well when the visual objects are clear in the videos, but fails when the visual objects are blurred, transformed or occluded.
- **Tracking-based** method is used for addressing both problems too. It can overcome the transformation issue. Initializing the tracking-based method with a visual object and its location, it will track the occurrences and perform linking as long as possible. However, it has two non-negligible drawbacks. The first is that the tracking-based method cannot stop linking when the target

visual object disappears, and the second is that the tracking-based method cannot link the visual objects across different videos. In summary, none of these intuitive methods alone is suitable for addressing all the problems.

Based on the above analysis, we propose a hybrid method to generate the unique object table and fix the occurrence table. The proposed method treats the matching-based and tracking-based methods as the atomic operations during the whole process. It is divided into three stages. The first is local merge. It begins with applying the detector to get the occurrence table for each video. Then, the local merge applies the matching-based method to link the visual objects to generate the unique object table for each video. The second stage is global merge. It is proposed because of the fact that the unique object tables from local merge may still have redundancy globally. As a result, the visual objects from local merge are merged again to form the final table of unique visual objects. The third stage is propagation. It applies the tracking-based method to track the unique objects from the local merge in each video. Then, the missing occurrences are tracked in the videos as much as possible. It is worth noting that not all the tracked occurrences are reserved in our method. Only those pass the check by the matching-based method will be left. After that, the links between the unique visual objects and their occurrences in the videos are established sufficiently. Our experiments show that the proposed hybrid method is more accurate than the three intuitive methods no matter in generating the unique object tables or in fixing the occurrence tables. Furthermore, our experiments show that the major fraction of running time is occupied by the detector. Our proposed method just use a little more time to accomplish the two tasks.

In summary, our contributions in this work are as follows:

1. We propose to leverage automatically detected objects to index the video database, and find that this requires to change the original top-down process into bottom-up process. The new process consists of two tasks, namely, generating the unique object table and fixing the occurrence table.
2. We study three intuitive methods, namely, recognizing-based, matching-based and tracking-based methods, for the bottom-up process. By mechanism analysis and experiments, we show that all of them are not suitable for the two tasks we proposed.
3. Based on above analysis, we propose a hybrid method to perform the two tasks. It absorbs the advantages of matching-based and tracking-based methods, and further divides the process into three stages, namely local merge, global merge and propagation. Through the evaluation in terms of accuracy and efficiency, our proposed method is shown to be better than the three intuitive methods.

2   Related Works

The video database has been studied for many years. The research mainly focuses on designing enhanced query types to make a better use of the video data. Until now, three enhanced query types have been proposed for the video database, namely, low-level query, spatial-temporal query and semantic query [5,20]. The low-level query is equivalent to content-based video retrieval [18]. Given some exemplar videos, the query leverages the low-level features to search the similar videos [9]. In recent years, the low-level query has been improved a lot by more advanced features [7,13,23]. However, since the low-level query only involves similarity calculation and ranking, it seldom builds index structure to guide the query. In addition to that, the granularity of low-level query is high. For instance, if a user wants to query some visual objects in the video database, the low-level query cannot provide accurate location and time information. As an improvement, the spatial-temporal query is proposed to achieve lower granularity [12]. On one hand, it allows more flexibility in query conditions: any combination of object directional relations, topological relations, appearance and trajectory [5]. On the other hand, the spatial-temporal query accepts the visual objects as the input and returns arbitrary segments of videos [5]. The low-level query and spatial-temporal query are powerful, but they require specialist skills to work properly. The semantic query is proposed to achieve the easy-to-use purpose. It is usually decomposed into some sub-queries which in fact can be regarded a combination of the low-level query and the spatial-temporal query [1,18]. With these query types, many systems have been proposed in the past, such as OVID [17], QBIC [6], AVIS [1], CVQL [14] and BilVideo [22]. We notice that some of them also design index structure for the enhanced query type. However, the visual objects and links for constructing index are obtained manually, which limits the scalability of these systems.

Recent years sees an increasing tendency in developing automatic algorithms instead of human to locate objects visually. They have achieved a significant improvement by Deep Convolution Neural Network (DCNN) [21]. However, these algorithms are oriented to images. When they are applied to videos, the accuracy

Table 1. Summary of notations

Notation	Definition
T_u	The table of unique visual objects
u	A unique visual object in T_u
T_o	The table of object occurrences
o	A unique visual object's occurrence in T_o
v	A video in the video database
t	The frame sequential number within a video
$frame$	A frame of a video, which is actually an image
img	An image to represent a visual object's appearance
mbr	A minimum bounding rectangle to locate a visual object on a frame
$frame[mbr]$	Crop an image from $frame$ by mbr
c	The recognized content of a visual object

drops significantly [15]. One possible reason is that the images for training have better quality than the frames from videos on average. To improve this, some researchers propose the track-by-detection algorithms to increase the accuracy of object localization [2]. Such series of algorithms first try to locate the objects on some frames then track them in the following frames. But the tracking-by-detection algorithm keeps on tracking even when the objects already disappear [2]. Such behavior is fine in the tracking benchmark [24] because the visual objects in the test videos seldom disappear for a long time. However, in the real world videos, simply performing such tracker will bring a lot of false visual objects into the video database [15].

3   Problem Definition

There are two tables in video database to store the necessary information for the enhanced query types. The first one is the unique object table denoted by T_u which collects all the unique visual objects found in the current videos. In T_u, each unique visual object o is represented by a tuple (uid, img, c) where uid is the primary key of T_u, img is an image to represent object's appearance and c is the a text to describe object. The second table denoted as T_o stores the occurrence information of the visual objects in the videos. For example, when one unique visual object u is located by a minimum bounding rectangle mbr in the tth frame of video v, this occurrence o will be recorded by a tuple $(oid, u.uid, v, t, mbr, img, c)$ where oid is the primary key of T_o. In previous works [1,4,6,12,14,17,22], the visual object table T_u is given at first. Then, the human annotators will annotate the visual objects in the videos. Afterwards, the video database will generate the occurrence table T_o. It is obvious that the whole process requires a great deal of labors, which limits the scalability of the previous video databases. That is why the automatic visual object detection is necessary to the video database.

However, the whole process is reversed and challenging when the automatic visual object detection is applied. Under this direction, the video database gets the occurrence table T_o at first. Then, the table of visual objects needs to be generated by the occurrence table. This brings one new challenge, that is, the occurrence table is incomplete. The incompletion problem consists of two aspects. The first is that the occurrences are not complete. In other words, it is possible for the detection to detect a visual object on some frames, but very hard for the detection to detect a visual object on all the frames it appears. To amend this, we need to improve the detection rate. The second is that the occurrence tuples do not have the element uid. This is because there is no visual object table available initially. To amend this, we need to design a method to discover unique visual objects from their occurrences and link them together. In summary, we need to accomplish two tasks for using visual objects to index the video database:

- **Task 1:** Generating the unique object table from the occurrence table given by the detector;
- **Task 2:** Fixing the occurrence table to recover the object occurrence as much as possible.

Three intuitive methods are the basis of our method. They are recognizing-based method, matching-based method and tracking-based method, which we denote them as f_{rec}, f_{match} and f_{track} respectively. In this paper, we focus on their inputs and outputs. The recognizing-based method accepts two visual objects' recognized contents as input:

$$f_{rec}(c_1, c_2) = \begin{cases} true & c_1 \text{ and } c_2 \text{ are same;} \\ false & \text{otherwise.} \end{cases}$$

The matching-based method accepts two visual objects' appearances as input:

$$f_{match}(img_1, img_2) = \begin{cases} true & img_1 \text{ and } img_2 \text{ are visually same;} \\ false & \text{otherwise.} \end{cases}$$

The tracking-based method is different from the above two methods. It requires initialization where the inputs are the unique visual object u we want to track, u's last mbr as well as the last frame it appears. For simplicity, we hide the initialization in this paper. We focus on the inputs and outputs after it is initialized. This time, the tracking-based method accepts object appearance img and the current frame as input, then returns the mbr to indicate u's current locations on the frame:

$$mbr \leftarrow f_{track}(u.img, frame)$$

In our implementation, we consider four factors that could improve the accuracy of these three methods. They are size, illumination, occlusion and transformation. We found that f_{match} is more robust than f_{rec}. Therefore, in our proposed method, we only depend on f_{match} and f_{track}.

4   Proposed Method

The proposed method accepts the occurrence table as input, then generates the unique object table and form the index structure in Fig. 1. It is further divided into three stages, namely, local merge, propagation and global merge. In this section, we will describe their functionalities and how they generate the table of the unique visual objects and fix the missing occurrences.

4.1   Local Merge

The first stage is local merge. It aims at generating the table of unique visual objects for each video from the occurrence table T_o of each video. The key idea is to use the matching-based method f_{match} to discover the pairs of same visual objects in T_o, then group the visual objects belonging to these pairs together to form the unique visual objects. The outline of local merge is described in Algorithm 1 where the visual objects in the occurrence table T_o are examined

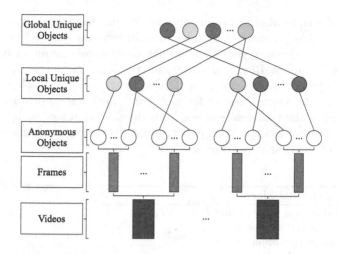

Fig. 1. The index structure after applying our proposed method.

Algorithm 1. Local Merge

Require: occurrence table T_o of video v
Ensure: unique object table T_u of video v
1: sort the visual objects in T_o by t in an ascending order
2: initialize $T_u \leftarrow \{\}$
3: **for all** visual object $o \in T_o$ **do**
4:     $flag \leftarrow false$
5:     **for all** visual object $u \in T_u$ **do**
6:         $flag \leftarrow f_{match}(o.img, u.img)$
7:         **if** $flag$ is $true$ **then**
8:             update T_o by $uid = u.uid$ where $oid = o.oid$
9:             **break**
10:     **if** $flag$ is $false$ **then**
11:         $T_u \leftarrow T_u \cup \{(uid, o.img, o.c)\}$
12: **return** T_u

---

one by one. When a visual object is not matched any visual object in T_u, it will be appended into T_u. Otherwise, the uid of the occurrence will be updated accordingly. It is worth noting that the unique visual objects found in this stage are just unique within a video, even though we already assign them with global $uids$ for facility. To make them unique globally, we introduce the global merge to obtain a global table of unique visual objects across videos.

4.2 Global Merge

The global merge is used to create the unique object table across different videos. Its outline is described in Algorithm 2. The key idea of global merge about merging is same to the local merge. But the global merge needs to update the column uid of each local object occurrence table when the visual objects in local unique object table are merged. Besides that, in local merge, we leverage some local properties of visual objects such as spatial-temporal continuity to enhance the matching accuracy. In global merge, all the local properties do not hold any more. To guarantee the matching results, the global merge selects the best appearances of the visual objects for matching. The best appearance can be selected by using the confidence scores from the detection tool or by averaging the merged visual objects. In this paper, the global merge uses the confidence scores to select the best appearances.

Algorithm 2. Global Merge

Require: local unique object table $T_{u_1}, \ldots, T_{u_k}$,
    local object occurrence table $T_{o_1}, \ldots, T_{o_k}$
Ensure: global unique object table T_{u_g}
  1: initialize $T_{u_g} \leftarrow \{\}$
  2: **for all** $i \in \{1, \ldots, k\}$ **do**
  3:    **for all** visual objects $u \in T_{u_i}$ **do**
  4:       $flag \leftarrow false$
  5:       **for all** visual objects $u' \in T_{u_g}$ **do**
  6:          $flag \leftarrow f_{match}(u.img, u'.img)$
  7:          **if** $flag$ is $true$ **then**
  8:             update T_{o_i} by $uid = u'.uid$ where $uid = u.uid$
  9:             **break**
10:       **if** $flag$ is $false$ **then**
11:          $T_{u_g} \leftarrow T_{u_g} \cup \{(u.uid, u.img, u.c)\}$
12: **return** T_{u_g}

4.3 Propagation

Fixing the occurrence table is another task our proposed method needs to deal with. The stage for fixing the occurrence table is called propagation in this paper. It leverages the unique object tables from local merge to recover their occurrences on the videos. The key idea is to track the unique visual objects and use the matching-based method to check whether the recover is correct. The check is necessary because tracking-based method can always track the visual objects in the videos even though they disappear. With the matching-based method, when the recovered occurrences do not match to the visual objects we want to track, these occurrences will be filtered out. Introducing the extra check will make decrease the false positives which are brought by the tracking-based method. The outline of propagation is described in Algorithm 3.

Algorithm 3. Propagation

Require: video v,
    unique object table T_u,
    occurrence table T_o
Ensure: occurrence table T_o
 1: **for all** frame $\in v$ **do**
 2:     **for all** visual object $u \in T_u$ **do**
 3:         $mbr \leftarrow f_{track}(u.img, frame)$
 4:         **if** $f_{match}(u.img, frame[mbr])$ is *true* **then**
 5:             $T_o \leftarrow T_o \cup \{(oid, u.uid, v.vid, frame.tid, mbr, u.c)\}$
 6: **return** T_o

5 Experiment

5.1 Data

The visual objects for the experiments are the car plates in this paper. They are detected by OpenALPR[1]. The collection consists of 393 videos which record the daily drives on the road. The durations of these videos are around 10 s and the fps is either 30 or 60. In addition to that, they are all recorded in FULL HD where the frame width is 1920 pixels and the frame height is 1080 pixels. The total size is 22 GB while the individual sizes range from 54 MB to 62 MB.

5.2 Evaluation Plan

The whole evaluation metrics are divided into two parts, namely, accuracy and efficiency. In details, for generating unique object tables, we will list the amount of unique objects each methods found and their overlap with the ground truth. The method which has closer amount and higher overlap rate to the ground truth is more accurate. For fixing object occurrence tables, we use precision and recall to evaluate the accuracy. The calculations of precision and recall are listed as follow:

$$Precision = \frac{\text{The amount of correct fixed occurrences}}{\text{The amount of total fixed occurrences}}$$

$$Recall = \frac{\text{The amount of correct fixed occurrences}}{\text{The amount of ground truth occurrences}}$$

Besides that, we also use F1 measure to provide a overall score. The calculation of F1 score we use is $F1socre = \frac{Precision*Recall}{Precision+Recall}$. In all these accuracy metrics, the method which obtains higher values is more accurate. The efficiency is further divided into evaluating the space and time cost. As comparison, we will list the space and time cost of detection as baseline.

---

[1] https://github.com/openalpr/openalpr.

5.3   Accuracy

We manually label the car plates appear in the videos and locate their occurrences. They are used as ground truth in the following experiments.

Table 2. The amount of unique objects generated by different methods

Method	#Visual objects
Ground truth	100
Recognizing	123
Matching	112
Our method	102

The first experiment evaluates the accuracy of unique visual object generation. In this experiment, we generate the local occurrence tables by the detection tools at first. Then applying local merge and global merge on these tables. During the merge we change the matching functions and record their results separately. In Table 2, the amount of global unique visual objects generated by different methods are listed. Since tracking-based method can only link visual objects within the videos, we do not list the amount of global unique visual objects that it generates. The results show that our method generate the closest amount to the ground truth than the other methods.

Table 3. The accuracy of different methods on occurrence fixing

Metric \ Method	Recognizing	Matching	Tracking	Our Method
Precision	0.9315	0.9372	0.3418	0.9225
Recall	0.2803	0.4046	0.9984	0.7490
F1 measure	0.2157	0.2826	0.2546	**0.4134**

The second experiment evaluates the accuracy of occurrence fixing. The inputs are the local unique object tables and the local object occurrence tables. Table 3 shows the accuracy of different methods on occurrence fixing. The results show that the recognizing-based method and the matching-based method achieve much higher precision than the tracking-based method, while the tracking-based method achieves much higher recall than the recognizing-based method and the matching-based method. Our method which takes the advantages of these intuitive methods achieve both high precision and recall at the same time. In the perspective of F1 score, our method has the highest accuracy.

5.4   Efficiency

Beside the accuracy, another aspect we focus on is the efficiency of local merge, global merge and propagation. In Table 4, we list the storage cost of the outputs from each stage and the average time cost of executing each stage per video. We notice that after the propagation the storage cost of occurrence table increases. Based on the accuracy analysis, this indicates our method discovers more object occurrences which the detection fails to discover. Our experiment also show that our proposed method which consists of local merge, global merge and propagation is faster than detection. It means that the time of generating unique object table and fixing object occurrences will be finished within acceptable time.

Table 4. Storage and time cost

Storage cost		Time cost	
Output	Volume	Process	Time
Initial occurrences	972.8 MB	Detection	437.98 s
Local objects	52.4 MB	Local merge	1.76 s
Global objects	22.4 MB	Global merge	34.21 s
Fixed occurrences	2334.7 MB	Propagation	304.64 s

We also want to show benefits after generating unique object table and fixing object occurrences. Fig. 2 summarizes two major benefits. The first benefit is the search space. If the unique object table does not exist, the video database needs to scan the occurrence table in a brute-force manner when the query is presented a visual object. Figure 2(a) shows the huge difference between scanning on the occurrence table and the unique object table. It indicates that the video database needs to pay much more time to scan the visual objects without the unique object table. Figure 2(b) displays the benefit after occurrence fixing. It shows that the amount of object occurrences increase significantly after propagation.

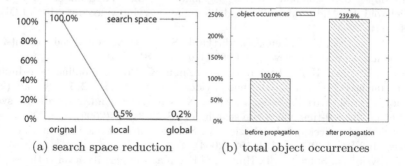

(a) search space reduction   (b) total object occurrences

Fig. 2. The effects of unique object generation and occurrence fixing

6   Conclusion

There are two basic structures for indexing video database by visual objects. They are unique object table and occurrence table. Previous methods obtain them by human labors. In this paper, we propose to obtain them automatically. Since the process is reversed compared to previous works [4,12,14,22], we propose a hybrid method to generate the unique object table and occurrence table. The proposed method consists of local merge, global merge and propagation. It is based on both matching-based and tracking-based methods. The experiments show that its accuracy and efficiency are better than the intuitive methods.

Acknowledgments. This research was jointly supported by the ARC project (Grant No. DP150103008) and the ARC DECRA project (Grant No. DE160100308).

References

1. Adali, S., Candan, K.S., Chen, S., Erol, K., Subrahmanian, V.S.: The advanced video information system: data structures and query processing. MMS **4**(4), 172–186 (1996)
2. Breitenstein, M.D., Reichlin, F., Leibe, B., Koller-Meier, E., Gool, L.J.V.: Robust tracking-by-detection using a detector confidence particle filter. In: ICCV, pp. 1515–1522 (2009)
3. Dönderler, M.E., Saykol, E., Arslan, U., Ulusoy, Ö., Güdükbay, U.: Bilvideo: design and implementation of a video database management system. MTA **27**(1), 79–104 (2005)
4. Dönderler, M.E., Ulusoy, Ö., Güdükbay, U.: A rule-based video database system architecture. Inf. Sci. **143**(1–4), 13–45 (2002)
5. Dönderler, M.E., Ulusoy, Ö., Güdükbay, U.: Rule-based spatiotemporal query processing for video databases. VLDB J. **13**(1), 86–103 (2004)
6. Flickner, M., Sawhney, H.S., Ashley, J., Huang, Q., Dom, B., Gorkani, M., Hafner, J., Lee, D., Petkovic, D., Steele, D., Yanker, P.: Query by image and video content: the QBIC system. IEEE Comput. **28**(9), 23–32 (1995)
7. Girshick, R.B., Donahue, J., Darrell, T., Malik, J.: Rich feature hierarchies for accurate object detection and semantic segmentation. In: CVPR, pp. 580–587 (2014)
8. Hjelsvold, R., Midtstraum, R.: Modelling and querying video data. In: VLDB, pp. 686–694 (1994)
9. Hu, W., Xie, N., Li, L., Zeng, X., Maybank, S.J.: A survey on visual content-based video indexing and retrieval. SMC C **41**(6), 797–819 (2011)
10. Huang, Z., Shen, H.T., Shao, J., Cui, B., Zhou, X.: Practical online near-duplicate subsequence detection for continuous video streams. TMM **12**(5), 386–398 (2010)
11. Huang, Z., Shen, H.T., Shao, J., Zhou, X., Cui, B.: Bounded coordinate system indexing for real-time video clip search. TOIS **27**(3), 17 (2009)
12. Köprülü, M., Cicekli, N.K., Yazici, A.: Spatio-temporal querying in video databases. Inf. Sci. **160**(1–4), 131–152 (2004)
13. Krizhevsky, A., Sutskever, I., Hinton, G.E.: Imagenet classification with deep convolutional neural networks. In: NIPS, pp. 1106–1114 (2012)
14. Kuo, T.C.T., Chen, A.L.P.: Content-based query processing for video databases. TMM **2**(1), 1–13 (2000)

15. Kuznetsova, A., Ju Hwang, S., Rosenhahn, B., Sigal, L.: Expanding object detector's horizon: incremental learning framework for object detection in videos. In: CVPR, pp. 28–36 (2015)

16. Le, T.-L., Thonnat, M., Boucher, A., Brémond, F.: A query language combining object features and semantic events for surveillance video retrieval. In: Satoh, S., Nack, F., Etoh, M. (eds.) MMM 2008. LNCS, vol. 4903, pp. 307–317. Springer, Heidelberg (2008)

17. Oomoto, E., Tanaka, K.: OVID: design and implementation of a video-object database system. TKDE 5(4), 629–643 (1993)

18. Petkovic, M., Jonker, W.: Content-Based Video Retrieval - A Database Perspective, Multimedia systems and applications, vol. 25. Springer (2003)

19. Ren, S., He, K., Girshick, R.B., Sun, J.: Faster R-CNN: towards real-time object detection with region proposal networks. In: NIPS, pp. 91–99 (2015)

20. Shen, H.T., Shao, J., Huang, Z., Zhou, X.: Effective and efficient query processing for video subsequence identification. TKDE 21(3), 321–334 (2009)

21. Szegedy, C., Toshev, A., Erhan, D.: Deep neural networks for object detection. In: NIPS, pp. 2553–2561 (2013)

22. Ulusoy, Ö., Güdükbay, U., Dönderler, M.E., Saykol, E., Alper, C.: Bilvideo video database management system. In: VLDB, pp. 1373–1376 (2004)

23. Wang, H., Schmid, C.: Action recognition with improved trajectories. In: ICCV, pp. 3551–3558 (2013)

24. Wu, Y., Lim, J., Yang, M.: Object tracking benchmark. PAMI 37(9), 1834–1848 (2015)

25. Yang, Y., Huang, Z., Shen, H.T., Zhou, X.: Mining multi-tag association for image tagging. WWWJ 14(2), 133–156 (2011)

Virtual Samples Construction Using Image-Block-Stretching for Face Recognition

Yingnan Zhao[1,2,3(✉)], Xiangjian He[3], and Beijing Chen[1,2]

[1] Jiangsu Engineering Center of Network Monitoring, Nanjing University of Information Science and Technology, Nanjing 210024, China
ann_zhao_99@163.com, nbutimage@126.com
[2] School of Computer & Software, Nanjing University of Information Science and Technology, Nanjing 210044, China
[3] School of Computing and Communications, University of Technology Sydney, Sydney, Australia
xiangjian.he@uts.edu.au

Abstract. Face recognition encounters the problem that multiple samples of the same object may be very different owing to the deformation of appearances. To synthesizing reasonable virtual samples is a good way to solve it. In this paper, we introduce the idea of image-block-stretching to generate virtual images for deformable faces. It allows the neighbored image blocks to be stretching randomly to reflect possible variations of the appearance of faces. We demonstrate that virtual images obtained using image-block-stretching and original images are complementary in representing faces. Extensive classification experiments on face databases show that the proposed virtual image scheme is very competent and can be combined with a number of classifiers, such as the sparse representation classification, to achieve surprising accuracy improvement.

Keywords: Face recognition · Virtual image · Sparse representation

1 Introduction

As one of the most active branches of biometrics, face recognition is attracting more and more attention [1, 2]. However, it is still faced with a number of challenges, such as varying illumination, facial expression and poses [3–6]. These appearance deformations lead to the fact that the same position of different images of the same object may have varying pixel intensities. It thereby increases the difficulty to recognize the object accurately [7]. Since more training samples are able to reveal more possible variation of the deformations, they are consequently beneficial for correct face classification. However, in real-world applications, there are usually only a limited number of available training samples, due to limited storage space or capturing training samples in a short time. In fact, non-sufficient training samples have become one bottleneck of face recognition [8, 9].

On feasible way is to construct virtual samples to obtain better face recognition result. The literatures have proposed several schemes to synthesize new samples. From the viewpoint of applications, they can be categorized into two kinds, i.e. 2D virtual

© Springer International Publishing AG 2016
M.A. Cheema et al. (Eds.): ADC 2016, LNCS 9877, pp. 346–354, 2016.
DOI: 10.1007/978-3-319-46922-5_27

face images and 3D ones. As for 2D virtual face images, Tang et al. [10] used prototype faces and an optic flow and expression ratio image based method to generate virtual facial expression. Thian et al. [11] used simple geometric transformations to generate virtual samples. Ryu et al. [12] exploited the distribution of the given training set to generate virtual samples. Beymer et al. [13] and Vetter et al. [14] synthesized new face samples with virtual views. Jung et al. [15] exploited the noise to synthesize new face samples. Sharma et al. [16] synthesize multiple virtual views of a person under different poses and illumination from a single face image and exploited extended training samples to classify the face. In order to overcome the small sample size problem of face recognition, Liu et al. [17] represented each single image as a subspace spanned by its synthesized (shifted) samples. Xu et al. [18–20] proposed a series of virtual sample construction methods based on symmetry, mirror samples and average samples. These virtual samples are easy to obtain, and the relative algorithms do yield attractive accuracy improvement in face recognition. On the other hand, 3D face synthesis also plays an important role in computer vision and virtual reality [21]. These studies offer good solutions to produce more available samples by exploiting the characteristics of face images.

In this paper, we propose to exploit the image-block-stretching method to generate new training samples and be combined with a sparse representation based method to perform face recognition. The new training samples indeed reflect some possible appearance of the face. The sparse representation based method simultaneously uses the original and new training samples to perform a face classification. This method also takes advantages of the score level fusion, which has proven to be very competent and is usually better than the decision level and feature level fusion.

Classification is a basic task in the fields of pattern recognition and computer vision. Various classification algorithms have been proposed with different viewpoints. For face recognition, sparse representation classification (SRC) algorithm is one of the best algorithms and has obtained very good performance [22]. However, Zhang et al. pointed out that it was collaborative representation (CR) rather than l_1 norm sparsity that contributes to the final classification accuracy [23]. Based on the non-sparse l_2 norm, Collaborate representation classification (CRC) [23] could lead to similar recognition but a significantly highly computational speed. It is noted that all the feature elements both in SRC and CRC share the same coding vector over their associated sub-dictionaries. This requirement ignores the fact that the feature elements in a pattern not only share similarities but also have differences. Therefore, Yang et al. presented a relaxed collaborative representation (RCR) [24] model to effectively exploit the similarity and distinctiveness of features. Liner Regression Classification (LRC) [25] can be referred as l_2 norm based on the linear regression model. Representation algorithms with the l_2 minimization have analytic solutions whereas conventional SRC algorithms should use iterative procedures to determine the solutions. Moreover, the formers always have lower computational costs than the conventional SRC algorithm [26].

With this paper, we present a novel scheme to generate virtual images for deformable faces. Our scheme first divides an original image into non-overlapping blocks. Then for two neighbored blocks in a row, we stretch them randomly. After all blocks are dealt with, we treat the new image as a virtual sample. The rationale of image-block-stretching in our scheme is based on the following fact. The deformation of the appearance of faces makes pixel intensities within the same block changeable. For example, different facial expressions of a person will lead to varying image block sat the same position of different face images. A slight smile will cause the pixel shift of some blocks in comparison with the original image. The experiments show that the proposed virtual images provide important information of deformable faces and can win very satisfactory accuracy for face recognition. This paper has the following contributes. (1) It proposes a kind of competent virtual images for deformable faces, which is very beneficial to reduce the data uncertainty. (2) The proposed scheme is simple, easy to understand and implement.

The rest of this paper is organized as follows. Section 2 first presents the proposed novel scheme to generate virtual images and interprets the reasonability, then describes the algorithm to integrate original images and virtual images. Section 3 performs experiments and Sect. 4 concludes the paper.

2   The Proposed Scheme and Algorithm

2.1   Scheme to Produce Virtual Samples Using Image-Block-Stretching

Suppose the size of each original training sample is $a \times b$. The proposed scheme to produce virtual samples includes the following steps.

1. Each original training sample is divided into non-overlapping image blocks, each block having the size of a' by b' pixels.
2. For all blocks within each training sample, an array $t[i,j]$ is generated, where $i = 1, 2, 3, \cdots m, j = 1, 2, 3, \cdots n$. Here, $m = [a/a']$ and $n = [b/b']$.
3. For each row of the array t, from the left to the right, the two neighbored blocks, denoted as $t[i,j]$ and $t[i,j+1]$, are stretched randomly. Suppose the stretching scale is s. Then the two stretching results is $a' \times (b'+s)$, $a' \times (b'-s)$ or $a' \times (b'-s)$, $a' \times (b'+s)$. Note that the interpolation method is the nearest neighbor algorithm.
4. When all the blocks are dealt with, we get a virtual sample. Every virtual image is converted into a vector.

Figure 1 shows some examples of original images and the corresponding virtual images. We see that the virtual image is still a natural image. Because it is a modification of the corresponding original image, it has obvious deviation from the original image and represents some possible variations of the object. As a result, the simultaneous use of the original images and virtual images allows more information of the object to be provided, and better recognition accuracy can be achieved. An intuitive explanation of the benefit of more available training samples is as follows. Because the object is deformable and the number of samples in the form of images is always smaller than the dimension of samples, the difference between images of the same object is usually great. Consequently, the test sample may be very different from the limited training samples of the same object. However, when virtual images are also used as training samples, the possibility that the training sample with the minimum distance to the test sample is from the same object as the test sample will be increased. As we know, if the training sample with the minimum distance to the test sample is from the same object as the test sample, we can obtain the correct recognition result of the test sample via the nearest neighbor classifier. Thus, more accurate recognition of objects can be obtained under the condition that the original images and virtual images are simultaneously used as training samples and the nearest neighbor classifier is exploited. We also say that the simultaneous use of the original images and virtual images enables the data uncertainty to be reduced.

Fig. 1. Examples of original images (the top row) and the corresponding virtual images (the bottom row).

2.2   Algorithm to Integrate Original Images and Virtual Images

First of all, it should be pointed out that our algorithm exploits three kinds of virtual training samples. The first kind of virtual training samples is the mirror face image presented in [19].The second kind of virtual training samples is obtained by applying our image-block-stretching scheme to the original face image. The third kind of virtual training samples is obtained by applying our image-block-stretching scheme to the mirror face image.

For a test sample, an arbitrary representation based classification algorithm is respectively applied to the original training samples and virtual training samples. Suppose that there are C classes. Let $d_1, \cdots, d_C$ denote the class residuals of the test sample with respect to the original training samples of the first to the C-th classes. Let $e_1, \cdots, e_C$ denote the class residuals of the test sample with respect to the first kind of virtual training samples of the first to the C-th classes. Let $f_1, \cdots, f_C$ and $g_1, \cdots, g_C$ denote the class residuals of the test sample with respect to the second and third kinds of virtual training samples.

The class residuals of the test sample are integrated using

$$q_j = t_1 d_j + t_2 e_j + t_3 f_j + t_4 g_j, \quad j = 1, \cdots, C. \tag{1}$$

Let $r = \arg\min_j q_j$. We assign the test sample to the r-th class.

As for the weight, we use an automatic procedure to determine it. The automatic procedure is as follows. Let $d_1, \cdots, d_C$ stand for the sorted results of $d_1, \cdots, d_C$ and suppose that $d_1 \leq \cdots \leq d_C$. Accordingly, $e_1, \cdots, e_C, f_1, \cdots, f_C$, and $g_1, \cdots, g_C$ have the similar meanings. We define $t_{10} = d_2 - d_1$ and $t_{20} = e_2 - e_1$. t_{30} and t_{40} are defined as $t_{30} = f_2 - f_1$ and $t_{40} = g_2 - g_1$. We respectively define

$$t_1 = \frac{t_{10}}{t_{10} + t_{20} + t_{30} + t_{40}}, \tag{1}$$

$$t_2 = \frac{t_{20}}{t_{10} + t_{20} + t_{30} + t_{40}}, \tag{2}$$

$$t_3 = \frac{t_{30}}{t_{10} + t_{20} + t_{30} + t_{40}}, \tag{3}$$

$$t_4 = \frac{t_{40}}{t_{10} + t_{20} + t_{30} + t_{40}}. \tag{4}$$

We would like to point out that a similar weight setting algorithm was proposed in [27], which is the first completely adaptive weighted fusion algorithm and obtained an almost perfect fusion result. However, the algorithm in [27] can fuse only two kinds of scores.

3   Experimental Results

In experiments, SRC and CRC are respectively used as the classifier. Because our scheme stretches the neighbored two blocks of an original image randomly, so we run it five times and take the average of the rates of classification errors as the final result. In the following tables CRC and SRC denote the algorithms of naïve CRC presented in [23] and SRC presented in [22], respectively. In the context "Our method integrated with CRC" is used to present the result of applying our method to improve CRC. "Our method integrated with SRC" is used to present the result of applying our method to improve SRC. Note that the stretching scale s is 1 for all the experiments.

3.1   Experiment on the FERET Database

We first use a subset of the FERET face dataset to perform an experiment. The subset contains 1400 face images from 200 subjects and each subject has seven different face images. This subset was composed of images in the original FERET face dataset whose names are marked with two-character strings: 'ba', 'bj', 'bk', 'be', 'bf', 'bd', and 'bg'. Every face image was resized to a 40 by 40 image. We respectively select the first 2, 3 and 4 face images of each subject as original training samples and use the remaining face images as testing samples. Figure 2 presents some cropped face images from the FERET face database. The size of the image block in our method is set to $a = 2$ and $b = 2$. Table 1 shows the rates of classification errors (%) of different methods on the FERET database. We see that our method outperforms naïve CRC and SRC.

Fig. 2. Some face images from the FERET face database.

Table 1. Experimental results on the FERET database.

Number of training samples per person	2	3	4
CRC	41.60	55.63	44.67
Our method + CRC	39.48	46.58	43.28
SRC	41.90	49.25	36.67
Our method + SRC	36.15	37.39	31.83

3.2 Experiment on the GT Database

In the GT face database there are 750 face images from 50 persons and every person provided 15 face images. The original images are color images with clutter background taken at the resolution of 640×480 pixels and show frontal and tilted faces with different facial expressions, lighting conditions and scales. We used the manually cropped and labeled face images to conduct the experiment. Figure 3 shows some of these images. We further resized them to obtain images with the resolution of 40×30 pixels. They were then converted into gray images. In our experiment the first 3–10 face images of every person are used as training samples and the other images are employed as test samples. The size of the image block in our method is set to $a = 2$ and $b = 4$. Table 2 shows the rates of classification errors (%) of different methods on the GT database. We see that our method also performs better than original CRC and SRC.

Fig. 3. Some face images from the GT face database.

Table 2. Experimental results on the GT database.

Number of training samples per person	3	4	5	6	7	8	9	10
CRC	54.67	52.91	51.20	44.44	41.50	40.57	39.00	36.00
Our method + CRC	51.63	48.52	46.38	42.04	36.29	35.83	32.40	29.78
SRC	55.33	53.82	51.20	44.44	41.75	41.43	37.33	34.40
Our method + SRC	54.18	52.26	49.37	41.60	38.44	37.62	34.50	33.48

3.3 Experiments on the ORL Database

The ORL face database includes 400 face images taken from 40 subjects each having 10 face images. In this database some images were taken at different times and with various lightings, facial expressions (open/closed eyes, smiling/not smiling), and facial details (glasses/no glasses). All images were taken against a dark homogeneous background with the subjects in an upright, frontal position (with tolerance for some side movement). We resized every image to form a 56 by 46 image matrix. Figure 4 shows some images from this ORL dataset. The first 1, 2, 3, 4, 5 and 6 face images of each subject were used as training samples and the others were exploited as test samples, respectively. The size of the image block in our method is set to $a = 4$ and $b = 8$. Table 3 shows rates of classification errors (%). It again tells us that our method is better than naïve CRC and SRC.

Fig. 4. Some images from this ORL dataset.

Table 3. Experimental results on the ORL database.

Number of training samples per person	1	2	3	4	5	6
CRC	31.94	16.56	13.93	10.83	11.50	8.13
Our method + CRC	31.49	16.46	13.40	11.29	11.41	8.81
SRC	32.78	17.81	16.07	12.50	12.00	10.00
Our method + SRC	30.18	16.62	15.38	11.27	10.82	9.63

4   Conclusions

In this paper, the proposed novel scheme not only can efficiently generate natural virtual images for deformable faces but also can lead to notable accuracy improvement. Moreover, the rationale of the proposed scheme is easy to understand. As shown earlier, image-block-stretching of an original image can reflect possible change of the appearance of a deformable face, so the obtained virtual image is a proper representation of the object. Besides the proposed scheme is applicable for faces, it can also provide useful representation for a general object because of the following factors. Within a small enough region of a general image, in most cases the difference of the values of pixels is usually little. As a result, image-block-stretching of a small region of an image will obtain reasonable and new representation of this region. It should be pointed out that a non-deformable object also usually has varying images owing to changeable illuminations, imaging distances and views. Thus, after we use the proposed scheme to produce virtual images for general objects, we can also integrate the virtual images and original images to obtain better classification performance.

Acknowledgments. This work is supported in part by the PAPD of Jiangsu Higher Education Institutions, Natural Science Foundation of China (No. 61572258, No. 61103141 and No. 51505234), and the Natural Science Foundation of Jiangsu Province (No. BK20151530).

References

1. Zhang, L., Chen, S., Qiao, L.: Graph optimization for dimensionality reduction with sparsity constraints. Pattern Recogn. **45**(3), 1205–1210 (2012)
2. Fan, Z., Xu, Y., Zhang, D.: Local linear discriminant analysis framework using sample neighbors. IEEE Trans. Neural Netw. **22**(7), 1119–1132 (2011)
3. Wang, S.J., Yang, J., Sun, M.-F., et al.: Sparse tensor discriminant color space for face verification. IEEE Trans. Neural Netw. Learn. Syst. **23**(6), 876–888 (2012)
4. Zhang, D., Song, F., Xu, Y., et al.: Advanced pattern recognition technologies with applications to biometrics. Medical Information Science Reference, New York (2009)
5. Zhang, X., Gao, Y.: Face recognition across pose: a review. Pattern Recogn. **42**(11), 2876–2896 (2009)
6. Kautkar, S.N., Atkinson, G.A., Smith, M.L.: Face recognition in 2D and 2.5D using ridgelets and photometric stereo. Pattern Recogn. **45**(9), 3317–3327 (2012)
7. Shahrokni, A., Fleuret, F., Fua, P.: Classifier-based contour tracking for rigid and deformable objects. In: British Machine Vision Conference, No. CVLAB-CONF-2005-014 (2005)
8. Zhang, P., You, X., Ou, W., et al.: Sparse discriminative multi-manifold embedding for one-sample face identification. Pattern Recogn. **52**, 249–259 (2016)
9. Sun, Z.L., Shang, L.: A local spectral feature based face recognition approach for the one-sample-per-person problem. Neurocomputing **188**, 160–166 (2016)
10. Tang, B., Luo, S., Huang, H.: High performance face recognition system by creating virtual sample. In: Proceedings of the International Conference on Neural Networks and Signal Processing, pp. 972–975 (2003)
11. Thian, N.P.H., Marcel, S., Bengio, S.: Improving face authentication using virtual samples. In: Proceeding of the IEEE International Conference on Acoustics, Speech and Signal Processing, pp. 6–10 (2003)
12. Ryu, Y.-S., Oh, S.-Y.: Simple hybrid classifier for face recognition with adaptively generated virtual data. Pattern Recogn. Lett. **23**(7), 833–841 (2002)
13. Beymer, D., Poggio, T.: Face recognition from one example view. In: Proceedings of the Fifth International Conference on Computer Vision, pp. 500–507 (1995)
14. Vetter, T.: Synthesis of novel views from a single face image. Int. J. Comput. Vis. **28**(2), 102–116 (1998)
15. Jung, H.-C.: Authenticating corrupted face image based on noise model. In: Proceedings of the Sixth IEEE International Conference on Automatic Face and Gesture Recognition, pp. 272–277 (2004)
16. Sharma, A., Dubey, A., Tripathi, P., et al.: Pose invariant virtual classifiers from single training image using novel hybrid-eigenfaces. Neurocomputing **73**(10–12), 1868–1880 (2010)
17. Liu, J., Chen, S., Zhou, Z.-H., Tan, X.: Single image subspace for face recognition. In: Zhou, S., Zhao, W., Tang, X., Gong, S. (eds.) AMFG 2007. LNCS, vol. 4778, pp. 205–219. Springer, Heidelberg (2007)
18. Xu, Y., Zhu, X., Li, Z., et al.: Using the original and 'symmetrical face' training samples to perform representation based two-step face recognition. Pattern Recogn. **46**(4), 1151–1158 (2013)
19. Xu, Y., Li, X., Yang, J.: Integrate the original face image and its mirror image for face recognition. Neurocomputing **131**, 191–199 (2014)
20. Xu, Y., Fagn, X., Li, X., et al.: Data uncertainty in face recognition. IEEE Trans. Cybern. **44**(10), 1950–1961 (2014)

21. Nguyen, H.T., Ong, E.P., Niswar, A., et al.: Automatic and real-time 3D face synthesis. In: VRCAI, pp. 103–106 (2009)
22. Wright, J., Yang, A.Y., Ganesh, A., Sastry, S.S., Ma, Y.: Robust face recognition via sparse representation. IEEE Trans. Pattern Anal. Mach. Intell. **31**(2), 210–227 (2009)
23. Zhang, D., Yang, M., Feng, X.-Ch: Sparse representation or collaborative representation: Which helps face recognition? In: 2011 IEEE International Conference on Computer Vision (ICCV), pp. 471–478 (2011)
24. Meng, Y., Zhang, D., Wang, S.: Relaxed collaborative representation for pattern classification. In: 2012 IEEE Conference on Computer Vision and Pattern Recognition (CVPR), pp. 2224–2231 (2012)
25. Imran, N., Togneri, R., Bennamoun, M.: Linear regression for face recognition. IEEE Trans. Pattern Anal. Mach. Intell. **32**(11), 2106–2112 (2010)
26. Zhang, Z., Xu, Y., Yang, J., et al.: A survey of sparse representation: algorithms and applications. IEEE Access **3**, 490–530 (2015)
27. Xu, Y., Lu, Y.: Adaptive weighted fusion: A novel fusion approach for image classification. Neurocomputing **168**, 566–574 (2015)

Miscellaneous

XB+-Tree: A Novel Index
for PCM/DRAM-Based Hybrid Memory

Lu Li[1], Peiquan Jin[1,3(✉)], Chengcheng Yang[1], Shouhong Wan[1,2],
and Lihua Yue[1,2]

[1] University of Science and Technology of China, Hefei, China
jpq@ustc.edu.cn
[2] Key Laboratory of Electromagnetic Space Information,
Chinese Academy of Sciences, Hefei, China
[3] Science and Technology on Electronic Information Control Laboratory,
Chengdu, China

Abstract. Phase Change Memory (PCM) has emerged as a new kind of future memories that can be used as an alternative of DRAM. PCM has a number of special properties such as non-volatility, high density, read/write asymmetry, and byte addressability. Specially, PCM has higher write latency than DRAM but has comparable read latency with DRAM. This makes it difficult to directly replace DRAM with PCM in current memory hierarchy. Thus, in this paper, we propose to construct hybrid memory architecture that involves both PCM and DRAM, which is a practical and feasible way to utilize PCM. Such hybrid memory architecture introduces many new issues for database researches, as existing algorithms have to be revised to be suitable for hybrid memory. In this paper, we study the indexing issue on PCM/DRAM-based hybrid memory and propose an improved version of the B+-tree called XB+-tree (eXtended B+-tree). The key idea of the XB+-tree is to detect the read/write tendency of the nodes in the tree index and organize write-intensive nodes on PCM while putting read-intensive nodes on DRAM. We propose a new node management and migration algorithm in the XB+-tree to effectively move nodes between DRAM and PCM. With this mechanism, we can reduce the read and write operations on PCM and improve the overall performance. We conduct trace-driven experiments and compare our proposal with three existing indices including the B+-tree, the OB+-tree (B+-tree with the overflow scheme), and the CB+-tree. The results in terms of PCM read/write count and run time suggest the efficiency of our proposal.

Keywords: B+-tree · Index · PCM · Hybrid memory

1 Introduction

In recent years, the advance of material science leads to the rapid development of new kinds of storage media, such as flash memory and Phase Change Memory (PCM). These new kinds of memories introduce new opportunities for big data storage, because they have many special properties that are better than traditional storage media. For example,

© Springer International Publishing AG 2016
M.A. Cheema et al. (Eds.): ADC 2016, LNCS 9877, pp. 357–368, 2016.
DOI: 10.1007/978-3-319-46922-5_28

Table 1. PCM vs. DRAM[1]

	DRAM	PCM
Durability	Volatile	Non-volatile
Read Latency	$20 \sim 50$ ns	$50 \sim 100$ ns
Write latency	50 ns	$100 \sim 500$ ns
Read bandwidth	$\sim$ GB/s	800 MB/s
Write bandwidth	$\sim$ GB/s	$50 \sim 100$ MB/S
Endurance	∞	10^8 for writes

PCM has the properties of byte addressability, non-volatility, low energy consumption, and high capacity. Table 1 shows a comparison between DRAM and PCM [1].

The special properties of PCM enable us to integrate PCM into the current memory architecture, which forms a kind of hybrid memory involving PCM and DRAM [1, 2], as shown in Fig. 1. In such hybrid memory architecture, the high density and non-volatility of PCM makes it possible to build non-volatile main-memory databases. Further, we can take the advantages of the two media to offer better performance for database systems.

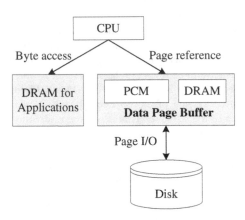

Fig. 1. Hybrid memory architecture consists of PCM and DRAM

In this paper, we focus on optimizing the B+-tree index for the PCM and DRAM based hybrid memory. The key idea of our approach is to keep write-intensive data on DRAM and read-intensive data on PCM. Thus, we can reduce the costly write operations on PCM and consequently improve the overall performance. The key issue is to determine the read/write tendency of data and to devise an efficient data migration algorithm between DRAM and PCM. We explore these issues in this paper and propose an efficient index structure called XB+-tree. In summary, we make the following contributions in this paper:

(1) We propose an efficient index scheme called XB+-tree (eXtended B+-tree) for DRAM and PCM based hybrid memory systems. XB+-tree improves the B+-tree

in three aspects. First, it is unbalanced because it enables leaf nodes consisting of overflow nodes. Second, the leaf nodes of XB+-tree can reside in DRAM or PCM. Third, XB+-tree supports data migration between DRAM nodes and PCM nodes.
(2) We conduct experiments to measure the performance of our proposal and make comparison with the traditional B+-tree, the overflow B+-tree, and the CB+-tree. The results suggest the efficiency of our proposal.

The rest of the paper is structured as follows. Section 2 introduces the background and the related work. Section 3 presents the structure of the XB+-tree as well as the operations on the index. Section 4 presents experimental results, and in Sect. 5 we conclude the paper.

2 Related Work

PCM is a new kind of non-volatile random access memory (NVRAM), which exploits the unique behavior of chalcogenide glass that enters two different states under different heating temperatures and durations. The two states, termed as amorphous state and crystalline state, have significantly different electrical resistivity. The high-resistance amorphous state represents "0"; and the low-resistance crystalline state represents "1" [3].

The properties of PCM lie on the gap between DRAM and NAND flash memory [4]. PCM has following properties: (1) It has high storage density, which is about 2 to 10 times higher than DRAM, that means it can store more information in less space, and improve the efficiency of many database applications; (2) The energy consumption of PCM is significantly decreased, and the idle energy consumption is almost 0; (3) The read latency of PCM is almost the same with DRAM; (4) Its non-volatile characteristic makes data persistent. However, PCM also has some disadvantages: (1) It has unbalanced read and write latency. Its write latency is 6 times higher than its read latency; (2) It has limited life time. A PCM cell will soon reach the upper limit of life when it is continuously written.

Therefore, PCM is considered to be a promising alternative to DRAM as main memory [5, 6] or a promising alternative to NAND flash as storage [7]. In this paper, we consider PCM and DRAM to be used for building hybrid main memory.

Index is one of the key technologies in modern database systems, search applications, and information retrieval. The B+-tree index, which has been widely used in database systems and file systems, is suitable for random and sequential access applications. However, it usually incurs high write costs.

As write operations have higher cost than reads on PCM, an intuitive idea of optimizing indices on PCM is to reduce write operations to PCM. Traditional B+-tree performs well in search operations, because nodes are ordered and binary search is used. But it causes many unnecessary writes to keep data in order when inserts and deletes are performed. The unsorted B+-tree [4] is helpful in reducing write counts to keep nodes in order, but it cannot avoid sort operations when splits and merge operations occur. As split operations on the B+-tree will lead to many page writes, incurring propagating updates from a leaf node to the root of B+-tree, a simple idea for solving this problem is to allow overflow pages for leaf nodes. Thus, newly inserted records can be put in overflow pages and splits will not be triggered by insertions, yielding an Overflow B+-tree (OB+-tree) [4, 8] as exemplified in Fig. 2. The OB+-tree is efficient in reducing write operations because split and merge operations are given priority within the overflow chain; thus it can reduce the write operation to the parent node. However, if the node size is small (less than 4 cache-line size) [4], the performance becomes poor. In addition, the overflow scheme introduces additional search costs, because we have to read all the overflow nodes in the targeted leaf node. The CB+-tree [9] adjusts the overflow chain to move the read-intensive leaf nodes out of the chain, which can reduce read operations on PCM.

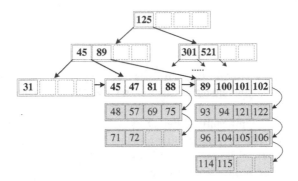

Fig. 2. Overflow B+-tree structure

In this paper, we also use the overflow policy to reduce PCM writes. However, we devise a new approach for forecasting the read/write tendency of data, which is used to conduct data migration between DRAM and PCM. Moreover, we propose a new algorithm to manage the nodes in DRAM.

3   XB+-Tree

The motivation of the XB+-tree comes from an investigation on the OB+-tree. The OB+-tree can effectively reduce writes on the traditional B+-tree because it can delay the split and merge operations. However, in case that a leaf node residing in PCM is frequently written (thus yields many overflow nodes), the great number of PCM writes will worsen the overall performance of the index. Therefore, a more efficient way is to organize write-intensive leaf nodes in DRAM and put those nodes that are not frequently written in PCM. This leads to the design of the XB+-tree.

3.1   Index Structure of the XB+-Tree

The structure of the XB+-tree is shown in Fig. 3. All the data is stored in leaf nodes. Every node of the XB+-tree consists of several cache lines [4]. One of them is *cacheline_h* and the others are *cacheline* type. We only store *<key, value>* in *cacheline*.

For *cacheline_h*, which stores auxiliary information, <*key, value*> pairs store in the remaining space. In the XB+-tree, auxiliary information includes *num_keys*, *is_leaf*, *parent*, and *brother*, in which *num_keys* denotes the number of <*key, value*> pairs, *is_leaf* indicates whether the node is a leaf node, *parent* and *brother* denote the parent and brother of the node. All these information takes up 16 bytes. Specially, we use DRAM to store the access information that will be used in the read/write tendency mechanism, because they are frequently updated.

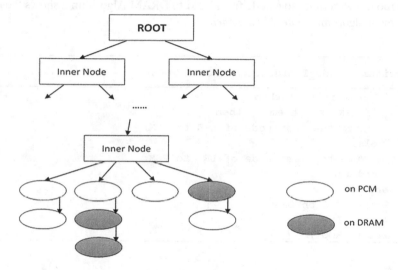

Fig. 3. Structure of the XB+-tree

3.2 Migration Between DRAM and PCM

As DRAM has higher performance than PCM, we move the write-intensive nodes to DRAM so as to reduce PCM writes. However, the limited capacity of DRAM makes it not possible to maintain a great number of write-intensive nodes. When DRAM reaches its capacity limit, we need to select the appropriate nodes to move back to PCM. The migration algorithm does not change the XB+-tree structure but change the storage location of each leaf node. In this section, we discuss how to manage the DRAM nodes to realize effective storage and migration.

We use two lists to manage the DRAM nodes. One is an LRU list that is used to manage all the nodes on DRAM. When a node is accessed (either read or write), we adjust this LRU list. The other one is an LRR (Least Recently Read) list that is used to manage the read-intensive nodes on DRAM. As write-intensive nodes may change with time into read-intensive nodes, we put read-intensive nodes into the LRR list and adjust the LRR list when these nodes are accessed by read requests. If read-intensive nodes become write-intensive ones, we remove them from the LRR list. Every time when a leaf node on PCM becomes write-intensive, it is moved to DRAM. If DRAM has free spaces, we put the write-intensive leaf node on DRAM and insert it into the head of the

LRU list. If DRAM is full, we select a victim node in DRAM and move it to PCM. When selecting the victim from DRAM, we select the last node in the LRR list as the victim if the LRR list is not empty. This is because only write-intensive nodes will be stored on DRAM; thus we prefer to choose a read-intensive node to move it out of DRAM. On the other side, if the LRR list is empty, the last node in the LRU list will be chosen as the victim, because in this case all the nodes in DRAM are all write-intensive and we evict out the coldest node.

Algorithm 1 shows the node immigration algorithm *Node_Immigration*, which is used to move a write-intensive node from PCM to DRAM Algorithm 2 shows the node management algorithm *Node_Management*.

Algorithm 1. Node_Immigration (ROOT, NODE)

```
1     if DRAM is full then
2       if LRR is not empty then
3         Move the last node of LRR to PCM;
4       else
5         Move the last node of LRU to PCM;
6       end if
7     end if
8     Write NODE on DRAM;
9     Adjust LRU;
10    return ROOT;
```

Algorithm 2. Node_Management (ROOT, NODE)

```
1     if NODE is on PCM and Tendency(NODE)=WRITE then
2       Node_Immigration (ROOT, NODE);
3     end if
4     if NODE is on DRAM then
5       Adjust LRU;
6       if Tendency(NODE)=READ then
7         Adjust LRR;
8       end if
9     end if
10    return ROOT;
```

The Tendency(NODE) routine in Algorithm 2 returns the read/write tendency of a leaf node. In order to determine the read/write tendency of a leaf node, we propose to calculate a write proportion *ratio* of a leaf node and use a threshold-based approach to determine read/write tendency. When *ratio* is below a given threshold, the node is regarded to have read tendency, otherwise it is regarded to have write tendency. In addition, we use a sliding time window to periodically calculate the *ratio*.

$$ratio = ratio_h * weight_h + ratio_c * weight_c \tag{1}$$

The *ratio* is calculated by (1). $ratio_h$ is the total write proportion and $ratio_c$ is the write proportion in current time window. $weight_h$ and $weight_c$ are the weights of $ratio_h$ and $ratio_c$, respectively. Such calculations can balance the historical information with recent access information. Recent access information has higher weight. Because the access characteristics of elements may change with time, it can avoid the situation that historical data misleads the current type judgment.

3.3 Operations of the XB+-Tree

Algorithm 3 shows the insert operation of the XB+-tree. When inserting *<key, value>*, we first find the target leaf node. If a split operation is needed, we create a new leaf node that is on the same media with the target node. Then, we adjust the overflow chain. Finally, the read/write tendency of the target leaf node is calculated and we trigger the migration mechanism if needed.

Algorithm 3. Insert(ROOT, key, value)

```
1    Insert <key, value> into the target leaf node NODE;
2    if NODE need to split then
3      Split NODE;/*new node and NODE are on the same media*/
4      Adjust the overflow chain;
5    end if
6    Update the access information for NODE ;
7    Node_Management(ROOT, NODE);
8    return ROOT;
```

Algorithm 4. Delete(ROOT, key)

```
1    Delete key and its value from the target leaf node NODE;
2    if NODE need to merge or distribute then
3      if NODE has a neighbor on the same media then
4        Find a neighbor;
5      else
6        Find a neighbor on different media;
7      end if
8    end if
9    Update the access information for NODE;
10   Node_Management(ROOT, NODE);
11   return ROOT;
```

Algorithm 4 shows the delete operation of the XB+-tree. When deleting *key*, we first find the target node. If merging and distributing is needed, we find a neighbor from the same media.

Algorithm 5 shows the search operation of the XB+-tree. When searching the key *key*, we find the target leaf node and search in it. Finally, we calculate its access tendency to see whether migration is needed.

Algorithm 5. Search(ROOT, key)

1	Find the target leaf node NODE in the overflow chain;
2	Search within NODE;
3	Node_Management(ROOT, NODE)
4	**return** ROOT;

---

4   Performance Evaluation

4.1   Experimental Setup

We conduct comparative experiments to compare our proposal with three existing indices, which are the traditional B+-tree, the OB+-tree [4, 8], and the CB+-tree [9]. The OB+-tree without the migration mechanism is used as the baseline index. The length of the overflow chain in the OB+-tree, the CB+-tree, and the XB+-tree is all set to 6. We run experiments on a computer with Ubuntu 14.04, a CPU of AMD Athlon II X2, 4 GB RAM, and 1 TB Seagate hard disk. In addition, we use DRAM to simulate PCM by artificially set the read/write latency according to Table 1.

We use the TPC-C workload to generate traces for the experiments. Particularly, we run the TPC-C workload on the open-source PostgreSQL, and use BenchmarkSQL [10] to collect page requests on the tables contained in the TPC-C. When running the TPC-C workload, 10 warehouses and 100 clients are configured. The TPC-C workload contains eight index files built on eight tables. The size of the tables is approximately 1 GB. We first perform 5.4 million insertion requests to build eight indices and get the original B+-tree index file that is about 135 MB. Then, we generate a trace containing about 3.8 million index requests including 74.2 % searches, 23.8 % insertions, and 1 % deletions. We run this trace on each index to measure the performance.

4.2   Results

We measure the read and write count on PCM of each compared index. We use two kinds of settings. In the first setting, we vary the node size of each index, while in the second setting, we vary the buffer size.

Figures 4 and 5 show the PCM read/write count when varying the node size and the cache size. Figures 4(a) and 5(a) show the read counts of each index. We can see that

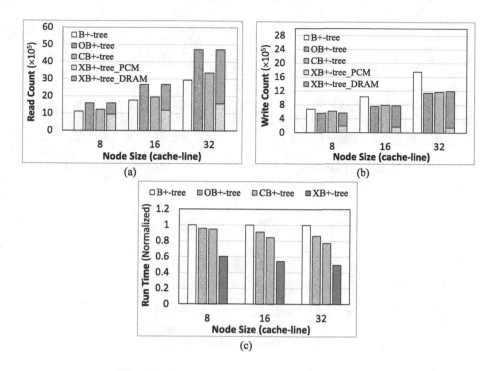

Fig. 4. PCM performance w.r.t. different node sizes

the OB+-tree has the highest read count because of its overflow chain. However, although the XB+-tree has the same overflow chain as the OB+-tree, it gets the least read count on PCM, which is even less than that of the traditional B+-tree. This is because that the migration mechanism of the XB+-tree makes the write-intensive nodes be stored on DRAM. Thus, the read requests to these nodes are all re-directed to DRAM rather than on PCM. This indicates that our migration design that moving write-intensive nodes to DRAM is helpful for reducing PCM read operations. The total read count of the XB+-tree on DRAM and PCM is the same as that of the OB+-tree, because the migration mechanism does not change the overflow structure.

Figures 4(b) and 5(b) show the write counts of the cache-lines. The traditional B+-tree has the highest write count because it does not use the overflow scheme. The write counts of the OB+-tree and the CB+-tree are almost the same but less than that of the traditional B+-tree. Owing to the migration mechanism that moves the majority of write operations to DRAM, the XB+-tree has less PCM writes compared with the B+-tree and the CB+-tree. The total write count of the XB+-tree is a bit higher than that of the OB+-tree because the node immigration algorithm in the XB+-tree causes extra write operations. Note that the XB+-tree has much less PCM reads than the OB+-tree. To this end, the XB+-tree has better overall performance than the OB+-tree, which is proven in

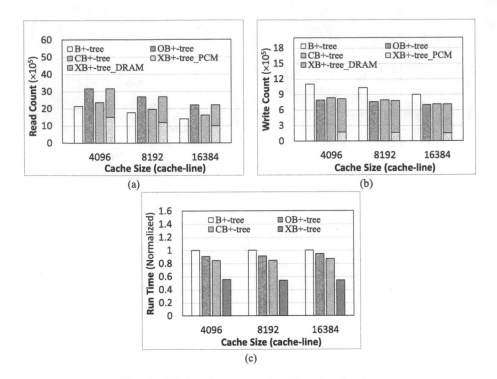

Fig. 5. PCM performance w.r.t. different cache sizes

Figs. 4(c) and 5(c). When the node size is set to 8 and 16, the XB+-tree performs better than CB+-tree, but when the node size gets larger, the write times of XB+-tree becomes bigger than the CB+-tree. We know that the more entries a node contains, the harder the access tendency can be determined. In other words, the access tendency of a node may change a lot with time. This can bring frequently migrate operations which cause writes to PCM.

Figures 4(c) and 5(c) show the run time of each index which is normalized according to the run time of the B+-tree, i.e., the run time of the B+-tree is always set to 1. The proposed XB+-tree has the best time performance. Although the PCM write count of the XB+-tree is a bit more than that of the OB+-tree, the XB+-tree has better read performance.

In addition, we record the number of the leaf nodes that are moved from PCM to DRAM when varying the node size. The result is shown in Fig. 6 All the write-intensive nodes that migrate to DRAM is only 2.17 %, 3.0 %, and 7.02 % among all the leaf nodes, respectively. This indicates that our migration mechanism has a small space cost.

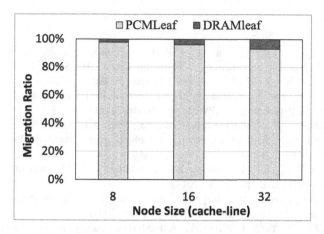

Fig. 6. The ratio of the leaf nodes moved from PCM to DRAM

5 Conclusions

PCM has been regarded as a new kind of future memories. Its non-volatility and high density makes it much suitable for non-volatile main-memory databases, where large volumes of data can be stored in PCM and can be operated by CPU directly. However, the use of PCM in non-volatile main-memory databases introduces many new issues. In this paper, we exploited the indexing problem of PCM-based hybrid memory systems. We proposed an efficient tree indexing approach called XB+-tree that is an improved version of the OB+-tree. We use DRAM to compensate the shortcomings of PCM, and proposed a node management algorithm for DRAM, which detects read/write tendency of nodes to store nodes with different kinds of access tendency on DRAM or PCM. We conducted experiments to compare our proposal with the traditional B+-tree, the OB+-tree, and the CB+-tree. The results in terms of PCM read/write count and run time showed that our proposal is efficient for PCM-based hybrid memory systems.

Our future work will concentrate on improving the method of predicting read/write tendency. The present method is a simple weighted summation that considers the portion of writes within the current time window as well as the portion of writes among all the operations during the measured time period. One possible improvement is to consider two or more time windows so that we can capture the read/write tendency more precisely. However, this will introduce additional maintaining costs. Therefore, we will investigate a k-window algorithm that can reach the best trade-off between maintaining costs and tendency prediction precision.

Acknowledgements. This work is partially supported by the National Science Foundation of China under the grant number 61472376, the Fundamental Research Funds for the Central Universities, and a fund from the Science and Technology on Electronic Information Control Laboratory.

References

1. Wu, Z., Jin, P., Yang, C., Yue, L.: APP-LRU: a new page replacement method for PCM/DRAM-based hybrid memory systems. In: Hsu, C.-H., Shi, X., Salapura, V. (eds.) NPC 2014. LNCS, vol. 8707, pp. 84–95. Springer, Heidelberg (2014)
2. Chen, K., Jin, P., Yue, L.: Efficient buffer management for PCM-Enhanced hybrid memory architecture. In: Proceeding of APWeb, pp. 29–40 (2015)
3. Jiang, L., Zhang, Y., Childers, B.R., Yang, J.: FPB: fine-grained power budgeting to improve write throughput of multi-level cell phase change memory. In: Proceeding of MICRO, pp. 1–12 (2012)
4. Chi, P., Lee, W.C., Xie, Y.: Making B+-tree efficient in PCM-based main memory. In: Proceeding of ISLPED, pp. 69–74 (2014)
5. Lee, B.C., Ipek, E., Mutlu, O., Burger, D.: Architecting phase change memory as a scalable dram alternative. In: Proceeding of ISCA, pp. 2–13 (2009)
6. Zhou, P., Zhao, B., Yang, J., Zhang, Y.: A durable and energy efficient main memory using phase change memory technology. In: Proceeding of ISCA, pp. 14–23 (2009)
7. Caulfield, A.M., De, A., Coburn, J., Mollow, T.I., Gupta, R.K., Swanson, S.: Moneta: a high-performance storage array architecture for next-generation, non-volatile memories. In: Proceeding of MICRO, pp. 385–395 (2010)
8. Jin, P., Yang, C., Jensen, C.S., Yang, P., Yue, L.: Read/Write-optimized tree indexing for solid state drives. VLDB J. (2016). doi:10.1007/s00778-015-0406-1
9. Li, L., Jin, P., Yang, C., Yue, L.: Optimizing B+-tree for PCM-based hybrid memory. In: Proceeding of EDBT, pp. 662–663 (2016)
10. BenchmarkSQL. http://sourceforge.net/projects/benchmarksql/

Effective Order Preserving Estimation Method

Chen Chen[1(✉)], Wei Wang[1], Xiaoyang Wang[2], and Shiyu Yang[1]

[1] The University of New South Wales, Sydney, Australia
{cchen,weiw,yangs}@cse.unsw.edu.au
[2] University of Technology, Sydney, Australia
xiaoyang.wang@uts.edu.au

Abstract. Order preserving estimation is an estimation method that can retain the original order of the population parameters of interest. It is an important tool in many applications such as data visualization. In this paper, we focus on the population mean as our primary estimation function, and propose effective query processing strategy that can preserve the estimated order to be correct with probabilistic guarantees. We define the cost function as the number of samples taken for all the groups, and our goal is to make the sample size as small as possible. We compare our methods with state-of-the-art near-optimal algorithm in the literature, and achieve up to 80 % reduction in the total sample size.

Keywords: Random sampling · Order guarantee

1 Introduction

Statistical method is essential for analyzing the massive data that arises in many applications. A synopsis of a large dataset captures vital properties of the original data while typically consuming much less resources. One of the most widely used properties is the population mean, which is the average value of a group of numeric data. The methods and criteria for a reliable population mean estimation are well developed in both statistical inference and computer science areas. However, when it comes to several groups of data, it is long neglected that an order estimation on the population means is of equal importance, since many applications require to know data trends or more specifically the relationships among variables [5,9,14]. Motivated by above observations, we study the order preserving estimation (OPE) problem in this paper. Given k groups of numeric data with unknown distribution along with $\delta \in (0,1)$, OPE returns an order estimate on the group average with a probabilistic guarantee $1 - \delta$. With the rapid development of modern computing ability, the offline sample processing time is less concerned in many applications. Instead, the sample size becomes a critical factor to be restrained. For example, the I/O cost of obtaining data from external or distributed storage is very high. Or in most clinical trials of new drugs, the number of human subjects is usually limited due to the risk or ethical issues. In both cases, it is not affordable to get as many samples as we want.

© Springer International Publishing AG 2016
M.A. Cheema et al. (Eds.): ADC 2016, LNCS 9877, pp. 369–380, 2016.
DOI: 10.1007/978-3-319-46922-5_29

Therefore, we consider the total sample size as the most important parameter in our algorithm design.

The state-of-the-art method [10] guarantees to output a correct order and is proved to be near optimal in the total sample size. Nevertheless, an important assumption for the near-optimality is that it does not allow the inactive confidence intervals to become to be active again. The assumption affects the logical strengths of the proof. Because the crude elimination of sampling outliers is unrealistic in real applications. Even though the assumption could be ignored without impairing the theoretical soundness, [10] tends to examine more samples than needed, which means their sample complexity upper bound is conservative empirically.

Contribution. Our improvements have two aspects, stop condition and sample strategy, respectively. We design two stop functions, named IntervalSeparation and PairwiseComparison. They utilize the relation among current sample means to make judicious decisions, in order to output a correct result while using as little samples as possible. We also propose a heuristic sample strategy to assign new sample points adaptively. Specifically, IntervalSeparation reduces order estimates to separating the underlying confidence intervals. Due to tail inequality, all true means can be bounded within their confidence intervals with high probabilites. Thus as long as the intervals do not overlap, the order can be guaranteed. Next in the PairwiseComparison method, we first show that minimizing the sample size in OPE is NP-hard by reducing from TSP problem on tournament graphs. In order to compute the failure probability, we introduce the prefix downsample technique to equalize the sample size between adjacent sample groups, where the unequal sample size is due to the proposd sample strategy.

The rest of the paper is organized as follows: Sect. 2 defines the problem and gives an overview of our algorithm framework. Section 3 introduces the interval separation algorithm and proves its correctness. Section 4 analyzes the pairwise comparison algorithm and the prefix downsample technique, along with the sample strategy. Section 5 shows our experimental results. Section 6 surveys related work and Sect. 7 concludes the paper.

2   Problem Definition and Proposed Framework

We first formulate the problem of order preserving estimation, then give an overview of our proposed framework.

2.1   Problem Definition

An important concept used in our algorithms is the confidence interval. Given $\epsilon, \delta \in (0, 1)$, for an estimated sample mean $\hat{\mu}$, let its confidence interval be $[\hat{\mu} - \epsilon, \hat{\mu} + \epsilon]$ with confidence level $1 - \delta$. It means that the population mean μ belongs to $[\hat{\mu} - \epsilon, \hat{\mu} + \epsilon]$ with at least $1 - \delta$ probability, which is denoted by $\mathbf{Pr}\left[|\mu - \hat{\mu}| \leq \epsilon\right] \geq 1 - \delta$.

In this paper, we use the empirical Bernstein-Serfling inequality proposed in [1] to derive robust and tight confidence intervals.

Definition 1 (Order Preserving Estimation). *Given k groups $g_1, g_2, \cdots, g_k$ of numeric values where the values in g_i are bounded by $[a_i, b_i]$. Denote the group population mean of elements in g_i as $\mu_i = \frac{1}{|g_i|}\sum_{e \in g_i} e$. Given $\delta \in (0, 1)$, an order preserving estimation for the group mean returns an order estimate on μ_i for $i \in [1, k]$, s.t. the order is correct with at least $1 - \delta$ probability. Correct order means if the population means satisfy $\mu_i > \mu_j$, the sample means also have $\hat{\mu}_i > \hat{\mu}_j$ for all $i, j \in [1, k]$.*

As justified in Sect. 1, the sample size is a major concern in various applications. In order to meet the ultimate user requirements, we model OPE as an minimization problem. The goal is to minimize the sample complexity. In such case, we consider the total sample size as the dominant factor in our analysis.

Definition 2 (Minimization Problem for OPE). *Let n_i be the number of samples taken from group i before the algorithm terminates, OPE aims to minimize $\sum_{i=1}^{k} n_i$, which is the total sample complexity of all the k groups.*

To achieve the minimization goal, we need to relate the sample complexity $\sum_{i=1}^{k} n_i$ to the user specified failure probability δ in the randomization framework. Therefore we adopt the empirical Bernstein-Serfling inequality for sampling without replacement [1], along with the union bound as our primary tools to solve the problem. The superiority lies in the fact that they does not make assumptions on the data distribution. As a consequence, OPE can return a correct order for *any* data distribution with a high probability.

2.2 Framework of OPE

In this section, we give an overview of our algorithms, followed by the intuition of our improvements.

The user issues a query about the order of k groups. An efficient **sample strategy** is designed to incrementally get random samples and update sample means. We use random sample without replacement throughout the paper, which will lead us to the population mean when the sample size equals population size. In the meantime, a **stop function** computes an order estimate and decides if it satisfies the probabilistic guarantee hence can stop. Above process is repeated until the stop condition is satisfied. The procedure is summarized in Algorithm 1.

Our key observation is that some groups are easy to be ordered correctly, as their population means have large difference. We call it easy case which only needs few samples to get a correct order. If relating easy case to the stop function, it will result in a small failure probability δ_i as the order estimate is unlikely to be incorrect. On the contrary, hard case refers to ordering the groups with small differences among their population means. It will cause large δ_i hence consuming a large amount of samples for a correct result. Therefore, we need to design a judicious sample strategy that can assign random samples *adaptively*, as well as effective stop functions that can allocate δ_i *dynamically*. We will materialize StopFunction in Algorithm 1 as IntervalSeparation (Sect. 3) and PairwiseComparison (Sect. 4). SampleStrategy (Line 5) is introduced at the end of Sect. 4.

Algorithm 1. OPE$(g_1, \cdots, g_k, \delta)$

Input: Data from $g_1, g_2, \cdots, g_k, \delta$.
Output: Order estimates $\hat{\mu}_1, \hat{\mu}_2, \cdots, \hat{\mu}_k$ on the group means.
1 $n \leftarrow 1$;
2 Draw n samples from g_i to produce initial sample mean $\hat{\mu}_i$, $i \in [1, k]$;
3 $\mathcal{F} \leftarrow$ StopFunction;   /* Estimate an order and compute the stop function */;
4 **while** $\mathcal{F} > \delta$ **do**
5 | Get new samples according to SampleStrategy then update $\hat{\mu}_i$;
6 | $\mathcal{F} \leftarrow$ StopFunction;   /* Update order and stop function */;
7 **return** Estimated order;

---

3   Interval Separation Method

We first give the intuition of IntervalSeparation, then define the stop function in Sect. 3.1 followed by its proof of correctness in Sect. 3.2.

Intuition. To order group g_i and g_j, we can examine the confidence intervals of their population means μ_i and μ_j. If both μ_i and μ_j are within respective confidence intervals with high probabilities, and the two intervals do not overlap, we can output an order that is probabilistic correct. When it comes to k groups, the idea is interleaving the interval boundaries $\{x_1, \cdots, x_{k-1}\}$ with k sample means that are ordered decreasingly, then deriving the total failure probability based on the union bound.

3.1   IntervalSeparation Stop Function

We will order the sample means decreasingly before evaluating the stop function $f_{\min}$. The ordered means are denoted by $\hat{\mu}_1, \cdots, \hat{\mu}_k$. To keep the notation clean, we use $\hat{\mu}_i$ as the i-th largest sample mean in the rest of Sect. 3, instead of the sample mean of i-th group used previously. As shown in Eq. (1), we first define function f, then minimize it in terms of $\mathbf{x} = \{x_1, \cdots, x_{k-1}\}$, where $x_i \in (\hat{\mu}_{i+1}, \hat{\mu}_i)$, and use $f_{\min}$ as the stop function. If $f_{\min}$ is no greater than the user specified δ, the algorithm are safe to stop. $f_{\min}$ in Eq. (1) will replace Line 3 & 6 of Algorithm 1. The stop condition is to decide if $f_{\min}$ exceeds δ at each round. Determining $f_{\min}$ is equivalent to finding the root of $\frac{\partial f}{\partial x_i}$ between $x_i \in (\hat{\mu}_{i+1}, \hat{\mu}_i)$. Since the closed-form expression for the root is unavailable, we use binary search to get an approximate result.

$$f(\mathbf{x}) = 5 \sum_{i=1}^{k-1} \left[\exp\left(-\left(\frac{\hat{\sigma}_i \sqrt{2\rho_i n_i} + \sqrt{2n_i(\rho_i \hat{\sigma}_i^2 + 2r_i \kappa(\hat{\mu}_i - x_i))}}{2r_i \kappa} \right)^2 \right) \right.$$
$$\left. + \exp\left(-\left(\frac{\hat{\sigma}_{i+1}\sqrt{2\rho_{i+1} n_{i+1}} + \sqrt{2n_{i+1}(\rho_{i+1}\hat{\sigma}_{i+1}^2 + 2r_{i+1}\kappa(x_i - \hat{\mu}_{i+1}))}}{2r_{i+1}\kappa} \right)^2 \right) \right],$$

$$f_{\min} = \min f(\mathbf{x}), \text{ subject to } x_i \in (\hat{\mu}_{i+1}, \hat{\mu}_i),$$

$$(1)$$

where for each group g_i, $\hat{\sigma}_i$ is the sample standard deviation, $\hat{\mu}_i$ is the sample mean, n_i and N_i are the sample and population size, $r_i = b_i - a_i$, $\kappa = \frac{7}{3} + \frac{3}{\sqrt{2}}$, and:

$$\rho_i = \begin{cases} 1 - \frac{n_i - 1}{N_i} & \text{, if } n_i \leq \frac{N_i}{2}. \\ (1 - \frac{n_i}{N_i})(1 + \frac{1}{n_i}) & \text{, if } n_i > \frac{N_i}{2}. \end{cases}$$

3.2   Analysis

We prove the correctness of IntervalSeparation in Theorem 1.

Theorem 1. *Replacing Line 3 and Line 6 of Algorithm 1 with Eq. (1) can return a correct order with at least $1 - \delta$ probability.*

Proof. Given a group of size N, assume each sample X_i is bounded by $[a, b]$, then we know the sample mean $\hat{\mu} = \frac{1}{n}\sum_{i=1}^{n} X_i$, and the population mean $\mu = \mathbf{E}[\hat{\mu}]$. Given ordered sample means $\hat{\mu}_1, \cdots, \hat{\mu}_k$, let $x_i \in (\hat{\mu}_{i+1}, \hat{\mu}_i)$, and define $\epsilon_i = |x_i - \hat{\mu}_i|$. According to the empirical Bernstein-Serfling Inequality [1], for the upper side of g_i, we have

$$\mathbf{Pr}\left[\mu_i \leq x_i\right] \leq 5\exp\left(-\left(\frac{\hat{\sigma}_i\sqrt{2\rho_i n_i} + \sqrt{2n_i(\rho_i\hat{\sigma}_i^2 + 2r_i(\hat{\mu}_i - x_i)\kappa)}}{2r_i\kappa}\right)^2\right).$$

Analogously, for the lower side of g_{i+1},

$$\mathbf{Pr}\left[\mu_{i+1} \geq x_i\right] \leq 5\exp\left(-\left(\frac{\hat{\sigma}_{i+1}\sqrt{2\rho_{i+1} n_{i+1}} + \sqrt{2n_{i+1}(\rho_{i+1}\hat{\sigma}_{i+1}^2 + 2r_{i+1}(x_i - \hat{\mu}_{i+1})\kappa)}}{2r_{i+1}\kappa}\right)^2\right).$$

Above derivations are applicable to all the boundaries x_i where i belongs to $[1, k - 1]$. Due to the continuity of f and the union bound, the correctness is concluded by

$$\mathbf{Pr}\left[\bigvee_{i=1}^{k-1}\left(x_i \in (\hat{\mu}_{i+1}, \hat{\mu}_i)\right)\right] \geq 1 - f_{\min} \geq 1 - \delta.$$

Remember that a confidence interval has two ingredients ϵ and δ. It can be interpreted as $\mathbf{Pr}[\text{error is within } \epsilon] \geq 1 - \delta$. Given k groups, the methods in [10] fix all the δ_i to be $\frac{\delta}{k}$, and derive the associated interval $[\hat{\mu}_i - \epsilon_i, \hat{\mu}_i + \epsilon_i]$ then test if there are overlaps among different intervals. The difference between IntervalSeparation and that in [10] is that we do it inversely. To be specific, we first place $\mathbf{x}$ as interval boundaries that satisfy the nonoverlapping condition. Then we compute the associated δ_i and define $\sum_{i=1}^{k-1} \delta_i$ as $f(\mathbf{x})$. Next we test if there exists $\mathbf{x}$ such that $f_{\min}$ is no greater than the given δ. It is equivalent to allocating δ_i dynamically. The advantage is that it assigns less δ_i to the easy cases while more δ_i to the hard cases for an early termination.

4 Pairwise Comparison Method

We first give the intuition of PairwiseComparison. Then we illustrate the steps for doing prefix downsample between adjacent groups with different sample sizes and explain its necessity. After that we define the stop function and prove it is NP-hard to solve the OPE minimization problem in Definition 2.

Intuition. Suppose we have values μ_a, μ_b and μ_c and want to order them decreasingly. A natural way is to show $\mu_a - \mu_b > 0$ and $\mu_b - \mu_c > 0$. Then it must have $\mu_a > \mu_b > \mu_c$ according to the transitivity of the greater than relation. We apply this simple yet effective idea with probabilistic constraint and propose the stop function for PairwiseComparison.

4.1 Prefix Downsample

This section considers prefix downsample method that can be maintained efficiently as the samples update. The sampling method we use is random sample without replacement.

Definition 3 (Prefix Downsample). *Given a group, denote its sample as an ordered[1] set $S = \{X_1, X_2, \cdots, X_n\}$, where n is the sample size. A prefix downsample of size m is defined as $S_{pre} = \{X_1, X_2, \cdots, X_m\}$, which is the first m elements of S.*

Given S and the corresponding S_{pre}, *prefix downsample mean* is the sample mean of m samples in S_{pre}, while full sample mean refers to the standard sample mean of S. It is easy to see that both of them are unbiased estimators of the population mean. Lemma 1 shows that a prefix downsample of S is a valid random sample for the same population.

Lemma 1. *Given a random sample S, any prefix S_{pre} of size m is a valid prefix downsample that can be used for evaluating the stop function.*

The proof is immediate by considering that S is induced from a random permutation. As a result, any prefix of size $m < n$ is also a random sample of the same population. Next we consider how to apply prefix downsample between two groups and define the relation "$>_{pre}$" between them.

Definition 4 (Binary Relation "$>_{pre}$" and "$-_{pre}$"). *Given two groups g_i and g_j with sample size n_i and n_j, denote their full samples as $\hat{\mu}_i$ and $\hat{\mu}_j$. Let the prefix downsample size m be $\min(n_i, n_j)$ for both groups. Then it holds $\hat{\mu}_i >_{pre} \hat{\mu}_j$ if the prefix downsample mean of g_i is greater than that of g_j. In addition, we use $\hat{\mu}_i -_{pre} \hat{\mu}_j$ to denote the difference between the prefix downsample means of g_i and g_j.*

---

[1] The order is induced from progressive sampling without replacement process. I.e., $Rank(X_i) < Rank(X_j)$, if $i < j$.

Prefix downsample is designed to equalize the sample size between adjacent groups to be compared in Eq. (4), in order to apply the empirical Bernstein-Serfling inequality. To be more specific, assume the adjacent groups are g_i and g_j. We will define $\hat{\mu}_i -_{pre} \hat{\mu}_j$ as an *individual* random variable to apply the inequality. It means the sample size of g_i and g_j must be identical due to the requirement of empirical Bernstein-Serfling inequality. Therefore, we need to find out the criteria for extracting a correct downsample that makes the most of available samples and can be maintained efficiently as S updates. Although we can use any size m prefix in terms of the algorithm correctness, we deem $m = \min(n_i, n_j)$ as the most effective one. Since it manages to take advantage of all the sample information in the adjacent groups with sample size n_i and n_j, respectively.

4.2   PairwiseComparison Stop Function

Before introducing the stop function h, we need to define candidate order as the input to the function.

Candidate Order. Given k groups and an arbitrary order $g_1, g_2, \cdots, g_k$, [2] g_i denotes the i-th rank in the order, and adjacent groups refer to (g_i, g_{i+1}), $i \in [1, k-1]$. If it holds $\hat{\mu}_i >_{pre} \hat{\mu}_{i+1}$ for all $i \in [1, k-1]$, we say that $g_1, g_2, \cdots, g_k$ is a candidate order.

Now suppose we have a candidate order with full sample means $\hat{\mu}_1, \cdots, \hat{\mu}_k$, we can define the stop function h for PairwiseComparison in Eq. (2) . h will replace Line 3 & 6 of Algorithm 1. The stop condition is to decide if h exceeds δ at each round.

$$h = \sum_{i=1}^{k-1} 5 \exp\left(-\left(\frac{\hat{\sigma}_i \sqrt{2 n_i \rho_i} + \sqrt{2 n_i \left(\rho_i \hat{\sigma}_i^2 + 2 r_i \kappa |\hat{\mu}_i -_{pre} \hat{\mu}_{i+1}|\right)}}{2 r_i \kappa} \right)^2 \right), \qquad (2)$$

where for each group g_i, we use $\hat{\mu}_i$ as the sample mean, n_i and N_i as the sample and population size, and:

$$\rho_i = \begin{cases} 1 - \frac{n_i - 1}{\min(N_i, N_{i+1})} & \text{, if } n_i \leq \frac{\min(N_i, N_{i+1})}{2}. \\ \left(1 - \frac{n_i}{\min(N_i, N_{i+1})}\right)\left(1 + \frac{1}{n_i}\right) & \text{, if } n_i > \frac{\min(N_i, N_{i+1})}{2}. \end{cases} \qquad (3)$$

$r_i = (b_i + b_{i+1}) - (a_i + a_{i+1})$, $\kappa = \frac{7}{3} + \frac{3}{\sqrt{2}}$, and $\hat{\sigma}_i$ is the sample standard deviation of $\hat{\mu}_i -_{pre} \hat{\mu}_{i+1}$.

Our observation is that determining whether the algorithm can stop is equivalent to deciding if there exists a candidate order with $h \leq \delta$. Nevertheless, it is NP-hard to solve the decision problem. It is because a given group g_i may have different prefix downsample means when downsampling with g_{i-1} and g_{i+1}, due to different downsample size $\min(n_{i-1}, n_i)$ and $\min(n_i, n_{i+1})$.

Given sample set S_i, full sample size n_i, full sample mean $\hat{\mu}_i$ and δ, the desicion problem for PairwiseComparison in OPE is defined as $\{\langle (S_i, n_i, \hat{\mu}_i)_{i=1}^k, \delta \rangle$: There exists a candidate order on the k groups, such that it holds $h \leq \delta\}$.

---

[2] Same as before we use g_i as the group with rank i in a candidate order in the rest of Sect. 4 in order to keep the notation clean.

Theorem 2. *PairwiseComparison in OPE is NP-hard.*

Proof Sketch. We will reduce from a known NP-hard problem "Traveling-Salesman Problem on Tournament graph (TSPT)" to prove the NP-hardness of OPE. The reduction can be summarized as follows. Without loss of generality, let n_i be arbitrary integers satisfying $n_1 < n_2 < \cdots < n_k$. For each vertex V_i in a tournament graph G, the reduction algorithm transforms it to group g_i in OPE. Then the cost function c_{ij} between V_i and V_j in TSPT is normalized and equated to Eq. (5). The sample set S_i hence $\hat{\mu}_i$ can be easily constructed from the equation. Finally the directed edge e_{ij} is transformed to "$>_{pre}$" relation between $\hat{\mu}_i$ and $\hat{\mu}_j$. This process can be done in polynomial time.

Consequently, we can use TSP algorithms to accelerate the process of finding the candidate order required.

4.3   Analysis

We prove the correctness of PairwiseComparison in Theorem 3.

Theorem 3. *Replacing Line 3 and Line 6 of Algorithm 1 with Eq. (2) can return a correct order with at least $1 - \delta$ probability.*

Proof. Given a candidate order with full sample mean $\hat{\mu}_i$ for $i \in [1, k]$, we examine the confidence interval of $\hat{\mu}_i -_{pre} \hat{\mu}_{i+1}$. According to the empirical Bernstein-Serfling inequality [1], given $\epsilon_i \in [0, \hat{\mu}_i -_{pre} \hat{\mu}_{i+1}]$, it has

$$\mathbf{Pr}\left[\mu_i - \mu_{i+1} > \hat{\mu}_i -_{pre} \hat{\mu}_{i+1} - \epsilon_i\right] \geq 1 - \delta_i. \tag{4}$$

The expression for δ_i is

$$\delta_i = 5\exp\left(-\left(\frac{\hat{\sigma}_i\sqrt{2n_i\rho_i} + \sqrt{2n_i\left(\rho_i\hat{\sigma}_i^2 + 2r_i\epsilon_i\kappa\right)}}{2r_i\kappa}\right)^2\right). \tag{5}$$

When ϵ_i decreases from $\hat{\mu}_i -_{pre} \hat{\mu}_{i+1}$ to 0, δ_i will increase monotonically. In order to achieve an early stop, we will let ϵ_i be $\hat{\mu}_i -_{pre} \hat{\mu}_{i+1}$. Fit into Eq. (4), we know

$$\mathbf{Pr}\left[(\mu_i - \mu_{i+1} > \hat{\mu}_i -_{pre} \hat{\mu}_{i+1} - \epsilon_i) \wedge (\epsilon_i = \hat{\mu}_i -_{pre} \hat{\mu}_{i+1})\right] \geq$$

$$1 - 5\exp\left(-\left(\frac{\hat{\sigma}_i\sqrt{2n_i\rho_i} + \sqrt{2n_i\left(\rho_i\hat{\sigma}_i^2 + 2r_i(\hat{\mu}_i -_{pre} \hat{\mu}_{i+1})\kappa\right)}}{2r_i\kappa}\right)^2\right).$$

Apply union bound to all the $k-1$ pairs of adjacent groups hence the correctness is concluded   □

Sample Strategy. We introduce our heuristic sample strategy in this section. As the data distribution is unknown to the algorithm, we cannot predict the gain of each sample. Therefore, we adopt the heuristic method to choose the next group to sample from, based on the failure probabilities at current sample round. The intuition is that if the population means of two groups are very close, it will need more samples to separate them than to separate the groups with large difference between their means.

Since our goal is to minimize the sample compelxity instead of the number of rounds the sampling proceeds, we will use the most conservative sampling scheme, i.e. at each round, add one more sample to one of the k groups. Specifically, the strategy will randomly choose the next sample from the group that introduces the most failure probability among all the groups at current round. The proposed sample strategy considers the distribution of all the groups' failure probabilities at current round. Large failure probability implies small difference between sample means, hence it is harder to get the correct order with a high confidence than the groups with far sample means between them. Moreover, the sample means of the groups with small failure probability are stable enough to draw a correct order. Therefore, we spend the rest samples on the groups with large failure probabilities.

5   Experiments

In this section, we present the results of a comprehensive performance study to evaluate the effectiveness of the proposed techniques in this paper.

5.1   Experiment Setup

In this paper we focus on reducing the total sample size required by OPE. Thus we report the sample ratio of the proposed methods by varying different parameters on both real and synthetic datasets.

Algorithms. We compare our methods with state-of-the-art algorithm IFOCUS in [10]. It stops sampling from a group as long as its confidence interval disjoints with all the other groups. To summarize, the algorithms evaluated in this section are listed below.

- **IFOCUS**. The state-of-the-art approach in [10].
- **SEP**. The Interval Separation method presented in Sect. 3.
- **COMP**. The Pairwise Comparison method introduced in Sect. 4.

Datasets. We use both real and synthetic datasets. For real datasets, we use the flight records in [6]. We utilize the "ActualElapsedTime" attribute, which denotes the actual travel time for each flight. The attribute represents the marketing preference of the flight company (e.g., short-haul fight or long-haul flight). We use two years' data, 2004 and 2005, which contain 7 million records for each, with 19 and 20 flight companies, respectively. For the synthetic datasets, we generate data from two distributions, uniform (Uniform) and mixture of truncated normal (MixNormal). For each group, we generate 1 million records. For Uniform we randomly select a mean from [0,50] and a data range from [70, 100]. For MixNormal, we select a set of truncated normal distributions in the following manner. Firstly we generate a number from $\{1, 2, 3, 4, 5\}$ indicating the number of truncated normal distributions that comprise the group. For each of the truncated normal distribution, we randomly select a mean from [0,100] and a variance from [1,16].

Workloads. In the experiments, we evaluate the methods by varying the failure probability δ from 0.05 to 0.2 with 0.05 as the default value. The number of groups k increases from 5 to 20 with 20 as the default value. For each setting we run 100 rounds and report the average performance.

Implementation Environment. All experiments are carried out on a PC with Intel Xeon 2.30 GHz and 96G RAM. The operating system is Redhat. All algorithms are implemented in C++ and compiled with GCC 4.8.2 with -O3 flag.

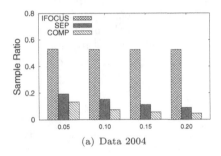
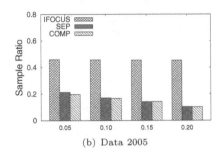

(a) Data 2004   (b) Data 2005

Fig. 1. Performance Evaluation on Real Datasets by Varying δ. (a) Data 2004. (b) Data 2005

5.2 Real Data Experiments

In Fig. 1, we report the sample ratio on the real datasets by varying δ from 0.05 to 0.2. As can be seen, the proposed algorithms, SEP and COMP, require less samples than IFOCUS and can achieve up to 80 % reduction in the total sample size. The gain of COMP is more significant on Data 2004 compared with SEP. In addition, we notice that IFOCUS is not sensitive to δ. By increasing δ, the sample size does not change much. This is due to the sample size upper bound in the IFOCUS algorithm explained in [10]. While for our proposed algorithms, increasing δ will make the sample size decrease. This is reasonable as due to the definition of δ, the correctness requirement relaxes with the increase of δ. Hypothetically, any reported order is deemed to be correct when δ equals 1 (fail probability is 100 %). If the algorithm is sensitive to δ, the user is able to make a trade off between the sample size and the accuracy by tuning δ. For the correctness of the reported order, IFOCUS always returns the correct order by consuming a large sample size. While for our algorithms, they report a wrong order only when δ equals 0.2. However, the percentage of the wrong order is less than 3 %, which still satisfies the requirement of $\delta = 0.2$ setting.

5.3 Synthetic Data Experiments

We will report the performance of the proposed methods on synthetic data by varying the number of data groups k. δ equals 0.05 for all the cases. As shown in Fig. 2, by increasing k, the sample size increases for all the algorithms. This

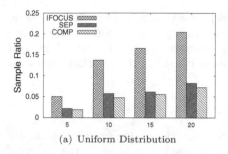

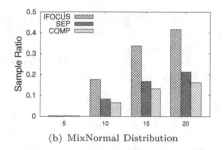

(a) Uniform Distribution   (b) MixNormal Distribution

Fig. 2. Performance Evaluation on Synthetic Datasets by Varying k. (a) Uniform Distribution. (b) MixNormal Distribution

is because when k increases, we need more samples to bound the correctness of the returned order, which is easy to verify from the stop functions (1) and (2). IFOCUS increases much faster than the algorithms proposed in this paper. The sample size needed for Uniform is much smaller than that of MixNormal, since uniform distribution is much easier to infer than normal distribution. COMP requires less samples than SEP, which coincides with the trend on the real datasets.

6   Related Work

Sampling is an important and long-studied method for approximate query processing in the database literature [3,4,7,12,13,15]. The most relevant method is the stratified sampling. Stratification is applied in [8] to increase the accuracy of join size estimation via random sampling. [3] uses query workload information to partition the data into strata in order to maximize the query accuracy. However, none of them considers to bring order into the aggregation for average among different strata. From the statistical point of view, one of the classic results in the stratified sampling theory is the Neyman allocation [11]. Given a fixed total sample size, it allocates sample size for each stratum in order to minimize the total variance. The optimization goal is different from our problem where the sample size is the major concern. Another related problem is hypothesis testing. There are standard methods for testing a single mean or the difference between means [2]. Nevertheless, they need to make assumptions on the distribution, which is not suitable for our setting of arbitrary distribution of the population.

Recently, [10] designs a visualization system that preserves the visual property of ordering population means. The authors improve the round-robin stratified sampling method, and prove the near optimality in the total sample size. The correctness of the order estimation is based on the reasonable estimates at each sample round. Specifically, for a newly added sample point, the updated confidence interval contains the true mean with probability more than $1 - \delta/k$, where k denotes the total number of groups. As all true means are bounded by their confidence intervals, ordering them correctly reduces to separating the underlying intervals. The merit is its simplicity for understanding. Nevertheless, it does not seem to be a tight bound for separating a set of values. Our IntervalSeparation improves the algorithm in [10] by dynamically allocating δ_i at each sample round. And PairwiseComparison instead considers the relation among all the sample means directly.

7   Conclusion

In this paper we propose two effective stop functions along with a heuristic sample strategy to solve the order preserving estimation problem. In the design of an effective stop function, we allocate the failure probability δ_i dynamically based on the current observed sample means. Given the total allowed δ, it amounts to allocating small portions of δ to the easy case while save the rest for the hard case. For the sample strategy, we prioritize the hard case to assign more samples, rather than give all the groups the same amount of samples. By conducting empirical evaluations on both synthetic and real datasets, we demonstrate the effectiveness of our proposed methods.

References

1. Bardenet, R., Maillard, O.A.: Concentration inequalities for sampling without replacement. Bernoulli 21(3), 1361–1385 (2015)
2. Casella, G., Berger, R.: Statistical Inference. Thomson Learning (2002)
3. Chaudhuri, S., Das, G., Narasayya, V.R.: Optimized stratified sampling for approximate query processing. TODS 32(2), 9 (2007)
4. Chaudhuri, S., Motwani, R., Narasayya, V.R.: On random sampling over joins. In: SIGMOD, pp. 263–274 (1999)
5. Cormode, G., Garofalakis, M.N., Haas, P.J., Jermaine, C.: Synopses for massive data: Samples, histograms, wavelets, sketches. Found. Trends Databases 4(1–3), 1–294 (2012)
6. DataExpo,: Flight records (2009). http://stat-computing.org/dataexpo/2009/the-data.html
7. Garofalakis, M.N., Gibbons, P.B.: Approximate query processing: Taming the terabytes. In: VLDB (2001)
8. Haas, P.J., Swami, A.N.: Sequential sampling procedures for query size estimation. In: SIGMOD, pp. 341–350 (1992)
9. Ilyas, I.F., Beskales, G., Soliman, M.A.: A survey of top-k query processing techniques in relational database systems. ACM Comput. Surv. 40(4) (2008)
10. Kim, A., Blais, E., Parameswaran, A.G., Indyk, P., Madden, S., Rubinfeld, R.: Rapid sampling for visualizations with ordering guarantees. PVLDB 8(5), 521–532 (2015)
11. Neyman, J.: On the two different aspects of the representative method: The method of stratified sampling and the method of purposive selection. J. Royal Stat. Soc. 97(4), 558–625 (1934)
12. Nirkhiwale, S., Dobra, A., Jermaine, C.M.: A sampling algebra for aggregate estimation. PVLDB 6(14), 1798–1809 (2013)
13. Piatetsky-Shapiro, G., Connell, C.: Accurate estimation of the number of tuples satisfying a condition. In: SIGMOD, pp. 256–276 (1984)
14. Sun, Y., Wang, W., Qin, J., Zhang, Y., Lin, X.: SRS: solving c-approximate nearest neighbor queries in high dimensional euclidean space with a tiny index. PVLDB 8(1), 1–12 (2014)
15. Vitter, J.S.: Random sampling with a reservoir. ACM TOMS 11(1), 37–57 (1985)

Joint Top-k Subscription Query Processing over Microblog Threads

Liangjun Song[1]([⊠]), Zhifeng Bao[1], Farhana Choudhury[1], and Timos Sellis[2]

[1] Computer Science & IT, RMIT, Melbourne, Australia
{liangjun.song,zhifeng.bao,farhana.choudhury}@rmit.edu.au
[2] Swinburne University of Technology, Melbourne, Australia
tsellis@swin.edu.au

Abstract. With an increasing amount of social media messages, users on the social platforms start to seek ideas and opinions by themselves. Publishers/Subscriber queries (a.k.a. pub/sub) are utilized by these who want to actively read and consume web data. Social media platforms give people opportunities to communicate with others. The social property is also important in the pub/sub while currently no other works have ever considered this property. Also, platforms like Twitter only allow users to post a short message which causes the short-text problem: single posts lack of contextual information. Therefore, we propose the microblog thread as the minimum information unit to capture social and textual relevant information. However, this brings several challenges to this problem: 1. How to retrieve the microblog thread while the stream of microblogs keeps updating the microblog threads and the results of subscription queries keep changing? 2. How to represent the subscription results while the microblog threads are frequently updated? Hence, we propose the group filtering and individual filtering to help to satisfy the high update rate of subscription results. Extensive experiments on real datasets have been conducted to verify the efficiency and scalability of our proposed approach.

1 Introduction

Increasingly, users seek others' opinions and follow the trends on the social networks. The publisher/subscriber (pub/sub) system over Web data is very useful in helping users acquire up-to-date information matching their subscriptions, and has attracted a lot research efforts [2,3,9,10]. As social media is becoming one of the major sources to seek the latest information, it is critical to build a pub/sub system that caters for social media data and its users. This problem emerges most recently and there are only two related works [1,9].

Compared to Web data, social media data has at least three distinguishing features which are also the challenges of this work:

(1) **Short Text** - a microblog has very short text (e.g., up to 140 characters per tweet), that alone cannot contribute enough contextual information to describe a storyline, and there will be a lot of near-duplicate entries for a certain query.

© Springer International Publishing AG 2016
M.A. Cheema et al. (Eds.): ADC 2016, LNCS 9877, pp. 381–394, 2016.
DOI: 10.1007/978-3-319-46922-5_30

(2) **User Interactions** - it is users' interaction (e.g., comment, reply, forward, etc.) that makes social media a unique channel to seek opinions and recommendations from friends. Leveraging the personal network to rank results will greatly improve the search experience as people tend to trust those who are socially closer [3–8].

(3) **High Update Rate** - take Twitter as an example, it has 284 million active users who post 500 million tweets per day[1].

By revisiting those closely related work on the top-k subscription query processing over social media data streams [1,9]. We find that: (1) Both of them focus on how to deal with the high-update rate in maintaining the up-to-date top-k results in Challenge 3, they still adopt a single microblog as the basic information unit of retrieval and indexing, leaving challenge 1 unfilled yet (i.e., such solution is not able to capture the storyline from a single microblog). In particular, Sharer et al. [9] studied how to find textually relevant top-k results while Chen et al. [1] studied a combined metric of textual relevance and diversity. (2) Both works ignore the social similarity part in finding the top-k results, while it has been pointed out in [6] that the computational cost on social similarity is much higher than its counterpart in textual dimension.

Therefore, we are motivated to fill in the gap by proposing a solution to address the above three challenges in one basket, in order to cater for users' various preference on these three dimensions unique to social network streaming data.

In particular, we first propose a tree-like data structure called "Microblog Thread" as the minimum information unit instead of individual microblog. On top of such a tree-like data structure, both the contents and the user interactions among microblogs (such as comment, reply, forward, etc.) can be inherently captured. Second, we define the joint top-k subscription query processing over the streaming Microblog Threads, and propose a series of pruning methods to continuously maintain the up-to-date top-k results for each subscription query as new microblog comes.

The above two components together form a novel framework in providing a continuous subscription query processing framework. In particular, we have made the following technical contributions.

- We propose Microblog Thread (MT) as the atomic information unit to capture user interactions on the microblogs along with the contents. Such a structure addresses challenge 1, as it consists of richer text and contexts. As user interactions over microblogs can naturally show how information propagates through the social network, MTs also solve challenge 2 by dynamically keeping track of user interactions within the microblog thread (Sect. 3).

- We define the joint top-k subscription query processing that takes social similarity, textual similarity and time recency into account on top of the microblog threads, then propose an efficient algorithm to provide the up-to-date top-k results of a massive amount of subscriptions against the high-velocity microblog streams in two steps: (i) filter the subscriptions that are

---

[1] https://about.twitter.com/company.

not affected by a newly arrived microblog, and (ii) update the contents and similarities of all valid MT candidates w.r.t. each query (Sect. 4).
- We conducted extensive experiments on a subset of Twitter dataset to evaluate the performance of both joint top-k query processing algorithm (Sect. 5).

2 Related Works

We focus on the social media data stream like Twitter and Facebook. We observe that different interpretations or focuses of the following settings can lead to different sub-problems and thereby various indexing and filtering techniques: (1) models of the social media data and the user query, (2) ranking metrics for top-k results.

The closest works are [1,3,9], and we would like to illustrate them in terms of the above two aspects.

Regarding data modeling, they assume each microblog is independent and take only the textual part of the social media data, resulting in a text stream. However, the microblogs inherently have relationships such as *forward, reply, comment, like* etc., as a result of user interactions within the social network, and it such social part should not be simply ignored. That also explains the output of the problem is top-k individual microblogs. In term of query modeling, they model users' subscription query as a set of keywords.

In term of ranking metrics of top-k results, Shraer et al. [9] adopts the freshness and textual relevance between the result and the query, while Chen et al. [1] take one more metric, the diversity of whole set of top-k results, into the ranking metric. In general, the ranking metric between a result candidate r and the query q looks like below: $S(q,r) = (\alpha \times A + (1 - \alpha) \times B) \times T(r)$. Where A is the textual relevance, B is the diversity value, and $T(r)$ is a decaying function to record the freshness of r. Since the user part is barely considered in the previous literature, existing solutions can provide little information of the social relevance between the query and the results. However, it has been pointed out that social relevance is important as people prefer to get updates from their close friends ahead of others [6].

Therefore, it is natural to exploit the linkage among microblogs for the problem setting, which essentially distinguishes this work from the literature. In particular, we propose the concept of microblog threads (which is a tree-structured set of microblogs) to capture such social user interactions and a storyline that individual microblogs could not tell, then we bring the social similarity between the query user and the microblog owner into the result ranking metrics (see Sect. 3).

Accordingly, this new problem demands for novel query indexing methods as well as filtering techniques, because (1) the result goes beyond an individual microblog and is a tree-like structure whose textual content and social relevance keep changing as new microblog flows in; however, in previous work, the result is somehow static in terms of its similarity score; (2) the computation cost on social similarity has been shown to be much more expensive than its counterpart

Fig. 1. Type of microblogs

at textual dimension [6], so it brings extra challenges on how to do candidate filtering in a social-first manner.

In term of index design and filtering techniques: Chen et al. [1] build a block based inverted file consists of inverted lists for the keywords of the queries, where each list stores the query ids containing that keyword. The queries are organized based on their top-k results. They observe that the same microblog may be in one of the top-k results of multiple subscription queries, so they propose a method to put those queries, which share the most number of results, together. They also prove that choosing a block to cover as many common documents as possible is the Minimum Set Cover problem, which is NP-hard.

3   Problem Formulation

3.1   Preliminaries

We model the social network as an undirected graph $G = (U, E)$, where U is the set of users and E is the set of edges representing the "friendship".

Definition 1. Microblog Stream. *A Microblog stream comprises a sequence of Microblogs, each represented as $mb = (\phi, ts, u)$, where ϕ is the set of keywords, ts is the timestamp, and u is the user who posts mb. A microblog can also be the "reply", "favorite", or "forward" of another microblog. Each microblog will appear after δt time, which is predefined.*

Intuitively, our system should take Microblog Stream as input, the results will be retrieved based on the relevant functions. We propose the following approaches to solve those challenges. First, we introduce Microblog Thread as minimum information unit to address challenges 1 and 2 mentioned in Sect. 1. To represent the form of Microblog Thread clearly, we have types of microblogs which are based on the original forms of microblog propagation shown in Fig. 1. Any microblog alone is a post, any pair of microblogs that have interactions is a pair, any group of microblogs that have interactions in a line is a dialogue, and any group of microblogs that have interactions connected with each other is a thread.

Definition 2. (Matching) Microblog Thread. *A Microblog Thread (MT) is represented as a tree structure $m = (N, E, root)$, where N is the set of microblogs*

$\{mb_1, mb_2, ..., mb_{|N|}\}$, E is the set of edges representing the interactions between two microblogs (e.g., reply, forward) and root is the original microblog. We say that an MTT matches a query q if all query keywords $q.\phi$ appear in the root node of m.

As Fig. 1 shows, the MT has different parts. The whole structure of it is called thread, a single node from MT is post, any two posts that have an interaction is a pair, the path from a post to the root is a dialogue.

3.2   Metrics

Metrics for Top-k Ranking. For any Microblog Thread m, we store a pair (k_i, tf_{k_i}) for each unique keyword appearing in m, where k_i is the keyword and tf_{k_i} is the frequency of k_i in m. The set of such pairs of all the keywords in m is called the signature of m, denoted as $Sig(m)$.

We use the following similarity function, $score(q, m)$ to compute the similarity between a Microblog Thread m and a query q. However, our proposed approaches are not limited to this measure, any other social-textual similarity measure can also be used. We adopt the state of art [1].

$$score(q, m) = STS(q, m) \times Recency(m) \tag{1}$$

The social and textual similarity measure $STS(q, m)$ balanced by the parameter λ, is computed as:

$$STS(q, m) = \lambda \times TS(q, m) + (1 - \lambda) \times SS(q, m) \tag{2}$$

where $STS(q, m)$ is the social textual similarity and $Recency(m)$ is the measure of freshness, adopted from the work [1] $\lambda \in [0, 1]$ is the preference parameter to balance between the textual and social similarities.

The textual relevance $TS(q, m)$ is computed as:

$$TS(q, m) = \begin{cases} \dfrac{\sum\limits_{k \in q.\phi \cap Sig(m).\phi} tf_k}{\sum\limits_{j \in Sig(m).\phi} tf_j}, & \text{if } q.\phi \subseteq m.root.\phi \\ 0, & \text{otherwise} \end{cases} \tag{3}$$

where, tf_k is the frequency of the keyword k in $Sig(m)$.

The social distance is evaluated by the number of "friends" of $q.u$ in an MT m. In social networks, although there is a well-known six-degrees-of-separation theory, the degree of acquaintance may even decrease super-exponentially, i.e., two persons may hardly know each other if they are just 3 hops apart [6]. Hence, we consider the percentage of 2-hop friends to measure the social similarity between $q.u$ and the microblog contributors of m.

$$SS(q, m) = \frac{|friend(q.u) \cap user(m)|}{|friend(q.u)|}, \tag{4}$$

where $user(m)$ is a set of users in nodes of microblog thread $m.N$ and $friend(q.u)$ is a set of users within 2-hop distance from $q.u$.

Here, we define the problem of joint top-k subscription query processing over microblog threads.

Definition 3. *Joint Top-k Query over Microblog Thread (TQMT).* *Given a set M of MT, a top-k query $q = \langle \phi, u, \lambda, k \rangle$, where ϕ is a set of query keywords,λ is the parameter for balancing the weights of social and textual information in the rank, u is the user who issues q, returns a ranked list of MTs M' such that $M' \subseteq M$, $|M'| = k$, $\forall t' \in M', m \in M, score(q, m') \geq score(q, m)$. $score(q, m)$ is a ranking function combining textual and social relevance, and the freshness of MT. Joint Top-k query is a set of such queries.*

4   Top-k Subscription Query Processing

4.1   Index for Social Dimension

To facilitate the pruning in social dimension, we introduce the concept of "Impact Region (IR)", such that if the contributor of a newly arrived microblog falls outside the IR of a query user, the top-k result of that query will not be affected.

Definition 4. *Impact Region (IR).* *Given a query user $q.u$, an Impact Region is a sub-graph containing all the users within a 2-hop away from $q.u$.*

Recall from the discussion of social relevance in Eq. 4 that the users beyond 2-hop distance barely know each other in social network. As the social relevance is computed by the number of 2-hop friends, we only need to consider those friends in IR.

Reverse Impact Region: By Definition 4, if a user u is in the IR of $q.u$, the microblogs posted by u may affect the top-k result of q. Thus, a Reverse Impact Region $RIR(u)$ of a user is the set of query users for which u belong to the impact region $IR(q.u)$.

Query-Aware User Grouping: Since there might be a few Impact Regions for different queries, they are likely to have some overlaps with each other. Intuitively, we should group the users based on the IR they belong to. Recall from Definition 4, the value of *SocialRelevance* will be exactly the same if two users have the same IR.

Example 1. Assume we have an example of four queries as shown in Table 1. Figure 2 shows a social graph where the dotted circles are the query users $\{u_1, u_{10}, u_{13}, u_{15}\}$ and each edge means they are friends with each other. Users u_1 and u_6 are both in the $IR(u_1)$ and $IR(u_{10})$. Therefore, they should be in the same user group, such as $G(u_1)$.

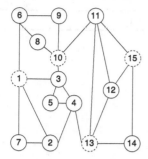

Fig. 2. User networks

G	$Members$	RIR
G_1	u_1, u_6, u_8, u_9	q_1, q_2
G_2	u_2	q_1, q_3
G_3	u_3, u_4, u_5	q_1, q_2, q_3
G_4	u_7	q_1
G_5	u_{10}	q_1, q_2, q_3, q_4
G_6	u_{11}, u_{13}, u_{15}	q_2, q_3, q_4
G_7	u_{12}	q_2, q_4
G_8	u_{14}	q_3, q_4

Fig. 3. User grouping

Therefore, we propose a heuristic algorithm to group the users, where users who are in the same IR, will be grouped together. In this approach, we randomly pick a user, and then heuristically search his neighbours in the social graph. If a user is already included in a group, then his friends have a higher chance to be included in that group too. We group all query users using this manner, as shown in Fig. 3, of which the users are in Fig. 2.

4.2 Baseline Approach

Recall Definition 3, a straightforward approach is to calculate the updated score of the corresponding MT w.r.t. each query, and update their results accordingly when a new microblog comes. However, the large number of subscriptions and the high update rate of social networks make this approach prohibitively expensive.

Therefore, when a new microblog mb_n comes, we want to quickly prune those queries whose top-k results cannot be affected by mb_n, in order to avoid unnecessary computations. As the similarity measure depends on both social and textual properties, we need to consider both dimensions while pruning. Li et al. [6] has shown that due to the complex graph structure of a social network, the social dimension has a more powerful pruning capacity. Therefore, we propose two solutions, (1) the baseline approach which adopts pruning on textual dimension only, and (2) the hybrid approach which primarily adopts pruning in social dimension on top of the baseline.

Both approaches consist of two steps: (1) *Group Filtering* and (2) *Individual Filtering*. In the group filtering step, we propose an approach where the pruning is done on both dimensions simultaneously with the help of our proposed "hybrid index", in order to alleviate the computational cost. Individual filtering is the same for both methods, where an upper bound of the score is computed for mb_n w.r.t. each query individually to decide whether the score is high enough to update the current top-k results.

The pseudocode of the workflow of both approaches is presented in Algorithm 1. When a newly arrived microblog mb_n is an original microblog containing all query keywords, it forms a new MT, as shown in Lines 2–3. Otherwise, if

---

Algorithm 1. Joint top-k pub/sub query processing

Input: a new Microblog from stream mb
Query set Q_s, the kth value of query $LB(q)$
Output: Top-k Results for all the queries

1 MT M;
2 **if** mb *is an original Microblog* **then**
3 | $M = mb$;
4 **else**
5 | $M = M.Append(mb)$;
6 $Q_g = GroupFilter(Q_s, m)$;
7 $Q_i = IndividualFilter(Q_g, m)$;
8 **for** $q \in Q_i$ **do**
9 | Update($Rel(q, M)$);

---

mb_n is a reply or forward, so it is added to its corresponding MT (Lines 4–5). If the top-k result of any subscription change due to the new microblog, we need to push the updates to those subscriptions.

A straightforward approach to solve the problem is to calculate the updated score of the corresponding MT with all the queries, and update their results accordingly. However, as an MT is incrementally built up, we check only those queries whose scores can be affected by mb_n through function $GroupFilter(q_s, m)$ (Line 6 of Algorithm 1). Now, we present two approaches to filter the irrelevant queries. The rest of the Algorithm 1, i.e., Lines 7–9 will be explained in Sect. 4.4.

Group Filtering Baseline. We present a technique that prunes queries using social and textual properties separately in the group filtering stage for comparison.

Index: For each MT m, we maintain a lookup table of the number of friends it contains for each query. This number is the numerator of Eq. 4. When a new microblog mb_n arrives, we need to update this number for the corresponding MT with respect to each query. According to Definition 4, we need to update for only the queries that are in the RIR of the contributor of mb_n. For textual similarity, for each keyword k_i we maintain an inverted list to store the queries that contain k_i.

Approach: Consider the example in Table 1, where the queries are q_1, q_2, q_3, and q_4. Let an MT consist of the nodes of users $\{u_1, u_3\}$, the signature of microblog thread m is $\{kobe : 0.3, ball : 0.1, nba : 0.4, usa : 0.1, best : 0.1\}$, then m is one of the top-k MTs of both q_2 and q_3. Let a new microblog mb_n is posted by u_{14}. Here, we only need to update the queries in the RIR of u_{14}. From Table 2, query q_2 is not in $RIR(u_{14})$, therefore can be pruned. However, q_3 is in $RIR(u_{14})$, so we update the score of m for q_3 in the social dimension.

Table 1. Example of queries

Q	U	IR	ϕ
q_1	u_1	$u_{1,2,3,4,5}$, $u_{6,7,8,9,10}$	nba, best
q_2	u_{10}	$u_{1,3,4,5,6,8,9}$, $u_{10,11,12,13,15}$	ball, best
q_3	u_{13}	$u_{2,3,4,5,10}$, $u_{11,13,14,15}$	usa, nba, ball
q_4	u_{15}	$u_{10,11,12,13}$, $u_{14,15}$	nba, usa

Table 2. Hybrid approach: Index

	G_1	G_2	G_3	G_4	G_5
nba	q_1	q_1, q_3	q_1, q_3	q_1	$q_{1,3,4}$
usa		q_3	q_3		q_3, q_4
best	q_1, q_2	q_1	q_1, q_2	q_1	q_1, q_2
ball	q_2	q_3	q_2, q_3		q_2, q_3

Each time a new microblog arrives, we need to update the number of friends with respect to each query in its RIR, so the complexity of social dimension update is $O(|RIR|)$.

We only need to check the queries that appear in the inverted list of keywords of mb_n. Let the size of inverted list be IL and number of new keywords be k_n. With both the social and textual index, the complexity of update is $O(|RIR| + \sum_{i=1}^{k_n} |IL_i|)$.

4.3 Hybrid Approach

Social-Aware Group Filtering. Recall Definition 4, we introduce RIR for the fast update of social relevance. In the baseline, the inverted lists consist of set of queries. However, it is not efficient if we want to find a group of queries in an inverted list. To solve this problem, we propose a *Hybrid Index* that groups the queries in inverted lists based on the query users' impact regions.

Index: We explain the index with the following example.

Example 2. We have the queries from Table 1. The user groups are constructed as mentioned at the beginning of this section. Table 2 shows the hybrid index, where, for each keyword, the queries are stored as these groups. For keyword "best", we have the inverted list "best" = $\{q_1, q_2\}$, for user group G_2, $RIR(G_2)$= q_1, q_3. Based on this, the query that can be affected by both "best" and G_2 is $\{q_1\}$. Then by calculating all these overlaps of pairs of user groups and keyword inverted lists, we can have this hybrid index.

The query-aware user groups and the index are computed once, and they do not change as long as the queries do not change.

Approach: When a new microblog mb_n comes, it can only come from one user. Therefore, in the hybrid index, we only need to check the user group where $mb_n.u$ belongs. The pruning strategy in the hybrid index is explained with the following example.

Example 3. Let a new microblog mb_n append to the MT m which is ranked in $\{q_1, q_2, q_3\}$, it has only the keyword "*usa*", and comes from a user belongs to the user group G_2. Then we only need to check the column for G_2 in the inverted list of *usa*. From Table 2, we find that mb_n could affect only query q_3, so the Individual Filter needs only one query to be considered.

The hybrid index combined the social dimension filter and textual filter, and the complexity is obviously $O((\sum_{i=1}^{k_n} |IL_i| \cap |RIR_G|))$.

4.4 Individual Filtering

The individual filtering approach is shown in Line 8 of Algorithm 1. We propose the notion of an upper bound value to further filter the irrelevant queries. Let $ScoreK(q)$ be the *Score* of the kth ranked MT of a query q. In computing the *STS* of Eq. 2, the social relevance is easy to calculate, as a new microblog can only come from one user, the number of friends with a query can increase at most 1. Computing the upper bound of *STS* is actually computing the upper bound of *TS*. Assume that our keywords of microblog is $mb.\phi$, in Eq. 3, the denominator is $|Sig(m)| + |mb.\phi|$, by adding the term frequencies of the newly arrived microblog. For the numerator, if we assume all the keywords are contained in $q.\phi$, then we could have the upper bound with it, because $|mb.\phi| \geq |q.\phi \cap mb.\phi|$. If this upper bound is smaller than the kth value, $ScoreK(q)$, then the new microblog cannot affect the top-k result of q. Otherwise, we compute the actual score to check if the result needs to be updated for q.

Table 3. Parameter setting

Parameter	Values
k	$\mathbf{5}, 10, 20, 30, 40$
Average # of shared keywords, $ANSK$	$\mathbf{0.3}, 0.6, 0.9, 1.2, 1.5$
Average percentage of shared friends, $APSF$	$\mathbf{0.1}, 0.2, 0.4, 0.6, 0.8$
# of queries, $|Q|$	$\mathbf{10K}, 50K, 100K, 200K$

5 Experiments

All methods are implemented in C++ and all experiments are run on a PC with Intel Core i5, 20 G memory, 500 GB SSD.

Dataset: The original Twitter dataset is collected from SNAP[2] with $17M$ users and $476M$ tweets. We extracted a subgraph of $500K$ users with $5.9M$ edges,

---

[2] http://snap.stanford.edu.

whose tweets have frequent follow up discussions. As a result, we form a set of 1.5 Million tweets.

Experiments Settings: We evaluate the problem of publisher subscriber in our system. Since this is the first work on joint top-k subscription query processing over MTs, we evaluate our proposed algorithm and compare with our developed baselines.

Subscription Query Generation: we generate two kinds of queries from the tweets: **random query set** and **feature query set**. For the random query set, we randomly selected 1–3 terms from each tweet as the query, and uses its owner as the query user. For the feature query set, we generate it to fulfil two features: the average number of shared keywords (ANSK) and the average percentages shared of 2-hop friends (APSF). For ANSK, we generate it from a random query set, repeat some keywords of its queries to simulate the real case, and assign a random user which will not affect another feature. The parameter of ANSK will affect the length of inverted lists, which could affect the group filtering in textual dimension. For the APSF, we use a set of queries that has a quite low value of APSF, then replace those users who are too far from others by closer ones. It could affect the length of social indexes, which affects the group filtering in social dimension.

Experiment Evaluations. Now we evaluate the performance by varying several parameters. Since the Baseline approach is for the group filtering stage, we separate the methods into (i) individual filtering + Hybrid group filtering as IGFilter-H, (ii) Baseline group filtering + individual as IGFilter-B, (iii) Hybrid group filtering alone as GFilter-H, and (iv) Baseline group filtering alone as GFilter-B. It could highlight the differences of group and individual filtering methods on different indexes. For each of following parameters, the runtime cost for processing it consist of two parts: (1) Time cost for finding the queries that include this microblog post and its microblog thread in this as results; (2) Time cost for index update, which includes the cost for updating query results tables, and update user groups.

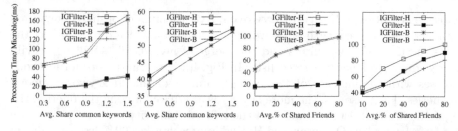

(a) Doc Processing (b) Query Insertion (c) Doc Processing (d) Query Insertion

Fig. 4. Various parameters on ANSK and APSF

See Table 3 for parameter settings, where those in bold are the default values. When one of the parameter varies, other parameters are setted as default. Here, "Shared Impact Region" indicates the average number of pairwise shared 2-hop friends of the queries, which is the number of common friends divided by the total number of friends. In all experiments, we use the default settings in the Table 3 in bold fronts and vary a single parameter to study the impact on average runtime. The first three parameters listed in Table 3 demonstrate the efficiency, and the last one is for scalability testing.

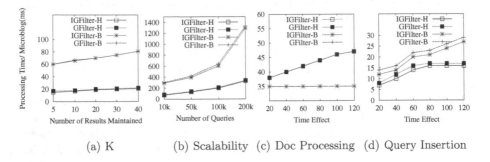

(a) K   (b) Scalability (c) Doc Processing (d) Query Insertion

Fig. 5. K, Scale and Time Effect

Number of Results Maintained: Number of results maintained affects the update of rankings. Larger number of results will cost more time during the ranking of results. Figure 5(a) shows that the runtime of all methods increase with the increases of the number of results maintained (k), as the cost of computing the scores and updating the ranks increase for higher values of k. As the hybrid approach reduces the number of computations compared to the baseline with the help of the hybrid index, it is 3 times faster than the baseline on average. However, on the level of group filtering and individual filtering, with or without individual filtering does not matter a lot in here because the scale of test set is not large enough.

Average Number of Shared Keywords: Average number of shared keywords will affect the average length of keywords inverted lists, which directly affect the processing time of group filters and the merge of inverted lists will be heavily affected from the longer inverted lists. In Fig. 4(a), the runtime of all methods increases with the increase of the average number of shared keywords among the queries. The reason is, the more keywords are shared among the queries, the longer the inverted lists become, and the textual similarity measure costs more time for a longer list. Our proposed Hybrid approach takes 60 % less runtime on average than the baseline, as the hybrid index reduces the lengths of the lists used in the filtering parts. In Fig. 4(b), it shows that the update cost increases because each update will deal with more lists in the index for both dimensions. However, the Hybrid index costs more because the user grouping part takes time.

Average Percentage of Shared Friends: Average percentage of shared friends will affect the size of RIR, which is part of the group filtering mechanism. Figure 4(c) shows that the runtime does not vary much with the percentage of shared friends among the queries, as we only need to update one RIR each time a new microblog comes, regardless of the sharing. The hybrid approach takes 70 % less runtime than the baseline, as the hybrid index exploits the shared friends among queries to reduce the computations. Moreover, the Fig. 4(d) shows that the cost for social index update is more than the cost for textual index update with the comparison of Fig. 4(b).

Scalability Test: The scalability of the approaches is shown in Fig. 5(b) by varying the number of queries. The hybrid approach outperforms the baseline by 4 times on average, where the benefit is higher for higher number of queries. The hybrid approach scales with the increase in the number of queries, as it utilizes the grouping of the queries, where the baseline approach processes them individually.

Time effect: In this set of experiments, each method runs for 120 min. To make sure we can handle the real case, we insert 1 microblog and 1 query each second. We report the average runtime of processing and the average runtime for query insertion during each period of 20 min. In Fig. 5(c), it shows the processing time for each Microblog when time evolves, after the number of documents increases the efficiencies are quite different between the hybrid and baseline. For the update of queries, in Fig. 5(d) shows that it does not change much for baseline, but the update of hybrid index will be affected a little bit because of the user grouping.

6 Conclusion

We consider the problem of Joint Top-k subscriptions query processing results over Microblog Thread. Each subscription query takes into account of text relevance, social relevance and document recency in evaluating the query result. We propose a novel approach to efficiently retrieve the microblog threads. The experimental results on real-world dataset show that our solution is able to achieve a reduction of processing time by 60 % compared with baselines.

References

1. Chen, L., Cong, G.: Diversity-aware top-k publish/subscribe for text stream. In: SIGMOD, pp. 347–362 (2015)
2. Guo, L., Zhang, D., Li, G., Tan, K.-L., Bao, Z.: Location-aware pub, sub system: When continuous moving queries meet dynamic event streams. In: SIGMOD, pp. 843–857 (2015)

3. Haghani, P., Michel, S., Aberer, K.: The gist of everything new: personalized top-k processing over web 2.0 streams. In: CIKM, pp. 489–498 (2010)
4. He, H., Yu, Z., Guo, B., Lu, X., Tian, J.: Tree-based mining for discovering patterns of reposting behavior in microblog. In: Motoda, H., Wu, Z., Cao, L., Zaiane, O., Yao, M., Wang, W. (eds.) ADMA 2013, Part I. LNCS, vol. 8346, pp. 372–384. Springer, Heidelberg (2013)
5. Jiang, J., Lu, H., Yang, B., Cui, B.: Finding top-k local users in geo-tagged social media data. In: ICDE, pp. 267–278 (2015)
6. Li, Y., Bao, Z., Li, G., Tan, K.-L.: Real time personalized search on social networks. In: ICDE, pp. 639–650 (2015)
7. Maniu, S., Cautis, B.: Network-aware search in social tagging applications: instance optimality versus efficiency. In: CIKM, pp. 939–948 (2013)
8. Schenkel, R., Crecelius, T., Kacimi, M., Michel, S., Neumann, T., Parreira, J.X., Weikum, G.: Efficient top-k querying over social-tagging networks. In: SIGIR, pp. 523–530 (2008)
9. Shraer, A., Gurevich, M., Fontoura, M., Josifovski, V.: Top-k publish-subscribe for social annotation of news. PVLDB 6(6), 385–396 (2013)
10. Yu, M., Li, G., Feng, J.: A cost-based method for location-aware publish/subscribe services. In: CIKM, pp. 693–702 (2015)

A Pattern-Based Framework for Addressing Data Representational Inconsistency

Bingyu Yi$^{(\boxtimes)}$, Wen Hua, and Shazia Sadiq

School of Information Technology and Electrical Engineering,
The University of Queensland, Brisbane, Australia
{b.yi1,w.hua}@uq.edu.au, shazia@itee.uq.edu.au

Abstract. Data representational inconsistency, where data has diverse formats or structures, is a crucial data quality problem. Existing fixing approaches either target on a specific domain or require massive information from users. In this work, we propose a user-friendly pattern-based framework for addressing data representational inconsistency. Our framework consists of three modules: pattern design, pattern detection, and pattern unification. We identify several challenges in all the three tasks in order to handle an inconsistent dataset both accurately and efficiently. We propose various techniques to tackle these issues, and our experimental results on real-life datasets demonstrate better performance of our proposals compared with existing methods.

1 Introduction

As a significant problem of data quality, data inconsistency arises everywhere. Data is generated and represented differently in various cultures, countries, companies, and contexts using different standards or formats. When this disparate data is integrated, data inconsistency becomes evident. Hence, this problem is attracting increasing attention from both industry and academia with data becoming more and more massive and heterogeneous [1].

In this paper, we focus on data representational inconsistency where data has diverse formats or structures. Current research on data representational inconsistency generally targets on a specific domain such as name, address, etc. [2,3]. Recent ETL (Extraction Transformation Loading) tools [4] can handle multiple domains, but require a large amount of information from users including metadata, transformation mappings and workflow definitions. To this end, we propose a user-friendly pattern-based framework for addressing data representational inconsistency in various domains. We define data format or structure as a pattern, namely a sequence of fields and separators represented using regular expressions, as illustrated in Table 1. Our framework consists of three modules:

- Pattern design - construct a pattern library for each domain.
- Pattern detection - recognize possible patterns for each data record.
- Pattern unification - transform all data records into a target pattern.

© Springer International Publishing AG 2016
M.A. Cheema et al. (Eds.): ADC 2016, LNCS 9877, pp. 395–406, 2016.
DOI: 10.1007/978-3-319-46922-5_31

Table 1. An example of data representational inconsistency.

BoardingStop	Pattern[a]	Consistent Data
Wynnum Plaza - Stop 58 [BT006135]	D - Stop SN [ID]	Wynnum Plaza - Stop 58 [BT006135]
A.&I.I.C.S. - 55/56 [BT005196]	D - SN [ID]	A.&I.I.C.S. - Stop 55/56 [BT005196]
Alison St - St 32 [BT002904]	D - St SN [ID]	Alison St - Stop 32 [BT002904]
Trouts/Redwood - 40 [BT002172]	D - SN [ID]	Trouts/Redwood - Stop 40 [BT002172]
	SN - D [ID]	
Griffith University Stop A [BT010434]	D Stop SN [ID]	Griffith University - Stop A [BT010434]
	D SN [ID]	

[a] D, SN, and ID are fields where D represents stop description with regular expression (|[A-Za-z0-9/&.])+, SN represents stop number with regular expression [A-Za-z0-9/]+, and ID represents stop ID with regular expression BT[0-9]+. { ,-,Stop,St,[,]} are separators between fields

Although the framework seems simple, challenges still abound in order to handle an inconsistent dataset both accurately and efficiently. First, the coverage and quality of patterns are critical. An incomplete pattern library will miss data records, while a badly-designed pattern library might cause conflicts between patterns. Therefore, an ideal pattern library should be complete and mutual exclusive. However, it requires extensive human efforts to construct such a pattern library from scratch. Second, pattern conflict means a data record can match multiple patterns. We observe two types of pattern conflict, namely field-field conflict where a substring maps to several fields (In Table 1, "Trouts/Redwood" in "Trouts/Redwood - 40 [BT002172]" can be a stop description as well as a stop number based on the regular expressions of D and SN), and field-separator conflict where a field covers a separator (In Table 1, we can regard "Griffith University" as a stop description and "Stop" as a separator in "Griffith University Stop A [BT010434]", but it is also possible to treat "Griffith University Stop" as a stop description). Hence, it is necessary to recognise such one-to-many mappings when conducting pattern detection. A straightforward approach is to adopt pairwise checking between data records and patterns which, however, is obviously very time consuming. Third, pattern unification is more complicated than string-based functions such as substring replacement. Consider "Alison St - St 32 [BT002904]" in Table 1 as an example. We cannot unify it to "Alison St - Stop 32 [BT002904]" simply by replacing "St" with "Stop". Instead, we need the semantic knowledge that "Alison St" as a whole denotes a stop description while the second "St" is a separator, and our goal is to unify only the separator "St" as "Stop". We tackle these challenges in this work.

- We propose an iterative and interactive approach to designing patterns, trying to achieve a complete and nearly mutual exclusive pattern library.
- We construct a Finite State Machine (FSM) to recognise all possible patterns for each data record and meanwhile avoid pairwise checking.
- We introduce a two-level pattern definition to combine both domain knowledge and regular expressions, and propose a spilt-transform-merge method to facilitate pattern unification.

– We conduct comprehensive evaluation on real-life datasets, and the experimental results verify the effectiveness and efficiency of our proposals.

The rest of this paper is organised as follows: we investigate related work on data consistency in Sect. 2, and formally define the problem of data representational inconsistency in Sect. 3. The details of our approaches and the evaluation results will be introduced in Sect. 4 and Sect. 5 respectively, followed by a brief conclusion and discussion about future work in Sect. 6.

2   Related Work

Data consistency is an important dimension of data quality and has been extensively studied. In this section, we review related work on three data consistency issues: data integrity, semantic consistency, and representational consistency.

Data integrity focuses on integrity constraints especially in relational models, and is supported by most commercial DBMSs (DataBase Management Systems). One way to guarantee data integrity is to declare integrity constraints together with the schema, and the DBMS will take care of database maintenance by rejecting transactions which might lead to a violation to the constraints [5]. Another way is to use triggers stored in the database, and the reaction to a potential violation is programmed as an action of the trigger [6].

Semantic consistency requires no contradiction between data items. Most of existing methods on semantic consistency are rule-based or CFD-based (Conditional Functional Dependency) [7,8]. Fan et al. [7] adopted CFDs to capture semantic inconsistency by enforcing bindings on semantically related values. This work mainly focused on detection of semantic inconsistency in data without providing fixing methods. In 2010, CFD-based editing rules [8] were introduced to repair data which, however, requires users to examine every tuple and hence is very expensive. Wang and Tang [9] proposed fixing rules to trigger repairing operations using both evidence patterns and negative patterns.

Data representational consistency, which is also the focus of this paper, examines whether a dataset contains unified formats or structures. This is usually regarded as a pre-processing step for other data processing tasks such as data search [10,11] and information extraction [12]. Existing works on data representational consistency are mostly domain specific. In other words, they target on a certain domain in a specific context. For example, Churches et al. [2] utilised Hidden Markov Models (HMMs) to format name and address data for record linkage. AddressDoctor [3] only deals with address inconsistency using a large dynamic address library. Our work, on the contrary, introduces a generalised framework for addressing representational inconsistency in domain agnostic data.

3   Problem Statement

In order to recognise representational inconsistency in data, we propose a two-level pattern definition in this work to reflect the format or structure of data.

Specifically, a pattern is defined in both semantic level and lexical level. At semantic level, a pattern can be regard as a sequence of fields and separators. While at lexical level, a pattern is a sequence of regular expressions.

Definition 1 (Pattern). *A pattern p is represented as a sequence of fields and separators, namely* $p = (f_1, s_1, f_2, s_2, f_3, ..., f_{t-1}, s_{t-1}, f_t)$ *where each* f_i *and* s_i *denote a field and a separator respectively, both of which are expressed as regular expressions.*

We denote the set of fields and the set of separators as $\mathbb{F} = \{f_1, f_2, ..., f_m\}$ and $\mathbb{S} = \{s_1, s_2, ..., s_n\}$ respectively. Consider the example in Table 1. The field set is $\mathbb{F} = \{D,SN,ID\}$, and the separator set is $\mathbb{S} = \{\sqcup,-,\text{Stop},\text{St},[,]\}$. From the field set and the separator set, we can construct several patterns such as "D - Stop SN $[ID]$", "D - SN $[ID]$", "D SN $[ID]$", etc. We denote the pattern library as $\mathbb{P} = \{p_1, p_2, ..., p_k\}$.

For each data record d from a dataset $\mathbb{D}$, we denote the set of patterns it maps to as $\mathbb{P}_d = \{p_i | p_i \in \mathbb{P} \wedge d \to p_i\}$ where $d \to p_i$ means that data record d can match pattern p_i. Hence, the set of patterns that dataset $\mathbb{D}$ contains can be denoted as $\mathbb{P}_\mathbb{D} = \bigcup_{d \in \mathbb{D}} \mathbb{P}_d$.

Definition 2 (Data Representational Inconsistency). *A dataset $\mathbb{D}$ is representational inconsistent when it contains multiple patterns, namely $|\mathbb{P}_\mathbb{D}| > 1$ where $|\mathbb{P}_\mathbb{D}|$ represents the size of (or number of patterns contained in) $\mathbb{P}_\mathbb{D}$.*

The goal of addressing data representational inconsistency is to unify an inconsistent dataset to a single pattern. Figure 1 demonstrates the framework proposed in this work, which consists of three modules: pattern design, pattern detection, and pattern unification.

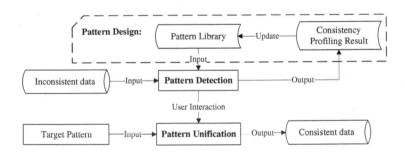

Fig. 1. Framework overview.

The pattern design module constructs a pattern library for each data domain. For example, Table 3 demonstrates the pattern library constructed for "BoardingStop". As discussed in Sect. 1, an ideal pattern library should be complete and mutual exclusive, in order to cover the entire dataset and eliminate pattern conflict. This incurs extensive efforts if we manually build the pattern library from scratch. We adopt an iterative and interactive approach to reduce human efforts. Starting with a seed set of patterns, we trigger the pattern detection module to handle the given dataset, and obtain the *consistency profiling result* which contains information about unrecognised data records and conflicting patterns. Based on the consistency profiling result, we design and add more patterns into the seed set and meanwhile revise conflicting patterns, and then recur the above process. The detailed pattern design process can be found in technical report [15].

Given a dataset $\mathbb{D}$ and the pattern library $\mathbb{P}$, the pattern detection module recognises possible patterns $\mathbb{P}_d$ for each data record $d \in \mathbb{D}$. We will discuss the details of pattern detection in Sect. 4.1. If the dataset $\mathbb{D}$ is representational inconsistent (i.e., $|\mathbb{P}_{\mathbb{D}}| > 1$), the pattern unification module will be triggered. It receives a target pattern p^* from the user, and transforms all data records into this pattern to make the dataset $\mathbb{D}$ consistent. We will discuss the details of pattern unification in Sect. 4.2. Note that when a data record d maps to multiple patterns, we need to pick a pattern p_i from the pattern set $\mathbb{P}_d$ and conduct unification based on the specific pattern p_i. This can be done randomly. However, if a semantically incorrect pattern is chosen as the unification pattern, it will cause the resulting data record to be semantically incorrect. Take "Trouts/Redwood - 40 [BT002172]" in Table 1 as an example. Assume "D - Stop SN [ID]" is the target pattern and we select "SN - D [ID]" as the pattern for this data record, then the unification result will be "40 - Stop Trouts/Redwood [BT002172]" which is obviously wrong. Hence, a better solution is to ask users to select the best pattern when multiple patterns are detected. But in order to reduce the amount of user interactions, it also requires the pattern library to be less conflicting.

4 Methodology

4.1 Pattern Detection

With the pattern library $\mathbb{P}$ at hand, pattern detection needs to recognise the possible patterns $\mathbb{P}_d$ for each data record $d \in \mathbb{D}$ from a specific domain. A straightforward approach is to check every record-pattern pair one by one to see whether the record conform to the pattern. The time complexity of such a method is obviously $O(|\mathbb{D}| \times |\mathbb{P}|)$, where $|\mathbb{D}|$ and $|\mathbb{P}|$ denote the size of the dataset and the pattern library respectively. In this work, we improve the efficiency by conducting pattern detection in a batch manner.

As defined in Sect. 3, a pattern is a sequence of fields and separators which are represented by regular expressions. Therefore, a pattern can also be regarded as a concatenation of all the regular expressions. In order to determine whether a data record (i.e., a string) satisfies a pattern, it suffices to check whether this record can be recognised by the concatenated regular expression. This is typically accomplished using a Finite State Machine (FSM) such as Nondeterministic Finite Automaton (NFA) and Deterministic Finite Automaton (DFA). Figure 2 illustrates the process of constructing an NFA to represent a pattern using Thompson's construction algorithm [13]. We first build NFAs for each field and separator based on their regular expressions, and then combine these NFAs using *series connection*. In order to detect patterns in a batch manner, we compile the entire pattern library into one NFA such that all possible patterns for a data record can be detected by scanning the record only once. Specifically, we employ parallel connection to combine the NFAs for each pattern into a large NFA.

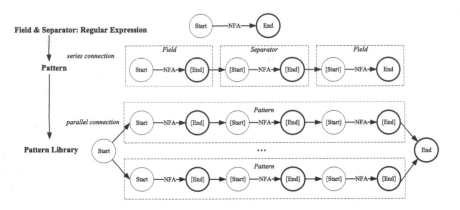

Fig. 2. Construction of NFA to encode the pattern library.

In the following we consider a simple pattern library as a running example. The field set and the separator set are $\mathbb{F} = \{f_1 = [ab1]+, f_2 = [a1]+\}$ and $\mathbb{S} = \{s1 = 11, s2 = a1\}$ respectively, and the pattern library is $\{p_1 = f_1 s_1 f_2 = [ab1] + 11[a1]+, p_2 = f_1 s_2 f_2 = [ab1] + a1[a1]+\}$. Figure 3 illustrates an NFA for this pattern library. In order to split each data record (e.g., "aba11a") into a collection of field values (e.g., "aba" and "a") and separator values (e.g., "11"), which is a prerequisite to pattern unification and will be discussed in Sect. 4.2, pattern detection should be able to determine both the matching patterns and the boundaries between fields and separators. Hence, we assign each final state in the NFA with a special label to notify which pattern has been detected when reaching the final state. We also assign each state corresponding to a separator with a special label (i.e., yellow states in Fig. 3) to differentiate fields and separators.

The NFA for a pattern library is usually huge, containing thousands of states with multiple choices on transitions. To reduce computational cost when processing on the NFA, we need to ensure that the constructed NFA has as few states as possible. A classic approach is to transform the original NFA into a DFA using subset construction algorithm and then minimise the DFA [14]. Figure 4 shows the minimised DFA of the original NFA in Fig. 3. There are three obvious drawbacks of the classic DFA: (1) it has no back-tracking and hence cannot recognise multiple patterns for a data record (e.g. "aba11a" → $\{p_1, p_2\}$, "abba1a" → $\{p_2\}$); (2) it cannot determine which pattern the data record maps to when reaching the

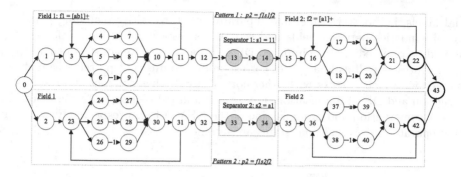

Fig. 3. NFA for $\{[ab1] + 11[a1]+, [ab1] + a1[a1]+\}$.

final state; (3) it cannot detect boundaries between fields and separators. The key problem is that classic subset construction and DFA minimisation algorithms combine and reduce states without considering their specific meanings, namely whether the states are fields or separators and which pattern the states correspond to. Hence, We introduce several modifications to the subset construction and DFA minimisation algorithms.

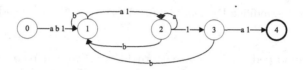

Fig. 4. Classic DFA for $\{[ab1] + 11[a1]+, [ab1] + a1[a1]+\}$.

One requirement of the modified DFA is to recognise multiple patterns by scanning the data record only once. In other words, after recognising one pattern, the modified DFA should enable back-tracking to check other patterns. Back-tracking can be easily supported by NFA since NFA allows for multiple next states given a specific input symbol. However, this is not the case in DFA. Therefore, we introduce a special state - branch state - into the modified DFA which can support multiple transitions on a single input symbol. Furthermore, in order to distinguish patterns, the information that different states lead to different patterns should be retained in the modified DFA. To this end, we first check which patterns the states lead to when merging states in subset construction. If the states represent multiple patterns, we then combine them into a single branch state and add a transition for each pattern. Branch states will not be merged with other states in DFA minimisation. The automatons in Fig. 5 show modified DFAs of the NFA in Fig. 3. The green states are branch states which transit to two next states corresponding to patterns p_1 and p_2 respectively.

Another requirement for the modified DFA is to determine patterns and pattern boundaries. As mentioned above, we assign states in the NFA with two special labels, i.e., pattern label and separator label, to notify matching patterns and boundaries between fields and separators respectively. These information should be retained in the minimised DFA. In particular, we constrain that separator states as well as final states corresponding to different patterns cannot be merged with each other or with other normal states when conducting subset construction and DFA minimisation. Figure 5(a) and (b) demonstrate the modified subset construction and minimisation of the NFA in Fig. 3 respectively, wherein the separator states and final states are retained.

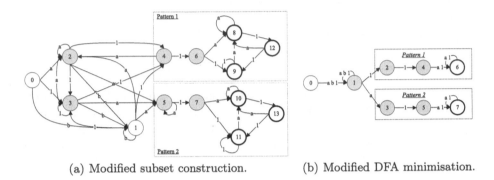

(a) Modified subset construction.   (b) Modified DFA minimisation.

Fig. 5. Modified DFA for $\{[ab1] + 11[a1]+, [ab1] + a1[a1]+\}$.

Given the modified DFA for a pattern library, we can recognise matching patterns for a data record accurately and efficiently. Consider the modified DFA in Fig. 5(b) and the data record "aba11a" as an example. At branch state 1, the automaton can go through states $1 \to 1 \to 1 \to 2 \to 4 \to 6$ to arrive at pattern p_1 (with "aba" and "a" as the field values, and "11" as the separator value), but it can also back-track and go through states $1 \to 1 \to 3 \to 5 \to 7 \to 7$ to arrive at pattern p_2 (with "ab" and "1a" as the field values, and "a1" as the separator value). The back-tracking ends when there is no next state for the branch state or the last character of the data record can not reach a final state.

4.2   Pattern Unification

Given a dataset $\mathbb{D}$ from a specific domain along with the detected pattern for each data record $p_d \in \mathbb{P}_d$, pattern unification transforms all data records into a target pattern p^*. As discussed in Sect. 1, pattern unification is more complicated than traditional string-based functions since it requires additional knowledge. In the example of "Alison St - St 32 [BT002904]" in Table 1, we need the semantic knowledge that the former "St" is part of the stop description filed while the latter "St" is a separator, in order to unify this data record to "Alison St - Stop 32 [BT002904]". Consider "2003/03/17" in Fig. 6 as another example. Both the semantic knowledge that the second "03" belongs to the month field and the domain knowledge that "03" as a month can be represented as "Mar" are necessary to transform "2003/03/17" to "2003-Mar-17". To this end, we adopt a two-level pattern definition, namely a pattern is a sequence of fields and separators (semantic level) expressed as regular expressions (lexical level), and propose a split-transform-merge approach for pattern unification. Figure 6 illustrates a running example of pattern unification.

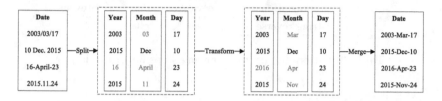

Fig. 6. An example of pattern unification.

Split. As discussed in Sect. 4.1, our approach to pattern detection can not only determine matching patterns but also recognise boundaries between fields and separators. Hence, we split each data record from a specific domain into a set of field values and separator values, based on the pattern detection result. In Fig. 6, each date (e.g., "2003/03/17") is divided into three fields namely year (e.g., "2003"), month (e.g., "03"), and day (e.g., "17"). For ease of representation, we ignore the separators in Fig. 6.

Transform. Given the set of field values and separator values, we conduct transformation according to the target pattern p^*. We introduce plenty of transformation functions for this task, many of which cannot be easily supported by traditional string-based functions. Table 2 presents some examples of our proposed transformation functions which can be roughly classified into two categories: structure change and value change. Structure change functions reorganise the order of fields and separators, while value change functions modify the actual values of fields and separators. We associate domain knowledge with some value change functions such as fielding mapping, abbreviation, etc. In Fig. 6, the month field value "03" is modified to "Mar".

Merge. After transformation, we merge the revised values of fields and separators together using string concatenation function to obtain the unified data record. In Fig. 6, the pattern unification result for data record "2003/03/17" is "2003-Mar-17".

5   Evaluation

We conducted extensive experiments to evaluate the accuracy and efficiency of our framework for addressing data representational inconsistency. All the algorithms were implemented in C#, and all the experiments were conducted on a

Table 2. Examples of transformation functions.

Category	Functions	Examples
Structure Change	field order change	"John Stevens" → "Stevens John"
Value Change	separator change	"2014/03/13" → "2014-03-13"
	field mapping	"03" → "Mar"
	measurement change	"10cm" → "0.1m"
	abbreviation	"Stevens John" → "Stevens J"
	string reverse	"abc" → "cba"

Table 3. Pattern library for boarding stops in gocard data.

Pattern[a]				
D - SN [ID]	D (Stop SN) [ID]	D - Platform SN [ID]	D - Zone SN [ID]	D (St SN) [ID]
D - Stop SN [ID]	D-SN [ID]	D Platform SN [ID]	D Zone SN [ID]	D (stop SN) [ID]
D Stop SN [ID]	D (SN) [ID]	D, platform SN [ID]	D -SN [ID]	D stop SN [ID]
D'SN' [ID]	Stop SN D [ID]	D - Platform'SN' [ID]		

  [a] D= (|[A-Za-z0-9/!,;'@#$%&*()-."])+, SN=[A-Za-z0-9/]+, ID=BT[0-9A-Za-z.]+

server with 2.90GHz Intel Xeon E5-2690 CPU and 192GB memory. Our evaluation was based on real-life datasets from various domains and demostrated that our approach is domain agnostic. Due to limitation of paper length, we only report experimental results on the public transportation data in this section. Readers can refer to our technical report [15] for results on other datasets inluding DBLP.

5.1 Accuracy

We downloaded a one-month (i.e., March 2013) snapshot of Translink gocard data in Brisbane, which consists of 4,329,128 records of gocard touch-on and touch-off information. We chose the noisy boarding stop domain to conduct our experiments. Using the iterative and interactive approach to pattern designing, we constructed a pattern library of 18 patterns, as listed in Table 3. We rank the patterns according to their coverage (i.e., number of matched records) in descending order, add them into the pattern library one by one, and then check the coverage and conflict of the resulting pattern library. Figure 7(a) illustrates the change of the number of unmatched records, single matched records and multiple matched records in log-scale when adding patterns into the library. We can see that the first pattern (i.e., D - SN [ID]) captures more than 25 % boarding stops in the gocard data. With more patterns inserted into the pattern library, its coverage increases gradually, at the cost of more conflict among patterns.

5.2 Efficiency

Given a pattern library and an inconsistent dataset, the naive solution for pattern detection is pairwise checking. In order to improve efficiency, we propose to

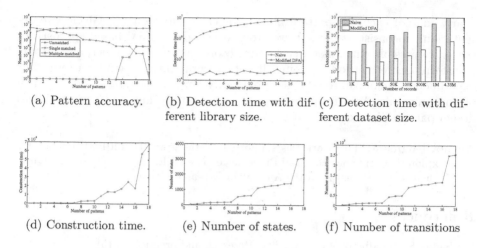

(a) Pattern accuracy.

(b) Detection time with different library size.

(c) Detection time with different dataset size.

(d) Construction time.

(e) Number of states.

(f) Number of transitions

Fig. 7. Accuracy and efficiency.

examine patterns in a batch manner. Specifically, we combine all patterns into an NFA and then utilise modified subset construction and DFA minimisation algorithms to reduce the original NFA into a modified DFA. In this part, we compare pattern detection time of the naive solution and the modified DFA. We change the size of pattern library and the size of dataset, and report the evolving of pattern detection time in Fig. 7(b) and (c) respectively. As we can see, the pattern detection time of both methods increases linearly with the growth of dataset size. But when more patterns are included in the pattern library, the detection time of naive solution still rises linearly while our modified DFA remains stable.

Although DFA construction can be accomplished offline, it can still be quite time consuming especially when the pattern library is huge. Therefore, we also evaluate the DFA construction time when varying the size of pattern library. From Fig. 7(d) we can see that the DFA construction cost grows when the number of patterns increases. To obtain a more insightful understanding of the rise of construction time, we present the number of states and transitions in the DFA respectively in Fig. 7(e) and (f). It is obvious that when the size of pattern library increases, the resulting DFA becomes larger, which in turn causes longer construction time.

6   Conclusion

In this paper, we target an important data quality problem - data representational inconsistency where data has diverse formats or structures. We propose a user-friendly pattern-based framework for addressing such inconsistency. We define a pattern as a sequence of fields and separators expressed as regular expressions. We adopt an iterative and interactive method to design a complete and nearly mutual exclusive pattern library for each data domain. Given a representational inconsistent dataset, we propose a modified FSM-based approach to recognise possible patterns for each data record in a batch manner, and a split-transform-merge method to unify all data records into a target pattern. Our experimental results on real-life datasets verify both the accuracy and efficiency of our proposals. As a future work, we plan to study automatic strategies to design pattern libraries.

Acknowledgment. This work was supported by the grant DP140103171 (Declaration, Exploration, Enhancement and Provenance: The DEEP Approach to Data Quality Management Systems) from the Australian Research Council.

References

1. Sadiq, S.: Handbook of data quality: Research and practice (2015)
2. Churches, T., Christen, P., Lim, K., Zhu, J.X.: Preparation of name and address data for record linkage using hidden markov models. BMC Med. Inf. Decis. Making **2**(1), 9 (2002)
3. GmbH, A.: Addressdoctor enterprise documentation - informatica (2014)
4. Rahm, E., Do, H.H.: Data cleaning: problems and current approaches. IEEE Data Eng. Bull. **23**(4), 3–13 (2000)
5. Türker, C., Gertz, M.: Semantic integrity support in sql: 1999 and commercial (object-) relational database management systems. VLDB J. **10**(4), 241–269 (2001)
6. Ceri, S., Cochrane, R., Widom, J.: Practical applications of triggers and constraints: Successes and lingering issues. In: VLDB, pp. 10–14 (2000)
7. Fan, W., Geerts, F., Jia, X., Kementsietsidis, A.: Conditional functional dependencies for capturing data inconsistencies. ACM Trans. Database Syst. (TODS) **33**(2), 94–115 (2008)
8. Fan, W., Li, J., Ma, S., Tang, N., Yu, W.: Towards certain fixes with editing rules and master data. Proc. VLDB Endowment **3**(1–2), 173–184 (2010)
9. Wang, J., Tang, N.: Towards dependable data repairing with fixing rules. In: Proceedings of the 2014 ACM SIGMOD International Conference on Management of Data, pp. 457–468 (2014)
10. Li, G., Zhou, X., Feng, J., Wang, J.: Progressive keyword search in relational databases. In: 2009 IEEE 25th International Conference on Data Engineering, pp. 1183–1186, March 2009
11. Luo, Y., Wang, W., Lin, X., Zhou, X., Wang, J., Li, K.: Spark2: Top-k keyword query in relational databases. IEEE Trans. Knowl. Data Eng. **23**(12), 1763–1780 (2011)
12. Huynh, D.T., Hua, W.: Self-supervised learning approach for extracting citation information on the web. In: Sheng, Q.Z., Wang, G., Jensen, C.S., Xu, G. (eds.) APWeb 2012. LNCS, vol. 7235, pp. 719–726. Springer, Heidelberg (2012)
13. Thompson, K.: Programming techniques: regular expression search algorithm. Commun. ACM **11**(6), 419–422 (1968)
14. Aho, A.V., Lam, M.S., Sethi, R., Ullman, J.D.: Compilers: Principles, techniques, and tools (2006)
15. Yi, B., Hua, W., Sadiq, S.: Technical report: Pattern-based framework for addressing data representation inconsistency (2016). https://drive.google.com/folderview?id=0B7vhn9TkNVEVYjN4WWhIclpLdTA&usp=sharing

Optimising Queue-Based Semi-stream Joins by Introducing a Queue of Frequent Pages

M. Asif Naeem[1]([⊠]), Gerald Weber[2], and Christof Lutteroth[2]

[1] School of Engineering, Computer and Mathematical Sciences,
Auckland University of Technology, Private Bag 92006, Auckland, New Zealand
mnaeem@aut.ac.nz
[2] Department of Computer Science,
The University of Auckland, Auckland, New Zealand
{gerald,christof}@cs.auckland.ac.nz

Abstract. Semi-stream joins perform a join between a stream and a disk-based table. These joins can easily deal with typical workloads in online real-time data warehousing in many scenarios and with relatively modest system requirements. The disk access is page-based. In the past, several proposals have been made to exploit skew in the distribution of the join attribute. Such skew is a common result of natural short- or long-tailed distributions in master data. Several semi-stream joins use caching strategies in order to improve performance. This works up to a point, but these algorithms still require relatively slow processing of stream data that matches with the infrequent tuples in the master data. In this work we explore the possibility of an additional strategy to exploit data skew: disk pages that are frequently accessed as a whole are accessed with priority. We show that considerable gain in service rate can be achieved with this strategy, while keeping memory consumption low. In essence we gain a three-stage approach to deal with skewed, unsorted data: caching plus our new strategy plus processing of the long tail of the distribution. We also present a cost model for our approach and validate our approach empirically.

Keywords: Semi-stream join · Performance optimisation · Indexing

1 Introduction

Stream-based joins are important operations in modern systems architectures, where just-in-time delivery of data is expected. We consider a particular class of stream-based join, a semi-stream join that joins a single stream with a disk-based table. Such a join can be applied, for example, in real-time data warehousing [5–7].

In this work we only consider one-to-many equijoins, as they appear between foreign keys in the stream and a referenced primary key in the master data table. This is obviously a very important class of joins, and they are a natural case of a join between a stream of updates and master data in a data warehousing

© Springer International Publishing AG 2016
M.A. Cheema et al. (Eds.): ADC 2016, LNCS 9877, pp. 407–418, 2016.
DOI: 10.1007/978-3-319-46922-5_32

context [5], online auction systems [2] and supply-chain management [15]. If a join is one-to-many and if high throughput is the aim, it is important to exploit this and to choose a one-to-many join operator, since it allows to expire stream tuples as soon as they have matched a master data record. Consequently, we do not consider here many-to-many joins, e.g. joins on categorical attributes in master data, such as gender.

Most semi-stream join algorithms keep recent stream tuples in main memory in order to amortize the expensive disk access cost over a large section of the stream [8–11,13,14]. As a consequence, many algorithms have a queue data structure that represents a section of the stream currently in processing. The queue keeps track of the stream tuples based on their loading timestamps. The queue typically stores join attribute values, and some of these algorithms [8,9,11] use these join attribute values in the queue (queue elements) as look up and load relevant disk-based master data through an index. In this paper we denote these queue elements *lookup elements*.

Several semi-stream algorithms have contributed caching strategies in order to exploit skew in the distribution of join attributes in the stream. Skewed data is the norm in many application scenarios [1], such as data warehousing for purchases. A skewed distribution commonly found in applications is the 80/20 rule, for example it is a rule of thumb in many markets that 80 % of sales are related to 20 % products [1]. However, caching is only effective down to a certain frequency of the join attribute, even if one allows for a tuple-level cache [9,11]. In our recent work [12] we have investigated whether one can further improve the disk-based part of the join operations. We have made some inroads by showing that by optimizing the choice of a lookup element from the queue, the service rate can be improved for a given skew. However, up to now it was not clear that this is an optimal strategy. This strategy was evaluated in the existing HYBRIDJOIN and CACHEJOIN algorithms [8,9]; in this paper we denote these algorithms once they are improved by the lookup strategy (and therefore modified) by HYBRIDJOIN-L and CACHJOIN-L respectively. In this paper we present a new analysis that shows that this lookup strategy suffers still from two drawbacks: (a) this position varies with the degree of skew in stream data and (b) at the lookup position we do not find a good lookup element each time we look, so the algorihm still performs suboptimal. Further details about these two issues are presented in Sect. 3.2.

In this paper we address these issues by presenting a new approach that replaces the early lookup element: it is a priority queue *PQ* of frequent pages. The *PQ* contains page IDs of the master data pages with a high stream probability. We added this new strategy again to the basic HYBRIDJOIN and CACHEJOIN, resulting in two new algorithms HYBRIDJOIN-PQ and CACHEJOIN-PQ. In this paper we compare them with HYBRIDJOIN-L and CACHEJOIN-L and can demonstrate a further improvement. This improvement can be explained because the new algorithms now access the frequent pages in an optimal time interval; in contrast the strategy of HYBRIDJOIN-L and CACHEJOIN-L achieved a preferential treatment of the high probability pages, but with the side effect of loading other pages before time. Further details about this are presented in Sect. 4.

Our main findings in this research can be summarized as follows:

Higher service rate: By implementing the new strategy, both HYBRIDJOIN-PQ and CACHEJOIN-PQ outperform existing HYBRIDJOIN-L and CACHEJOIN-L, respectively, for skewed data.

Adaptability: By using the new strategy both HYBRIDJOIN-PQ and CACHEJOIN-PQ to adapt to changes in the stream data, e.g. the value of skew in stream data. By contrast, due to the fixed position for the optimal lookup element in the queue, HYBRIDJOIN-L and CACHEJOIN-L cannot optimally adapt themselves to skew variation in the stream data.

Three-stage approach to skewed data: The new algorithm CACHEJOIN-PQ can be said to employ three different stages to deal with a skewed distribution. For very frequent tuples, an in-memory cache is employed, for tuples that are not in themselved frequent but cluster on pages that then get frequent as a page (e.g. due to locality effects), we employ the in-memory priority queue of frequent pages. Finally, for all other tuples we use direct lookup.

The rest of the paper is structured as follows. Section 2 presents related work. Section 3 describes background and based on that formulate the problem statement. Section 4 presents the solution. Section 5 describes our experiments and finally Sect. 6 concludes the paper.

2 Related Work

In this section, we present an overview of the previous work that has been done in the area of semi-stream joins, focusing on those that are closely related to our problem domain.

A seminal algorithm MESHJOIN [13, 14] has been designed especially for joining a continuous stream with disk-based master data, like in the scenario of active data warehouses. The MESHJOIN algorithm is a hash join, where the stream serves as the build input and the disk-based relation serves as the probe input. A characteristic of MESHJOIN is that it performs a staggered execution of the hash table build in order to load in stream tuples more steadily. To implement this staggered execution the algorithm uses a queue. The algorithm makes no assumptions about data distribution or the organization of the master data, hence there is no master data index. The algorithm always removes stream tuples from the end of the queue, as they have been matched with all master data partitions.

R-MESHJOIN (reduced Mesh Join) [10] clarifies the dependencies among the components of MESHJOIN. As a result the performance is improved slightly. However, R-MESHJOIN implements the same strategy as the MESHJOIN algorithm for accessing the disk-based master data, using no index.

Partitioned Join [4] improved MESHJOIN by using a two-level hash table, attempting to join stream tuples as soon as they arrive, and using a partition-based wait buffer for other stream tuples. The number of partitions in the wait buffer is equal to the number of partitions in the disk-based master data. The algorithm uses these partitions as an index, for looking up the master data. If a partition in a wait buffer grows larger than a preset threshold, the algorithm loads the relevant partition from the master data into memory. The algorithm allows starvation of stream tuples as tuples can stay in a wait buffer indefinitely if the buffer's size threshold is not reached.

Semi-Streaming Index Join (SSIJ) [3] was developed recently to join stream data with disk-based data. In general, the algorithm is divided into three phases: the pending phase, the online phase and the join phase. In the pending phase, the stream tuples are collected in an input buffer until either the buffer is larger than a predefined threshold or the stream ends. In the online phase, stream tuples from the input buffer are looked up in cached disk blocks. If the required disk tuple exists in the cache, the join is executed. Otherwise, the algorithm flushes the stream tuple into a stream buffer. When the stream buffer is full, the join phase starts where master data partitions are loaded from disk using an index and joined until the stream buffer is empty. This means that as partitions are loaded and joined, the join becomes more and more inefficient: partitions that are joined later can potentially join only with fewer tuples because the stream buffer is not refilled between partition loads. By keeping the stream buffer full and selecting lookup elements carefully the performance could be improved.

CACHEJOIN [9] is an extension of HYBRIDJOIN, which adds an additional cache module to cope with Zipfian stream distributions. This is similar to Partitioned Join and SSIJ, but a tuple-level cache is used instead of a page-level cache to use the cache memory more efficiently. CACHEJOIN is able to adapt its cache to changing stream characteristics, but similar to HYBRIDJOIN, it uses the last queue element as a lookup element for tuples that were not joined with the cache. SSCJ [11] is an improved version of CACHEJOIN, which optimizes the manipulation of master data tuples in the cache module. While CACHEJOIN uses a random approach to overwrite tuples in the cache when it is full, SSCJ overwrites the least frequent tuples. However, both SSCJ and CACHEJOIN use the same suboptimal strategy to access the queue.

3   Background

Semi-stream joins which implement staggered execution of stream data mostly use a queue data structure [8–11,13,14]. The key role of this queue component is to keep track of every stream tuple in memory with respect to loading time. The other purpose of the queue is to ensure that a stream tuple which enters into memory will certainly be processed. Moreover, some of these semi-stream joins [8,9,11] also use these queue values as lookup values to load master data into memory via an index.

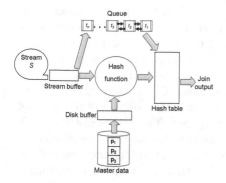

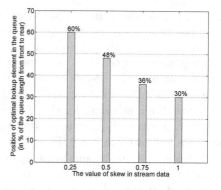

Fig. 1. Data structures and architecture of HYBRIDJOIN

Fig. 2. Analysis of position of the optimal lookup element in the queue using different values of the skew

3.1 HYBRIDJOIN-L and CACHEJOIN-L

We recently presented HYBRIDJOIN [8], an index-based semi-stream join with a simple architecture as shown in Fig. 1. HYBRIDJOIN [8], addresses the issue of accessing disk-based master data efficiently.

Since HYBRIDJOIN uses the oldest tuple of the queue as lookup element before a partition is loaded again after a join, new tuples matching that partition need to move all the way from the beginning to the end of the queue. For frequent partitions (partitions that have more matches) in master data are loaded not much more often than the common partitions. Therefore, the choice of the oldest tuple of the queue as lookup element is not optimal. Recently, we presented our work to determine a position for optimal lookup element from the queue [12]. This lookup element is then used as an index to load the master data partition in memory.

We provided an explanation of the achievement by introducing the concepts of stream probability and load probability. Based on these concepts it was empirically determined that the position at about 30 % of the queue size contains a lookup element that results in a particularly high service rate for a skew of 1. This improved the performance of HYBRIDJOIN, and we refer to this new improved algorithm as HYBRIDJOIN-L.

We also demonstrated this concept by applying to our existing CACHEJOIN [9] yielding CACHEJOIN-L. We have shown in the past that this still delivers an improvement over both CACHEJOIN and HYBRIDJOIN-L [12]. This means, that even with a cache, the careful optimization of master data access of frequent pages brings advantages.

3.2 Limitations and Problem Definition

HYBRIDJOIN-L and CACHEJOIN-L perform significantly better than their counterparts using only the oldest element in the queue for lookup. However, there are limtations. In [12] we were concerned with the optimal lookup element position solely for the skew value of 1. However, in further experiments we have confirmed that the optimal position of the lookup element is quite variable depending on the data, and specifically depending on the skew in the data. We present here the results of our experiment, and we use HYBRIDJOIN-L to illustrate the effect. We ran HYBRIDJOIN-L algorithm with synthetic stream data with different values for the skew and measured the position of the optimal lookup element in the queue, maximising throughput. The results of the experiment are presented in Fig. 2. From the figure we can observe that the position of the optimal lookup element in the queue moves forward to the young end of the queue (where new stream tuples are inserted) as the skew increases, reaching a position of 30 percent for a skew exponent of 1. For less pronounced skew with an exponent of 0.25 the optimal value is nearer to the other end of the queue.

Therefore both algorithms HYBRIDJOIN-L and CACHEJOIN-L would need further adaptations to perform optimally. Furthermore, even in the optimal position, the early lookup does not make a good choice every time. In a substantial fractions of the lookup, the element found is not on a frequent page and hence the early lookup is inefficient. The advantage only materializes in the average behavior. Therefore the question arises, whether we can avoid the bad cases and load a frequent page, at the right time, every time. This is what we will propose now.

4   Priority Queue of Frequent Pages

The improvement first offered in HYBRIDJOIN-L is based on the observation that unmatched tuples in the queue take up space. Therefore it is best for achieving a high service rate to load frequent pages as soon as a certain number of tuples in the queue is expected to match with this page. In HYBRIDJOIN-L this is achieved with the early lookup element: this element has been shown to be preferentially from a frequent page. However if we know which pages are frequent, we can use a more direct approach. A page is frequent if it has a high stream probability: This is the probability that if we pick a random stream tuple, that stream tuple belongs to that page [12]. We denote this value for the page here as *page.frequency*. This value can be easily observed at the time the page is first looked up: the observed number of matches divided by the queue length gives a good estimate of the frequency due to the law of large numbers: cases where an infrequent page is just accidentally matched by many stream tuples (and therefore this infrequent page enters the queue) are rare. For a frequent page, if we wait longer with loading that page we get more matches with a single lookup. However the many tuples that accumulate use up space in the hashtable, and they do that for a longer time. Therefore there is a tradeoff value at which it makes sense to load the page, we call this *pageThreshold*. Since new incoming stream tuples belong to random pages (but we cannot tell which tuple belongs to which page without accessing the disk) we know that after a certain number of new stream tuples have been entered into the queue, there is an expected number of tuples matching a given page. In fact for every page (not only the frequent ones), the expected number of stream tuples matching that page increases linearly with the number of stream tuples entered into the queue. Therefore if this expected number is larger as $pageThreshold$, then we should load the page. This is the case after $pageThreshold * \frac{h_S}{page.frequency}$ new loaded stream tuples. Hence the best strategy is to load each frequent page after a period of so many newly loaded stream tuples. The straightforward way to implement this is to keep a running counter of how many stream tuples have been processed and to keep a priority queue. The pages are inserted into the queue at the next counter value when they should be loaded. As soon as the counter value for the top page in the priority queue has been reached, this page is popped from the priority queue, the page is loaded and processed according to the algorithm for the disk phase, and finally the page is inserted with the new counter value $ptc + pageThreshold * \frac{h_S}{page.frequency}$. We do not discuss here issues of resetting the counter in order to prevent overflow since these are trivial exercises. We also do not discuss here removal of pages from the priority queue. This is trivial and can be based on the observed number of matches. As soon as the number of matches is below a certain threshold the page can be removed from the queue. Likewise the frequency of a page can steadily be adapted, by using a gliding average. A gliding average with an exponential decay model simple and does not require any further datastructure: Each time the page is loaded, its current number of matches c is observed, and we update according to $page.frequency = (page.frequency * d) + c * (1 - d)$ with a parameter d that controls the decay, e.g. $d = 0.9$ would be a typical value that smoothens out accidental changes in page frequency. This gliding average was however not installed in the algorithms used in our apparatus, since it would introduce an element of arbitrariness in the measurements (which decay parameter should we choose?) and the test data that we use here do not employ a drift of the skew. Nevertheless the algorithm is adaptive as it is now: if it is run with data with different skew values, it will adapt to whatever skew is present, without any configuration necessary.

5   Experiments

In this section we present an extensive experimental study of our algorithms. We compare the performance of all the algorithms using synthetic, TPC-H, and real-life datasets.

5.1   Experimental Setup

Hardware and software specifications: We performed our experiments on a *Intel-i5* with 8 GB main memory and 250 GB Solid Sate Drive (SSD) as secondary storage. We implemented our experiments in Java using the Eclipse IDE. As join attribute values can be duplicated in stream data due to the attribute being a foreign key, a hash table is needed that can store multiple values against one value of the master data. The hash table provided by the Java library does not support this feature, therefore `org.apache.MultiHashMap` was used.

Table 1. Data specifications for synthetic dataset

Parameter	Value
Total allocated memory M	50 MB to 250 MB
Size of disk-based relation R	0.5 million to 8 million tuples
Size of each disk tuple v_R	120 *bytes*
Size of each disk page p_{size}	$8 * 2^{10}$ *bytes*
Size of each stream tuple v_S	20 *bytes*
Size of each node in the queue	12 *bytes*
Size of each node in PQ	16 *bytes*
Data set	based on Zipf's law (skew value from 0.25 to 1)

Measurement strategy: The service rate of the join is measured by calculating the number of tuples processed in a unit second. In each experiment, the algorithm first completed their warmup phase before starting the actual measurements. These kinds of algorithms normally need a warmup phase to tune their components with respect to the available memory resources, so that each component can deliver a maximum service rate. For each service rate measurement, we calculated the 95% confidence interval. The calculation of the confidence intervals is based on 1000 to 4000 measurements for one setting. During the execution of the algorithm, no other application was running in parallel. We used a constant stream arrival rate throughout a run in order to measure the service rate for all algorithms.

Data specifications: We analyzed the service rate of the algorithms using synthetic, TPC-H, and real-life datasets. The relation R was stored on disk using a MySQL database. Both the algorithms read master data from the database. To measure the I/O cost more accurately, we set the fetch size for `ResultSet` equal to the disk buffer size.

Synthetic data. The stream dataset we used is based on a Zipfian distribution. We tested the service rate of all algorithms by varying the skew value from 0.25 (lightly skewed) to 1 (highly skewed). The master data we used was unsorted and had an index. Moreover, the size of the master data could be changed online. We used memory, master data, disk tuple, stream tuple and queue *pointer* sizes similar to original CACHEJOIN-L and HYBRIDJOIN-L. The detailed specifications of our synthetic dataset are shown in Table 1.

TPC-H. We also analyzed the service rate of all algorithms using the TPC-H dataset, which is a well-known decision support benchmark. We created the dataset using a scale factor of 100. More precisely, we used the table `Customer` as master data and the table Order as stream data. In table Order there is one foreign key attribute custkey, which is a primary key in the Customer table, so the two tables can be joined. Our Customer table contained 20 million tuples, with each tuple having a size of 223 bytes. The Order table contained the same number of tuples, with each tuple having a size of 138 bytes. The plausible scenario for such a join is to add customer details corresponding to an order before loading the order into the warehouse.

Real-life data. We also compared the service rate of all algorithms using a real-life dataset[1]. This dataset basically contains cloud information stored in a summarized weather report format. It was also used to evaluate the original MESHJOIN, CACHEJOIN-L and HYBRIDJOIN-L. We consider our master data table by combining meteorological data corresponding to months April and August, while consider the stream data by combining data files from December. The master data table contains 20 million tuples, while the streaming data table contains 6 million tuples. The size of each tuple in both the master data table and the stream data table is 128 bytes. Both tables are joined using a common attribute, longitude (LON). The domain of the join attribute is the interval [0,36000].

5.2 Service Rate and Costs Analysis

We conduct the service rate analysis with respect to three key parameters: the size of the master data table R, the total memory available, and the value of skew in the Zipfian distribution. For the sake of brevity, we restrict the discussion for each parameter to a one-dimensional variation, i.e. we vary one parameter at a time.

Analysis by varying size of memory: In this experiment we compared the service rate of all algorithms while varying the memory size from 50 MB to 250 MB, with the fixed size of R equal to 2 million tuples and skew value in stream data is equal to 1. Figure 3(a) presents the results of the experiment. From the figure it can be noted that for even a small memory size (50 MB) both our algorithms perform noticeably better than the existing algorithms and this improvement increases with the increase in the memory size. Furthermore in case of HYBRIDJOIN-PQ the scale of improvement is even better as because of no cache component, PQ takes the full advantage of the skew in stream data.

Analysis by varying size of R: In this experiment we compared the service rate of all the algorithms with different sizes of R. We keep fixed memory size (50 MB) and skew value is equal to 1. The results of the experiment are shown in Fig. 3(b). From the figure it can be seen that again both the new algorithms perform considerably better than the existing algorithms. Also in case of increasing the size of R the service rate of both CACHEJOIN-PQ and HYBRIDJOIN-PQ does not decrease with that rate as it decreases in CACHEJOIN-L and HYBRIDJOIN-L respectively. The plausible reason behind this behaviour is

---

[1] This dataset is available at: http://cdiac.ornl.gov/ftp/ndp026b/.

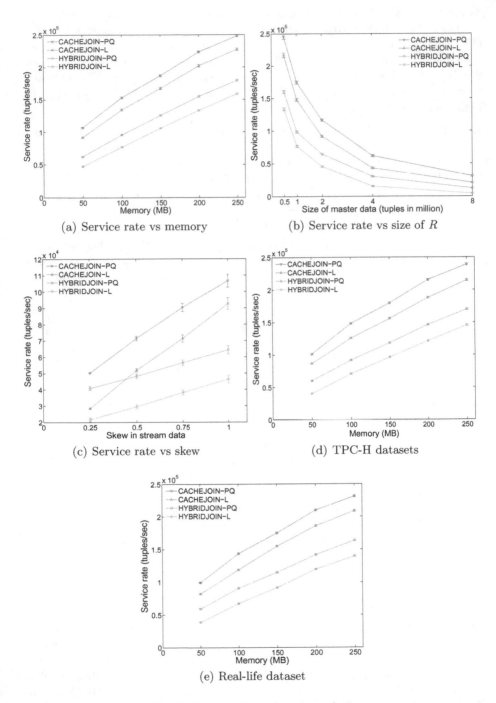

(a) Service rate vs memory

(b) Service rate vs size of R

(c) Service rate vs skew

(d) TPC-H datasets

(e) Real-life dataset

Fig. 3. Service rate and costs analysis

that, in the existing algorithms the position of optimal lookup element changes by increasing the size of R.

Analysis by varying skew value: In this experiment we compared the service rate of all the algorithms while varying the skew value in streaming data. To vary the skew, we varied the Zipfian exponent from 0.25 to 1. At 0.25 the input stream S has a light skew, while at 1 the stream has a strong skew. The size of R was fixed at 2 million tuples and the available memory was set to 50 MB. The results presented in Fig. 3(c) show that both CACHEJOIN-PQ and HYBRIDJOIN-PQ perform significantly better than CACHEJOIN-L and HYBRIDJOIN-L respectively. Particularly for small values of skew this difference of improvement is more prominent. This is an evidence for our argument that by changing the skew value in stream data the position of optimal lookup element in the queue also changes. We do not present data for skew values larger than 1, which would imply short tails. Also we do not consider fully uniform stream data (i.e. Zipfian exponent is equal to 0) as this is very unlikely in supermarket transactional data.

TPC-H and real-life datasets: In these experiments we measured the service rate produced by all the algorithms at different memory settings. The results of using both TPC-H and real-life datasets are shown in Fig. 3(d) and (e) respectively. From the both figures it can be noted that under all memory settings both CACHEJOIN-PQ and HYBRIDJOIN-PQ outperform existing CACHEJOIN-L and HYBRIDJOIN-L respectively .

6   Conclusions

Recently we published the strategy of finding a position of optimal lookup element from the queue. However, the approach suffers with two drawbacks (a) it is not necessary that the lookup element always gives a frequent page from the master data, a page that has frequent number of matches with stream tuples in memory (b) the position of the lookup element can vary by varying the nature of stream input, e.g. the value of skew in stream data. In this paper we addressed these issues by introducing a new component called priority queue. The priority queue keeps the record of frequent disk pages by storing their page IDs. While the criteria of deciding that a page is frequent is based on the total number of matches in whole queue size stream data against that page. In this way unlike to the existing approach, every element in the priority queue gives a frequent page of master data and this does not affect by changing the skew in stream data. To validate our argument we implemented our new strategy to the existing CACHEJOIN-L and HYBRIDJOIN-L algorithms and named them CACHEJOIN-PQ and HYBRIDJOIN-PQ. We provided experimental results that show that the both new algorithms perform significantly better than the existing ones for all synthetic, TPC-H and real-life datsets. We also provide a cost model for our CACHEJOIN-PQ and validate that empirically.

In the future, we will extend our work to consider many-to-many equijoins and certain classes of non-equijoins.

References

1. Anderson, C.: The Long Tail: Why the Future of Business Is Selling Less of More. Hyperion, New York (2006)
2. Arasu, A., Babu, S., Widom, J.: An abstract semantics and concrete language for continuous queries over streams and relations. Technical Report 2002–57, Stanford InfoLab (2002)
3. Bornea, M., Deligiannakis, A., Kotidis, Y., Vassalos, V.: Semi-streamed index join for near-real time execution of ETL transformations. In: IEEE 27th International Conference on Data Engineering (ICDE 2011), pp. 159–170, April 2011
4. Chakraborty, A., Singh, A.: A partition-based approach to support streaming updates over persistent data in an active datawarehouse. In: IPDPS 2009: Proceedings of the 2009 IEEE International Symposium on Parallel and Distributed Processing, pp. 1–11. IEEE Computer Society, Washington, DC, USA (2009)
5. Golab, L., Johnson, T., Seidel, J.S., Shkapenyuk, V.: Stream warehousing with datadepot. In: SIGMOD 2009: Proceedings of the 35th SIGMOD International Conference on Management of Data, pp. 847–854. ACM, New York, NY, USA (2009)
6. Karakasidis, A., Vassiliadis, P., Pitoura, E.: ETL queues for active data warehousing. In: IQIS 2005: Proceedings of the 2nd International Workshop on Information Quality in Information Systems, pp. 28–39. ACM (2005)
7. Naeem, M.A., Dobbie, G., Weber, G.: An event-based near real-time data integration architecture. In: EDOCW 2008: Proceedings of the 2008 12th Enterprise Distributed Object Computing Conference Workshops, pp. 401–404. IEEE Computer Society, Washington, DC, USA (2008)
8. Naeem, M.A., Dobbie, G., Weber, G.: HYBRIDJOIN for near-real-time data warehousing. Int. J. Data Warehouse Min. (IJDWM) 7(4), 24–43 (2011)
9. Naeem, M.A., Dobbie, G., Weber, G.: A lightweight stream-based join with limited resource consumption. In: Cuzzocrea, A., Dayal, U. (eds.) DaWaK 2012. LNCS, vol. 7448, pp. 431–442. Springer, Heidelberg (2012)
10. Naeem, M.A., Dobbie, G., Weber, G., Alam, S.: R-MESHJOIN for near-real-time data warehousing. In: DOLAP 2010: Proceedings of the ACM 13th International Workshop on Data Warehousing and OLAP. ACM, Toronto, Canada (2010)
11. Naeem, M.A., Weber, G., Dobbie, G., Lutteroth, C.: SSCJ: a semi-stream cache join using a front-stage cache module. In: Bellatreche, L., Mohania, M.K. (eds.) DaWaK 2013. LNCS, vol. 8057, pp. 236–247. Springer, Heidelberg (2013)
12. Asif Naeem, M., Weber, G., Lutteroth, C., Dobbie, G.: Optimizing queue-based semi-stream joins with indexed master data. In: Bellatreche, L., Mohania, M.K. (eds.) DaWaK 2014. LNCS, vol. 8646, pp. 171–182. Springer, Heidelberg (2014)
13. Polyzotis, N., Skiadopoulos, S., Vassiliadis, P., Simitsis, A., Frantzell, N.: Supporting streaming updates in an active data warehouse. In: ICDE 2007: Proceedings of the 23rd International Conference on Data Engineering, pp. 476–485. Istanbul, Turkey (2007)
14. Polyzotis, N., Skiadopoulos, S., Vassiliadis, P., Simitsis, A., Frantzell, N.: Meshing streaming updates with persistent data in an active data warehouse. IEEE Trans. Knowl. Data Eng. 20(7), 976–991 (2008)
15. Wu, E., Diao, Y., Rizvi, S.: High-performance complex event processing over streams. In: Proceedings of the 2006 ACM SIGMOD International Conference on Management of Data, SIGMOD 2006, pp. 407–418. ACM, New York, NY, USA (2006)

Exploiting Hierarchies for Efficient Detection of Completeness in Stream Data

Simon Razniewski[1]([✉]), Shazia Sadiq[2], and Xiaofang Zhou[2]

[1] Free University of Bozen-Bolzano, Bolzano, Italy
`razniewski@inf.unibz.it`
[2] School of ITEE, The University of Queensland, Brisbane, Australia
`{shazia,zxf}@itee.uq.edu.au`

Abstract. In big data settings, the data can often be externally sourced with little or no knowledge of its quality. In such settings, users need to be empowered with the capacity to understand the quality of data sets and implications for use, in order to mitigate the risk of making investments in datasets that will not deliver. In this paper we present an approach for detecting the completeness of high volume stream data generated by a large number of data providers. By exploiting the inherent hierarchies within database attributes, we are able to devise an efficient solution for computing query specific completeness, thereby improving user understanding of implications of using query results based on incomplete data.

1 Introduction

Recent years have seen an unprecedented investment in big data initiatives. As companies intensify their efforts to get value from big data, the growth in the amount of data being managed continues at an exponential rate, leaving organizations with a massive footprint of unexplored, unfamiliar datasets. On February 8th, 2015, a group of global thought leaders from the database research community outlined the grand challenges in getting value from big data [2]. The key message was the need to develop the capacity to understand *how the quality of data affects the quality of the insight we derive from it*. Organizational big data investment strategies regarding what data to collect, clean, integrate, and analyze are typically driven by some notion of perceived value. However, even in traditional IS projects that follow carefully crafted analysis and design lifecycles, the misalignment between perceived and actual value has been a notoriously difficult problem leading to dismal return on investment [14]. In big data projects, where the data can often be externally sourced with little or no knowledge of its schemata, quality, or expected utility, the risk of making investments that will not deliver is greatly heightened.

It is widely known that the value of the dataset is inescapably tied to the underlying quality of the data. While data quality has been studied against a large number of so-called data quality dimensions [11], in this paper we focus on the completeness dimension. Completeness can be interpreted in multiple ways including (1) missing values, (2) missing records, and (3) sufficiency of the data

© Springer International Publishing AG 2016
M.A. Cheema et al. (Eds.): ADC 2016, LNCS 9877, pp. 419–431, 2016.
DOI: 10.1007/978-3-319-46922-5_33

for the task at hand. Unlike missing values which are easy to detect, missing records are not visible and can only be detected by using additional reference sources or meta data outside of a database e.g. from provenance information [10], master data or from business processes [18]. At the same time, when dealing with large and uncertain data sets, a user might experience long query processing times, only to realize that the provided result is based on incomplete records. Thus a key challenge in this regard is efficient assessment of record completeness in order to empower users with improved understanding of the implications of using query results based on incomplete data.

2   Motivating Scenario

Consider a fictional regional transportation agency, which we refer to by its acronym RTA in the following[1]. Use of RTA vehicles is by a smartcard, which passengers need to scan when boarding and alighting vehicles. RTA vehicles are fitted with readers that transmit scans in near real time to a central server. RTA uses the smartcard scan data in various decisions, such as:

1. Dynamic service scaling: If the demand for specific services is exceptionally high, the RTA tries to dispatch additional vehicles.
2. Connecting services: Especially at night time, when between service waiting times are long, managing the impact of delayed connections is an efficiency and safety issue.
3. Bus bunching detection and handling: The natural tendency of vehicles to cluster, called bus bunching, is a classical problem of public transport [1], which the RTA aims to predict or detect as early as possible.

For all three scenarios, if any part of the data is not complete (due to faults in data transmission, or technical failures that lead to card scans not being recorded at all), it may lead to incorrect query results and misinformation on the real status of the transport network. For example, the service control center might wrongly believe that a service is not full, and guaranteeing a delayed connection is not necessary. To avoid these misconceptions, knowledge of the completeness of data within a query result is necessary. The above example provides a high volume, high velocity big data setting. In this paper, we address the challenge of efficient data completeness assessment in such settings. Our approach is based on the notion of inherent hierarchies within data (such as RTA smartcard scan data) which we exploit to efficiently compute data completeness for specific queries. We will use RTA smartcard data as a running example throughout the paper to illustrate our approach.

Consider a single table *cardscan* in the RTA database, which contains the following attributes:

VehicleType, Line, VehicleID, DateTime, Stop, SmartcardID.

---

[1] There are actually 14 corporations in the transit sector with that acronym listed on Wikipedia.

Furthermore, suppose that VehicleID functionally determines Line, and Line functionally determines VehicleType, i.e., every vehicle can belong to only one line, and every line can run vehicles of only one type (bus/train/ferry/..). We consider a time window of 10 days, with days labelled from 1 to 10, accordingly. Some sample data is shown below.

VehicleType	Line	VehicleID	DateTime	Stop	SmartcardID
Bus	1	1B	Day 4 - 1:37 PM	Dutton Park	48789397
Bus	1	1A	Day 7 - 9:11 AM	Fairfield	39150370
Bus	2	2A	Day 2 - 6:50 PM	Yeronga	61096215
Bus	1	1B	Day 9 - 2:07 PM	South Bank	93367832
...	...	...	...	...	...

For illustration purposes, let us assume that the RTA has only 4 vehicles, all buses, with IDs 1A, 1B, 2A, and 2B, organized in two lines, 1 (vehicles 1A and 1B) and 2 (vehicles 2A and 2B). We also assume that RTA has knowledge on completeness of the data such as "the data for all vehicles is complete on all days, except for vehicle 2A on Day 3". The assumption of availability of completeness information is reasonable given the increasing functionality of complex onboard electronics [3]. Suppose now that the RTA wants to know how many people used each vehicle on Day 3. They could issue the following SQL query $Q_{VehiclesDay3}$:

```
SELECT VehicleID, COUNT(*)
FROM cardScan
WHERE CAST(DateTime as DATE) = 'Day3'
GROUP BY VehicleID
```

A hypothetical result to this query shown below could be colored with completeness information as shown in Table 1 (left side). White stands for correct rows (based on complete data), and gray stands for possibly incorrect rows (based on possibly incomplete data). A third color, light-gray is also possible, as the next query shows. Suppose now the RTA also wants to know how many people were riding each line in the last 14 days. They could issue the query Q_{Lines}:

```
SELECT Line, COUNT(*)
FROM cardScan
GROUP BY Line
```

The result to this query could be presented as shown in Table 1 (right side). Clearly, the count for Line 1 is complete, because vehicles 1A and 1B are complete, thus this row is colored in white. For Line 2, the result is not complete, but we propose to color the row in light gray, because there are two possible specializations of that row that are complete: (1) The count for Line 2 on days other than Day 3, and (2) The count of Line 2 for rides on vehicle 2B.

Table 1. Results for Queries $Q_{VehiclesDay3}$ and Q_{Lines}.

VehicleID	Count(*)
1A	823
1B	712
2A	152
2B	357

Line	Count(*)
1	4823
2	2712

Suppose now that instead of 4, there are 1000 vehicles each sending data and completeness statements every 5 s, and that there are 1000 queries that the operation control would like to pose in real-time. Naively, this scenario would require 1000×1000, that is, 1 million recomputations of query completeness every 5 s, or 200 k per second. Timely computation of completeness information for query results as demonstrated in the above examples is the key aim of the approach presented in this paper.

3   Related Work

The problem of data completeness assessment over partially complete databases was studied in [12,15,17]. Motro [15] used views to describe complete parts of databases, while Levy [12] used local completeness statements, similar to views. Biswas et al. [4] discussed the tradeoff between completeness and resource consumption in sensor networks, while in [7], the influence of probabilistic keys on the completeness of probabilistic databases was discussed. In the present work we rely on patterns, a less expressive formalism of selections on tables, introduced in [17], which allows completeness statements to be expressed in the same schema as the base data. Subsequently, this work introduced an algebra for manipulating these statements, similar to relational algebra.

The idea of promoting completeness information was first discussed in [16], where conditional finite domain constraints (CFDCs) were used. Conditional functional dependencies (CFDs) are well-known database constraints that originated in data cleaning applications [6]. They are more specific than functional dependencies, as they apply only to records that satisfy certain selection conditions, for instance, only for customers in the UK, the ZIP code determines the street. In [16], it was also shown how CFDCs can be used in promoting completeness statements using disjunctive logic programming. Promotion was also discussed in [17], although the approach there is a database-instance dependent one: Information about join values actually present in the database is used to assess whether incomplete join results already cover all possible parts of a join result. However this will not provide a way to derive the conclusions made by hierarchical promotion as presented in Sect. 2.

Our work is also related to data streams because of the setting with a large number of data providers (i.e. RTA vehicles) emitting data at high frequency. Streams are time-evolving series of data [9]. Completeness of parts of data streams has received attention since early works by Tucker et al. introducing the notion of *punctuations*, which are special symbols in data streams that indicate that parts of the stream have been fully transmitted [19]. Later work by Johnson and Golab [8] investigates the semantics of queries over partially complete streams. The focus of this work is on stream warehouses, where base tables come from individual streams, tables are partitioned by time, and views may be built upon base tables. It then investigates which time partitions of views are already trustworthy, or which ones can be quickly recomputed, even if data in the base tables is still missing.

4 Foundation Concepts

Completeness Descriptions. Incompleteness in databases is commonly understood as follows: The available database D is considered to be a subset of an unknown ideal database D^i, which describes the facts that hold in reality. For example in reality it might be the case that there were 823/712/902/357 card scans in vehicles 1A/1B/2A/2B on Day 3, while in the database D, only 823/712/*152*/357 are recorded. Completeness statements are then used to describe in which parts the available database D corresponds to the ideal database D^i. In the example above, we could (1) state that the data for Vehicle 1A is complete on Day 3, while stating (2) that the data for Vehicle 2A is complete would be incorrect, as 750 records are missing. Completeness statements are tuples that may contain wildcards, and where the constants describe selection conditions. Prior work [17] considered completeness statements of arbitrary shape, i.e., constants and wildcards could appear at arbitrary positions in the statement. This is not the case in our scenario.

Example 1. Vehicle 1A publishes statements such as

- *(*, *, 1A, Day 2 - 2pm–4pm, *, *)*,
- *(*, *, 1A, Day 3 - 9am–1pm, *, *)*.
- *(*, *, 1A, Day 8 - 7am–8pm, *, *)*.

Notably, the statements differ only in the *DateTime* field. In the following, we assume that all data and completeness statements come from a known set of objects, called *data-generating entities* (DGEs). Each DGE has a *descriptor pattern* that may consist of constants ("Bus", "B1"), wildcards ("*"), and underscores ("_"). For all attributes in the DGE descriptor that are not underscores, any completeness statement produced by the DGE must have the same values as the descriptor pattern. For instance, the descriptor pattern for Vehicle 1A is *(*, *, 1A, _, *, *)*.

Example 2. Consider again the scenario from Sect. 2, where data from all vehicles besides Vehicle 2A on Day 3 was complete. We could describe this with the following completeness statements for the table *cardScan*:

VehicleType	Line	VehicleID	DateTime	Stop	SmartcardID	
Bus	2	2A	Day 4 - 1:37 PM	Dutton Park	48789397	} Data
...	...	...	...	...	...	
*	*	1A	*	*	*	}
*	*	1B	*	*	*	Completeness
*	*	2A	Days 1-2, 4-14	*	*	Statements
*	*	2B	*	*	*	}

Query Completeness Reasoning. Completeness statements can also be used to describe complete parts of query results. For $Q_{VehiclesDay3}$, the statements (1A,*), (1B,*) and (2B,*) hold, which we used before to color the corresponding rows in the result in white. A core problem is to find the statements that can be derived for a query result, given statements for database tables. This problem is called *query completeness reasoning.*

Example 3. Consider that the statements (*, *, 1A, *, *, *) and (*, *, 1B, *, *, *) hold for *cardScan*, i.e., that the table contains all scans for vehicles 1A and 1B. If this is the case, then, based on the information that 1A and 1B are the only vehicles of Line 1, the table contains also all card scans for it, i.e., the former statements logically entails (1,*) for Q_{Lines}, thus the white coloring of that row in Table 1.

Existing work reduces completeness reasoning to query rewritability [15], query independence of updates [12], query containment [17] and logic programming [16], but does not take into account hierarchic data, as discussed next.

Hierarchies. Hierarchies within attributes can be visualized using trees, where each level corresponds to an attribute, each node in a level represents a possible attribute value, and edges link possible attribute value combinations. Figure 1 shows the hierarchy for our use case.

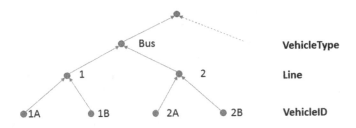

Fig. 1. Hierarchy of our use case.

Technically, hierarchies can be seen as a stronger version of conditional finite domain constraints (CFDCs) [16]. CDFCs are database constraints used to assure that a certain value in a certain field limits the possible values of another field, e.g., that for records with *Line=1* it holds that *VehicleID* is either 1A or 1B. Our setting requires also the converse implication: If *VehicleID* is 1A or

1B, *Line* must be 1. We therefore introduce an extended formalism that we call *conditional two-way dependencies* (CTWDs). Formally, a CTWD has the form $R\{A_1 = v\}[A_2] = \mathcal{W}$, where $A_1 = v$ is a selection condition on the relation R, and $\mathcal{W}$ describes the possible values of attribute A_2 in records that satisfy $A_1 = v$. Their semantics is as follows: For any record r that has $r[A_1] = v$, it must hold that $r[A_2]$ is in $\mathcal{W}$ (like CFDCs), and conversely, if $r[A_2]$ is in $\mathcal{W}$, then $r[A_1]$ must have the value v. A hierarchy tree can then be decomposed into a set of CTWDs, with each parent-children-pair in the tree corresponding to one CTWD. For example, the parent-children-pair (1, (1A, 1B)) corresponds to the CTWD *cardScan*{Line=1}[VehicleID]={1A, 1B}.

5 Query Completeness Computation with Hierarchies

In this section we present our approach to data completeness assessment through efficient computation of query completeness. The hierarchical data model introduced in the previous section has a twofold impact on completeness computation:

Enabling Efficient Promotion. Promotion in [17] requires access to the database content, and to apply it in our use case we would need to normalize the database and to rewrite single-table queries into join queries, even if none of the attributes of the joined table is in the result, which may be counter-intuitive. Promotion in [16] requires to solve Π_2^P-complete logic programming problems, which is generally not scalable. Our first challenge is thus to perform efficient promotion on a single table.

Pruning Irrelevant DGEs. Consider again the 1000 vehicles sending statements every 5 s, and the 1000 real-time queries, which would naively require 200k recomputations of query completeness per second. Most likely, most statements do not affect the completeness of most queries, i.e., they are irrelevant for the respective queries. Our second challenge is thus to devise techniques to determine whether completeness statements from a DGE are relevant to a query or not.

5.1 Enabling Efficient Promotion

In this section we discuss how to do efficient promotion on a single table. Our idea is to use the CTWDs to promote completeness information, that a set of siblings has in common, recursively upward.

Definition 1 (Hierarchical Promotion). *Let R be a table and $F = R\{A_1 = v\}[A_2] = \mathcal{W}$ be a CTWD. Then if there exists a set S of completeness statements for R that is unifiable except for the attribute A_2, and if $\{\pi_{A_2}(s) \mid s \in S\} = \mathcal{W}$, we can assert the unifier of S, with attribute A_2 being promoted and attribute A_1 being replaced by v, for R.*

Example 4. Consider the CTWD *goCardRecords*{Line=1}[VehicleID]={1A, 1B} and the completeness statements *(*, *, 1A, Days 3-5, *, *)* and *(*, *, 1B, Days*

4-10, *, *). Then, since these statements cover all possible values of the CTWD, and are unifiable, we can assert the unified statement *(*, 1, *, Days 4-5, *, *)* for Line 1.

If we repeat this procedure bottom-up, we will eventually derive all statements that hold for a table. As a preparation, we have to arrange completeness statements in a hierarchy, analogous to the hierarchy for the attribute values. Whenever a DGE transmits a new completeness statement s for a node n in this hierarchy, the method *processStmt(n, s)* notifies the parent, which in turn checks the applicability of promotion and recursively notifies further ancestors. The next theorem proves that the procedure *processStmt* is complete for all statements that can be inferred over a single table. We say that a set of completeness statements S is maximal, if all statements entailed by S are in S.

Theorem 1. *Consider a maximal set of completeness statements S stored in a hierarchy of nodes N, and a completeness statement s for a node n in N. Then processStmt(n, s) produces all completeness statements that are entailed by $S \cup \{s\}$.*

Instead of doing promotion repeatedly at query time (for the same or different queries), it is done here only once for every statement. Unlike in [17], where promotion requires repeated database access to retrieve relevant values, here no database access is needed as long as hierarchies can be kept in memory. Hierarchical promotion can also be applied to tables that satisfy multiple orthogonal hierarchies, e.g., if lines are associated with operators, and operators and vehicle types are orthogonal, that is, an operator may operate both buses and ferries, or trains can be operated by different operators. In such situations, the attributes would not be arranged in a tree but in a directed acyclic graphs with a single sink, and in the promotion, an OR-computation would take place: All vehicles would be complete if all operators OR all vehicle types were complete.

5.2   Pruning Irrelevant DGEs

Not all completeness statements are relevant to all queries. For instance, for a query for ferry traffic, completeness of the bus line 1 is clearly irrelevant. Recomputing query completeness only on the arrival of relevant statements may potentially yield huge savings. Recall the example at the beginning of this Section, where 1000 queries were posed and 1000 statements were published every 5 s. Let us assume that on average, each statement is relevant to only 5 queries (which makes sense if many queries concern few vehicles, and only few queries concern many vehicles). Then, ideally, it would be necessary to recompute only 5000 queries per 5 s, or 1000 queries per second, instead of 200 k per second. To find out whether a query needs to be recomputed, we can check whether a source is relevant to a query, using query independent of update (QIU) techniques. Solutions for general QIU problems exist in [5,13]. In our case, what is needed to be checked is whether a query is independent of updates adhering to a DGE descriptor. If this is the case, statements produced by this DGE are relevant to the query.

Example 5. Consider a query Q_{F+B} for cards that were used both on a ferry and a bus. As a conjunctive query, this is written as

$$Q_{F+B}(c)\text{:-}cardScan(Ferry, x_1, x_2, x_3, x_4, c), \; cardScan(Bus, y_1, y_2, y_3, y_4, c).$$

Furthermore, consider the following two DGEs:

1. $d_1 = $ *(Ferry, F2, 7B, _, * , *)*
2. $d_2 = $ *(Train, T3, 9A, _, * , *)*

We can see that d_1 is relevant for this query, as its descriptor pattern syntactically matches the first atom of Q_{F+B}. In contrast, d_2 is not relevant for this query, because it matches neither atom of Q_{F+B}.

To check relevancy, we need to expand DGE descriptors using the CTWDs derived from the hierarchy. Although the descriptor *(*,*,9A,*,*)* is semantically equivalent to d_2 from above, it matches both atoms of Q_{F+B} and hence one might wrongly conclude that its statements are relevant to the query. Computationally, descriptor pattern expansion is linear in the height of the hierarchy, and relevancy checking is linear in the length of the query, as it involves only pairwise comparison of constants and variables. If we cannot find any atom in a query to which the source description can be mapped, we say that the source is not relevant for the query. The following proposition formalizes that statements deemed as irrelevant can be ignored in query completeness computation.

Proposition 1. *Let Q be a conjunctive query and p be a DGE descriptor that is not relevant to Q. Then*

1. *For no database, adding data records that match p changes the result to Q.*
2. *For no set of completeness statements, adding completeness statements that match p changes the completeness statements that hold for Q.*

5.3   Implementation of Query Completeness Reasoning

The core data component is the *completeness statement hierarchy*, which is both a data structure that allows to store completeness statements based on a hierarchy, and an object that implements the hierarchical promotion. For each database table, one such component would be deployed. DGEs would submit statements to this component, while queries would subscribe to the nodes in the hierarchy that are relevant to them. The completeness statement hierarchies would then be the bridge between DGEs and queries. On arrival of new relevant completeness statements, they would notify the subscribed queries. Once completeness of a query is recomputed (which need not happend immediately after notification, depending on load), the query can pull the completeness statements from the nodes of the completeness statement hierarchy that are relevant to it. The full interaction is illustrated in Fig. 2. In our motivating example, data from all DGEs is relevant to the queries $Q_{VehiclesDay3}$ and Q_{Lines}, hence, they would subscribe to all nodes in the completeness statement hierarchy.

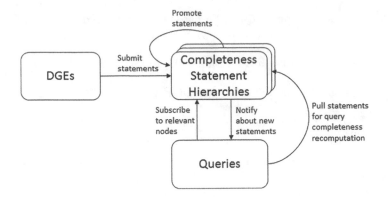

Fig. 2. Interaction between DGEs, completeness statement hierarchies and queries.

6   Experiments

We have experimentally analyzed the scalability of completeness statement computation using our approach. The factor of interest is how fast query completeness can be recomputed based on changes to completeness statements. Three parameters are relevant:

1. The size of the hierarchy. We have chosen a hierarchy with 4 attributes, and vary its size by changing the branching factor (default branching factor 10, resulting in 10,000 DGEs).
2. The queries. We start with elementary queries that select data from a node in the hierarchy, one per node in the hierarchy, and obtain complex queries by joining elementary queries on the *SmartcardID* field.
3. The frequency with which completeness statements are submitted. As default we use 0.2 Hz (every 5 s).

From all the completeness statements generated in the experiments, we randomly dropped 1 %, and set the arrival time of the others to a random timepoint in the next period (for instance, statements for the interval 10–15 s arrive randomly somewhere between second 15 and 20). Our experiments were performed on a desktop machine with an Intel i7 processor with 3.4 GHZ and 16 GB RAM. Our implementation was done as a single-threaded Java program.

In the first experiment, we compare completeness computation with a baseline, where for any query, completeness is recomputed whenever a new completeness statement arrives. The results are shown in Fig. 3 (left). The results indicate that the delay of the baseline method exceeds reasonable amounts already for less than 10,000 DGEs, while computation with hierarchies is below 2 s delay even with 40,000 DGEs. Second we increase the number of joins. As one can see in Fig. 3 (right), computation for queries with 3 joins is possible for up to 20,000 DGEs within a second. In the third experiment we compare the delay in completeness statement processing and query completeness recomputation for

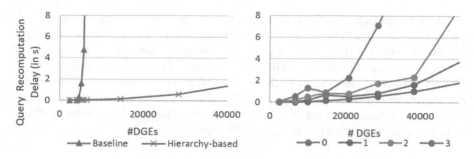

Fig. 3. Recomputation delay in our hierarchical approach compared with a naive baseline (left), and in relation to query length (right).

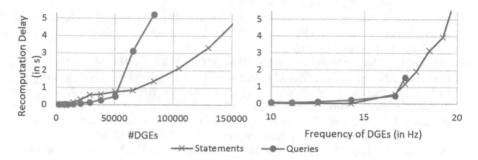

Fig. 4. Recomputation delay for queries and completeness statements, depending on the number of DGEs (left) and on the frequency of statement submission (right).

elementary queries. The results shown in Fig. 4 (left), indicate that the delays for both are below one second for up to 50,000 DGEs. Afterwards, the query computation delay increases steeply, while the completeness processing delay increases gradually, as completeness statements have priority in our implementation. In the last experiment, we vary the period of data submission of the DGEs between 100 ms and 50 ms (10–20 Hz), finding that the recomputation delays are reasonably small until a period of 60 ms (17 Hz). At higher frequencies, all compute time is used for hierarchical promotion, and queries are never recomputed (Fig. 4 right).

7   Discussion and Outlook

In the experiments, we have seen that the framework can be applied for elementary queries for up to 50,000 DGEs submitting statements every 5 s, or that up to 10,000 DGEs could submit data every 60 milliseconds. We believe that in practical applications, the frequency and the number of DGEs will be much lower[2], thus, the observed bounds validate the applicability of our framework.

---

[2] http://goo.gl/x2kZD5, for instance, cites 1200 buses in Brisbane in 2012.

We envisage that the efficient completeness annotation we have proposed is applicable in several domains besides transport, where a large number of DGEs submit data in real-time, and data loss occasionally happens. For example in *logistics* to manage inventory and delivery cycles; and similarly in *smart factories* where a variety of parameters such as state of processes and energy consumption needs continuous monitoring, and where deviations from desired states may require just-in-time interventions. In all such scenarios timely knowledge about incomplete data is critical to make accurate decisions.

References

1. https://en.wikipedia.org/wiki/Bus_bunching
2. Abiteboul, S., Dong, L., Etzioni, O., Srivastava, D., Weikum, G., Stoyanovich, J., Suchanek, F.M.: The elephant in the room: getting value from big data. In: WebDB, pp. 1–5. ACM (2015)
3. Ashton, K.: That 'internet of things' thing. RFiD J. **22**(7), 97–114 (2009)
4. Biswas, J., Naumann, F., Qiu, Q.: Assessing the completeness of sensor data. In: Li Lee, M., Tan, K.-L., Wuwongse, V. (eds.) DASFAA 2006. LNCS, vol. 3882, pp. 717–732. Springer, Heidelberg (2006)
5. Blakeley, J.A., Coburn, N., Larson, P.: Updating derived relations: Detecting irrelevant and autonomously computable updates. In: VLDB (1986)
6. Bohannon, P., Fan, W., Geerts, F., Jia, X., Kementsietsidis, A.: Conditional functional dependencies for data cleaning. In: ICDE (2007)
7. Brown, P., Link, S.: Probabilistic keys for data quality management. In: Zdravkovic, J., Kirikova, M., Johannesson, P. (eds.) CAiSE 2015. LNCS, vol. 9097, pp. 118–132. Springer, Heidelberg (2015)
8. Golab, L., Johnson, T.: Consistency in a stream warehouse. In: CIDR, pp. 114–122 (2011)
9. Tamer Özsu, M., Golab, L.: Issues in data stream management. ACM Sigmod Rec. **32**(2), 5–14 (2003)
10. Hartig, O., Zhao, J.: Using web data provenance for quality assessment. In: CEUR Workshop Proceedings (2009)
11. Jayawardene, V., Sadiq, S., Indulska, M.: The curse of dimensionality in data quality. In: ACIS, pp. 1–11 (2013)
12. Levy, A.Y.: Obtaining complete answers from incomplete databases. In: VLDB, pp. 402–412 (1996)
13. Levy, A.Y., Sagiv, Y.: Queries independent of updates. In: Proceedings of the VLDB, pp. 171–181 (1993)
14. McAfee, A.: Mastering the three worlds of information technology. Harvard Bus. Rev. **84**(11), 141 (2006)
15. Motro, A.: Integrity = Validity + Completeness. ACM TODS **14**(4), 480–502 (1989)
16. Nutt, W., Paramonov, S., Savkovic, O.: Implementing query completeness reasoning. In CIKM, pp. 733–742 (2015)

17. Razniewski, S., Korn, F., Nutt, W., Srivastava, D.: Identifying the extent of completeness of query answers over partially complete databases. In: SIGMOD, pp. 561–576 (2015)
18. Razniewski, S., Montali, M., Nutt, W.: Verification of query completeness over processes. In: Daniel, F., Wang, J., Weber, B. (eds.) BPM 2013. LNCS, vol. 8094, pp. 155–170. Springer, Heidelberg (2013)
19. Tucker, P., Maier, D., Sheard, T., Fegaras, L., et al.: Exploiting punctuation semantics in continuous data streams. TKDE 15(3), 555–568 (2003)

Demo Papers

Visualization-Aided Exploration
of the Real Estate Data

Mingzhao Li[1][(✉)], Zhifeng Bao[1], Timos Sellis[2], and Shi Yan[1]

[1] RMIT University, Melbourne, Australia
Mingzhao.Li@rmit.edu.au
[2] Swinburne University of Technology, Melbourne, Australia

Abstract. An efficient analysis of the real estate data is critical for buyers to understand the real estate market and seek appropriate properties to live in or rent. In this paper, we first collect data from different channels which are not provided in existing commercial real estate systems, and integrate them to build a location-centred comprehensive real estate dataset, including information other than the house itself, such as education profile, transportation profile and regional profile. Then we develop HouseSeeker, a visualization-aided system for buyers to explore the real estate data, find appropriate properties based on their individual requirements, and compare properties/suburbs from different aspects to discover the strengths and weaknesses of each property/suburb. We demonstrate the effectiveness of our system based on a real-world dataset in Melbourne metropolitan area: it is able to help zero-knowledge users better understand local real estate market and find preferred properties based on their individual requirements. A preliminary implementation of the system is available at http://115.146.89.158/.

1 Introduction

We have noticed four main problems when using existing commercial systems, in particular *Real Estate Australia*[1] and *Domain*[2], to search for real estate properties. **First**, data in those systems are merely about properties, while a lot of useful information that is critical to users' choices is not captured, e.g., the regional, educational and transportational profiles (see Sect. 2). **Secondly**, users are requested to pre-define their requirements, and then the corresponding housing properties are displayed in the form of a list or map. However, neither SQL nor Top-K based search adopted by those commercial systems, manages to provide users with enough candidate properties based on their individual requirements. **Thirdly**, they only show locations of properties on the map and do not support presentation and comparison of properties in multiple aspects. **Last**, those commercial systems fail to provide a way for users to learn about the real estate knowledge and truly explore the real estate market data. We introduce HouseSeeker, an interactive real estate visualization system to augment

---

[1] http://www.realestate.com.au/.
[2] http://www.domain.com.au/.

© Springer International Publishing AG 2016
M.A. Cheema et al. (Eds.): ADC 2016, LNCS 9877, pp. 435–439, 2016.
DOI: 10.1007/978-3-319-46922-5_34

current commercial systems and assist users in understanding the real estate market, finding candidate properties that satisfy their requirements, and visually comparing properties/suburbs in those aspects at users' own preference. A preliminary demo system and a demonstration video are available at http://115. 146.89.158/ and https://youtu.be/uhp_a7_1mpY, respectively.

2   System Architecture

As shown in Fig. 1, we collect the location-centered data as summarized in the following profiles, and store them in a relational database MySQL.

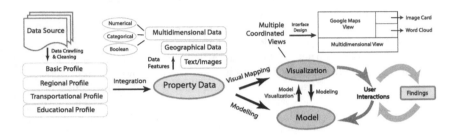

Fig. 1. The framework of HouseSeeker.

- **Basic Profile**: it contains basic information of properties crawled from Real Estate Australia.
- **Regional Profile**: it includes the mapping from each property to a specific SA1[3] (Statistical Area Level 1, the smallest unit for the processing and release of Australian census data) and census statistics based on each SA1.
- **Transportational Profile**: it includes the walking time from each property to its nearest train station and the travel time between each pair of train stations. We are then able to compute efficiently about how long it takes from each property to a user's working place at runtime.
- **Educational Profile**: it contains an exact public secondary school associated to each property and the ranking of all the schools[4].

Then, based on different characteristics, we divide all attributes associated with housing properties into several categories (Fig. 1), as geographical, numerical, categorical and boolean attributes, etc.

Finally, based on our analysis tasks and data characteristics, we propose a visualization design with four multiple coordinated views [1]. **The Google Maps view** (Fig. 2(b)) maps properties as polygons on the top of Google Maps

---

[3] http://www.abs.gov.au/ausstats/abs@.nsf/Lookup/2901.0Chapter23102011.
[4] https://bettereducation.com.au/results/vce.aspx.

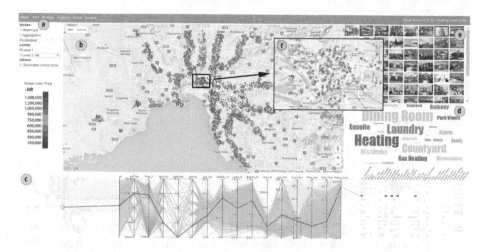

Fig. 2. The interface of HouseSeeker, with all properties that are within 15-min walk to the nearest train station displayed, and a particular apartment highlighted in the multidimensional view. (Color figure online)

and use retinal variables such as colours and shapes to visualize additional information of properties. **The multidimensional view** (Fig. 2(c)) directly connects parallel coordinates (middle) with a geo-coded scatter plot (left) and a coloured boolean table (right). The design is effective to visualize different categories of multiple attributes and also allows users to link the elements in this view with that in the Google Maps view (the design problem of multiple views [3,4]). We also design **the Image Card view** (Fig. 2(e)) and **the Word Cloud view** (Fig. 2(d)) to display additional image and textual information.

Please refer to our previous work [2] for details about the process of data integration, data characteristic analysis and visualization design.

3   Demonstration Scenarios

The dataset generated for our demonstration contains 52,896 housing properties sold in Melbourne metropolitan area from January 2013 to October 2015.

We demonstrate our system based on three exploration queries, and illustrate how our system effectively helps non-knowledgeable users understand local real estate patterns, find their preferred candidate properties, and compare properties in detail to find strengths and weaknesses of each property. None of those exploration queries is supported in existing commercial systems.

Specifically, our demonstration is based on the scenario that a user, *John*, is trying to find a property to live in Melbourne. However, as he does not have much knowledge about Melbourne real estate market, he is not sure of what kind of properties that he could afford or which area he should live in.

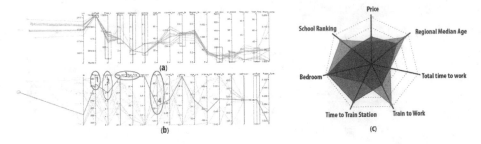

Fig. 3. (a) Interactive exploration with parallel coordinates; (b) detailed comparison in parallel coordinates; (c) detailed comparison with spider chart. (Color figure online)

Exploration Query 1. Personalized Overview at Different Granularities of Details. Our first exploration query gives users an overview from different aspects and helps them quickly understand property/suburb characteristics.

When John enters the main page of HouseSeeker, he first observes from the Google Maps view that the price of properties in East Melbourne and Southeast Melbourne is higher than that in North Melbourne, and the price of that in West Melbourne is the lowest. He also notices that the price seems has an approximately negative correlation with the distance from the Melbourne CBD (central business district); the properties around the CBD are quite expensive, but those exactly in the CBD are actually much cheaper. After further zooming in, he finds that the reason behind this is that there are more 1- and 2-bedroom apartments in the CBD (Fig. 2(f)). John is guided by animations to use parallel coordinates to filter out the properties; by observing the changes in the Google Maps view, he understands how price is affected by different factors, and where some properties (such as those are within school zones of good secondary schools) locate.

Exploration Query 2. Interactive Exploration. After users know better about the local real estate market, our second exploration query helps them further understand their individual requirements, and helps them find some candidate properties based on the requirements.

John now knows that he needs a house that has three or four rooms, within the school zone of a top 30 % secondary school, and within half an hour to his working place. He interacts with parallel coordinates by selecting the corresponding values from each dimension according to his requirements, all other views (e.g. the Google Maps view) change simultaneously to allow him understand every step of his interaction. Finally, There are about 20 properties that fully satisfy John's requirements (as shown in Fig. 3(a)).

Exploration Query 3. Detailed Comparison of Selected Properties. When users are satisfied with the filtering results, or if they want to compare the properties that they have saved, our third exploration query provides a detailed comparison of the properties and allows users to discover strengths and weaknesses of each property.

As John is satisfied with the current result, he would like to know the detailed comparison of those selected properties. With visual assistance from our system, he drags those factors that he cares about more, such as the price, travelling time and school ranking, in front of other dimensions; he also reverts the coordinate axes of some dimensions and makes sure the best values of important dimensions are represented at the top and front of parallel coordinates. As shown in Fig. 3(b)), the highlighted property has a lower price (b-1), is very close to train stations (b-2) and John's working place (b-3); while it has a relatively worse school comparing to other candidate properties (b-4). John also selects those factors that he concerns about and compares the properties in the Spider Chart. As shown in Fig. 3(c), he clearly observes that the red coloured property has a better school ranking and less time to his working work, but the other factors are worse than the blue coloured one.

References

1. Javed, W., Elmqvist, N.: Exploring the design space of composite visualization. In: IEEE PacificVis (2012)
2. Li, M., Bao, Z., Yan, S., Sellis, T.: A visual analytics framework for the housing estate data. In: IEEE PacificVis (Poster) (2016)
3. Sun, G., Liang, R., Qu, H., Wu, Y.: Embedding spatio-temporal information into maps by route-zooming. IEEE Trans. Vis. Comput. Graph. (2016). Early Access
4. Turkay, C., Slingsby, A., Hauser, H., Wood, J., Dykes, J.: Attribute signatures: dynamic visual summaries for analyzing multivariate geographical data. IEEE Trans. Vis. Comput. Graph. **20**(12), 2033–2042 (2014)

A Real Time System for Indoor Shortest Path Query with Indexed Indoor Datasets

Zhou Shao[1(✉)], Joon Bum Lee[2], David Taniar[1], and Yang Bo[3]

[1] Monash University, Melbourne, Australia
zsha14@student.monash.edu, david.taniar@monash.edu
[2] Gachon University, Seongnam, Gyeonggi, South Korea
leejb905@gc.gachon.ac.kr
[3] Melbourne, Australia
ben83feel@gmail.com

Abstract. In comparison to query processing in spatial road networks, indoor query processing is not well studied due to limited number of available indoor datasets. In this paper, we present a system to represent floor plans of indoor spaces which enables users to find the shortest path between two indoor locations. Indoor spaces, such as shopping malls and office buildings, contain a large number of rooms and doors, which cause some difficulties to people to find their ways inside the building such as a large shopping centre. We have developed a system that is used to index floor plans and allow users to specify any two given locations and calculate the shortest path.

1 Introduction

People spend a large part of their lives in an indoor environment, such as office buildings, shopping malls and universities. However, location-based services (LBS) are not available inside buildings. The reason is that the dominant technology such as global positioning system (GPS) only applicable to an outdoor environment. For indoor spaces, locations cannot be identified using GPS accurately. Another reason is that for to calculate shortest path in an outdoor space, spatial road networks can be downloaded from the internet, such as OpenStreetMap. The network is basically a graph $G=(V, E)$, where V is a set of vertices and E is a set of edges. According to G, the shortest path between two points can be calculated. However, for indoor spaces, the main challenge is that there are no available indoor datasets. For example, a user is located in Menzies Building in Monash University and wants to find a shortest path from his current location to Room 935, this cannot be done unless an indoor data of this building is given.

An indoor space is totally different from a outdoor space because it consists of indoor entities such as rooms, hallways, doors and walls. Hence, the notion of shortest path in indoor is different from that in outdoor. The shortest distance between two indoor points is the *minimum indoor walking distance* [2]. It is restricted by the floor plan of an indoor space. In the given floor plan (see Fig. 1), two query points are represented by two red points p_{source} and $p_{destination}$.

© Springer International Publishing AG 2016
M.A. Cheema et al. (Eds.): ADC 2016, LNCS 9877, pp. 440–443, 2016.
DOI: 10.1007/978-3-319-46922-5_35

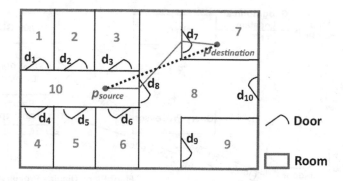

Fig. 1. Minimum indoor walking distance

The black dashed line is the Euclidean distance which is the straight line between the two indoor points. It cannot be used as the shortest path because users are restricted to walk through doors and spaces inside rooms, not through walls. Therefore, the minimum indoor walking distance is used which is indicated by the red path passing through d_8 and d_7 to reach to the destination point.

In this paper, we introduce how to index an indoor space into dataset files. Based on these files, we designed a system to represent the floor plans of indoor spaces. Additionally, a demonstration is given that calculates the shortest path between two given indoor locations.

2   System Overview

The framework of the proposed system is shown in Fig. 2, which contains 3 main parts. The first part is Google Maps Geocoding API [1] which takes the geographic location from the user. Based on this location, the address is retrieved and sent to the database consisting of all available indoor datasets. If the indoor dataset of the given address exists, the databases returns the files to the user GUI in order to display floor plans of the given building. If a user invokes a shortest path query, the query processor will return the shortest path to user GUI based on the floor plan files as well as two given query points.

Geocoding API. This component mainly provides the transformation from a geographic location to an address. At first, it displays Google Maps for users to choose any location which is basically a point inside a building area on the map. After that, the point consisting of latitude and longitude coordinates is transferred into a human-readable address. This address is used to get the corresponding indoor dataset from the database.

Indoor Datasets. This is an important component of the proposed system as it maintains the data for the other parts such as query processor and user GUI. As we have mentioned before, indoor datasets are not available at this stage, hence, constructing the indoor dataset has to be done manually. Take the given floor plan in Fig. 1 as an example, two files are created to store the information

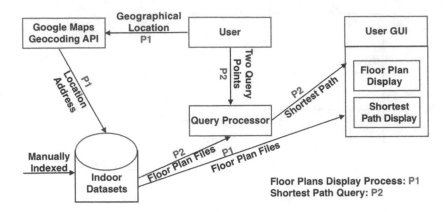

Fig. 2. System overview

of an indoor space: *door.txt* and *room.txt*. For a door, it consists of a 4-tuple $(ID, Room_{from}, Room_{to}, Coordinate)$. ID is an unique ID of the door, and the second and third are the IDs of from and to rooms of this doors. The last one is the coordinate of this door. For example, 4-tuple of d_1 is $(1, 1, 10, (2, 14))$. For a room, it is a 7-tuple $(ID, T_{room}, Coordinate_{topleft}, Coordinate_{bottomright}, Level, S_{doors}, S_{coordinates})$. ID is an unique ID of the room and T_{room} is the type of this room. Room type is either rooms, hallways or stairs. The next 2 coordinates are used to represent the top left and bottom right points of the room, based on the coordinates, a rectangular room can be retrieved. After that, $Level$ shows which level this room is located. S_{doors} is a set of doors contained in this room, while $S_{coordinates}$ are used by hallways only because hallway is not always a rectangle, hence, hallway is partitioned into several rectangles and the coordinates of each rectangle are stored in $S_{coordinates}$.

Query Processor. The query processor component implements the shortest path query algorithm. This algorithm, it is a room based expansion method. At first, rooms containing source and destination points are located. Based on the source room, it expands to the adjacent rooms which have at least one common door with the source room. Meanwhile, the distance between the doors inside the same room is calculated in order to get the current nearest door to the source point. All possible paths which have been expanded are stored in a candidate set, together with their distances. The expansion process keeps expanding to the rooms which contains the current nearest door to the source point until the destination room is reached and the current shortest distance is shorter than any distance in the candidate set. In order to accelerate the processing speed, unnecessary rooms and doors are removed after source and destination rooms are located. An unnecessary room is a room which contains only one door and it is not a source or destination room. The shortest path will never enter into these rooms as the entrance and exit doors are the same.

3   Demonstration

We demonstrate the system that we have developed that is able to display the the floor plans of the building given by the user. The first picture in Fig. 3 shows the outdoor map for part of Monash University Clayton Campus. The red shaded area is the user-clicked building. The system will then show the floor plan refer to the second picture. The user is then allowed to specify two locations on the indoor floor plan, the shortest path is then shown (see the third picture).

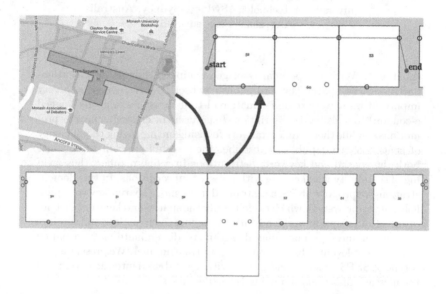

Fig. 3. Demonstration

4   Conclusion

In this paper, we developed a system to show the floor plans of a user-specified building. This system gives users q useful query, indoor shortest path query. It enables users to find the shortest path from its current location to any destination inside the building easily. This is very useful with more indoor datasets are available in this system. Whenever people go to some indoor buildings they are not familiar with, such as foreign airports, hospitals, etc., the system gives people the details of the building, as well as finding their ways in the building.

References

1. https://developers.google.com/maps/documentation/geocoding/intro
2. Lu, H., Cao, X., Jensen, C.S.: A foundation for efficient indoor distance-aware query processing. In: IEEE 28th International Conference on Data Engineering (ICDE), pp. 438–449. IEEE (2012)

SKPS: Towards Efficient Processing of Spatial-Keyword Publish/Subscribe System

Xiang Wang[1](✉), Shiyu Yang[1], and Ying Zhang[2]

[1] The University of New South Wales, Sydney, Australia
{xiangw,yangs}@cse.unsw.edu.au
[2] University of Technology Sydney, Sydney, Australia
ying.zhang@uts.edu.au

Abstract. With the popularity of geo-equipped devices and location-based services, spatial-keyword publish/subscribe has emerged as a very important framework to disseminate real-time messages (e.g., geo-tagged e-coupon) to registered subscriptions (e.g., users interested in nearby promotions). While there are several work focusing on improving the efficiency of spatial-keyword publish/subscribe, their techniques fail to consider both the spatial and keyword distributions in a fine manner, thus lacking of scalability when coping with massive subscriptions. In this demonstration, we propose SKPS, a centralized in-memory spatial-keyword publish/subscribe system, which exploits fully the spatial and keyword distributions of subscription workload during indexing construction, and employs an efficient message matching algorithm to disseminate each incoming message to relevant subscriptions in a real-time manner. We present a prototype of SKPS which provides users with a web-based interface to explore the message dissemination in publish/subscribe system.

Keywords: Publish/subscribe system · Spatial-keyword query · Boolean range query · Stream

1 Introduction

Content-based Publish/Subscribe (Pub/Sub for short) system has attracted a lot of attention since the last decade [1,2,5,7]. Subscribers can register their interest as subscriptions, and publishers issue messages which need to be delivered to all the *relevant* subscribers in a real-time manner. Recently, due to the proliferation of *User Generated Content* and geo-equipped devices, there is a vast amount of data with both spatial and textual information, referred to as *geo-textual* data; they often come in a rapid streaming fashion in many important applications such as social networks (e.g., Facebook, FourSquare and Twitter) and location-based services (e.g., iAd[1]), thus leading to the *location-awareness* of subscribers. For instance, a user may want to retrieve all the tweets discussing the recent movie *Birdman* in her home city *Sydney*.

---

[1] http://advertising.apple.com.

© Springer International Publishing AG 2016
M.A. Cheema et al. (Eds.): ADC 2016, LNCS 9877, pp. 444–447, 2016.
DOI: 10.1007/978-3-319-46922-5_36

To make sense of streaming geo-textual data and satisfy the increasing location-aware demand of subscribers, it is critical to develop efficient location-aware Pub/Sub system. In this paper, we investigate the problem of processing spatial-keyword subscriptions in a location-aware Pub/Sub system; that is, efficiently delivering a stream of incoming geo-textual messages issued by publishers to all the *relevant* spatial-keyword subscriptions registered by a large amount of subscribers. This problem plays a fundamental role in a variety of applications such as information dissemination [6], location-based recommendation [4] and sponsored search [3].

Challenges. There are two key challenges in efficiently processing spatial-keyword subscriptions. *Firstly*, a massive number of subscriptions, typically in the order of millions, are registered in many applications, and hence even a small increase in efficiency results in significant savings. *Secondly*, the streaming geo-textual messages (e.g., geo-tagged tweets) may continuously arrive in a rapid rate which also calls for high throughput performance for better user satisfaction.

In this paper, we propose a novel system, called SKPS, to efficiently support spatial-keyword Pub/Sub. The system consists of two main components, one of which is a spatial-keyword indexing structure built upon subscriptions to facilitate the dissemination of incoming messages and the other is a front-end web interface built on top of Django² framework for user exploration. As to the spatial-keyword index, we explore the spatial and keyword distributions of subscription workload to build a novel adaptive indexing structure, which greatly improves the system throughput. In terms of front-end, users can input a geo-textual message manually and the system will visualize all the subscriptions which are relevant to it. Our principal contributions are summarized as follows:

- A novel and efficient indexing structure is built in the back-end to support real-time dissemination of streaming messages.
- A visualization web interface is designed to enable user exploration with the help of back-end indexing structure.

2   SKPS System

Figure 1 demonstrates our system architecture. Our system is built upon Django, which is a high-level Python Web framework. There are two main components of our system: the back-end and the front-end.

Back-end. The back-end is where the subscription index lies. Our subscription index is essentially a multi-ary tree structure, which consists of three types of tree nodes, i.e., s-node, k-node and l-node. s-node is an intermediate node which employs spatial partition, such as Quadtree and Grid, to partition the subscriptions based on their query ranges. k-node is an intermediate node which utilizes keywords to divide subscriptions according to their keywords. Finally, l-node is

---

² https://www.djangoproject.com/.

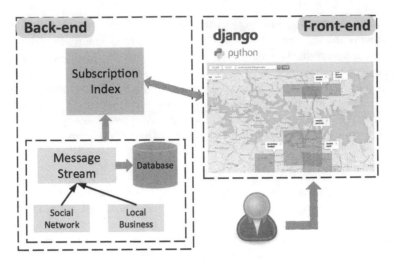

Fig. 1. System framework

a leaf node containing all the subscriptions which satisfy the spatial and keyword constraints imposed by the intermediate nodes along the path from root node. By taking into consideration both the spatial and keyword distributions of subscriptions, the subscription index can partition the subscriptions in a fine granularity, facilitating the dissemination upon an incoming message. The backend component also builds a stream collector, which digests messages from either social network or local business and feeds the message to the subscription index continuously.

Front-end. The front-end is a web interface which can be used to explore the message dissemination procedure. The web interface shows the keywords and query ranges of all the subscriptions on Google Maps with Google Maps API[3]. A user can issue a message as a query, and the system will return all the matching subscriptions (i.e., the subscription keywords are fully covered by message keywords and the message location is inside the subscription range) being highlighted in Google Maps for user exploration. Note that, since the number of subscriptions is often at million scale, the web interface only shows the subscriptions whose query ranges are within current map window to reduce system overhead.

3   Demonstration

Figure 2 depicts the web interface of our system. Figure 2(a) is the initial web interface, where a user can specify a message and submit this message to our system to identify all the matching subscriptions. For example, a user issues a message with location *Sydney* (latitude: -33.88, longitude: 151.21) and a set

---

[3] https://developers.google.com/maps/.

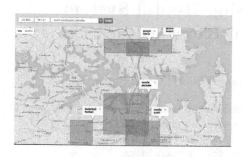 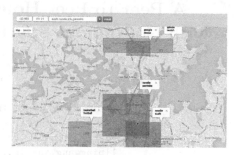

(a) Initial Interface   (b) After submitting a message

Fig. 2. Performance over various datasets (Color figure online)

of keywords *sushi, noodle, tofu* and *pancake*. After submitting this message to our system, all the matching subscriptions will be highlighted with red color in Fig. 2(b). Users can also drag the map to explore some other regions they are interested in.

4 Conclusion

In this paper, we present SKPS, a spatial-keyword publish/subscribe system which continuously feeds a message stream to a massive number of subscriptions. A subscription index is built to accelerate the message dissemination upon subscriptions by exploring both spatial and keyword information of subscriptions. A web interface is also demonstrated to facilitate users to explore the message dissemination in an interactive manner.

References

1. Aguilera, M.K., Strom, R.E., Sturman, D.C., Astley, M., Chandra, T.D.: Matching events in a content-based subscription system. In: PODC, pp. 53–61 (1999)
2. Fabret, F., Jacobsen, H.A., Llirbat, F., Pereira, J., Ross, K.A., Shasha, D.: Filtering algorithms and implementation for very fast publish/subscribe systems. In: ACM SIGMOD Record, vol. 30, pp. 115–126. ACM (2001)
3. Konig, A., Church, K., Markov, M.: A data structure for sponsored search. In: ICDE, pp. 90–101 (2009)
4. Park, M.-H., Hong, J.-H., Cho, S.-B.: Location-based recommendation system using bayesian user's preference model in mobile devices. In: Indulska, J., Ma, J., Yang, L.T., Ungerer, T., Cao, J. (eds.) UIC 2007. LNCS, vol. 4611, pp. 1130–1139. Springer, Heidelberg (2007). doi:10.1007/978-3-540-73549-6_110
5. Shraer, A., Gurevich, M., Fontoura, M., Josifovski, V.: Top-k publish-subscribe for social annotation of news. PVLDB (2013)
6. Yan, T.W., García-Molina, H.: Index structures for selective dissemination of information under the boolean model. TODS (1994)
7. Zhang, D., Chan, C.Y., Tan, K.L.: An efficient publish/subscribe index for e-commerce databases. PVLDB **7**(8), 613–624 (2014)

A Peer to Peer Housing Rental System with Continuous Spatial-Keyword Searching

Yingqian Hou, Xinyin Wang, Liping Wang[(✉)], and Junjie Yao

Shanghai Key Lab for Trustworthy Computing,
East China Normal University, Shanghai, China
hyqat1994@gmail.com, xywang281@163.com,
{lipingwang,junjie.yao}@sei.ecnu.edu.cn

Abstract. Apart from traditional intermediary companies, online housing rental systems as their convenience are gaining their popularity. Search is a critical function in these systems, but it does not always meet users' satisfaction. For example, spatial and keyword distributions are generally be considered separately and the user cannot submit a continuous query requirement before he/she rent a satisfied house. In this paper, we develop a peer to peer housing rental system (P2PHRS) based on Django Framework of Python. We propose an efficient house searching algorithm called Quad-tree plus Inverted List (QIL) to filter housing resources for users according to their spatial and keyword requirements. P2PHRS is designed to be adaptive to a variety of front-end clients, like Web, Android and iOS platform etc. We show the advantages of P2PHRS by several spatial-keyword query demonstration scenarios.

Keywords: House renting · Continuous spatial-keyword searching · Django Framework

1 Introduction

With the acceleration of urbanization, a large number of people have started to migrate to metropolis. Renting a house has been a primary issue for people nowadays. Online housing rental systems or APPs are popular for tenants to rent houses. After studying traditional renting systems like fang.com and anjuke.com, we find three fundamental drawbacks. Firstly, it is inconvenient for landlord to release a new house resource. And during renting process, landlords and tenants rely too much on house agencies, whose agency fee constitutes a high proportion of rents. Secondly, spatial and keyword distributions of the queries cannot be considered at the same time. When spatial (keyword)-first indexing method is adapted, keyword (spatial) factor cannot be well exploited. Thirdly, if query results do not satisfy user requirements, the user will have to give up the query and stop searching.

In this paper, we design and implement P2PHRS to overcome above three drawbacks. Different from traditional renting systems, in P2PHRS landlords

© Springer International Publishing AG 2016
M.A. Cheema et al. (Eds.): ADC 2016, LNCS 9877, pp. 448–452, 2016.
DOI: 10.1007/978-3-319-46922-5_37

and tenants can release and search houses online without participation of house agencies. We introduce a new role called manager to verify the authenticity of housing resources reliably. One of advantages of P2PHRS is that spatial and keyword distributions can be considered at the same time. Moreover, subscribe function [3] is designed for users to handle concurrent queries continuously.

One of the core challenges in P2PHRS lies on how to efficiently implement continuous spatial-keyword searching over housing stream. The problem can be stated as follows. Given a continuous spatial-keyword query set Q, for each incoming housing object o from streaming spatial-textual data O, we aim to rapidly deliver o to all the matched queries. Note that traditional spatial-keyword search methods [2] are not applicable to our system, for we focus on continuous queries. Recently AP-Tree [4] and IQ-Tree [1] were proposed for continuous query problem, but building cost models is dependent on complex probability computation, which leads to the fact that these two methods cannot be directly applied to our system. To tackle this problem, we introduce a novel quad-tree plus inverted list (QIL) in the searching algorithm. The other challenge is to make P2PHRS adapt to different front-end clients. Thus we make the design of API in server part complies with the Django RESTful framework.

The paper is organized as follows. In Sect. 2.1, we overview the system and examine how it works. Additionally, the continuous searching algorithm is introduced as the key technology in Sect. 2.2. After this, we demonstrate our system in Sect. 3.

2 System Overview

2.1 System Framework

Figure 1 shows the system framework of P2PHRS. There are three kinds of user in the system: tenant, landlord and system manager. In P2PHRS, landlords

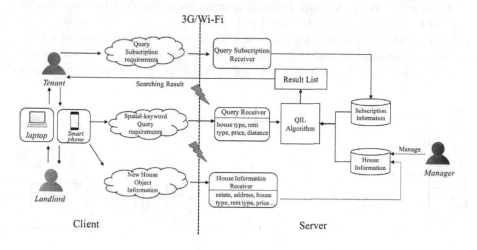

Fig. 1. System framework for P2PHRS

can upload their houses from the laptop or mobile phone. The system manager has the right to review and manage housing resources in order to ensure their reliability and integrity. Tenants submit their query with spatial and keyword information and then the query is sent to the server which has deployed QIL searching. Finally, the searching results are shown to tenants by setting markers on the Baidu Map. In case of unsatisfied results, we also offer a subscribe function in the system that allows tenants to subscribe their queries.

In server part Django Framework of Python is utilized. The functions of searching, uploading and deleting houses etc. are encapsulated as Django API. All house information is stored in MySQL database. QIL, which is stored in sever memory, is constructed according to house objects and queries. The server communicates with different kinds of clients by JSON message.

In the front end Baidu Map is introduced as the interactive interface. We implement one kind of clients as an Android application based on Android SDK 4.4.

2.2   Continuous Spatial-Keyword Search

QIL algorithm is divided into two processes: index building and object matching. Figure 2 is the flow chart of building the index.

We firstly build index for both space and keyword in QIL so that we can manage spatial and keyword distributions. To build index for space, we build a quad-tree to divide the whole space into four child space recursively. Each node has its geo-location information and a query list, which contains all the queries in the node. By comparing the area of query q and node n, we can store q in the node in the lowest level that contains the range of q. In this way, spatial index is built. The building of index for keyword is completed by an inverted list, which organizes the keywords of all the queries in this node. In the inverted list, each keyword is followed by the queries that contains it. The quad-tree and inverted lists will be updated if a new query comes.

Let $o = (\psi, loc)$ represents a spatial-textual house object, where $o.\psi$ denotes a set of keywords and $o.loc$ is a geo-location. $q = (\psi, \gamma)$ records a query, where $q.\psi$ is a set of user-specified keywords and $q.\gamma$ indicates spatial region of interest to user. An incoming spatial-textual house object o matches a continuous spatial-keyword query q if and only if $o.\psi \supseteq q.\psi$ and $o.loc \in q.\gamma$. Thus when it comes

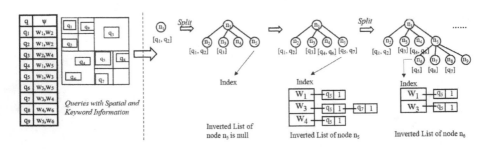

Fig. 2. An example of building QIL index

to object matching, we first find the node set N that o locates in. For each node n_i belongs to node set N, we retrieve the query list of node n_i and get the query set Q_1 matching o spatially. Then we search inverted list of node n_i to get query set Q_2 matching o in terms of keyword. The intersection of Q_1 and Q_2 is the result set of n_i.

3   Demonstration

We collect about 15,000 housing resources as objects from fang.com. The performance of P2PHRS is evaluated in Fig. 3 where the number of queries grows from 3,000 to 30,000. The result shows that P2PHRS is scalable to the number of queries. Moreover, continuous spatial-keyword query processing is demonstrated and two test scenarios are designed. In the first context, after choosing searching terms, the search results are shown in Fig. 4 and the house details can be browsed in Fig. 5. The second context demonstrates the subscribe function. As depicted in Fig. 6(a), when there is no satisfied results, the user subscribe the query (See Fig. 6(b)). When new objects that match the query appear, they will be listed in "My Subscription" - "Subscription Details" (See Fig. 6(c)). In P2PHRS, it is very convenient for user to subscribe more than one query, which will be registered at the server end and managed by QIL continuously until the user find satisfied houses.

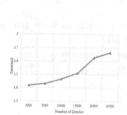

Fig. 3. Performance experiment

Fig. 4. Result of searching by (entire rent, 2 rooms, 4000–5000 RMB, 3 km)

Fig. 5. House details of Fig. 4

(a) Searching by (Entire Rent, 3 Rooms and above, 3000-4000RMB, 5km), no results

(b) Subscribe Query

(c) Subscription Result List - new objects matching the query

Fig. 6. Subscribe a query

Acknowledgments. This work was partially supported by NSFC 61232006, NSFC 61401155 and NSFC 61502169.

References

1. Chen, L., Cong, G., Cao, X.: An efficient query indexing mechanism for filtering geo-textual data. In: SIGMOD 2013, pp. 749–760. ACM (2013)
2. Chen, L., Cong, G., Jensen, C.S., Wu, D.: Spatial keyword query processing: an experimental evaluation. VLDB **6**(3), 217–228 (2013)
3. Wang, X., Zhang, Y., Zhang, W., Lin, X., Huang, Z.: Skype: top-k spatial-keyword publish/subscribe over sliding window. VLDB **9**(7), 588–599 (2016)
4. Wang, X., Zhang, Y., Zhang, W., Lin, X., Wang, W., Ap-tree: Efficiently support continuous spatial-keyword queries over stream. In: ICDE 2015, pp. 1107–1118. IEEE (2015)

SCHOLAT: An Innovative Academic Information Service Platform

Feiyi Tang[1], Jia Zhu[2(✉)], Chaobo He[3], Chengzhou Fu[2], Jing He[1], and Yong Tang[2]

[1] College of Engineering and Science, Victoria University, Melbourne, Australia
[2] School of Computer Science, South China Normal University, Guangzhou, China
jzhu@m.scnu.edu.cn
[3] School of Information Science and Technology,
Zhongkai University of Agriculture and Engineering, Guangzhou, China

Abstract. We present a system called SCHOLAT, which is implemented as a scholar-oriented social network that aims to form an academic community to let users establish connection with other researchers. SCHOLAT provides two novel professional services that are useful to researchers, namely, **XPSearch** and **XSRecom**. **XPSearch** is a service that provides vertical searching of research papers with author name disambiguation. **XSRecom** uses a topic community-based method to provide users with a list of "Recommend Scholars", which can help them find potential collaborators who share the same research interests and may be interested in building a cooperative relationship. These two services boost the efficiency of searching papers and discover research opportunities for scholars.

1 Introduction

In the era of Web2.0, Social Networking Sites (SNS) have become an important part of our daily life. These platforms have well satisfied the needs of people in different fields. However, many professionals, particularly scholars, still find themselves being flooded with too much information. More importantly, they expect SNSs to provide them not only with the chance to know new people or share moments with family and friends, but also with academic services for conducting research and expanding their academic networks. Thus, we propose an innovative scholar oriented social network (SOSN) platform called SCHOLAT. In addition to common social network functions, such as message, status update, and so on, SCHOLAT provides two creative and useful services for scholars, namely, **XPSearch** and **XSRecom**, which will be introduced in this demo:

XPSearch: This service provides vertical searching of research papers with author name disambiguation. Users can not only input keywords to search articles but also to search an author's name and retrieve all his/her publications with visualized name disambiguation results. We use information from neighboring pages to improve the performance of name disambiguation.

M.A. Cheema et al. (Eds.): ADC 2016, LNCS 9877, pp. 453–456, 2016.
DOI: 10.1007/978-3-319-46922-5_38

XSRecom: This service provides users with a list of recommended scholars that is based on topic community discovered by combing user link and content information. This tool can help SCHOLAT users to find potential collaborators who share the same research interests and may be interested in building up a cooperative relationship. The novelty of XSRecom is we fuse user link and content information to make scholars recommendation more appropriate.

2   SCHOLAT Architecture

As showed in Fig. 1, SCHOLAT is built on a layered architecture. The core layer in SCHOLAT is the application layer, which include XPSearch and XSRecom applications.

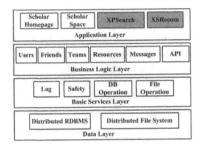

Fig. 1. Overview of SCHOLAT architecture

3   Technical Details

3.1   XPSearch

In SCHOLAT, we design an academic search engine that provides publication search and scholar search services to users with author name disambiguation. Unlike existing works, apart from using common information such as co-authorship and paper title [4], we focus on retrieving more information from web pages, such as personal homepages. Therefore, the challenge is finding a way to identify which pages are personal genre pages as not all web pages are useful, which is our main contribution.

We propose a framework that not only uses the information on the page but also considers information in neighboring pages to identify the genre of a page because neighboring pages can handle the pages with few textual content [3]. Assume that the original page and its neighboring pages are plotted in a directed graph $G = (V, E)$, $PATH(u, v)$ is the set of all possible paths from vertex u to vertex v. We then calculate the recommendation score $Score(u, v)$ of page u that propagates to v through $PATH(u, v)$ according to [1]. We then have the similarity score of two pages based on the $Score(u, v)$. The information of top K neighboring pages with the highest similarity scores will be used to construct features. More technical details and how we use information from neighboring pages for name disambiguation can be found in our previous works [3,4].

3.2   XSRecom

The overall framework of XSRecom is shown in Fig. 2. The first phase constructs the user link matrix and the content feature matrix after data extraction and pre-processing. The second phase mines the topic community via the joint non-negative matrix factorization (NMF) model, and user membership can be determined according to his/her strength distribution to a given community. The third phase computes pairwise user similarities in each community to generate a list of candidate scholars. We then combine these candidate lists to obtain the Top-K recommendation scholars for each target user. More technical details and experimental results can be found in our previous works [2].

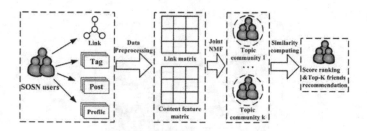

Fig. 2. XSRecom framework

4   Demonstration Scenarios

4.1   Author Name Disambiguation

As Fig. 3 shown, the XPSearch disambiguation result of "Wei Wang" when users search a scholar' name in SCHOLAT. Unlike other systems, SCHOLAT displays the authors with the same name in different circles. As the Fig. 3 shown, we have two scholars called "Wei Wang" with two circles consists of "Wei Wang" co-authors. Users can either click the scholar' portrait to obtain more information, such as his/her research interest, affiliation, and so on, or click the edge in the graph to retrieve a list of papers between the co-author and "Wei Wang". By this design, users can find the scholar with his/her publications in which they are interested.

4.2   Recommend Scholars

XSRecom will recommend potential scholars to the user using the topic community-based method. Figure 4 displays the list of recommended scholars by XSRecom to the target user "Wei Wang". Every scholar has a collaboration score computed using the community-based similarity measure as described in Sect. 3.2. The user can also click the scholar portrait to obtain more his/her information. Compared to other systems that only displays co-authorship, our system can easily let users know how closed the recommended scholars to them.

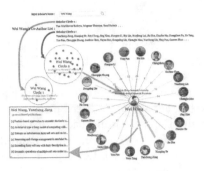

Fig. 3. Author disambiguation result

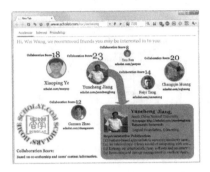

Fig. 4. Recommend scholars

In addition, users can get more recommendations by clicking the "Interest" or "Friendship" tab, which generates results based on the keyword we extracted from scholars' published paper and co-authorship, respectively.

Acknowledgments. This work was supported by the S&T Projects of Guangdong Province (No. 2015B010109003, 2016A030303055, 2016A030303058, 2016B030305004).

References

1. Arasu, A., Cho, J., Garcia-Molina, H., Paepcke, A., Raghavan, S.: Searching the web. ACM Trans. Internet Technol. **1**(1), 2–43 (2001)
2. He, C.B., Li, H.C., Fei, X., Tang, Y., Zhu, J.: A topic community-based method for friend recommendation in online social networks via joint nonnegative matrix factorization. In: The 3rd International Conference on Advanced Cloud and Big Data, pp. 28–35 (2015)
3. Zhu, J., Xie, Q., Yu, S.I., Wong, W.H.: Exploiting link structure for web page genre identification. Data Min. Knowl. **30**(3), 550–575 (2015)
4. Zhu, J., Yang, Y., Xie, Q., Wang, L.W., Hassan, S.: Robust hybrid name disambiguation framework for large databases. Scientometrics **98**(3), 2255–2274 (2014)

Data-Driven Prediction and Visualisation of Dynamic Bushfire Risks

Laura Rusu[✉], Hoang Tam Vo, Ziyuan Wang, Mahsa Salehi, and Anna Phan

IBM Research, 204 Lygon Street, Carlton, VIC 3053, Australia
laurusu@au1.ibm.com

Abstract. The potential impact of bushfires is a significant concern for communities and fire response agencies, and the ability to predict the fire risk timely and accurately is critical. However, that cannot be achieved without accessing and processing very large amounts of data in almost real time. We demonstrate a data-driven fire risk prediction system that leverages big geospatial and meteorological data, where the results are visualised and made available to communities and fire agencies for risk mitigation strategies.

1 Introduction

Risk management for natural disasters such as bushfires is an important area of concern to the government, its responsible agencies, and communities. However, the existing solutions have several limitations: some systems [1] do not capture the dynamics of data sources while others do not work at the individual property level [2]. In this demonstration, we present a solution that makes use of large and dynamic datasets (e.g. weather), together with large but sometimes less dynamic datasets (e.g. vegetation, terrain and cadaster) to predict the risk of damage due to bushfires in a given area, at individual property level. The risk information is then used for implementing mitigation strategies at individual, community or fire agency level.

2 System Overview

Figure 1 shows a high level overview of our proposed system that uses big geospatial and meteorological data to predict bushfire risk. The following sections briefly describe the four modules of the system. They will highlight the volume, velocity and variety of the data that are input into the system.

2.1 Sensing Bushfire Risk Conditions Using Outlier Detection

This first module predicts the dynamic level of bushfire risk in the study area. As data input for this module, meteorological forecasts from Bureau of Meteorology (BOM) could be used. However, we chose to conduct a high-resolution meteorological reanalysis that resolves smaller phenomena to provide more detailed

© Springer International Publishing AG 2016
M.A. Cheema et al. (Eds.): ADC 2016, LNCS 9877, pp. 457–461, 2016.
DOI: 10.1007/978-3-319-46922-5_39

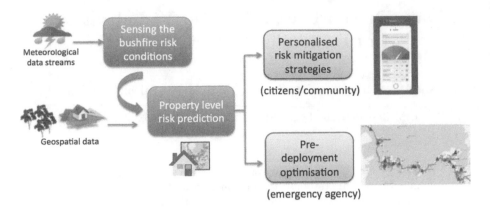

Fig. 1. High-level system overview.

data than would otherwise be possible from observations. The simulation was conducted with the Weather and Research Forecasting model (WRF-ARW) [3]. The model outputs include temperature, precipitation, pressure, relative humidity and wind speed. The study area (Blue Mountains, NSW, Australia) covers an area of 275 km × 275 km centred on the town of Blackheath (-34.80, 138.90) and has a grid spacing of 5 km, resulting in a 54×54 grid and roughly 3000 grid cell observations. The reanalysis was run for 15 days in October 2013 and the simulation output is at a 1-min frequency, hence a total of approximately 65,000,000 observations to analyse.

We employ an outlier detection method that is completely unsupervised so it can be easily applied in other areas of interest, using other big weather datasets. For technical details on the outlier detection approach please refer to [4]. Being an ensemble learning approach, scalability is achieved by parallelisation. The results of this module are risk values for each grid cell in the study area and for each timestamp, as shown in Fig. 2(a). Seen as a timeline, it gives a dynamic risk profile for all properties located in each grid cell, as shown in Fig. 2(b).

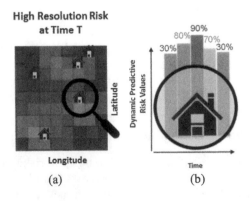

Fig. 2. Results of the "sensing the bushfire risk conditions" module.

2.2   Property-Level Risk Prediction

Once the areas (i.e. grid cells) at high bushfire risk have been identified, the system utilises a risk model to predict the dynamic bushfire risk at individual household level, considering their particularities in terms of position and terrain slope, surrounding vegetation, as well as construction type and other factors. The input data to the risk model is summarised in Table 1. For details on the risk model components please refer to [5].

Table 1. Overview of input data to the risk model

Data	Source	Type/format	Resolution	Update rate
Address	Open	Point/shapefile	NA	Static
Elevation	Open	Raster/GTiff	25 m	Static
Vegetation	Open	Polygon/shapefile	NA	>Yearly
Weather	BOM	Grid	1–5 km	1–15 min
Drought factor	ADFD	Grid/NetCDF	3–6 km	Seasonally
House specification	User	JSON	NA	Semi-static

The results of this component are risk values for each property and for each timestamp in the study period, as exemplified in Fig. 3 for households in Blue Mountains, NSW. For risk visualisation we use a range of colours, from light blue for lowest risk to dark red for highest risk (see Fig. 3(c) for the full legend).

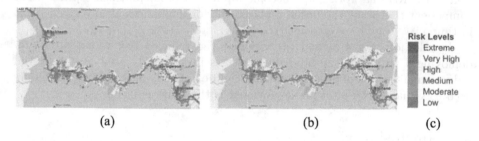

(a)   (b)   (c)

Fig. 3. Properties fire risk in Blue Mountains at 2 pm (a) and 5 pm (b) on 17 Oct 2013. (Color figure online)

2.3   Personalised Risk Mitigation Recommender and Pre-deployment Optimisation

Using the forecasted risk values from the previous module, our system is able to: (i) enable individuals to take proactive actions to reduce the risk to their homes

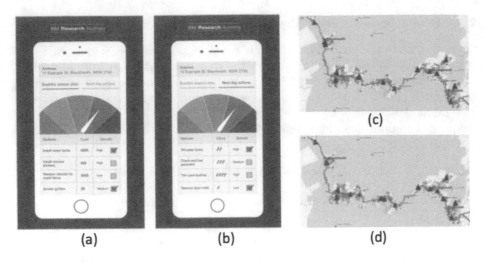

Fig. 4. Applications of dynamic risk prediction.

and (ii) enable emergency agencies to take proactive actions to prepare resource to defend communities.

Figure 4(a) and (b) show two screenshots of a mobile app designed to help people reduce risk via long-term and short-term actions. Once people select actions that could reduce their risk, the system responds in real time with a recalculation of the predicted risk. As part of this module, the large dataset of household-level predicted risk values, together with the number of possible stopping locations for fire-fighting appliances, their availability and capacity, as well as the road network graph become input in a pre-deployment optimisation component that finds the optimal placement for appliances at any given point in time.

In addition, Fig. 4(c) and (d) visualise the reduced risk values after pre-deploying a number of 20 fire-fighting assets in the study area. A substantial reduction in the risk of property damage can be noticed, and the range of colours used to visualise risk after pre-deployment is the same as in Fig. 3.

3   Conclusion

This demonstration will present an application using big geospatial and meteorological data (with high volumes, velocity and variety) to solve real word problems. We continue to develop the system by leveraging more dynamic data (e.g., sensor data and satellite imagery) and cognitive capabilities into the system, to increase accuracy and timeliness of predictions that can save lives.

References

1. NSW Rural Fire Service: BAL risk assessment application kit. New dwellings and alterations and additions to existing dwellings (2012)
2. McArthur, A.G.: Fire behaviour in eucalypt forests. Comm. of Australia For. & Timber Bur. Leaflet No. 107 (1967)
3. Dee, D., Uppala, S., Simmons, A., Berrisford, P., et al.: The era-interim reanalysis: configuration and performance of the data assimilation system. Q. J. R. Meteorol. Soc. **137**(6565), 553–597 (2011)
4. Salehi, M., Rusu, L.I., Lynar, T., Phan, A.: Dynamic and robust wildfire risk prediction system: an unsupervised approach. In: ACM SIGKDD Conference on Knowledge Discovery and Data Mining (SIGKDD) (2016)
5. Wang, Z., Roberts, M.E., Rusu, L.I.: Dynamic and personalised recommendations for proactive bushfire risk management. In: Australian and New Zealand Disaster and Emergency Management Conference (ANZDMC), pp. 207–222 (2016)

EDMS: A System for Efficient Processing Distance-Aware Influence Maximization

Xiaoyang Wang[1(✉)], Chen Chen[2], and Ying Zhang[1]

[1] University of Technology Sydney, Ultimo, Australia
{xiaoyang.wang,ying.zhang}@uts.edu.au
[2] The University of New South Wales, Sydney, Australia
cchen@cse.unsw.edu.au

Abstract. As a key problem in viral marketing, influence maximization has been widely studied in the literature. It aims to find a set of k users in a social network, which can maximize the influence spread under a certain propagation model. With the proliferation of geo-social networks, location-aware promotion is becoming more and more necessary in real applications. However, the importance of the distance between users and the promoted locations is underestimated in the existing work. For example, when promoting a local store, the owner may prefer to influence more people that are close to the store instead of people that are far away. In this demonstration, we propose EDMS, a centralized system that efficiently processes the distance-aware influence maximization problem. To meet the online requirements, we combine different pruning strategies and the best first search algorithm to significantly reduce the search space. We present a prototype, which provides users with a web interface to issue queries and visualize the search results in real time.

Keywords: Influence maximization · Distance-aware · NP-Hard

1 Introduction

Given a social network G, a positive integer k and a certain propagation model (*e.g.,* independent cascade model), influence maximization problem aims to find a set of k users, called seed set, that can lead to the largest spread of the information. This problem has many real applications [5]. For example, a company plans to promote its new product and wants to select certain influential users. By giving them some incentives (*e.g.,* discounts or VIP cards), the company expects that they can propagate the news about the product from friends to friends through their social networks. Consequently, it can achieve a large exposure of the news. The propagation leverages the benefit of the word-of-mouth property, which is shown to be more effective than traditional channels, such as TV advertisements [2]. As the proliferation of location enabled devices, many users are associated with a location. Thus, it is necessary to consider users' locations when conducting a location-aware promotion. There are some research

© Springer International Publishing AG 2016
M.A. Cheema et al. (Eds.): ADC 2016, LNCS 9877, pp. 462–466, 2016.
DOI: 10.1007/978-3-319-46922-5_40

papers [4, 7] that consider the location information in the influence maximization problem. However, they overlook the importance of the distance between users and the promoted location. Usually, users that are close to the promoted location, such as a local restaurant, are more likely to attend. While users that are far away may not consider it due to the distance factor, even if they are influenced.

As a consequence, it is necessary to consider the distance information when selecting the influential users. In this paper, we study the distance-aware influence maximization problem (DAIM) [6], in which uses are assigned weights based on their distances to the promoted location. The closer a user is to the location, the larger weight the user will have. DAIM aims to find a set of k users that can lead to the largest influence spread under the weighted model. In this paper, we focus on the independent cascade model to simulate the information propagation, which is widely adopted in the literature [3]. For different query locations (*i.e.,* promoted stores), the weights of users are different. Thus the final selected influential users will be different. We aim to develop a system that can efficiently process DAIM query in real time.

Challenges. The challenges of processing DAIM queries efficiently lie in two aspects. Firstly, the problem is NP-Hard, and the calculation of influence spread is #P-Hard. Even based on the simple greedy algorithm, the cost is still very high. Secondly, there might be a large number of queries issued. So it is not an easy task to provide a real time response, especially when the social network is large.

To handle these two challenges, we propose the EDMS system in this paper. The system consists of two parts, a back-end and a front-end. In the back-end, it utilizes the MIA model [1] as an approximation to estimate the influence of users, and builds novel index structures for the geo-social network. By using the index structures, we propose powerful pruning strategies and search algorithms to offer efficient response for each query. In the front-end, it provides a web interface for users to issue the queries. We not only show the result of returned seed set, but also provide the interface for users to explore the spatial distribution of the users influenced by the selected seed set.

2   EDMS System

Figure 1 demonstrates our system architecture. The system consists of two parts, a back-end and a front-end. The front-end provides users with a web interface, which is built on Django[1], a high-level Python Wed framework.

Back-end. In the back-end, we store the index structures for the geo-social networks crawled from the web. The index structures have four main components, MIA index, anchor points index, region index and pivots index. The MIA index is used to approximate the users' influence in a tree model. In this case, we can efficiently evaluate the importance of each user. The anchor points index

---

[1] https://www.djangoproject.com/.

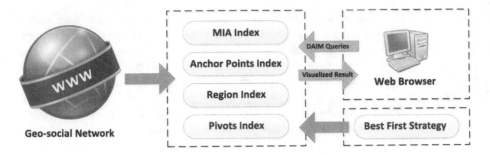

Fig. 1. EDMS system framework

and region index are used to estimated the upper bound and lower bound of users' influence and marginal influence. Since the objective function of DAIM is monotonic and submodular. Due to the hardness of the problem, we can use the greedy algorithm to obtain a result with a bounded ratio. During the search, we can safely prune the nodes whose (marginal) influence upper bounds are small. Thus the search cost and update cost are both reduced. The anchor points index is used for all users to obtain a loose bound by randomly sampling some locations (*i.e.*, anchor points) in space as the query points, and stores their influence according to the samples. When the query is given, we can obtain the bounds based on the value stored in the nearest anchor point. While the region index is used to obtain tighter bounds for those more influential nodes by further partitioning the users they can influence. The pivot index stores the result seed set for some pre-sampled queries. By using the spatial properties, we can obtain the approximation ratio of these results when a query is given. Then we can use them to further prune those less promising results. Another component in the back-end is the best first search strategy, which accesses the nodes based on their (marginal) influence upper bounds. By combining these powerful indexes, we can further reduce the search space.

Front-end. The front-end provides users with a web interface. It uses the Google Map API[2] and shows the distributions of user locations on the Google Map. Users can submit a query by inputing the promoted locations (*i.e.*, latitude and longitude) and the number of users they want to select (*i.e.*, k). By submitting the query, the system will present the seed set returned on the Google Map, and the seeds are marked by green rectangles and their user id in the system. In addition, by conducting randomly simulations from the seed set, the system also presents the distribution of the users that are influenced by the seed set on the map. Then users can have a better experience by knowing which user will be influenced. Users can also utilize this feedback to decide whether to query another location or to change the value k. Users can zoom-in or zoom-out to check the information in the preferred manner.

---

[2] https://developers.google.com/maps/.

3   Demonstration

Figure 2 presents the web interface of the system. Figure 2(a) shows the initial interface. User can submit a DAIM query by inputing the location of the query and specifying the number of users they want to select. Initially, all the users in the social network are presented on the map based on the spatial locations. Each user is marked by a small yellow rectangle. As shown in Fig. 2(b), after submitting a query (latitude is 36.340378, longitude is -78.785995 and k equals 2), the system visualizes the two users selected by the system with id 36 and 457, and they are marked by large green rectangles. In addition, the system also shows all the users influenced by the two users with red rectangles while the users uninfluenced are removed. The shade of the red color denotes the probability that the user can be influenced under the independent cascade model.

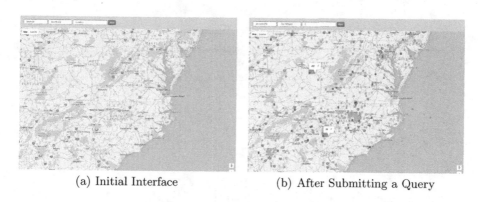

(a) Initial Interface   (b) After Submitting a Query

Fig. 2. System demonstration

4   Conclusion

In this paper, we present EDMS, a system that efficiently processes distance-aware influence maximization problem for online requirement. Novel index structures and search strategies are embedded to significantly reduce the search space. We present a friendly front-end to allow users issue queries and visualize the results in details.

References

1. Chen, W., Wang, C., Wang, Y.: Scalable influence maximization for prevalent viral marketing in large-scale social networks. In: KDD, pp. 1029–1038 (2010)
2. Chen, W., Wang, Y., Yang, S.: Efficient influence maximization in social networks. In: KDD, pp. 199–208 (2009)
3. Kempe, D., Kleinberg, J.M., Tardos, E.: Maximizing the spread of influence through a social network. In: KDD, pp. 137–146 (2003)

4. Li, G., Chen, S., Feng, J., Tan, K., Li, W.: Efficient location-aware influence maximization. In: SIGMOD 2014, pp. 87–98 (2014)
5. Tang, Y., Shi, Y., Xiao, X.: Influence maximization in near-linear time: a martingale approach. In: SIGMOD, pp. 1539–1554 (2015)
6. Wang, X., Zhang, Y., Zhang, W., Lin, X.: Distance-aware influence maximization in geo-social network. In: ICDE (2016)
7. Zhu, W., Peng, W., Chen, L., Zheng, K., Zhou, X.: Modeling user mobility for location promotion in location-based social networks. In: KDD, pp. 1573–1582 (2015)

Linking News and Tweets

Xiaojie Lin[1(✉)], Ye Gu[1], Rui Zhang[1], and Ju Fan[2]

[1] Department of Computing and Information Systems, The University of Melbourne, Melbourne, Australia
{xiaojiel1,yeg1}@student.unimelb.edu.au, rui.zhang@unimelb.edu.au
[2] Renmin University of China, Beijing, China
fanju1984@gmail.com

Abstract. In recent years, the rise of social media such as Twitter has been changing the way people acquire information. Meanwhile, traditional information sources such as news articles are still irreplaceable. These have led to a new branch of study on understanding the relationship between news articles and social media posts and fusing information from these heterogeneous sources. In this paper, we present a system that is able to effectively and efficiently link news and relevant tweets. Specifically, given a news stream and a tweet stream, the system discovers tweets that are relevant to each news in the news stream.

1 Introduction

Nowadays, a real world event such as a traffic accident or a criminal activity not only is covered by news articles, but also stimulates ordinary people to post their comments on social media such as Twitter[1], Facebook[2] and Weibo[3].

The strong relationship between news and social media interests researchers. Many studies on analyzing these two types of information sources together have been carried out. For example, Yang et al. [4] use relevant social media posts to summarize and extract highlights from news articles; Minkyoung et al. [2] use the relationship to analyze the characteristics of different types of media and the diffusion pattern of news events.

These studies and applications require reliable system to link news articles and relevant social media posts. In this paper, we present such a system which effectively and efficiently links news and relevant tweets.

2 The Linking System

The structure of the system is shown in Fig. 1. The input of the system is a news stream and a tweet stream. When a news article is received, it will be first preprocessed (e.g. tokenization and stemming) and then stored in a buffer for D

---

[1] https://twitter.com/.
[2] https://www.facebook.com/.
[3] The most popular microblogging platform in China. https://weibo.com.

© Springer International Publishing AG 2016
M.A. Cheema et al. (Eds.): ADC 2016, LNCS 9877, pp. 467–470, 2016.
DOI: 10.1007/978-3-319-46922-5_41

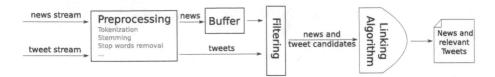

Fig. 1. System structure

days. Also, some indices will be created for the news in the buffer, which facilitate the following filtering and linking processes. When a tweet arrives, it will also be preprocessed, and then the system will use an efficient filtering algorithm (e.g. BM25 with a minimum threshold) to determine if the tweet should be added to the *tweet candidate set* of a news article in the buffer. When the filtering module has processed a certain amount of tweets, it will output a set of news along with their tweet candidates. The more expensive linking algorithm (discussed below) will now do the linking and output the final results — news and their relevant tweets.

We use an SVM classifier for the final linking. For each pair of a news and a tweet, a feature vector is extracted, and SVM will predict if the tweet is relevant to the news. The most important features we used are as follows:

BM25. BM25 computes a relevance score for a document and a query. In our case, we treat the news in the buffer as the document corpus and each tweet as a query.

Time. For a news and tweet published at time t_1 and t_2 respectively, the time feature is computed as $1/(t_2 - t_1 + 1)$. Note that we only consider tweet published after the news, so $t_2 - t_1 > 0$.

Named Entity. We extract named entities and calculate a TF-IDF score for each of them. The named entity feature is computed as:

$$\max_{n \in NE(a) \cap NE(t)} tfidf(n) \, ,$$

where $NE(a)$ and $NE(t)$ is the named entities extracted from news a and tweet t respectively.

Event Phrase. We use a dependency parser[4] to extract relations and noun phrases from news and tweets. Collectively, we call them event phrases since they can describe the essence of an event. Examples of extracted event phrases is shown in Fig. 2. We train another SVM classifier to generate a confidence score for each event phrase. The score indicates how well the event phrase describes a news article. For a news and a tweet, the event phrase feature is calculated as: $\max_{e \in EP(t)} confidence(e, a) \, ,$ where $EP(t)$ is the set of event phrases extracted from tweet t and a is a news.

---

[4] http://www.cs.cmu.edu/~ark/TweetNLP/#tweeboparser_tweebank.

> – Sigh. What a <u>9 year old</u>'s <u>Uzi accident</u> tells us about gun rights in America
> – That time a 12-year-old girl shot dead her assailant. Worth remembering with 9-year-old Uzi accident...
> – <u>Child</u> accidentally <u>shoots</u> and kills gun <u>instructor</u>
> – A 9 yr old <u>child</u> fatally <u>shoots</u> gun <u>instructor</u> with an UZI..a Submachine Gun!! #NRA whats next?! Tossing live grenades

Fig. 2. Event phrases extracted from tweets relevant to the event of "A 9-year-old girl accidentally shoots and kills her gun instructor with an automatic Uzi".

3   Experiments

We use a dataset derived from Guo's dataset [1], which contains 12,704 news and 34,888 tweets. In the gold standard, a tweet and a news article are considered relevant if the tweet contains a URL pointing to the news article. URLs in the tweets are removed before conducting experiments.

Guo's dataset does not contain the full content of news articles. Also, most of the news articles do not have any relevant tweets. Therefore, we identify the news articles with no less than 20 relevant tweets and download the full contents. A small amount of news are also removed because of download or parsing errors. The final dataset contains 381 news with full contents and all the 34,888 tweets.

Some of the news in the dataset are about the same event, and they are very similar to each other. For example, the news "Scores Dead as Fire Sweeps Through Nightclub in Brazil" and "Hundreds killed in Brazil nightclub fire" are about the same accident. Therefore, we also conduct extra experiments on a clustered version of the dataset, which contains 240 news clusters.

We test a wide range of unsupervised approaches along with ours. The results are shown in Table 1. The unsupervised approaches include the model of Tsagkias et al. [3] which is based on the language model (LM), BM25 using news as document corpus (BM25-news), BM25 using tweets as document corpus (BM25-tweets), cosine similarity of TF-IDF word vectors and the WTMF-G model [1].

For the unsupervised approaches, 5-fold cross-validation is used to determine the cut-off thresholds which maximizes the F_1 score. Precision and recall are reported under the same threshold. For our supervised approaches, the same 5-fold cross-validation is used for training/testing.

As shown in Table 1, our approach "SVM with event phrase features" performs the best in both the unclustered and clustered versions of the dataset. Note that Tsagkias's model (LM) does not work well in a binary classification setting because the relevance scores generated for different news are very different, so we are not able to find a reasonable cut-off threshold, and the reported metric values are very poor.

4   Demonstration

We build an online news service based on our system. After tweets are linked to news, we also use the relevant tweets to analyze the popularity and trending of each news. Our news service can be accessed via our website, Android client or REST API. Screenshots of the website and Android client are shown in Fig. 3.

Table 1. Performance of different approaches

Approaches	Unclustered			Clustered		
	Precision	Recall	F_1 Score	Precision	Recall	F_1 Score
LM	0.0016	0.0315	0.0030	0.0035	0.6641	0.0069
BM25 (news)	0.4635	0.5379	0.4979	0.6699	0.5658	0.6135
BM25 (tweets)	0.2210	0.3693	0.2765	0.4132	0.3757	0.3936
Cosine similarity	0.4810	0.4830	0.4820	0.6051	0.5539	0.5784
WTMF-G	0.5147	0.4797	0.4966	0.6897	0.5770	0.6284
SVM	0.6474	0.5634	0.6025	0.8183	0.5327	0.6453
SVM (event phrases)	0.6726	0.5544	**0.6078**	0.8145	0.5540	**0.6595**

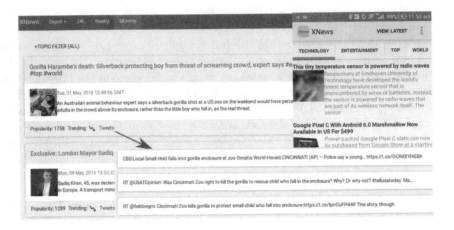

Fig. 3. Website and Android client

References

1. Guo, W., Li, H., Ji, H., Diab, M.T.: Linking tweets to news: a framework to enrich short text data in social media. In: ACL, vol. 1, pp. 239–249. Citeseer (2013)
2. Kim, M., Newth, D., Christen, P.: Trends of news diffusion in social media based on crowd phenomena. In: Proceedings of the Companion Publication of the 23rd International Conference on World Wide Web Companion, pp. 753–758. International World Wide Web Conferences Steering Committee (2014)
3. Tsagkias, M., de Rijke, M., Weerkamp, W.: Linking online news and social media. In: Proceedings of the Fourth ACM International Conference on Web Search and Data Mining, pp. 565–574. ACM (2011)
4. Yang, Z., Cai, K., Tang, J., Zhang, L., Su, Z., Li, J.: Social context summarization. In: Proceedings of the 34th International ACM SIGIR Conference on Research and Development in Information Retrieval, pp. 255–264. ACM (2011)

A Data-Packets-Saving Multi-agent Scheduling of Large Online Meetings

Usman Ali and Guangyan Huang[✉]

School of Information Technology, Deakin University,
Burwood, Melbourne, VIC 3125, Australia
{au,guangyan.huang}@deakin.edu.au

Abstract. Communication among a lot of people remotely distributed all over the world using various devices often becomes complex in a large online meeting and this challenges the meeting coordination which involves sending out invitation, waiting for answer and negotiating for the meeting time. To resolve this challenge in the large online meeting, this work implements a novel multi-agent scheduling system, which is demonstrated to be able to save data packets, reduce energy consumption and increase the meeting acceptance rate.

Keywords: Multi-agent · Scheduling · Large online meetings · Energy efficiency

1 Introduction

With the extensive range of multimedia applications, digital communication is very popular but becomes more complex in large online meetings, since it must handle various devices such as cell phones, PDAs and tablets [1]. This challenges traditional systems. This paper adopts a multi-agent scheduling to make the online meeting among multiple people more effective: to reduce data packet overhead for improving communication quality while reducing energy dissipation. To effectively schedule multi-agent, some protocols must be developed to ensure that the agent can understand the meaning of the message received and ignore the useless message so they can prevent the communication deadlock among all the agents. Proposed multi-agent scheduling for large online meetings is demonstrated effective to reduce the number of communication data packets, save energy and increase the meeting acceptance rate. Also, the proposed scheduling is more efficient than the traditional scheduling when the meeting is larger.

2 The Proposed Method

We developed a novel multi-agent meeting scheduling system based on the approach in [2] to resolve the challenge of scheduling large online meetings. The techniques implemented in the proposed system include a communication protocol, an internal architecture of the schedule, a state transition diagram and a mechanism for error correction and detection. Particularly, we have designed a software agent which makes decision based on a calendar and a scheduler and communicates with other software agent. These agents

© Springer International Publishing AG 2016
M.A. Cheema et al. (Eds.): ADC 2016, LNCS 9877, pp. 471–475, 2016.
DOI: 10.1007/978-3-319-46922-5_42

interact with each other; and when the meeting time is shared each agent changes the available time. If a free slot is available, the scheduler confirms the time and the notification is sent to the other agents. Due to intelligent cooperation among multiple agents, a high probability that around 90 % of the meeting time shared in organisation will be accepted.

The main components of the proposed scheduling comprise a state transition diagram and an internal architecture of a scheduler as shown in Fig. 1. There are five states in Fig. 1(a): *begin, waiting, search new proposals, evaluation* and *end.* At the *begin* state, the user is required for the meeting or received meeting time. In the *waiting* state, the agent waits for the approval of the time or checks the calendar so that the proposal time can be approved or rejected. If the proposed meeting time is not confirmed, the state is transferred to *search new proposals* and the agent send the new meeting time to the other agents. Once the time is approved, the *evaluation* is conducted and the information is shared with all the other agents which are involved in the meeting.

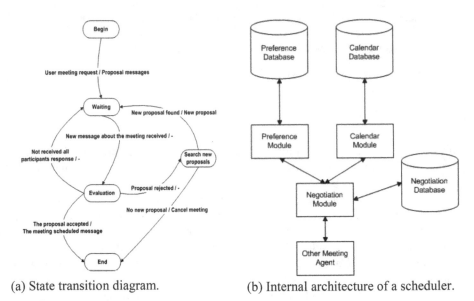

(a) State transition diagram.   (b) Internal architecture of a scheduler.

Fig. 1. Agent scheduling.

The processes continuous until all the agents or maximum number of agent accept or reject the proposed meeting time. The state transition diagram has been utilized to describe state evolution of agent during the communication and coordination protocol. For each transition, there is an input and output message. When a software agent is in the initial state, it receives a meeting scheduling request from its user. It sends a proposal message to the invitee agent and the agent enters in *waiting* state on a newly received message. The agent passes it into an evaluated and the presence of certain condition are checked it all the reply are not received the agent goes back in *waiting* state. If message from the entire attendant are received, the proposed time is accepted and it sends the confirmation message to the entire attendant and goes to the *end* state. If the proposed is rejected, then *it will go to search new proposals state*.

3   Experimental Study

We have adopted an open source tool, NETLOGO [3], to simulate the proposed solution. NetLogo is a multi-agent programmable modeling environment. We also compared the performance of the proposed solution with a traditional solution in [4] in terms of data packet overhead, energy and computation time.

We have deployed forty agents based on computing nodes in the simulation environment. In the agent node the computing processing units have been simulated to validate the solution up to maximum extends. We set the simulation area of the agent nodes to "256 × 256". The resources share was performed with the nodes which were in the neighbour agent computing nodes. Each agent computing nodes can be considered as the set of resources for performing the computing task. We adopted the multiple listing techniques for the communication and coordination protocol in [4]. The experiment has been conducted on one hundred transactions.

At the initial stages, the traditional overhead remains the same; but as the number of the transaction increases, the overhead also increases. It has been observed in Fig. 2(a) that the communication and coordination protocol of the proposed solution have less overhead than the traditional solution while transmitting the data packet. As the data packet size is smaller the less energy is required to transmit the message or vice versa, so the energy consumption is reduced in our method as shown in Fig. 2(b). However, the computation time of our proposed method is less only when the number of transactions greater than 70. This demonstrates that our method is more suitable for scheduling

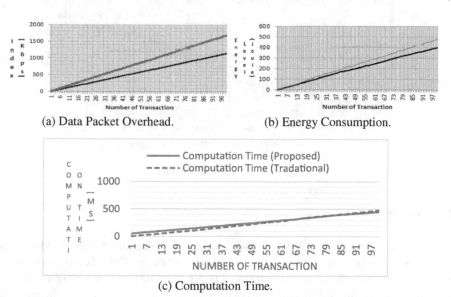

(a) Data Packet Overhead.   (b) Energy Consumption.

(c) Computation Time.

Fig. 2.   Energy efficiency achieved by reducing data packet overhead.

larger meetings. Meanwhile, we observed that the proposed solution averagely increases the meeting acceptance rate, as shown in Fig. 3; this also demonstrates the effectiveness of our solution.

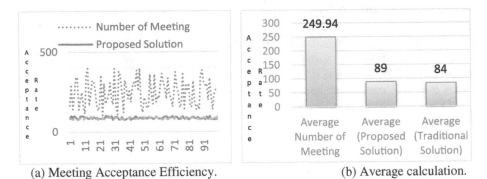

(a) Meeting Acceptance Efficiency.   (b) Average calculation.

Fig. 3. Meeting acceptance efficiency.

It has been investigated that this algorithm within context of scheduling required less resources as compared to the traditional algorithm. We have utilized the behaviour model of the NETLOGO to calculate the time complexity of our proposed scheduling, which achieves a good scale of $O(n)$, as shown in Fig. 4.

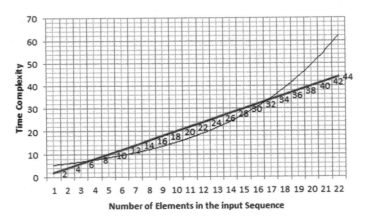

Fig. 4. Time Complexity: $O(n)$.

4   Conclusion

In this work, we have provided a multi-agent meeting scheduling, which reduces the data packet overhead in communication and saves power. The effectiveness of the proposed scheduling has been demonstrated in NETLOGO simulator, which performs better than the existing solution in terms of data packets overhead, energy efficiency and meeting acceptance rate while spending similar computation time (actually the proposed scheduling is more efficient when the meeting is larger). The proposed scheduling can be applied to general distributed environments, such as cloud computing.

References

1. Niazi, M.A., et al.: Verification & validation of agent based simulations using the VOMAS (Virtual Overlay Multi-agent System) approach. In: The MAS&S 2009 at Multi-agent Logics, Languages, and Organisations Federated Workshops, Torino, Italy (2009)
2. Bryman, A.: Integrating quantitative and qualitative research: how is it done? Qual. Res. **6**(1), 97–113 (2012)
3. Wilensky, U.: (1999). http://ccl.northwestern.edu/netlogo/
4. Khattak, A.S., Malik, S., Hayat, K., Sanam, S.R.: Verification & validation of a multi agent meeting scheduling simulation model. J. Comput. Sci. **2**(1), 47–64 (2014)

A Classifier Hub for Imbalanced Financial Data

Chirath Abeysinghe[1], Jianguo Li[1,2](✉), and Jing He[1]

[1] College of Engineering and Science, Victoria University, Melbourne, Australia
lvchirathdevinda@gmail.com, {Jianguo.Li,Jing.He}@vu.edu.au
[2] School of Computer Scinence, South China Normal University, Guangzhou, China

Abstract. We design and implement a classifier hub that can explore the detailed information on the imbalanced dataset and classify the dataset into two classes. Against the data imbalance, through setting imbalance ratio, it can adjust the proportion of majority and minority class. In this hub, we also implement Decision Tree, KNN and Random Forrest machine learning classifiers based on Python and Java. In the experiments, we use 30,000 loan records from an online P2P system as the dataset to demonstrate the functions of the classifier hub. The influences of different imbalanced ratio on classification performance have been compared through Decision Tree, KNN and Random Forrest algorithms.

1 Introduction

DUE to the widespread adoption of Internet banking, especially online P2P loan, risk control became more difficult and different for the Internet banking industry. P2P loan allows personal loans without banks' involvement and provides access to a wide range of loan types. Loans can be made between individual users with lower, fixed rate and transparent fee structures. Extracting effective knowledge from the multidimensional loan data records and avoiding loss due to lender's loan defaults have enormous profit potential for the investors that is described in [1]. Our main target for financial dataset is to reduce the number of instance that fail to make the loan repayment (that's false positive rate) and at the same time guarantee enough loans are granted (that's true positive rate, also called sensitivity). If we just predict all instances positive, we also get accuracy over 90 % because of imbalance. But the predication is meaningless, so we use AUC (the area under the Receiver Operating Characteristic curve) in [2] to evaluate different classifiers. ROC curve has a good characteristic of keeping stable when proportion of positive and negative samples in training set is changing especially imbalance appearing.

2 Classifier Hub Overview

Classifier Hub has friendly interfaces and is comprised of three parts: Data Explorer, Balance Adjusting and Machine Learning Classifiers.

© Springer International Publishing AG 2016
M.A. Cheema et al. (Eds.): ADC 2016, LNCS 9877, pp. 476–479, 2016.
DOI: 10.1007/978-3-319-46922-5_43

2.1 Data Explorer

It has been designed to explore each dataset features with a statistical description such as minimum, maximum, mean and standard deviation values along with a basic status information include number of instances, number of features and file name. It supports only CSV file format which consists of rows of comma separated values. In addition, it depicts the class distribution through a bar chart and it provides user a better visualization of the class information. This bar chart shows each class count as Y-coordinate and its category name as X-coordinate.

2.2 Balance Adjusting

Class balance adjusting is used for skewed class distribution as an over-sampling technique using SMOTE algorithm in order to increase minority class examples. When performing SMOTE in the dataset, three parameters being considered. First parameter is defined by the minority class and the second one is the percentage of minority class samples user wants to increase and the percentage value is ranging from 1 to 100. Next, the third parameter is K-Neighbors per instance. Based on this K value, for each instance, the number of K nearest neighbors being selected. This process iterates through each minority samples which has already been determined by the first and second parameters. Then, randomly picked one of the neighbor and compute its attribute value different with the corresponding instance attribute value. This difference then is multiplied by a random number between 0 and 1 and finally this value is added to the instance attribute value. This computation is repeated through all the attributes for each instance in the minority class. This newly generated array of samples now can be called as Synthetic samples. If a user wants to increase more synthetic samples, then current SMOTED dataset can be saved and re-loaded into the Classifier Hub and then SMOTE can be re-applied to itself. Following this way, user can create many datasets which have different class ratios.

2.3 Machine Learning Classifiers

Decision Tree, KNN, and Random Forest are an example of classical machine learning algorithms and they have been implemented here by using Python [4]. As a way of performance evaluation, Stratified 10-fold cross validation is used and confusion matrix displays the total of each folds classification and miss-classification counts. Furthermore, classifier hub displays performance metrics in details such as Accuracy, True Positive Rate (TPR), False Positive Rate (FPR), Precision and F1 score for both positive and negative class with ROC area along with AUC score, after applying each algorithm. In addition, it visualizes performed algorithms metrics in the same plot with appropriate axis coordination as a way of enabling user to compare performances of different algorithms. All the plots have extra functionalities such as print, save, zooming and chart properties for user convenience. As an additional feature, software log file viewer has been integrated and the user can check logs where logs display any errors with a timestamp and it is useful for troubleshooting purposes.

3   Demonstration and Experiments

Classifier Hub has been developed and tested on laptop with Intel Core (TM) i5 CPU @ 2.67 GHz, 4.00 GB RAM, 64-bit and Windows 10 Operating system. Apart from the system requirements, python interpreter must be used to execute scripts as a software requirement. Furthermore, python Sklearn machine learning library in [4] has used to implement default algorithms and Java is being used to execute these Python scripts through Java Virtual Machine (JVM). Figure 1 demonstrates the system interfaces built in Classifier Hub.

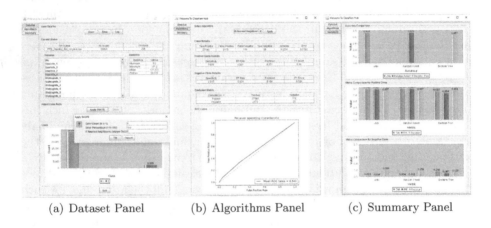

(a) Dataset Panel   (b) Algorithms Panel   (c) Summary Panel

Fig. 1. Software interfaces

Our dataset that is from an Internet financial P2P loan platform contains 30,000 instances and 225 features overall about lender's personal information, network behavior information, social network information and other web information. Among them, 27802 positive instances repay money on time and 2198 negative instances fail to pay back their loans. The proportion of positive instances and negative ones is 12.65. We call this value as imbalance ratio (IR) in [5] that is defined as the ratio of the number of instances in the majority class to the number of examples in the minority. As the following Fig. 2 illustrate, increasing of minority class samples heavily impact on the classifiers performances. Decision Tree, KNN and Random Forest classifiers are performing poorly on the dataset which has 12.65 class imbalance ratio for its negative class (minority class). After reducing the class ratio up to 3.16, classifiers have improved their performances and according to AUC score, it has increased from 0.53 to 0.83, 0.54 to 0.95 and 0.61 to 0.92 for Decision Tree, KNN and Random Forest classifiers respectively.

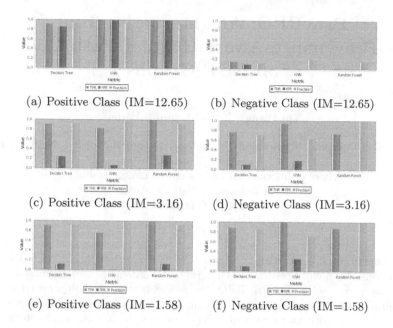

(a) Positive Class (IM=12.65)   (b) Negative Class (IM=12.65)

(c) Positive Class (IM=3.16)   (d) Negative Class (IM=3.16)

(e) Positive Class (IM=1.58)   (f) Negative Class (IM=1.58)

Fig. 2. TPR, FPR and Precision comparison for different imbalance ratio

This is a significant result over a reduction of class skewness. Finally, class distribution retained as 1.58 imbalance ratio in which Random Forest and KNN are better classifiers for this dataset with an AUC score of 0.96.

Acknowledgment. This research has been funded by the Guangzhou Science and Technology Plan Project "Collaborative Innovation Project Oriented Big Data Security Industry Chain" (No. 201508010067).

References

1. He, J., Zhang, Y., Shi, Y., Huang, G.: Domain-driven classification based on multiple criteria and multiple constraint-level programming for intelligent credit scoring. IEEE Trans. Knowl. Data Eng. **22**(6), 826–838 (2010)
2. Fawcett, T.: An introduction to ROC analysis. Pattern Recogn. Lett. **27**(8), 861–874 (2006)
3. Chawla, N.V., Bowyer, K.W., Hall, L.O., Kegelmeyer, W.P.: SMOTE: synthetic minority over-sampling technique. J. Artif. Intell. Res. **16**, 321–357 (2002)
4. Pedregosa, F., et al.: Scikit-learn: machine learning in Python. J. Mach. Learn. Res. **12**, 2825–2830 (2011)
5. Garcìa, V., Sànchez, J.S., Mollineda, R.A.: On the effectiveness of preprocessing methods when dealing with different levels of class imbalance. Knowl.-Based Syst. **25**(1), 13–21 (2012)

ProvRPQ: An Interactive Tool for Provenance-Aware Regular Path Queries on RDF Graphs

Xin Wang[1,2(✉)] and Junhu Wang[2]

[1] School of Computer Science and Technology, Tianjin University, Tianjin, China
wangx@tju.edu.cn
[2] School of Information and Communication Technology,
Griffith University, Nathan, Australia
j.wang@griffith.edu.au

Abstract. *Regular Path Queries (RPQs)* are building blocks for expressing navigations over RDF graphs. We demonstrate an interactive query tool, called ProvRPQ, for *provenance-aware* RPQs on RDF graphs. In contrast to merely showing pairs of source and target nodes in the conventional answers of RPQs, our tool, with users' exploratory interaction, could clearly justify how paths conforming to RPQs can be navigated from source to target resources in RDF graphs.

Keywords: Regular path queries · Provenance-aware · Interactive query tool · RDF graphs

1 Introduction

The *Resource Description Framework (RDF)* is the standard format for publishing and sharing data on the Semantic Web. The RDF data model is essentially a graph model in which resources are organized as *RDF graphs.* Let U and L be the sets of URIs and literals, respectively. An RDF graph T is a set of triples, with each *triple* $(s, p, o) \in U \times U \times (U \cup L)$ stating the fact that the resource s has the relationship p to the resource $o \in U$, or the resource s has the property p with the value $o \in L$, where s is called the subject, p the predicate (or property), and o the object. In recent years, inspired by the *Linked Data* campaign, more and more RDF graph data from various domains have been openly published on the Web. The class of *regular path queries*, or RPQs, is widely recognized as an essential mechanism for expressing *navigations* on RDF graphs, which has been reinforced by the introduction of *property paths* to the latest version of the W3C RDF query language, SPARQL 1.1 [1].

An RPQ $Q = (x, r, y)$ over an RDF graph T asks for a set of pairs of resources (s, o) such that there exists a path ρ in T from s to o with the label of ρ, denoted by $\lambda(\rho)$, satisfying the regular expression r in Q. For example, Fig. 1(a) depicts an RDF graph T_1 excerpted from the DBpedia dataset, which shows

© Springer International Publishing AG 2016
M.A. Cheema et al. (Eds.): ADC 2016, LNCS 9877, pp. 480–484, 2016.
DOI: 10.1007/978-3-319-46922-5_44

predecessor and father relationships among seven British monarchs. The RPQ $Q_1 = (x, (\text{dbo:predecessor}|\text{dbp:father})^+, y)$ on T_1 asks to find pairs of monarchs (s, o) such that s can navigate to o via a path in T_1 that consists of one or more dbo:predecessor or dbp:father edges. The answers to Q_1 are shown in Fig. 1(b) (prefix dbr: omitted). However, since it is unrealistic to see the entire view of large-scale RDF graphs in real-world settings, given an answer, e.g., (dbr:Elizabeth_II, dbr:Queen_Victoria), of Q_1, users cannot know how the paths from dbr:Elizabeth_II to dbr:Queen_Victoria are routed to satisfy the regular expression (dbo:predecessor|dbp:father)$^+$. To this end, *provenance-aware* answers of RPQs have been introduced [2,3], which are actually subgraphs of an RDF graph consisting of "witness triples" for RPQs.

More formally, given an RPQ $Q = (x, r, y)$ over an RDF graph T, the conventional answer set of Q is defined as $[\![Q]\!]_T = \{(s, o) \mid \exists$ a path ρ in T from s to o s.t. $\lambda(\rho) \in L(r)\}$, whereas the provenance-aware answer set of Q is defined as $[\![Q]\!]_T^{prov} = \{(s, p, o) \mid \exists$ a path ρ in T s.t. (s, p, o) is an edge in ρ and $\lambda(\rho) \in L(r)\}$. Obviously, $[\![Q]\!]_T^{prov}$ is a subset of T.

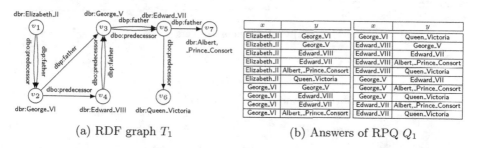

x	y
Elizabeth_II	George_VI
Elizabeth_II	George_V
Elizabeth_II	Edward_VIII
Elizabeth_II	Edward_VII
Elizabeth_II	Albert,_Prince_Consort
Elizabeth_II	Queen_Victoria
George_VI	George_V
George_VI	Edward_VIII
George_VI	Edward_VII
George_VI	Albert,_Prince_Consort

x	y
George_VI	Queen_Victoria
Edward_VIII	George_V
Edward_VIII	Edward_VII
Edward_VIII	Albert,_Prince_Consort
Edward_VIII	Queen_Victoria
George_V	Edward_VII
George_V	Albert,_Prince_Consort
George_V	Queen_Victoria
Edward_VII	Albert,_Prince_Consort
Edward_VII	Queen_Victoria

(a) RDF graph T_1 (b) Answers of RPQ Q_1

Fig. 1. An example RDF graph T_1 and answers of RPQ Q_1

We have developed an interactive query tool, called *ProvRPQ*, which can guide users to explore the provenance-aware answers of RPQs step by step from their conventional answers to justify how source resources can reach target resources by navigating through "witness triples". The algorithm behind ProvRPQ is a form of interactive graph search in provenance-aware answers of RPQs, which finds all the edges in $[\![Q]\!]_T^{prov}$ such that each of them occurs in some path from the given source resource s to the target o. To the best of our knowledge, ProvRPQ is the first query tool that interactively shows the provenance-aware answers of RPQs from their corresponding conventional answers in an exploratory way.

2 Demonstration

We demonstrate the interaction of ProvRPQ on synthetic and real-world RDF graphs, i.e., LUBM and DBpedia, respectively.

2.1 Provenance-Aware RPQs on LUBM

Figure 2 shows screenshots of the steps in exploring the provenance-aware answer of the RPQ $Q_2 = (x, \text{lubm:subOrganizationOf}^+/\text{rdf:type}, y)$ on LUBM by using ProvRPQ. In Fig. 2(a), we can see five conventional answers represented as blue directed edges from source resources (blue nodes) to target resources (red nodes). When expanding a selected answer (in red in Fig. 2(a)), we get Fig. 2(b) with two "witness edges" (in grey) being introduced. When hovering the mouse on a source resource, in Fig. 2(c), both the conventional (in purple) and provenance-aware (in orange) answers are highlighted. Figure 2(d) shows that a "witness node" http://www.University1.edu (in green) is added after expanding the conventional answer (http://www.Department0.University1.edu, http://swat.cse. lehigh.edu/onto/univ-bench.owl#University). Finally, the complete provenance-aware answer of Q_2 is shown in Fig. 2(e).

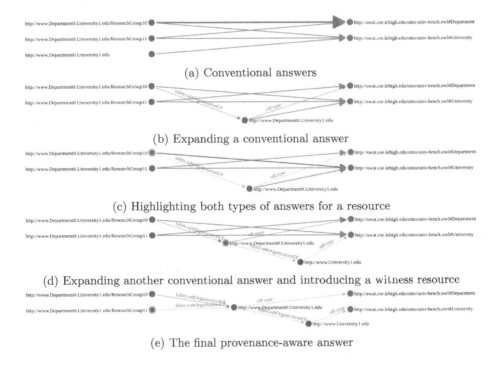

(a) Conventional answers

(b) Expanding a conventional answer

(c) Highlighting both types of answers for a resource

(d) Expanding another conventional answer and introducing a witness resource

(e) The final provenance-aware answer

Fig. 2. Exploring a provenance-aware RPQ on LUBM by using ProvRPQ (Color figure online)

2.2 Provenance-Aware RPQs on DBpedia

Figure 3 shows the initial and final steps in exploring the provenance-aware RPQ $Q_3 = (\text{dbr:Elizabeth\_II}, (\text{dbo:predecessor}|\text{dbp:father})^+, y)$ on DBpedia. Two conventional answers, i.e., (dbr:Elizabeth_II, dbr:Queen_Victoria) and (dbr:Elizabeth_II, dbr:Albert, _Prince_Consort), to Q_3 are shown in Fig. 3(a) with images from the corresponding Wikipedia articles being attached to the resources. Most likely, users may wonder how dbr:Elizabeth_II is related to dbr:Queen_Victoria and dbr:Albert,_Prince_Consort as answers to RPQ Q_3. After expanding the two conventional answers by using ProvRPQ, one can clearly view the whole picture of relationships among the resources involved, which is shown in Fig. 3(b). In this example, we can also observe that the provenance-aware answer of Q_3 can use a graph with 7 edges (3 edges with two relationships) in Fig. 3(b) to efficiently encode all 20 conventional answers of Q_1 in Fig. 1(b).

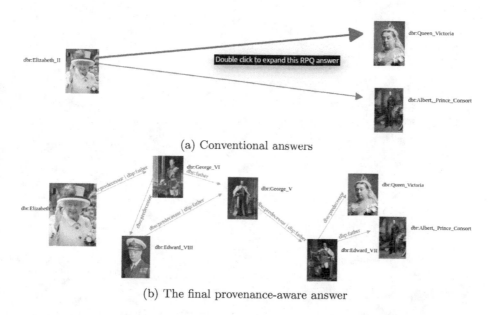

(a) Conventional answers

(b) The final provenance-aware answer

Fig. 3. Exploring a provenance-aware RPQ on DBpedia by using ProvRPQ

3   Conclusion

We have developed an interactive query tool, ProvRPQ, for provenance-aware RPQs on RDF graphs. With the help of ProvRPQ, users could clearly justify how conventional answers of RPQs are formed by navigating paths from source to target resources in the corresponding provenance-aware answers.

Acknowledgments. This work is supported by the National Natural Science Foundation of China (61572353) and the Australia Research Council (ARC) Discovery grants DP130103051.

References

1. Harris, S., Seaborne, A.: SPARQL 1.1 query language. W3C Recommendation (2013)
2. Dey, S., Cuevas-Vicenttín, V., Köhler, S., Gribkoff, E., Wang, M., Ludäscher, B.: On implementing provenance-aware regular path queries with relational query engines. In: EDBT/ICDT 2013, pp. 214–223. ACM (2013)
3. Wang, X., Ling, J., Wang, J., Wang, K., Feng, Z.: Answering provenance-aware regular path queries on rdf graphs using an automata-based algorithm. In: WWW 2014 (companion), pp. 395–396. ACM (2014)

Author Index

Printed in the United States
By Bookmasters